FLORA ZAMBESIACA

Flora terrarum Zambesii aquis conjunctarum

T0289734

VOLUME SEVEN: PART THREE

APOCYNACEAE (PART 2)

FLORA ZAMBESIACA

MOZAMBIQUE, MALAWI, ZAMBIA,

ZIMBABWE, BOTSWANA

VOLUME SEVEN: PART THREE
APOCYNACEAE (PART 2)

Authors

DAVID J. GOYDER (*Royal Botanic Gardens, Kew, U.K.*)

MICHAEL G. GILBERT (*Royal Botanic Gardens, Kew, U.K.*)

H. JOHAN T. VENTER (*University of the Free State, Bloemfontain, South Africa*)

Edited by

MIGUEL A. GARCÍA
Royal Botanic Gardens, Kew
Real Jardín Botánico (CSIC), Madrid, Spain

Published by the Royal Botanic Gardens, Kew
for the Flora Zambesiaca Managing Committee
2020

Citation: Goyder, D. J., Gilbert, M. G. & Venter, H. J. T. (2020). Apocynaceae (part 2). In: M. A. García (ed.), *Flora Zambesiaca*, Vol. 7(3). Royal Botanic Gardens, Kew.

First published in 2020 by
Royal Botanic Gardens, Kew,
Richmond, Surrey, TW9 3AB, UK
www.kew.org

Distributed on behalf of the Royal Botanic Gardens, Kew in North America by the University of Chicago Press, 1427 East 60th Street, Chicago, IL 60637, USA

ISBN 978-1-84246-713-8
eISBN 978-1-84246-714-5

British Library Cataloguing in Publication Data
A catalogue record for this book is available from the British Library

Design and page layout: Christine Beard
Production management: Jo Pillai

Printed in the UK by Short Run Press Limited

For information or to purchase all Kew titles please visit shop.kew.org/kewbooksonline or email publishing@kew.org

Kew's mission is to inspire and deliver science-based plant conservation worldwide, enhancing the quality of life.

Kew receives approximately one third of its running costs from Government through the Department for Environment, Food and Rural Affairs (Defra). All other funding needed to support Kew's vital work comes from members, foundations, donors and commercial activities including book sales.

CONTENTS

NEW TAXA AND COMBINATIONS PUBLISHED IN THIS VOLUME

7. APOCYNACEAE (Part 2)

by David J. Goyder, Michael G. Gilbert & H. Johan T. Venter[1]

Notes on floral structures within the more derived subfamilies of Apocynaceae

The five stamens are fused apically to the expanded stylar head and together form a compound structure called the gynostegium. The ovary is therefore almost entirely concealed within a ring of stamens (the staminal column) whose filaments are usually fused into a tube, but free in subfam. Periplocoideae. Only the sterile apex of the stylar head remains visible, level with or extending beyond the anthers; the receptive portions of the stylar head are on its underside behind the frequently sclerified margins of the anthers which form a chamber for the deposition of pollen. The pollen transfer apparatus is formed by secretions from the stylar head and, in periplocoid genera, generally consists of a spathulate translator onto which pollen tetrads, or more rarely pollen masses (pollinia), from adjacent anthers are shed; in the two remaining subfamilies, the secretions form a central corpusculum linked by caudicles or translator arms to 2 or 4 pollinia of adjacent anthers – the unit is transferred in its entirety from one flower to another and is called a pollinarium.

The corona presents many diagnostic characters for generic and specific recognition. Following the system devised by Liede & Kunze in Pl. Syst. Evol. **185**: 275–284 (1993), it may be corolline (derived from the corolla) or gynostegial (from the staminal column). Corolline coronas can be divided into those occurring in the corolla lobe sinuses and those forming an annulus in the corolla tube. Gynostegial coronas again have two basic elements, a staminal corona attached dorsally to the stamens, and an interstaminal corona, and these elements can be combined in a number of ways: e.g. staminal corona lobes only; a fused ring of staminal and interstaminal lobes; fused staminal and interstaminal corona with additional staminal lobes.

Key to the derived subfamilies of Apocynaceae (formerly Asclepiadaceae) – see Fig. 7.3.**120**.

1. Staminal filaments free; anther lacking sclerified margins; pollen in tetrads and appearing granular, or occasionally aggregated into pollinia, deposited onto a spathulate pollen carrier or translator situated on the stylar head, between adjacent anthers. . . . . .subfam. **Periplocoideae**
   - Staminal filaments, when present, united into a tube or annulus; anthers with sclerified margins (anther wings or guide rails), adjacent anther wings adnate to each other, with a groove between them; pollen of each anther cell aggregated into 1 or 2 pollinia, the pollinia from adjacent anther locules attached directly or indirectly to a corpusculum situated on the stylar head, at the top of the guide rails 2
2. Corpusculum pale, minute; pollinia 4 to each corpusculum, attached directly or on short stalks . . . . subfam. **Secamonoideae**
   - Corpusculum hard, horny, mostly dark brown or black; pollinia 2 per corpusculum, attached by variously structured translator arms. . . . . . .subfam. **Asclepiadoideae**

[1] D.J. Goyder: introductory section; Secamonoideae (genus 39); Asclepiadoideae – tribes Fockeeae (genera 40–41, Marsdenieae (genera 42–43), Ceropegieae in part (genera 44–47 & 49–57), Asclepiadeae (genera 58–77).

M.G. Gilbert: Asclepiadoideae – tribe Ceropegieae (Ceropegiinae) (genus 48).

H.J.T. Venter: Periplocoideae (genera 30–38).

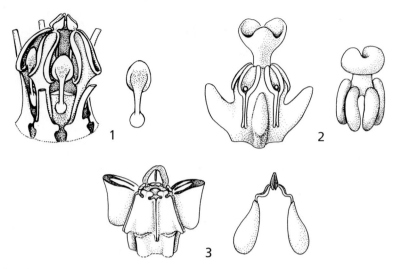

Fig. 7.3.**120**. Stylised gynostegia and pollen transport arrangements (to be used in conjunction with the key to subfamilies): 1, PERIPLOCOIDEAE: corona lobes cut away to reveal stamens depositing granular pollen onto the spoon-like translator. 2, SECAMONOIDEAE: adjacent anther wing margins sclerified and closely adnate with a groove between, pollinarium with four pollinia attached ± directly to the minute soft corpusculum. 3, ASCLEPIADOIDEAE: gynostegium with two corona lobes removed to expose anthers, again with sclerified margins to the anther wings and a groove between, pollinarium of two pollinia attached indirectly to the hard corpusculum my translator arms (but note Fockeeae; see Fig. 7.3.**133**-1 – two pollinia attached directly to the corpusculum). Drawn by Margaret Tebbs. Reproduced from Flora of Tropical East Africa (2012).

Subfam. **PERIPLOCOIDEAE**

Periplocoideae (Apocynaceae) Endl., Gen. Pl. **8**: 587 (1838). —Schumann in Engler & Prantl, Nat. Pflanzenfam. **4**(2): 209 (1895). —Hutchinson & Dalziel in F.W.T.A. **2**(1): 50–54 (1931). —Venter & Verhoeven in Taxon **46**: 705 (1997); in Ann. Missouri Bot. Gard. **88**: 550–568 (2001). —Venter in Hedberg & Edwards, Fl. Ethiopia Eritrea 4(1): 102 (2003); in Fl. Somalia **3**: 135 (2006); in F.T.E.A., Apocynaceae (part 2): 117 (2012).

Perennial climbers, shrubs or herbaceous geophytes with white or occasionally red or orange latex. Roots often tuberous. Interpetiolar stipules of simple lines, dentate colleterate ridges or a collar of frills. Leaves opposite, rarely alternate, whorled or fascicled, petiolate to sessile, axils with dentate and/or hairy colleters; blades simple, venation pinnate, often with colleters on main veins, margin entire. Inflorescences terminal and/or axillary, simple cymose or paniculate cymose, few- to many-flowered. Flowers actinomorphic, pentamerous, semi-epigynous, variously coloured, glabrous or hairy, sometimes aromatic. Sepals free, often with paired colleters at inner bases. Petals fused, tube saucer-shaped, campanulate, cylindrical or reflexed. Corona corolline, alternating with petals, 1 or 2-whorled; lower whorl mostly present, inserted on inner wall of corolla tube, always above stamen whorl, of 5 lobes, lobes simple or segmented, free or fused, mostly basally fused with staminal filaments forming coronal feet, coronal feet sometimes also fused to interstaminal nectaries forming a coronal annulus; upper whorl, when present, arising from corolla lobe sinuses, pocket-like or pocket-lobular with lobe from rim of pocket. Stamens epipetalous from base to near mouth of corolla tube, alternating with corolla lobes, glabrous or hairy; filaments free or fused to coronal feet and/or laterally to interstaminal nectaries; anthers free, inner bases fused to style-head forming a gynostegium, laterotrorse; pollen in rhomboidal, decussate or linear tetrads or in pollinia, shed onto translators. Nectaries epipetalous at base of tube, opposite corolla lobes, interstaminal, surrounding upper half of ovary, lobular or pocket-like, free or fused with coronal feet. Ovaries 2, free, semi-inferior, styles 2, apically fused into style-head; style-head pentangular, ovoid, deltoid or rarely cylindrical-ovoid, bearing translators on upper surface in between adjacent anthers. Translator of receptacle (spoon), stalk (present or absent) and basal sticky disc (viscidium). Gynostegium concealed in corolla tube or exposed from it. Follicles paired or single. Seeds compressed, mostly obliquely ovate, with coma or collar (rarely) of hairs.

Thirty-three genera and approximately 190 species occur in Africa, Madagascar, Asia, South-east Asia and Australia. Of the 18 genera found in Africa, 16 are endemic to the continent. Approximately 100 species are endemic. Nine genera and 44 species are present in the Flora Zambesiaca area.

1. Corolla tube campanulate or cylindrical 2
–   Corolla rotate with tube shallow and saucer-shaped 7
2. Stamens arising at base of corolla tube 3
–   Stamens arising at mouth of corolla tube 6
3. Corona lobes arising from corolla lobe sinuses; lobes matchstick-like . . . .
    . . . . **34. Stomatostemma**
–   Corona lobes arising from about middle of or lower down in corolla tube; lobes variously shaped, but never matchstick-like 4
4. Flowers large, corolla 3 cm or longer . . . . **33. Cryptostegia**
–   Flowers small, corolla less than 2 cm long 5
5. Corolla with mouth and inside of lobes papillate; corona lobes of channeled radiating processes with upper part bisegmented, the outer segment subulate to rounded, the inner concavely hood-shaped; plant parts turning black when dried . . . . **31. Parquetina**
–   Corolla tube and lobes not papillate; corona lobes clavate, subulate or filiform, plant parts never turning black when dried. . . . .**32. Cryptolepis**
6. Pollen in pollinia, lower corona lobes with bases fused into a collar around stamens, style terete, glabrous. . . . . **38. Schlechterella**
–   Pollen in tetrads, lower corona lobes with bases free, if fused into a collar then style ribbed and hairy . . . . **37. Raphionacme**
7. Interpetiolar collar of green frills, flowers large, 1.5–2.5 cm in diameter . . . .
    . . . . **36. Mondia**
–   Interpetiolar lines or dentate ridges without frills, flowers less than 1 cm in diameter 8
8. Stamens and inside of corolla lobes hairy . . . . **30. Periploca**
–   Stamens and inside of corolla lobes glabrous . . . . **35. Tacazzea**

30. **PERIPLOCA** L.[2]

Periploca L., Sp. Pl.: 211 (1753). —Browicz in Arbor. Kórnickie **11**: 5–104 (1966). —Venter in S. African J. Bot. **63**: 123–128 (1997); in Hedberg & Edwards, Fl. Ethiopia Eritrea **4**: 103–106 (2003); in Fl. Somalia **3**: 138–139 (2006); in F.T.E.A., Apocynaceae (part 2): 118–120 (2012).

Shrub, scrambler or liana. Latex white. Leaves persistent or soon falling, opposite, broadly elliptic, linear-elliptic, broadly ovate or linear-ovate, sometimes bracteate. Inflorescences compact to lax, few to many-flowered. Sepals mostly with paired colleters at inner base. Corolla tube saucer-shaped; lobes glabrous to villous, inside with or without dark coloured glandular centres and densely puberulous white spots. Corona inserted at corolla mouth, lobes simple, undivided or bi- or tri-segmented; undivided lobes filiform, linear or ovate, apically simple or

---

[2] by H.J.T. Venter

D.E.

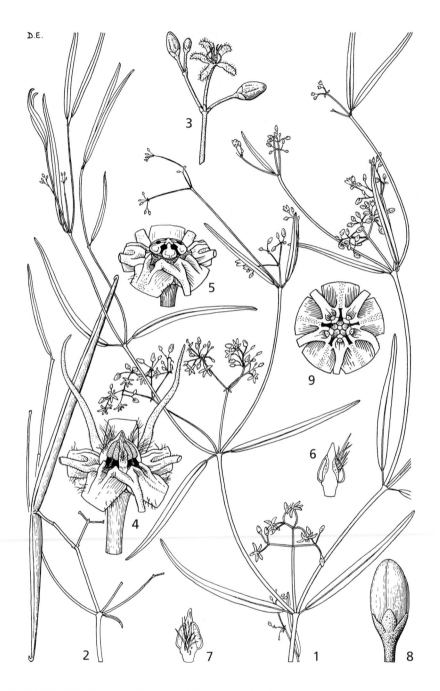

Fig. 7.3.**121**. PERIPLOCA LINEARIFOLIA. 1, stem with flowers (× 1); 2, stem with follicles (× 1); 3, part of inflorescence (× 3); 4, 5, flower (× 15); 6, adaxial view of anther (× 25); 7, abaxial view of anther with hairs (× 25); 8, young bud (× 8); 9, flower from above (style removed) (× 8). Drawn by D. Erasmus. Reproduced from Flora of Tropical East Africa (2012).

bifid; bi-segmented lobes subulate; tri-segmented lobes with central segment filiform or linear, apically simple or bifid, lateral segments fleshy, ovoid-deltoid and fused to inside base of corolla lobes, basally fused with staminal filaments. Stamens inserted directly below corona lobes, anthers villous or hirsute, pollen in tetrads. Interstaminal nectaries fused with staminal filaments, lobular. Style-head ovoid to broadly ovoid, translator spoon ovate, stalked. Gynostegium exposed from corolla. Follicles paired, linear-ovoid to very narrowly ovoid, narrowly to horizontally divergent.

Periploca includes 12 species which are widely distributed over Africa, Europe and Asia. One species occurs in the F.Z. area.

Periploca linearifolia Quart.-Dill. & A. Rich. in Ann. Sci. Nat., Bot., ser. 2, **14**: 263 (1840). —Venter in F.T.E.A., Apocynaceae (part 2): 118 (2012). Type: Ethiopia, Adowa, *Quartin-Dillon & Petit* s.n. (P holotype, K). FIGURE 7.3.**121**.

    Periploca linearis Hochst. in Flora 24(1, Intelligenzb.): 25 (1841), superfluous name. Type as for *P. linearifolia.*
    Periploca refractifolia Gilli in Ann. Naturhist. Mus. Wien **77**: 17 (1973). Type: Tanzania, Livingstone Mountains, Madunda, *Gilli* 396 (W holotype).

Large liana. Stems up to 20 m long or more, glabrous, bark greyish-brown. Leaves glabrous, shortly petiolate, leathery; blade 3–7(9) cm × 2–5 mm, linear, linear-ovate to very narrowly ovate, long attenuate, base cuneate or obtuse. Inflorescences paniculate, usually many-flowered; peduncles 1–2 cm long; pedicels 2–6 mm long; bracts ovate, ± 1 mm long. Sepals 1 × ± 1 mm, broadly ovate to sub-orbicular, apex obtuse. Corolla sub-herbaceous; tube ± 0.5 mm long; lobes 3–4 × 1–2 mm, linear-ovate to very narrowly ovate, obtuse to retuse, above violet with dark glabrous centre, white-villous on margin and apex. Corona lobes violet, puberulous, tri-segmented; central segment 3–4 mm long, filiform, puberulous, lateral segments fleshy ovoid-deltoid and fused to inner base of corolla lobes. Stamens ± 1 mm long, anthers hastate. Style ± 0.5 mm long, style-head ± 1 mm long. Follicles horizontally divergent, 6–12(16) cm × 4–5 mm, cylindrical-ovoid. Seeds black, 7–9 mm long, coma 2.5–3 cm long.

Zambia. N: Nyika Plateau, Chowo, fl. i.1981, *Dowsett-Lemaire* 98 (K). **Malawi**. N: Chipita Dist., Misuku Hills, fl. & fr. 23.v.1989, *Goyder et al.* 3274 (K, MAL, PRE). C: Dedza, Chongoni Forest, 10.i.1961, *Chapman* 1152 (K, SRGH). S: Zomba Dist., lower slopes of Mount Zomba, fl. 23.i.1966, *Agnew & Jones* 162 (MAL).

Also in Burundi, D.R. Congo, Ethiopia, Kenya, Sudan, Tanzania and Uganda. In bamboo and podocarp forest, riparian scrub forest, open dry montane forest and semi-deciduous dry forest; 700–2800 m.

The milksap is fed to cows to enhance lactation; very strong twine, rope and fish nets are woven from the stem bark; tea is brewed from the roots.

31. PARQUETINA Baill.[3]

Parquetina Baill. in Bull. Mens. Soc. Linn. Paris **2**: 806 (1889); Hist. Pl. **10**: 294 (1891). —Schumann in Engler & Prantl, Nat. Pflanzenfam. 4(2): 218 (1895). —Venter in S. African J. Bot. **75**: 557–559 (2009); in F.T.E.A., Apocynaceae (part 2): 122–123 (2012).
    Omphalogonus Baill. in Bull. Mens. Soc. Linn. Paris **2**: 812 (1889). —Brown in F.T.A. 4(1): 256 (1902).

Lianas with white latex; stems woody, twining. Leaves glabrous, petiolate, leathery; blade broadly ovate to elliptic, obtuse to cuspidate, base cordate, above glossy, green, below pale green, secondary veins divaricate. Inflorescences axillary, many-flowered, glabrous. Sepals glabrous, with colleters at their inner bases. Corolla coriaceous, deep crimson, purplish or pink. Corolline corona with coronal feet fused to stamens and nectaries forming a collar; stamens arising directly

---

[3] by H.J.T. Venter

below corona lobes. Gynostegial corona absent. Style terete; style-head deltoid, obtuse; translator with receptacle, stalk and sticky disc. Follicles paired, horizontally opposite.

A genus of 2 species, endemic to tropical Africa.

Parquetina has had a checkered history, with recent molecular analyses placing it firstly with *Periploca*, then more recently with *Cryptolepis*. It is maintained here as a separate genus for consistency with other regional tropical Floras.

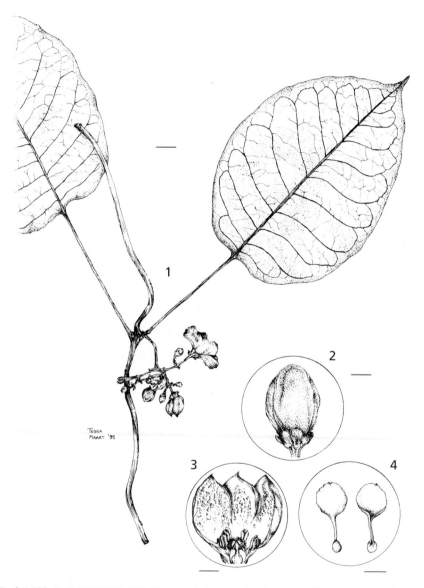

Fig. 7.3.**122**. PARQUETINA CALOPHYLLA. 1, stem with leaves and flowers; 2, flower bud; 3, flower opened showing the corona lobes, hairy stamens and pistil; 4, translators, posterior and anterior views. Scale bars: 1 = 10 mm; 2, 3 = 2.5 mm; 4 = 0.3 mm. Drawn by T. Maart. Reproduced with permission from South African Journal of Botany (1996).

Parquetina calophylla (Baill.) Venter in S. African J. Bot. **75**: 558 (2009). Type: Tanzania, Zanzibar, *Boivin* 1008A (P holotype). FIGURE 7.3.**122**.

Omphalogonus calophyllus Baill. in Bull. Mens. Soc. Linn. Paris **2**: 812 (1889); Hist. Pl. **10**: 300 (1891). —Brown in F.T.A. 4(1): 256 (1902). —Venter & Verhoeven in S. African J. Bot. **62**: 24 (1996).

Omphalogonus nigritanus N.E. Br. in Bull. Misc. Inform. Kew **1912**: 279 (1912). Type: Southern Nigeria, *Thomas* 1011 (K holotype).

Periploca calophylla (Baill.) Roberty in Bull. Inst. Franç. Afrique Noire **15**: 1429 (1953), illegitimate name, non (Wight) Falc. (1841).

Parquetina nigrescens sensu (Afzel.) Bullock in Kew Bull. **15**: 205 (1961), in part.

Cryptolepis calophylla (Baill.) L. Joubert & Bruyns in Taxon **65**: 498 (2016).

Large liana. Roots unknown. Stems up to 20 m long, glabrous, pale brown, verrucose, flaky. Leaf petiole (2)5–7 cm long; blade broadly ovate, 9–12(14) cm long, 5–11 cm wide, veins pale green above, dark green beneath. Inflorescences 10–30-flowered; peduncles 1–3 cm long, pedicles 4–5 mm long; bracts acicular, ± 2 mm long. Sepals sub-orbicular, 2 mm long, 2 mm wide. Corolla glabrous outside; tube campanulate, 5–6 mm long, inside of mouth papillate; lobes broadly elliptic to broadly ovate, obtuse, spreading, 6–7 mm long, 6–7 mm wide, inside papillate, pink, maroon, dark red or violet to deep violet. Coronal collar borne in lower half of corolla tube; lobes violet to maroon, 2 mm long and fused to corolla tube's inner face for ± 2 mm, radiating into corolla tube cavity, radiating processes channelled, upper part bisegmented, outer segment subulate to obtuse, inner segment concavely hood-shaped. Stamens green; filaments terete, ± 0.5 mm long, glabrous; anthers narrowly hastate-ovate, ± 0.5 mm long, pubescent to rarely glabrous on back. Interstaminal nectaries sub-orbicular, erect around style. Style terete, ± 1 mm long, style-head ± 1 mm long; translators ± 1 mm long. Gynostegium concealed in corolla tube. Follicles vary narrowly deltoid-ellipsoid to deltoid-ovoid, 2-edged, obtuse-acute, 11–25 cm long, 2–4 cm diam., glabrous. Seeds narrowly elliptic to obliquely narrowly elliptic, dark brown, 6–8 mm long, reticulate to warty; coma 3–4 cm long.

Mozambique. N: Nampula Province, between Lurio and Namapa, 17.iii.1948, *Pedro & Pedrogão* 4846 (LMA) MS: Inhamitanga Forest Reserve, 30.iii.2017, *Wursten* 1670 (BR).

Sterile material of what is probably this species have also been seen from Gonarezhou in SE Zimbabwe (Bart Wursten pers. comm.).

Also in Tanzania, Kenya, Uganda, Sudan, the Central African Republic and across West Africa to Mali and Senegal. In swampy forest, riverine forest and coastal thicket and grassland; 0–1100 m.

The roots are used as an aphrodisiac. The flossy cortical fibres are spun into fish nets and bow strings.

32. **CRYPTOLEPIS** R. Br.[4]

Cryptolepis R. Br., On the Asclepiadeae: 58 (1810). —Brown in F.T.A. 4(1): 242 (1902); in Fl. Cap. 4(1): 526 (1907). —Hutchinson & Dalziel in F.W.T.A. **2**: 53 (1931). —Venter in Fl. Somalia **3**: 136–138 (2006); in F.T.E.A., Apocynaceae (part 2): 126–135 (2012).

Leposma Blume, Bijdr. Fl. Ned. Ind.: 1049 (1826).

Lepistoma Blume, Fl. Javae: 7 (1828), illegitimate name.

Cryptolobus Steud., Nomencl. Bot., ed. 2 **1**: 450 (1840), non Spreng. (1826).

Curroria Benth. in Hooker, Niger Fl.: 457 (1849).

Ectadiopsis Benth. in Bentham & Hooker, Gen. Pl. **2**: 741 (1876).

Mitolepis Balf.f. in Proc. Roy. Soc. Edinburgh **12**: 78 (1883).

Cochlanthus Balf.f. in Proc. Roy. Soc. Edinburgh **12**: 78 (1883), illegitimate name, non Benth. (1852).

Socotranthus Kuntze in Post & Kuntze, Lex. Gen. Phan.: 523 (1903).

---

[4] by H.J.T. Venter

Mesophytic climbers or sclerophyllous shrubs, scramblers or small trees, interpetiolar ridges and leaf axils with reddish dentate colleters, latex white or rarely yellow-orange to orange. Leaves opposite on normal shoots, sub-fascicled on stunted shoots, petiolate to sessile; blade linear to broadly ovate or obovate to orbicular, herbaceous, leathery or succulent, secondary and tertiary veins sometimes invisible. Inflorescences few to many-flowered or flowers solitary. Flower buds with corolla apex helically twisted and deltoid to apiculate. Corolla, creamy yellow, mauve, red or violet-red; tube campanulate and shorter than lobes, rarely cylindrical and longer than lobes, mostly with elliptic papillose spot below every corona lobe; lobes linear to narrowly ovate to ovate. Corona 1 or 2-whorled; lower whorl of lobes inserted around middle of corolla tube, mostly concealed; lobes clavate, acicular, peg-like or filiform, glabrous, free; upper whorl, when present, of simple - or lobed pockets in corolla lobe sinuses. Stamens inserted near base of corolla tube; filaments short, mostly dilated; anthers deltoid to hastate, glabrous or rarely hairy; pollen in tetrads. Insterstaminal nectariferous pouches at base of corolla tube. Style-head deltoid, broadly deltoid, ovoid or broadly ovoid, translators with receptacle concavely narrowly elliptic and sessile on sticky disc, rarely triangular and stalked on sticky disc. Gynostegium concealed at bottom of corolla tube. Follicles paired, widely to narrowly divergent, cylindrical-ovoid to narrowly ovoid.

A genus of 27 species in Africa, 3 in Asia and 1 in Arabia.

1. Corolla pink, pink-violet, violet or brownish-red. 2
   – Corolla white, creamy yellow, yellow, greenish-white or greenish-yellow. 3
2. Erect shrublet; leaves oblong, oblong-elliptic or oblong-ovate with base obtuse to cuneate, 2–6 cm long, 1–6 mm wide, light green on both sides; flowers solitary; corona lobes filiform . . . . **2.** *decidua*
   – Climber; leaves elliptic, broadly elliptic, broadly obovate or sub-orbicular, base cordate or rarely obtuse, 6–16 cm long, 4–10 cm wide, above green, beneath glaucous; flowers in cymose racemes or panicles; corona lobes ovoid with curved apex . . . . **5.** *hypoglauca*
3. Flowers with a single corona (only a lower corona of clavate, deltoid or acicular lobes) 4
   – Flowers with two corona whorls (a lower corona of clavate lobes more or less at middle of corolla tube plus an upper corona of pockets or lobed pockets in corolla lobe sinuses) 6
4. Leaves 1–2.5 cm long, 4–12 mm wide, puberulous; corolla 3–6 mm long; follicles 3–5 cm long . . . . **6.** *delagoensis*
   – Leaves 5–17 cm long, 2–7 cm wide, glabrous; corolla 7–21 mm long; follicles 15–20 cm long 5
5. Leaf blade leathery; anthers glabrous; follicles horizontally divaricate, narrowly cigar-shaped . . . . **4.** *apiculata*
   – Leaf blade herbaceous; anthers villous; follicles narrowly divaricate, very narrowly cylindrical, falcate and nodose . . . . **7.** *capensis*
6. Erect shrubs or shrublets. . . . . **1.** *oblongifolia*
   – Climbers 7
7. Leaf blade leathery, apex rotund-apiculate and recurved; bud apex acute to rotund, half-turn helically twisted; anthers villous; follicles ovoid-apiculate to narrowly ovoid-apiculate . . . . **3.** *cryptolepioides*
   – Leaf blade herbaceous, apex obtuse-mucronate; bud apex full-turn helically twisted; anthers glabrous; follicles cylindrical-ovoid and attenuate. . . . . . **8.** *obtusa*

1. **Cryptolepis oblongifolia** (Meisn.) Schltr. in J. Bot. **34**: 315 (1896). —Venter in F.T.E.A., Apocynaceae (part 2): 130 (2012). Type: South Africa, KwaZulu-Natal, Umgeni, xi.1839, *Krauss* 132 (K-Bentham lectotype, K-Hooker, BM, MO), lectotypified by Venter in F.T.E.A., Apocynaceae (part 2): 130 (2012). FIGURE 7.3.**123**.

    Ectadium oblongifolium Meisn. in London J. Bot. **2**: 542 (1843).

Secamone acutifolia Sond. in Linnaea **23**: 76 (1850). Types: South Africa, Gauteng, Magaliesberg, *Zeyher* 1182 (B† syntype, G, K, P, S); Port Natal, *Gueinzius* 431 (B† syntype, S).

Ectadiopsis acutifolia (Sond.) B.D. Jacks. in Index Kew. **1**: 822 (1893).

Ectadiopsis oblongifolia (Meisn.) B.D. Jacks. in Index Kew. **1**: 822 (1893).

Cryptolepis brazzaei Baill. in Bull. Mens. Soc. Linn. Paris: **2**: 803 (1889). Type: Angola, Pungo Andongo Dist., *Welwitsch* 4197 (P holotype, COI, G, K, LISU).

Ectadiopsis lanceolata Baill. in Bull. Mens. Soc. Linn. Paris: **2**: 803 (1889). Type: Angola, Huila, *Welwitsch* 4207 (P holotype, BM, COI).

Ectadiopsis myrtifolia Baill. in Bull. Mens. Soc. Linn. Paris: **2**: 803 (1889). Type: Angola, Huila Dist., near Lopollo, *Welwitsch* 4206 (P holotype, BM, G, K, LISU).

Ectadiopsis welwitschii Baill. in Bull. Mens. Soc. Linn. Paris: **2**: 802 (1889). Type: Angola, Huila, *Welwitsch* 4203 (P holotype, BM, COI, G, K, LISU).

Cryptolepis sizenandii Rolfe in Bol. Soc. Brot. **11**: 86 (1893). Type: Angola, Malange, *Marques* 185 (COI holotype, K).

Ectadiopsis buettneri K. Schum. in Engler & Prantl, Nat. Pflanzenfam. **4**(2): 219 (1895). Type: D.R. Congo, Kinshasa (Leopoldville), *Buettner* s.n. (BR holotype?).

Cryptolepis welwitschii (Baill.) Schltr. in J. Bot. **33**: 301 (1895).

Cryptolepis welwitschii var. *luteola* Hiern, Cat. Afr. Pl. **1**: 677 (1898). Type: Angola, Huila Province, near Huila, *Welwitsch* 4205 (BM syntype, K, LISU).

Cryptolepis angolensis Hiern in Cat. Afr. Pl. **1**: 677 (1898). Type: Angola, Huila Dist., near Lopollo, *Welwitsch* 4204 (BM syntype, K, LISU).

Cryptolepis myrtifolia (Baill.) Hiern in Cat. Afr. Pl. **1**: 677 (1898).

Ectadiopsis suffruticosa K. Schum. in Bot. Jahrb. Syst. **28**: 453 (1900). Type: Tanzania, Uhehe, Uringa, Weru Area, *Goetze* 665 (B† holotype).

Cryptolepis hensii N.E. Br. in F.T.A. **4**(1): 246 (1902). Type: D.R. Congo, Ntombi River near Lutete, 15.ii. 1888, *Hens* 227 (K holotype, BR, L, P, Z).

Cryptolepis baumii N.E. Br. in F.T.A. **4**(1): 247 (1902). Type: Angola, Amboella Dist., Longa River below Napalanka, *Baum* 577 (K holotype, BM, BR, COI, G, HBG, M, Z).

Cryptolepis producta N.E.Br. in F.T.A. **4**(1): 247 (1902); Venter in F.T.E.A., Apocynaceae (part 2): 136 (2012). Type: Angola, Amboella, Kubango River above Kinimarva, *Baum* 457 (K holotype, BM, W).

Cryptolepis suffruticosa (K. Schum.) N.E. Br. in F.T.A. **4**(1): 251 (1902).

Cryptolepis baumii Schltr. in Warburg, Kunene-Sambesi Exped.: 340 (1903), illegitimate name, non N.E. Br. (1902). Type: Angola, Kubango, *Baum* 457 (B† holotype, E, HBG, K, M, W, Z).

Cryptolepis linearis N.E. Br. in Bull. Misc. Inform. Kew **1908**: 408 (1908).

Ectadiopsis oblongifolia (N.E. Br.) Bullock in Kew Bull. **11**: 278 (1956), superfluous name.

Ectadiopsis producta (N.E.Br.) Bullock in Kew Bull. **11**: 278 (1956).

Erect, procumbent or scrambling shrub of up to 0.3–1.25 m high. Roots woody, rootstock perennial. Stems up to 1.5 m × 4–7 mm, reddish-brown, scaberulous, sometimes lenticulate. Leaves opposite, leathery, glabrous, sessile, sub-sessile or petiole up to 5 mm long; blade linear-ovate to broadly ovate, narrowly obovate, obovate, linear elliptic or elliptic, obtuse-mucronate, acute or acuminate, base cuneate to obtuse, (2)4–9(12) × (0.2)1–2(3) cm, above dark green, beneath pale green to glaucous, glabrous, scabredulous or scabrous, main and secondary veins visible, secondary veins arched to divaricate and looped, 5–17 per side. Inflorescences axillary and terminal, usually compact, few to many-flowered, glabrous or scabrous; peduncles 0.2–4 cm long, pedicels 2–6 mm long, bracts very narrowly ovate to very narrowly triangular, ± 1.5 mm long. Buds ovoid with apex deltoid, half-turn helically twisted. Flowers sweet-scented. Sepals narrowly triangular to ovate to broadly ovate, 1.5–2.5 × 1 mm, green tipped purple, glabrous or scabredulous, with margins fimbriate. Corolla creamy white, yellow or pale lemon-yellow inside, pale yellow tinged red or pale brownish-green to purplish outside, glabrous; tube campanulate, 1.5–4 mm long, with papillose spots below corona lobes; lobes oblong-ovate to obliquely ovate, obtuse, 2–8 × 1–2 mm, glabrous. Corona double: lower whorl of clavate lobes, ± 1 mm long; upper whorl of sinus pockets, pocket-rim lobed. Stamens glabrous, filaments ± 0.5 mm long; anthers hastate, acuminate, ± 1 mm long, sometimes slightly hairy. Style terete, ± 0.5 mm long; style-head conical to broadly conical, obtuse to bifid, ± 1 mm long, translators narrowly elliptic.

Fig. 7.3.**123**. CRYPTOLEPIS OBLONGIFOLIA. 1, flowering shoot (× 2/3); 2, fruiting shoot, (× 2/3); 3–6, variation in leaf shape (× 1); 7, inflorescence (× 4); 8, half-flower (× 8); 9, flower from above (× 4); 10, 11, stamen from beneath and from above (× 27); 12, 13, translators with pollen deposited (× 38). 1, 7–13 from *Thomas* 3909; 2 from *Purseglove* 2240; 3, 5 from *Milne-Redhead & Taylor* 9760; 4 from *Richards* 195; 6 from *Michel* 4643. Drawn by D. Erasmus. Reproduced from Flora of Tropical East Africa (2012).

Follicles single or paired, when paired ± 30° divergent, narrowly ovoid, acute to blunt acute, ribbed, glabrous, (4)8–12(16) × 0.5–1.3 cm. Seeds obliquely ovate to oblong-ovate, 2–7 mm long, warty, dark brown; coma 3–4 cm long, whitish

Caprivi. Mashi/Caprivi, fl. 5.xi.1962, *Fanshawe* 7140 (K, NDO). **Botswana**. N: Aha Hills, fr. 25.iv.1980, *P.A. Smith* 3438 (K, PSUB, SRGH). SE: Lobatse, fr. 4.iv.1981, *Woollard* 957 (SRGH). **Zambia**. B: 3 miles S of Kalabo, fl. 16.xi.1959, *Drummond & Cookson* 6561 (K, SRGH). N: Isoka Dist., 18 km from Tunduma on road to Mbala, fl. 10.i.1975, *Brummitt & Polhill* 13701 (BR, K, SRGH). W: Mwinilunga Dist., near source or R. Matonchi, fl. 7.x.1937, *Milne-Redhead* 2631 (K). C: Mkushi Dist., Fiwila, fl. 3.i.1958, *Robinson* 2595 (K, M, SRGH). E: Chipata, 10 km NE, fl. 11.i.1959, *King* 451 (SRGH). S: Kalomo Dist., Tara P.F.A. Siachitema Chieftancy, fl. 26.xi.1963, *Bainbridge* 915 (K, NDO, SRGH). **Zimbabwe**. N: Urungwe Dist., Urungwe Nature Reserve, Zwipani Camp, fl. 30.xi.1957, *Goodier* 422 (K, SRGH). W: Matobo Dist., Hope Fountain Mission, fl. 3.xi.1973, *Norrgrann* 433 (K, SRGH). C: Marondera Dist., Dombi Estate, 20 km N of Marondera, fl. 13.xii.1970, *Biegel* 3436 (K, LISC, SRGH). E: Chimanimani, Chimanimani Mountain Forest Reserve, fl. 15.xi.1967, *Mavi* 645 (K, LISC, SRGH). S: Bikita Dist., Turgwe River, fr. 6.v.1969, *Pope* 95 (K, LISC, SRGH). **Malawi**. N: 8 km E of Mzuzu, fr. 7.vi.1975, *Pawek* 9673 (K, MAL, SRGH, UC). C: Dedza to Mphunzi road, fr. 18.iii.1968, *Jeke* 159 (K, LISC, SRGH). S: Blantyre Dist., Kirk Range, 10 km along Thambani road from Mwanza, fl. 16.i.1992, *Goyder & Paton* 3524 (BR, K, MAL, MO, PRE). **Mozambique**. N: Mission Catholique de Unango, fl. xii.?1932, *Sousa* 1038 (K, ?LMA). Z: Entrance to Gurué and Ile, fr. 5.iv.1943, *Torre* 5076 (LISC, LMA). T: Angónia Dist., Ulongue, fl. 2.xii.1980, *Macuacua* 1368 (K, LMA, WAG). MS: Manica e Sofala, Edmundium Copper Mine, fr. 29.iii.1966, *Chase* 4812 (BR, K, LISC, SRGH).

Also found in Angola, Benin, Burkina, Burundi, Cameroon, Central African Republic, Chad, D.R. Congo, Ghana, Guinea, Guinea Bissau, Ivory Coast, Kenya, Mali, Namibia, Nigeria, Rwanda, Senegal, South Africa, Swaziland, Tanzania, Togo and Uganda. In open rocky miombo woodland, *Uapaca* woodland on grey sandy loam, riverbank woodland, grassland, common in disturbed ground; 800–2400 m.

Chimpanzees eat this species apparently when feeling unwell. The woody roots are used as aphrodisiac. Roots are boiled for 15–20 minutes and the extract is drunk for coughs and tuberculosis. Vernacular name "Nyashinda" in the Chimanimani Region.

2. **Cryptolepis decidua** (Benth.) N.E. Br. in F.T.A. **4**(1): 243 (1902). Type: Angola, *Curror* s.n. (K holotype).

> *Curroria decidua* Benth. in Hooker, Niger Fl.: 457 (1849). —Engler in Bot. Jahrb. Syst. **10**: 244 (1889). —Schumann in Engler & Prantl, Nat. Pflanzenfam. 4(2): 218 (1895). —Huber in Merxmüller, Prodr. Fl. Sudwestafr. 113: 3 (1967).

A suffrutescent dwarf-shrub. Roots unknown. Stems erect, up to 0.5–1.0 m × 5 mm, smooth, glabrous, interpetiolar ridges villous, branchlets often stunted. Leaves opposite or fascicled, axils villous, sub-sessile or petiole up to 2 mm long; blade oblong, occasionally oblong-elliptic or oblong-ovate, obtuse to acute and occasionally emarginate, mucronulate to mucronate, base cuneate to obtuse, 1–4.5 × 0.1–0.6 cm, slightly succulent, glabrous, light green, occasionally with purple spots, veins invisible. Flowers solitary on stunted branchlets; pedicels 5–10(19) mm long; bracts when present acicular, 0.5–1.0 mm long, fimbriate with long silky trichomes. Buds oblong-ovoid, acute, slightly turned to half-turn helically twisted. Sepals oblong-ovate, acute, 0.5–2.5 × 0.5–1.0 mm, margin fimbriate. Corolla violet, glabrous; tube shortly campanulate, 2–3 mm long; lobes spreading, linear-oblong, obtuse, 6–9 × 1 mm. Corona of lower whorl only, slightly exserted, lobes filiform, simple or occasionally bifid, 2–3 mm long. Stamens sessile, anthers triangular, acute, glabrous, 0.7–1.0 mm long. Style terete, ± 0.3 mm long; style-head deltoid, obtuse to acutely bifid, 0.7–1.0 mm long; translator triangular. Follicles erect, widely divaricate to horizontal, narrowly ovoid, attenuate, bases cuneate, 2.5–6.5 × 0.3–1 cm, finely ribbed, light brown. Seeds oblong to broadly elliptic, 3–6 × 1–2 mm; coma white to yellowish-white, 1–2 cm long.

Botswana. SE: Kanye, Pharing, fr. xi.1947, *Miller* B527 (K).

Also found in Angola, Namibia and South Africa. Widespread in the southern African desert and semi-desert regions. Grows on rocky hillsides, on plains and along ravines, usually in sandy or calcareous soils; 550–1500 m.

3. **Cryptolepis cryptolepioides** (Schltr.) Bullock in Kew Bull. **10**: 281 (1955). Type: South Africa, Botsabelo, 29.xii.1893, *Schlechter* 4082 (B† holotype, BM, BOL, BR, G, GRA, K, NBG, NH, PRE, Z).

> *Ectadiopsis cryptolepioides* Schltr. in Bot. Jahrb. Syst. **19**(Beibl. 51): 10 (1895).
> *Cryptolepis transvaalensis* Schltr. in J. Bot. **34**: 315 (1896), illegitimate name. —Brown in Fl. Cap. **4**(1): 528 (1907). Type as for *C. cryptolepioides*.

A climber or scrambler. Roots unknown, Stems slender, densely branched, twining, up to 2 m × 5 mm, slightly verrucose, glabrous, interpetiolar ridges villous. Leaves decussate opposite, coriaceous, axils villous; petiole purplish-red, glabrous, 4–9(20) mm long; blade broadly obovate, broadly elliptic or orbiculate, apex rotund, rarely emarginate, apiculate, acuminate or cuspidate, recurved, base obtuse, 2–4(6) × 2–4(5) cm, coriaceous, rugose, glabrous, dark green above, pale green beneath, bright purple spots occasionaly present along abaxial side of main vein, main -, secondary - and tertiary veins visible, secondary veins divaricate and looped, 12–20 per side. Inflorescences axillary, compact, many-flowered; peduncles semi-frail, primary 4–10(25) mm long, secondaries 2–6(11) mm long; pedicels 2–3(12) mm long; bracts densely packed, acicular, 1.5 mm long, margin fimbriate. Buds ovoid, acute to rotund, half-turned helically twisted. Sepals broadly ovate, ± 1.5 × 1 mm, margin fimbriate. Corolla greenish, deep cream or pale yellow, glabrous; tube campanulate, 1.5–2.0 mm long; lobes spreading, oblong, rotund to obtuse, 2–3 × 1 mm. Corona double; lower lobes clavate with conical apices, 0.8–1.0 mm long, glabrous, greenish to yellowish; upper lobes pocket-like, glabrous. Stamens subsessile; anthers hastate, attenuate, ± 0.5 mm long, outside villous. Style ± 0.3 mm long, style-head angular deltoid, ± 0.5 mm long, acutely bifid, translator narrowly elliptic. Follicles erect, widely divaricate, ovoid to narrowly ovoid and apiculate, base obtuse, finely ribbed, 3–6 × 0.6–1.5 cm. Seeds obliquely obovate to obovate, 4–9 × 2–4 mm, warty; coma white to yellowish-white, 2–3 cm long.

Zimbabwe. N: Harare, Dales Estate, fl. 2.ii.1966, *Simon* 670 (K, LISC, SRGH). W: Matobos, Besna Kobila farm, fl. 30.i.1973, *Grosvenor* 795 (K, SRGH). C: Honzo Mountain, Kukwanisa Training Farm, fl. 5.i.1968. E: Mutare, Dora Farm, fl. 17.xii.1950, *Chase* 3439 (SRGH). S: Masvingo, Lake Kyle, fl. ii.1979, *Burrows* 1338 (SRGH). **Mozambique**. N: Serra Mepáluè, 7.ix.1968, *Macedo & Macuácua* 3561 (LMA).

Common in bushveld, scrub and ravine forest, on mountain slopes, cliffs and plateaus. This species is associated with sandy soil on granite or sandstone outcrops in moist areas along river beds and around dams; 750–1500 m.

4. **Cryptolepis apiculata** K. Schum. in Engler, Pflanzenw. Ost-Afrikas **C**: 320 (1895). —Venter in F.T.E.A., Apocynaceae (part 2): 128 (2012). Type: Tanzania, Amboni, vi.1893, *Holst* 2564 (B† holotype, M lectotype, K, COI), lectotypified by Venter in F.T.E.A., Apocynaceae (part 2): 128 (2012).

Many-stemmed climber. Roots unknown. Stems twining, up to 3 m × 2–4 mm, brown, glabrous, sparsely lenticulate. Leaves opposite, leathery, glabrous; petiole 0.5–3 cm long; blade narrowly elliptic or oblong-elliptic, obtuse-attenuate, base obtuse to cuneate, (5)7–13(17) × (2)3–5(7) cm, glossy dark green above, pale green beneath, main and secondary veins visible, secondary veins arched and looped, 7–12 per side. Inflorescences axillary, open, few-flowered, glabrous; peduncles sturdy, primary 3–5 cm long, secondaries 3–6 cm long; pedicels sturdy, 0.8–1 cm long; bracts narrowly ovate, ± 1 mm long. Bud ovoid, long apiculate, full-turn helically twisted. Flowers glabrous. Sepals narrowly ovate to narrowly triangular, 2–3 × 1 mm. Corolla white, creamy yellow or pale greenish-yellow; tube campanulate, 3–5 mm long, with papillose spots; lobes linear, rounded, 12–16 × 1–2 mm. Corona: lower whorl of clavate or acicular lobes, ± 1 mm long; upper whorl of sinus pockets. Stamens: filaments ± 1 mm long; anthers narrowly triangular, attenuate,

± 1 mm long, pale yellow, glabrous. Style ± 0.5 mm long; style-head broadly angular-deltoid, ± 1 mm long, attenuate, translator narrowly elliptic. Follicles paired, horizontally opposite, cylindrical-ellipsoid, attenuate, 15–16 × 0.8–1.0 cm, brown, glabrous. Seeds obliquely narrowly elliptic, 8 mm long, 2 mm wide, coma white, 3–4 cm long.

Zambia. S: Kaloma, Fl, 18.xii.1962, *Mataundi* 17/30 (LISC). **Zimbabwe**. E: Mutare Dist., waterfall S of Drumfad, fl. 2. xi.1952, *Chase* 4696 (BM, K, SRGH). **Mozambique**. MS: Chimoia, Belas, fr. 2.iv.1948, *Garcia* 854 (LISC, BR).

Also in Kenya, Tanzania and D.R. Congo. Lowland forest and shrubby thickets in grassland; 30–350 m.

5. **Cryptolepis hypoglauca** K. Schum. in Engler, Pflanzenw. Ost-Afrikas **C**: 320 (1895). —Venter in F.T.E.A., Apocynaceae (part 2): 129 (2012). Type: Tanzania, Amboni, vi.1893, *Holst* 2728 (B† holotype, K lectotype) lectotypified by Venter in F.T.E.A., Apocynaceae (part 2): 129 (2012).

Many-stemmed climber. Roots unknown. Stems twining, up to 6 m × 2–5 mm, pale brown to reddish brown, glabrous, lenticulate. Leaves opposite, sessile or sub-sessile, glabrous; blade elliptic, broadly elliptic, broadly obovate or sub-orbicular, retuse or obtuse, mucronate or apiculate, base cordate to obtuse, (6)8–10(16) × (4)5–8(10) cm, leathery, green above, glaucous beneath, violet spotted or blotched, main and secondary veins visible, secondary veins arched and looped, 5–7 per side. Inflorescences axillary, rarely terminal, open, monochasial branches 3–6-flowered, glabrous; peduncles sturdy, reddish brown, primary 2–3(10) cm long, secondaries 1–3 cm long, densely packed with floral bracts and scars; bracts broadly ovate to triangular, clasping, ± 1 mm long; pedicels sturdy, 1–5 mm long, reddish brown. Buds sub-globose, short apiculate, half-turn helically twisted. Sepals narrowly ovate to narrowly triangular, ± 2 × ± 0.5 mm, green, reddish blotched, glabrous, margins fimbriate. Corolla dull purple, brownish-red, pinkish-purple or pink, glabrous; tube broadly campanulate, 2.5–3 mm long, with papillose spots; lobes ovate to broadly ovate, acute to obtuse, 3–4 × 1–2.5 mm. Corona double; lower whorl of pale green to yellow ovoid lobes with curved apex, ± 1 mm long; upper whorl of sinus pockets. Stamens glabrous, filaments ± 0.5 mm long, anthers triangular, ± 1mm long, whitish. Style ± 0.5 mm long; style-head conical, long attenuate, ± 1.5 mm long, translators narrowly elliptic. Follicles pendulous, paired, horizontally opposite, dark reddish-brown, narrrowly cylindrical and long attenuate, 25–30 × 0.5–0.8 cm. Seeds very narrowly sub-elliptic to very narrowly sub-ovate, brown, 10 × 2 mm; coma white, 3–5 cm long.

Mozambique. N: Cabo Delgado Province, below NW escarpment of Namacubi Forest, W of Quiterajo, 29.xi.2008, *Goyder et al.* 5075 (K, LMA).

Also in Kenya, Tanzania, Cameroon and Central African Republic. In coastal dry forest; sea level.

6. **Cryptolepis delagoensis** Schltr. in Bot. Jahrb. Syst. **38**: 26 (1905). Type: Mozambique, Maputo [Lourenço Marques], xii.1897, *Schlechter* s.n. (B† holotype); neotype: South Africa, KwaZulu-Natal, Tembe Elephant Park, *Venter* 9335 (PRE neotype, BLFU), neotypified by Bester & Joubert in Bothalia **41**: 200 (2011).

A climber. Roots unknown. Stems slender, twining, up to 4 m × 3.5 mm, greyish-brown, slighty verrucose, glabrous, interpetiolar ridges glabrous. Leaves decussate opposite, petiole 1–3 mm long; blade elliptic to broadly elliptic, ovate or rarely obovate, obtuse-apiculate, cuspidate, acute or acuminate, base obtuse, 1.0–2.5 × 0.4–0.7(1.2) cm, coriaceous, puberulous and yellowish-green above, papillate and pale green beneath, margin revolute, secondary and tertiary veins indistinct. Inflorescences compact, few-flowered, glabrous; primary peduncles 1–3(10) mm long, secondaries 1–2 mm long; pedicels 1–3 mm long; bracts densely packed, acicular, 0.5–1.0 mm long, glabrous to fimbriate. Buds ovoid, acute to apiculate, full-turn helically twisted. Sepals broadly ovate, acute, ± 1 × ± 0.5 mm, glabrous, occasionally fimbriate. Corolla white to creamy yellow, glabrous; tube campanulate, 1.5–2.0 mm long; lobes spreading, linear-oblong, obtuse, 1.5–3.5 × ± 0.7 mm. Corona of lower whorl only, concealed, lobes oblong to clavate, emarginate to acute, fleshy, 0.3–0.5 mm long, glabrous. Stamens subsessile, anthers hastate, attenuate, villous outside, 0.7–0.9 mm long. Style terete, 0.2–0.3 mm long, style-head deltoid, acute, 0.5–0.6 mm long, translator narrowly elliptic. Follicles erect, widely to horizontally divaricate, narrowly ovoid, apex acute and recurved, 3–5 × 0.5–0.6 cm, dark brown. Seeds narrowly elliptic, 8–11 × 2.5 mm, dark reddish-brown, smooth; coma white, 1.3–1.7 cm long.

Mozambique. M: Maputo, Licuati Forest Reserve, fl. & fr. 7.xii.2001, *Goyder* 5034 (K, LMU).

Also found in South Africa (KwaZulu-Natal and Mpumalanga). Grows in sand forest and bushveld, mostly associated with sandy soil; 50–900 m.

7. **Cryptolepis capensis** Schltr. in Verh. Bot. Vereins Prov. Brandenburg **35**: 47 (1893). Type: South Africa, KwaZulu-Natal, Inanda, i.1881, *Wood* 1583 (B† holotype, BOL, K, MO).

A deciduous climber. Roots unknown. Stems slender, twining, up to 7 m × 3.5 mm, verrucose, glabrous, interpetiolar ridges villous. Leaves opposite, herbaceous, glabrous, axils villous; petiole 8–13(20) mm long; blade elliptic, rarely obovate, acuminate, cuspidate or mucronate, rarely obtuse or retuse, base obtuse, 5–6(11) × 2–3(5) cm, dark green above, pale green beneath, main and secondary veins visible, secondary veins arched and looped, 5–6 per side. Inflorescences axillary, open, few-flowered, glabrous; peduncles frail, primary 2–5(8.5) cm long, secondaries 1.5–3.5(9) cm long; pedicels 0.5–2.5 cm long; bracts acicular, 1.5 mm long, glabrous. Buds oblong, apiculate, full-turn helically twisted. Flowers glabrous. Sepals ovate, acute, 3 × 1 mm. Corolla white to greenish-white; tube campanulate, 2–5 mm long; lobes reflexed, linear-ovate, acute, 5–11 × 3 mm. Corona of lower whorl only, lobes deltoid to awl-shaped, 0.8–1.0 mm long, white to greenish-white. Stamens subsessile; anthers hastate, attenuate, ± 1 mm long, outside villous. Style ± 1 mm long; style-head angular-deltoid, ± 0.5 mm long, acutely bifid, translator narrowly elliptic. Follicles pendulous, narrowly divaricate, cylindrically ovoid, falcate, attenuate, 12–20 × 0.3–0.4 cm. Seeds oblong-obovate, 4–9 × 1.5–2.0 mm, finely ribbed or slightly warty, coma white, ± 2 cm long.

This species is most probably also present in Mozambique as it was collected in the Lebombo Mountain on the border of South Africa with Mozambique. Fairly common in the eastern region of South Africa. Afromontane and coastal forests, in margins and clearings; 50–1000 m.

8. **Cryptolepis obtusa** N.E. Br. in Bull. Misc. Inform. Kew **1895**: 110 (1895). —Venter in F.T.E.A., Apocynaceae (part 2): 132 (2012). Type: Mozambique, between Tete and sea coast, iii.1860, *Kirk* s.n. (K lectotype), lectotypified by Bullock in Kew Bull. **10**: 281 (1955). FIGURE 7.3.**124**.

    Cryptolepis obtusa K. Schum. in Engler, Pflanzenw. Ost-Afrikas **C**: 320, 424 (1895), illegitimate name. Type: East Africa (Tanzania?), *Stuhlmann* 7827 (B† holotype).

Many stemmed climber. Stems slender, twining, 3–5 m × 2–4 mm, brown, glabrous. Leaves opposite, glabrous, herbaceous; petioles 4–10 mm long; blade oblong-elliptic, oblong-obovate, elliptic or obovate, obtuse-mucronate, mucro 1–2 mm long, base truncate, 3–7(9) × 0.2–0.4 cm, bright green above, pale green beneath, veins all visible, secondary veins divaricate and looped, 9–14 per side. Inflorescences axillary, lax to sub-compact panicles with 3–7 monochasiums, few to many-flowered, glabrous; peduncles frail, primary 5–15 mm long, secondaries 5–10 mm long; pedicels 2–4 mm long; bracts acicular, ± 1 mm long. Buds ovoid to ellipsoid, apiculate, full-turn helically twisted. Flowers glabrous. Sepals broadly ovate, 1.5–2 × 1–1.5 mm. Corolla creamy yellow; tube campanulate, 1.5–2 mm long, with papillose spots; lobes linear to linear-ovate, 6–8 × ± 1 mm, obtuse. Corona double; lower whorl of clavate lobes, ± 1 mm long; upper whorl of sinus pockets. Stamens sub-sessile; anthers hastate, ± 0.8 mm long, attenuate, glabrous. Style terete, ± 0.5 mm long; style-head broadly deltoid, 0.5–1 mm diam., translators narrowly elliptic. Follicles horisontally opposite, cylindrical-ovoid, attenuate, 8–15 × 0.4–0.7 cm. Seeds obliquely narrowly ovate, 5–7 mm long, dark brown, coma 30–40 mm long.

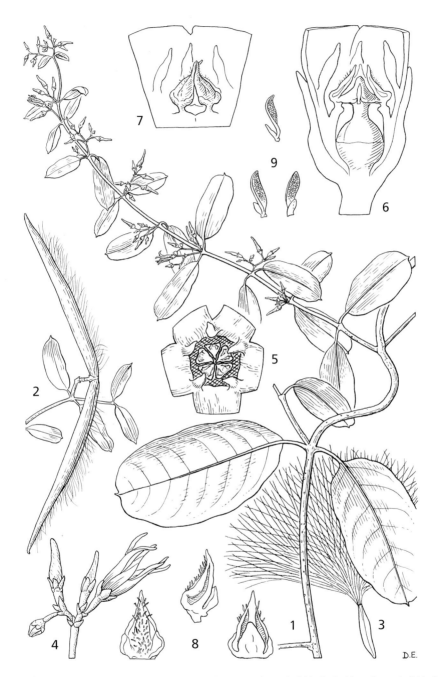

Fig. 7.3.**124**. CRYPTOLEPIS OBTUSA. 1, flowering shoot (× ²/₃); 2, fruiting shoot (× ²/₃); 3, seed with coma of hairs (× 2); 4, inflorescence (× 3); 5, centre of flower from above (× 8); 6, half-flower (× 12); 7, portion of flower showing gynostegium and 3 of the 5 corolline corona lobes (× 12); 8, stamens in dorsal, lateral and ventral views (× 20); 9 translators (× 27). 1, 4–9 from *Balsinhas & Marrime* 385; 2, 3 from *Faulkner* Kew 460. Drawn by D. Erasmus. Reproduced from Flora of Tropical East Africa (2012).

Zambia. N: Mpika Dist., 26 km S of Mfuwe, fr. 22.vii.1969, *Astle* 5704 (K, SRGH). C: Lusaka Province, Luangwa (Feira) Dist., Kingfisher Lodge, fr. 16.viii.1997, *Bingham* 11541 (K, MRSC). E: Luangwa Valley, Lupande area, Munkanya, fl. & fr. 19.iv.1968, *Phiri* 184 (K, UZL). S: Dist. Namwala, near Kafue River, fr. 11.vi.1949, *Hornby* 3017 (K, SRGH). **Zimbabwe**. N: Kaitano, Chiswiti Tribal Trust Land, fl. 8.iv.1965, *Bingham* 1442 (K, SRGH). E: Chimanimani, 5 km upstream from Lusitu, fl. 23.iv.1973, *Ngoni* 218 (BR, K, MO, SRGH). S: Gwanda, Doddieburn Ranch, 6.v.1972, *Pope* 651 (MO, SRGH). **Malawi**. N: Rumphi Gorge, 1.6 km E of Rumphi, fr. 15.viii.1977, *Pawek* 12883 (BR, K, MAL, MO, SRGH, UC, WAG). S: Zomba Dist., 3 km W of Lake Chilwa at Katchoka, fr. 1.vi.1970, *Brummitt & Williams* 11203 (BR, K, SRGH). **Mozambique**. N: Nampula, 7 km from town, fl. 13.iv.1961, *Balsinhas & Marrime* 385 (BM, COI, K, LISC, LMA). Z: Lugela-Mocuba Dist., Namagoa Estate, Lugela, fl. 22.v.1949, fr. 26.viii.1949, *Faulkner* 460 (BR, COI, K, SRGH). T: Baroda Dist., Sisitso, Zambezi River, fr. 15.vii.1950, *Chase* 2228 (K, SRGH). MS: Gorongosa Game Reserve, fr. 14.vii.1957, *Chase* 6623 (BR, COI, K, SRGH). GI: Gaza, between Chibuto and ponte de Cicacati, fr. 12.vi.1960, *De Lemos & Balsinhas* 90 (BM, COI, K, LMA, SRGH). M: Maputo, Inhaca Island, fr. 29.vii.1980, *De Koning & Nuvunga* 8382 (K, LMU, K, SRGH, WAG).

Also in South Africa (Provinces of Limpopo, Mpumalanga and KwaZulu-Natal). Found in bushland, savanna, wasteland, grassland and low shrubby vegetation; sea coast–500 m.

Used as medicinal plant. Vernacular names: "Munhambane" in Manica e Sofala Dist.; "Busali" (Ronga people, Maputo Region).

33. **CRYPTOSTEGIA** R. Br.[5]

Cryptostegia R. Br., Bot. Reg. **5**: t.435 (1820). —Marohasy & Forster in Aust. Syst. Bot. **4**: 571–577 (1991). —Klackenberg in Adansonia, ser. 3 **23**: 212 (2001). — Goyder in Hedberg & Edwards, Fl. Ethiopia Eritrea **4**: 112 (2003). —Venter in F.T.E.A., Apocynaceae (part 2): 120–122 (2012).

Liana or scrambling shrub with white latex. Stems twining; interpetiolar ridges with hair-like colleters, axillary colleters acicular. Leaves petiolate; blade elliptic to orbicular, leathery, main vein prominent beneath, secondary veins patent to divaricate. Inflorescences terminal, few-flowered, pedunculate. Corolla pinkish, purplish or white, glabrous; tube campanulate to funnelform; lobes narrowly ovate to ovate, obtuse. Corona of 5 entire or bifid lobes from about middle of corolla tube; each lobe with basal swollen foot fused to corolla tube and staminal filament base. Stamens from near base of corolla tube; anthers narrowly ovate to hastate, acuminate, completely fertile, pollen in tetrads. Interstaminal nectar pockets at base of corolla tube, fused sideways to filament bases. Style terete, style-head narrowly ovoid to ovoid. Gynostegium concealed within corolla-tube. Follicles paired, widely divergent to reflexed, narrowly ovoid with apiculate apex. Seeds oblong, comose, hairs white.

A genus of two species endemic to Madagascar. Both species were introduced into Africa as ornamental garden plants, that have escaped into the wild. Both species are present in the Flora area and may become invasive in seasonally dry climes.

Corona lobes bifid; follicles 10–16 cm long; leaves always glabrous; stems densely lenticellate . . . .**1.** *grandiflora*
Corona lobes entire; follicles up to 9 cm long; leaves sometimes hairy; stems sparsely lenticellate . . . . **2.** *madagascariensis*

---

[5] by H.J.T. Venter

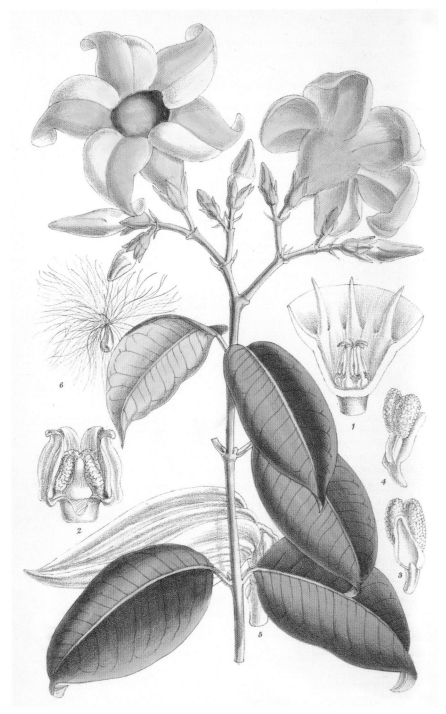

Fig. 7.3.**125**. CRYPTOSTEGIA MADAGASCARIENSIS. 1, flowering shoot, with one of the two follicles shown behind; 2, basal portion of corolla tube cut away to show gynostegium and entire corolline corona lobes; 3, gynostegium with one stamen removed to expose stylar head; 4, seed with coma of hairs; 5, 6, posterior and anterior views of translator with pollen tetrads. Drawn by M. Smith. Reproduced from Curtis's Botanical Magazine (1904).

1. **Cryptostegia grandiflora** R. Br. in Bot. Reg. **5**: t.435 (1820). Type: Bot. Reg. **5**: t.435 (1820) lectotype, designated by Marohasy & Forster in Austral. Syst. Bot. **4**: 574 (1991).

   Nerium grandiflorum Roxb., Fl. Ind. **2**: 10 (1832). Type: not designated.

Stems densely lenticellate, glabrous, up to 10 m long. Leaves glabrous; leaf petiole 0.5–1.5 cm long; blade 7–10 × 3–8 cm, elliptic, broadly elliptic or ovate, acuminate to cuspidate, base round, secondary veins 8–13 per side. Inflorescence glabrous, peduncles 4–10 mm long, pedicels 3–7 mm long, bracts 2–7 mm long. Sepals 10–20 × 5–10 mm, ovate, acuminate, glabrous. Corolla violet, pink or white, glabrous; tube 2–3 cm long; lobes 2–4.5 × 1.3–2.3 cm. Corona lobes 8–11 mm long, shallowly or deeply bifid. Staminal filaments 1–2 mm long; anthers oblong-hastate, 3–5 mm long. Ovaries 3–4 mm long; style 1–3 mm long, style-head ovoid, 3–4 mm long; translators elliptic. Follicles 10–16 × 2–4 cm, glabrous. Seeds 5–10 mm long; coma white, 2–4 cm long.

Botswana. N: Thamalakane River E of Maun, fl. 2.iii.2016, *A. & R. Heath* 2678 (GAB, K, PSUB). **Zimbabwe**. C: Harare [Salisbury] Dist., fl. 29.xi.1976, *Biegel* 5408 (K, SRGH). **Mozambique**. MS: Gorongoza, *Vasse* 436 (P).

Native to Madagascar. Cultivated and escaped locally. In coastal grassland and woodland; 0–1500 m.

2. **Cryptostegia madagascariensis** Decne. in Candolle, Prodr. **8**: 492 (1844). —Venter in F.T.E.A., Apocynaceae (part 2): 120 (2012). Type: Madagascar, cultured in Mauritius, *Bojer* s.n. (P holotype). FIGURE 7.3.**125**.

Liana or shrub-like. Stems sparsely lenticellate, glabrous or hairy, up to 10 m long. Leaves glabrous or hairy; leaf petiole 3–10 mm long; blade (2)4–9(11) × 2–5 cm, oblong-elliptic, elliptic, ovate or sub-orbicular, acuminate, base round, secondary veins 14–16 pairs. Inflorescences with peduncles 5–15 mm long, glabrous; pedicels 3–7 mm long, usually pubescent; bracts 2–7 mm long, narrowly triangular, usually pubescent on outside. Sepals 6–13 × 3–5 mm, narrowly ovate, narrowly triangular or elliptic, attenuate, pubescent on outside. Corolla pink, pubescent on outside, tube 1.0–2.5 cm long, lobes 2–4 x 1–3 cm. Corona lobes 6–9 mm long, acicular, entire. Staminal filaments 2–3 mm long, anthers 3–5 mm long, narrowly ovoid. Ovaries 2–3 mm long, style ± 1 mm long, style-head 3–4 mm long; translators elliptic. Follicles 5–9 × 1–3 cm diam., glabrous to puberulous. Seeds 6–9 mm long, brown with black base, warty; coma 2–3 cm long.

Zimbabwe. C: Harare [Salisbury] Dist., fl. & fr. 2.xii.1971, *Biegel* 3657 (K, SRGH). **Malawi**. S: Nasamo Bay, 12 km SE of Monkey Bay, fl. 11.xii.1986, *Brummitt* 18290 (K, MAL). **Mozambique**. N: Mogincual, fl. 26.vii.1948, *Pedro & Pedrogão* 4666 (LMA). M: Maputo, Vasco da Gama Municipal Garden, fl. 25.x.1946, *Gomes e Sousa* 3458 (K, LMA).

Native to Madagascar. A common garden ornamental from where the species may have escaped into the wild. Found in grassland and woodland; 0–1500 m.

34. STOMATOSTEMMA N.E. Br.[6]

Stomatostemma N.E. Br. in F.T.A. **4**(1): 252 (1902). —Brown in Fl. Cap. **4**(1): 530 (1907). —Venter & Verhoeven in S. African J. Bot. **59**: 52 (1993).

Climber or shrub with white latex. Roots tuberous. Stems perennial, woody, bark glabrous to verrucose, with reddish dentate interpetiolar ridges. Leaves petiolate, lanceolate or linear. Inflorescences terminal and axillary, few to many-flowered, glabrous. Sepals with colleters at inner base. Corolla campanulate, white to creamy white to yellowish green; tube with inner face swollen and nectiferous below filament bases, fluted towards nectary pockets at base of tube; lobes longer than tube, narrowly ovate to oblong ovate. Corona of 5 free clavate lobes from rim

[6] by H.J.T. Venter

of sinus pockets between corolla lobe bases. Stamens arising near base of corolla tube; pollen in tetrads. Gynostegium at base in corolla tube. Follicles paired.

A genus of two species, *S. monteiroae* widely distributed over Botswana, Mozambique, South Africa, Swaziland and Zimbabwe and *S. pendulina* restricted to east-central Mozambique.

Climber (rarely a shrubby bush); leaves narrowly lanceolate, lanceolate, elliptic or obovate; corolla broadly campanulate, creamy white with centre maroon to purple; corona lobes stout, 2–3 mm long . . . . **1.** *monteiroae*
Virgate shrub with pendulous branches, leaves linear; corolla campanulate, uniformly white; corona lobes slender, 5 mm long. . . . .**2.** *pendulina*

1. **Stomatostemma monteiroae** (Oliv.) N.E. Br. in F.T.A. 4(1): 252 (1902). Type: Mozambique, Delagoa Bay, *Monteiro* s.n. (K holotype, P). FIGURE 7.3.**126**.

   Cryptolepis monteiroae Oliv. in Hooker's Icon. Pl. **16**: t.1591 (1887).

Perennial climber. Root tubers numerous, spherical, 6–10 cm diam.. Stems woody, up to 10 m × 2.5 cm, twining, decussately branched, bark brownish, glabrous to verrucose. Leaves glossy dark green above, pale green beneath, semi-coriaceous, glabrous; petiole 0.1–10(3) cm long; blade 3–7(10) × 1–3 cm, narrowly ovate, ovate, elliptic or obovate, acute, acuminate or cuspidate, base round, midrib prominent beneath, lateral veins arched. Inflorescences with few–25 spreading flowers, primary peduncles 0.5–2.5 cm long; secondaries 0.5–1.5 cm long, bracts 2–3 mm long, very narrowly ovate to subulate, naviculate, margins fimbriate. Sepals 2–3 × 2 mm, narrowly triangular to narrowly ovate, acute to obtuse, margins fimbriate. Corolla creamy white to yellowish green with centre maroon to brownish purple; tube 5–6 mm long, broadly campanulate; lobes 12–16 × 5–6 mm, narrowly ovate to ovate, obtuse. Corona lobes glossy, green, 2–3 mm long, sturdy. Staminal filaments filiform, 2 mm long; anthers 2–3 mm long, angular-ovate, acuminate. Nectary pockets dark green. Ovaries 2 mm long; style 1–2 mm long, terete; style-head 1–2 mm long, ovoid, obtuse to acute; translators 2 mm long, spoon broadly angular-ovate, apex split, viscidium bifid. Follicles paired, often fused along inner face, narrowly divergent when free, stout, angular ovoid, obtuse, 6–9 × 1.5–2.5 cm. Seed 8–10 × 3 mm, brown, smooth, warty at margin, coma 2.5 cm long.

Botswana. N: Goha Hill, fr. 19.v.1977, *P.A. Smith* 2054 (K, MO, PRE, PSUB, SRGH). SE: 130 km WNW of Francistown on Maun road, fr. 29.iv.1957, *Drummond* 5281 (K, SRGH). **Zambia**. S: Livingstone, by gorge lip, fr. 22.iii.1961, *Fanshawe* 6456 (K, NDO). **Zimbabwe**. N: Near Kariba Dam wall, fl. iii.1960, *Goldsmith* 50/60 (K, SRGH). W: Victoria Falls, 3rd gorge, fl. 30.xii.1976, *Moyo* 6 (K, MO, SRGH). C: Que Que Sable Park, fl. 33.i.1976, *Stephens* 373 (MO, SRGH, Z). E: Chipinga Dist., west end of Mwangazi Gap, fl. 29.i.1975, *Pope et al.* 1440 (BR, K, MO, SRGH). S: Beitbridge Dist., Tchshiturapadsi [Chiturupazi], fl. 22.ii.1961, *Wild* 5330 (K, SRGH). **Malawi**. S: Litchenya Hill NW of Mulanje, fl. 12.i.1984, *Patel* 1414 (K, MAL). **Mozambique**. T: Tete, Monte Inanga, fr. 30.ii.1972, *Macêdo* 5123 (LISC, LMA, LMU, SRGH). MS: Manica e Sofala, Barue, fl. 20.xii.1965, *Torre & Correira* 13747 (LISC); GI: entre Dindiza e Nawali, fr. 15.i.1980, *de Koning* 8144 (K, LMU). M: Porto Enrique to Chengalene, fl. 18.i.1949, *Gomez e Sousa* 3949 (COI, K, MO).

Also in South Africa and Swaziland. Dry woodland of rocky habitats of hills, mountains or cliff sides in association with species of *Commiphora*, *Combretum*, *Brachystegia*, *Sclerocarya*, *Terminalia* and acacia, but also in sand forest with *Boscia*, *Croton*, *Diospyros*, *Ochna* and *Catunaregam*; sea coast–1500 m.

Flowering from November to March, peaking in January. The flowers are strongly scented and the plant is a magnificent sight when in full bloom.

2. **Stomatostemma pendulina** Venter & D.V. Field in Bot. J. Linn. Soc. **99**: 397 (1989). Type: Mozambique, Northern Dist., 17.6 km east of Namina, *Leach & Schelpe* 11441 (K holotype, SRGH).

   Cryptolepis pendulina (Venter & D.V. Field) P.I. Forst in Austrobaileya **3**: 280 (1990).

Fig. 7.3.**126**. STOMATOSTEMMA MONTEIROAE. 1, stem with flowers and fruit; 2, flower opened revealing the clavate corona lobes in the sinuses between corolla lobes and the cone of stamens covering the stigma head at the base of the corolla tube. From *Venter* 9084. Reproduced with permission from South African Journal of Botany (1993).

A virgate glabrous shrub up to 2.5 m high. Stems woody, erect with terminal branchlets pendulous, with dentate interpetiolar ridges. Leaves shortly petiolate; blade linear, acute, base cuneate, 10–11 × 1 mm, midrib prominent beneath. Inflorescences pendulous, with numerous flowers, primary, secondary and tertiary peduncles slender, each 3–3.5 cm long; pedicels frail, 1–1.5 cm long; bracts 2 mm long, linear. Sepals 1.5–2 × 0.8, narrowly triangular. Corolla uniformly white; tube 2–3 mm long, campanulate; lobes very narrowly oblong-ovate, obtuse, 6–7 × 1 mm. Corona lobes 5 mm long, terete, clavate. Staminal filaments ± 1.3 mm long; anthers ± 1 mm long, narrowly triangular, acuminate; translators 0.2 mm long. Ovaries ± 0.5 mm long, style-head quadrately rhomboid, 0.5 mm long. Follicles single or paired, obovoid, pendulous, c.4 × 1–2 cm. Seeds not known.

Mozambique. N: Nampula, Ribáuè, Monte Napitu, fl. & fr. 21.ix.1968, *Macedo & Macuácua* 3630 (LMA). Z: Ile, 26.vi.1943, *Torre* 5590 (LISC).

Granite hills with *Euphorbia, Xerophyta*, sedges and grasses; 500–1500 m.

35. TACAZZEA Decne.[7]

Tacazzea Decne. in Candolle, Prodr. **8**: 492 (1844). —Bentham in Bentham & Hooker, Gen. Pl. **2**(2): 745 (1876). —Schumann in Engler & Prantl, Nat. Pflanzenfam. **4**(2): 215 (1895). —Brown in F.T.A. **4**(1): 260 (1902); in Fl. Cap. **4**(1): 540 (1907). —Bullock in Kew Bull. **10**: 350 (1954). —Huber in Merxmüller, Prodr. Fl. Sudwestafr. **113**: 7 (1967). —Venter *et al.* in S. African J. Bot. **56**: 95 (1990).

Large perennial lianas or erect shrubs with white latex. Stems perennial, woody, bark brown, often verrucose. Leaves mostly opposite, sometimes alternate, rarely whorled, petiolate, interpetiolar ridges with reddish colleters; blade linear-ovate to broadly ovate or elliptic to broadly elliptic, puberulous. Inflorescences terminal and sub-terminal panicles with many-flowered, dichasial and monochasial branches, bracteate. Sepals broadly ovate, margin ciliate, with pairs of colleters at inner bases. Corolla rotate, tube saucer-shaped, lobes broadly ovate, oblong-ovate or oblong. Corona arising from corolla mouth, coronal feet, staminal filaments and nectaries fused into a wavy annulus, lobes filiform or narrowly ovate. Stamens from inner apex of coronal feet, glabrous; filaments free, anthers glabrous, pollen in tetrads. Interstaminal nectaries lobular, cone-shaped around style. Style-head subsessile, broadly ovoid; translators rhomboid to broadly ovate, apex split. Gynostegium exposed above corolla. Follicles paired, divergent, cylindrical-ovoid to ovoid.

Tacazzea is widely spread throughout the tropics and sub-tropics of Africa, from South Africa in the south through the equatorial forests and highlands of eastern and central Africa to Ethiopia in the north-east and westwards to south-western and western Africa. *Tacazzea* has 5 species, three are lianas found in swamp or riverine forests, and two are shrub-like and associated with water courses in more arid regions.

1. Erect shrubs; leaves linear ovate to narrowly ovate . . . . **3**. *rosmarinifolia*
– Lianas; leaves ovate, broadly ovate to elliptic or broadly elliptic 2
2. Leaf blade apex obtuse-acuminate to obtuse or emarginate, mucronate and base obtuse to cordate; panicle peduncles and pedicels slender and frail; follicles 180° divergent, narrowly ovoid with apex long-apiculate, 3–9 cm long. . . . . **1**. *apiculata*
– Leaf blade apex acute to acuminate and base cuneate or obtuse-tapering; panicle peduncles and pedicels sturdy; follicles slightly divergent, cylindrical-ovoid (cigar-shaped) with apices obtuse-acute, 7–20 cm long . . . . **2**. *conferta*

[7] by H.J.T. Venter

1. **Tacazzea apiculata** Oliv. in Trans. Linn. Soc. London **29**: 108 (1875). —Schumann in Engler, Pflanzenw. Ost-Afrikas **C**: 320 (1895). —Venter *et al.* in S. African J. Bot. **56**: 107 (1990). —Venter in Hedberg & Edwards, Fl. Ethiopia Eritrea **4** (1): 106 (2003); in Fl. Somalia **3**: 141 (2006); in F.T.E.A., Apocynaceae (part 2): 140 (2012). Type: Uganda, on the banks of the Madi Stream, xii.1862, *Speke & Grant* 711 (K lectotype), lectotypified by Venter *et al.* in S. African J. Bot. **56**: 97 (1990). FIGURE 7.3.**127**.

Tacazzea welwitschii Baill. in Bull. Mens. Soc. Linn. Paris **2**: 807 (1889). Type: Angola, Pungo Andongo, between Luxillo and Gazella, *Welwitsch* 4208 (P lectotype), lectotypified by Venter *et al.* in S. African J. Bot. **56**: 97 (1990).

Tacazzea thollonii Baill. in Bull. Mens. Soc. Linn. Paris **2**: 807 (1889). Type: D.R. Congo, Ogôone, *Savorgon* 507 (P holotype).

Tacazzea kirkii N.E. Br. in Bull. Misc. Inform. Kew **1895**: 248 (1895). Type: Mozambique, Zambesi Region, Lupata, iii/vi.1859, *Kirk* s.n. (K lectotype), lectotypified by Venter *et al.* in S. African J. Bot. **56**: 97 (1990).

Tacazzea bagshawei S. Moore in J. Bot. **44**: 88 (1906). Type: Uganda, Entebbe, 8.ix.1905, *Bagshawe* 745 (BM holotype).

Tacazzea bagshawei var. *occidentalis* Norman in J. Bot. **67**(Suppl. 2): 92 (1929). Type: Angola, Cuanza Region, Quilela Camabatela, *Gossweiler* 8467 (BM holotype).

Giant liana. Stems up to 20 m long and 15 cm diam., bark reddish brown, glabrous. Leaves opposite, semi-coriaceous; petiole 1–6 mm long; blade 3–22 × 2–12 cm, ovate to broadly ovate or elliptic to broadly elliptic, obtuse-acuminate to obtuse or emarginate, mucronate, base rounded to cordate, glabrous to sparsely tomentose above, tertiary veins conspicuously netted, beneath whitish to greyish tomentose. Inflorescences frail, peduncles 0.3–6 cm long, whitish tomentose; pedicels 0.2–1.5 cm long, puberulous, bracts ± 1 mm long, ovate, puberulous. Sepals 1.5 mm long, broadly ovate, pale green to brownish-red. Corolla pale green to pale yellow or reddish, glabrous, tube 0.5–1 mm long; lobes 5–7 × 2 mm, ovate, obtuse to acute. Coronal feet ± 1 mm long, lobes 5–11 mm long, filiform, sometimes 2- or 3-fid, yellowish or reddish. Stamens 1.5–2.0 mm long, filaments terete; anthers broadly oblong. Interstaminal nectaries squarish, emarginate or bifid. Style-head ± 1 mm long; translators ± 1 mm long, spoon rhomboidal. Follicles horizontally divergent, 3–9 × 0.5–1.5 cm, narrowly ovoid, long-apiculate, hirsute or tomentose to softly puberulous or glabrous. Seeds obliquely ovate, 3–9 mm long, coma 2–5 cm long.

Caprivi. Katima Mulilo, banks of Zambezi River, fl. 8.i.1959, *Killick & Leistner* 3317 (K, M, PRE). **Botswana**. N: Chobe River, Kasane, fl. 21.i.1972, *Gibbs et al.* 1355 (K, SRGH). **Zambia**. B: Barotseland, Mongu, fr. 5.viii.1965, *Van Rensburg* Z3028 (K, SRGH). N: Mweru Wa Ntipa Dist., Kaputa, fr. 18.x.1949, *Bullock* 1313 (K). W: Mwinilungwa Dist., edge of Mayowa Plains, fl. 3.x.1952, *White* 3443 (BR, K, FHO). C: Mulungushi Dam, fr. iii.1958, *Fanshawe* 4363 (K, NDO). E: Luangwa Valley, Kasenengwa, Munkanya, fr. 18.iv.1968, *Phiri* 175 (K). S: Livingstone Dist., Katambora, fl. 13.i.1956, *Gilges* 542 (K, SRGH). **Zimbabwe**. N: Kariba Dist., Chirundu, Zambezi River, fl. 15.iii.1966, *Simon* 714 (K, LISC, SRGH). W: Hwange [Wankie], fr. 18.vi.1934, *Eyles* 8068 (K, SRGH). C: Mutare Golf Course, *Chase* 7501 (K, LISC, M, MO). E: Lower Sabi Dist., Rupisi Hot Springs, fl. 29.i.1948, *Wild* 2388 (K, SRGH). S: Ndanga, Umtilikwe R., fl. 26.i.1949, *Wild* 2773 (K, SRGH). **Malawi**. N: Rumphi Dist., north side of Rumphi Gorge, 5.ii.1978, *Pawek* 13790 (BR, K, MAL, MO, UC, SRGH). S: Lengwe Game Reserve, fr. 30.vii.1970, *Hall-Martin* 873 (K, SRGH). **Mozambique**. N: Cabo Delgado, Rovuma riverbank, fl. 15.xi.2008, *Q. & P. Luke* 13807 (EA, K, LMA, P). Z: Mocuba Dist., Namagoa, fl. i.1944 *Faulkner* 52 (K, PRE). T: Tete, Mazoe River, Kabankanywa Kraal, fr. 22.ix.1948, *Wild* 2600 (K, SRGH). GI: Limpopo-Nuanetsi Rivers, fr. vii.1932, *Smuts* P308 (K, PRE). M: Maputo, Moamba, Chinhanguanine, left bank of Rio Incomati, fl. 14.xii.1979, *de Koning* 7760 (K, LMU).

Also in Angola, Burundi, Cameroon, Central African Republic, Chad, D.R. Congo, Ethiopia, Gabon, Ghana, Guinea, Guinea Bissau, Ivory Coast, Kenya, Liberia, Mali,

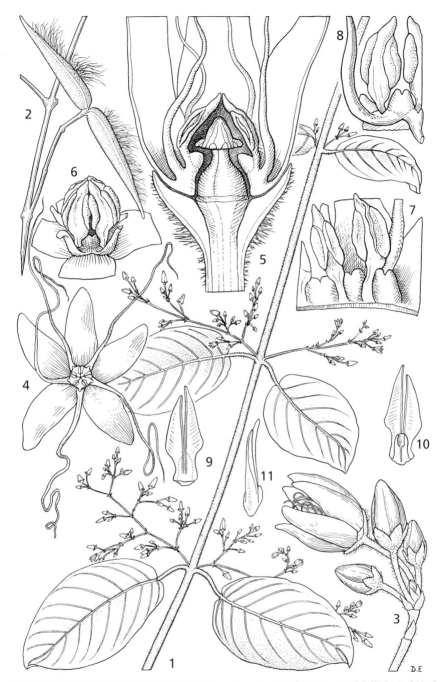

Fig. 7.3.**127**. TACAZZEA APICULATA. 1, flowering shoot (× 2/3); 2, paired follicles (× 2/3); 3, part of inflorescence (× 4); 4, flower from above (× 4); 5, half-flower (× 10); 6, centre of flower with corona lobes cut off to show gynostegium (× 8); 7, 8, portions of corolla showing relative positions of stamens, coronal feet and interstaminal nectaries (× 10); 9–11, translators (× 23). 1, 3–11 from *Drummond & Hemsley* 4263; 2 from *Jarrett* 461. Drawn by D. Erasmus. Reproduced from Flora of Tropical East Africa (2012).

Namibia, Niger, Nigeria, Senegal, Sierra Leone, Somalia, South Africa (Provinces of Limpopo, Mpumalanga and KwaZulu-Natal), Sudan, Swaziland, Tanzania and Uganda. In swamp, stream and lake bank forests; sea coast–2000 m.

Tacazzea apiculata is supposedly very poisonous. Vernacular name: Botswana: 'Sirozinama litiella'.

2. **Tacazzea conferta** N.E. Br. in Bull. Misc. Inform. Kew **1895**: 247 (1895); in F.T.A. 4(1): 265 (1902). —Venter *et al.* in S. African J. Bot. **56**: 107 (1990). —Venter in Hedberg & Edwards, Fl. Ethiopia Eritrea 4(1): 106 (2003); in F.T.E.A., Apocynaceae (part 2): 142 (2012). Type: Ethiopia, Efat, *Roth* 407 (K holotype).

> *Tacazzea floribunda* K. Schum. in Bot. Jahrb. Syst. **30**: 381 (1901). —Robyns in Fl. Parc Nat. Albert **2**: 84 (1947). Type: Tanzania, Poroto Mountains, Ngosi, *Goetze* 1289 (B†, K lectotype), lectotypified by Venter *et al.* in S. African J. Bot. **56**: 104 (1990).
>
> *Tacazzea galactagoga* Bullock in Kew Bull. **9**: 358 (1954). Type: Tanzania, Tanana in Ulugurus Mountains, *Bruce* 757 (K holotype, BM).

Giant liana; stems up to ± 20 m long, young stems tomentose, older stems glabrous, verrucose. Leaves petiolate; petiole 1–4 cm long, puberulous, adaxially with 2 fleshy colleters at apex; blade 6–16 × 3–10 cm, elliptic to oblong or oblong-ovate, acute to acuminate or drip-tipped-acuminate, base cuneate to obtuse, glabrous, or above glabrous to sparsely puberulous and beneath puberulous. Inflorescences of many-flowered panicles at terminal 4–6 nodes, puberulous; peduncles sturdy, 0.5–4 cm long; pedicels 0.4–2 cm long; bracts 1–3 mm long, broadly ovate to ovate. Sepals reddish-brown, ± 1.5 × ± 1.5 mm, broadly ovate, outside puberulous, margin ciliate. Corolla pale green to pale yellow or reddish; tube ± 1 mm long, sparsely puberulous outside; lobes 5–6 × 2–3 mm, broadly oblong, obtuse. Corona lobes 5–9 mm long, bases narrowly deltoid becoming filiform, helically twisted and sometimes bifid. Stamens 2–3 mm long, filaments terete, anthers ovate, acute. Interstaminal nectaries lobular, ± 1 mm long. Style-head ± 1 mm long; translators ± 1 mm long, spoon broadly ovate. Follicles slightly divergent, pendulous, 7–20 × 0.7–1.6 cm, cylindrical-ovoid, obtuse-acute. Seeds 0.8–1.4 cm long, obliquely ovate, coma 1.5–5 cm long.

Zambia. E: Nyika, fl. 26.vi.1966, *Fanshawe* 9762 (K, NDO); Nyika Plateau, Chowo Forest, fl. 9.ix.1976, *Pawek* 11783 (K), fl. 22.viii.1977, *Pawek* 12914 (MO). **Malawi**. N: Nyika Plateau, ± 2 km outside gate of National Park, fl. 9.vii.1970, *Brummitt* 11857 (K, MAL). S: Mlanje Dist., Luchenya Plateau, Mlanje Mt., fl. 10.vi.1962, *Richards* 16619 (K).

Also in Burundi, Central African Republic, D.R. Congo, Ethiopia, Ivory Coast, Kenya, Rwanda, Tanzania and Uganda. Found in bamboo or mixed *Podocarpus* afromontane forests, in moist bushland on basement complex or in riparian woodland; 1500–3000 m.

The vanilla-scented roots are used to enhance lactation in human females and cows.

3. **Tacazzea rosmarinifolia** (Decne.) N.E. Br. in F.T.A. 4(1): 263 (1902). Type: Angola, without locality, 1804, *J.J. da Silva* s.n. (P holotype, K).

> *Aechmolepis rosmarinifolia* Decne. in Candolle, Prodr. **8**: 493 (1844). —Schumann in Engler & Prantl, Nat. Pflanzenfam. 4(2): 220 (1895).
>
> *Tacazzea salicina* Schltr. in Warburg, Kunene-Sambesi-Exped.: 339 (1903). Type: Angola, Amboella Dist., Nambali (Kubango) River, *Baum* 245 (B† holotype, G lectotype, BM, COI, K, W), lectotypified by Venter *et al.* in S. African J. Bot. **56**: 107 (1990).
>
> *Tacazzea oleander* S. Moore in J. Bot. **50**: 338 (1912). Type: Angola, Cubango River near Fort Princeza Amelia, *Gossweiler* 2310 (BM holotype, COI, K).
>
> *Tacazzea venosa* Decne. subsp. *rosmarinifolia* (Decne.) Bullock in Kew Bull. **9**: 353 (1954).

Erect, virgate shrub up to ± 3 m high. Stems pendulous, glabrescent, bark verrucose. Leaves alternate, opposite or whorled with 3 leaves/whorl; petiole 4–10 mm long, sparsely puberulous; blade 7–12 × 1–4 cm, linear-ovate to narrowly ovate, attenuate-mucronate to obtuse-acute, base cuneate, above dark green, glabrous, with colleters on midrib, beneath greyish green, puberulous on veins. Inflorescences paniculate at terminal ± 10 nodes, puberulous; peduncles slender, 2–5 cm long; pedicels 7–15 mm long; bracts 2–3 mm long, ovate, acuminate. Sepals 2–3 × 2–3 mm, broadly ovate, acuminate to acute, outside puberulous, reddish, inside yellowish, margin ciliate. Corolla on outside reddish-violet and sparsely puberulous, inside yellow; tube ± 1 mm long; lobes 4–5 × 2 mm, oblong-ovate to ovate, acute to obtuse. Corona lobes 2–3 mm long, subulate to narrowly ovate, acuminate or bifid, fleshy, incurved. Staminal filaments ± 1.5 mm long; anthers ± 1 mm long, ovate. Ovaries ± 1 mm long; style-head ± 1 mm long, bifid; translators ± 1 mm long, spoon rhomboidal. Follicles slightly divergent, narrowly ovoid, acuminate, ± 6 × 1 cm. Seed ± 7 × 3 mm; coma ± 8 mm long.

Caprivi. Okavango River, east bank, Poppa Falls, 1.vii.2005, *Venter* 10739a (BLFU, PSUB).

Also in Angola and probably in Barotseland, Zambia as well. Common on sandbanks or rooted in rock crevices along rivers; 1000–1250 m.

36. **MONDIA** Skeels[8]

Mondia Skeels in Bull. Bur. Pl. Industr. U.S.D.A. **223**: 45 (1911).

Chlorocodon Hook.f. in Bot. Mag. **97**: t.5898 (1871), illegitimate name, non Fourr. (1869). — Schumann in Engler & Prantl, Nat. Pflanzenfam. **4**(2): 215 (1895). —Brown in F.T.A. **4**(1): 254 (1902); in Fl. Cap. **4**(1): 541 (1907). —Venter *et al.* in S. African J. Bot. **75**: 456–465 (2009).

Perennial climbers with white latex. Roots non-tuberous, woody. Stems twining, interpetiolar stipules frill-like. Leaves herbaceous to sub-leathery; petiole fluted, with reddish colleters; lamina ovate to elliptic to sub-orbicular, up to 25 × 20 cm, bright-green, glabrous to puberulent; main vein sunken above, with reddish colleters towards base, prominent beneath; main and secondary veins visible, secondary veins arching and looping. Inflorescences of lax axillary paniculate cymes, bracts acicular. Corolla rotate, coriaceous; tube saucer-shaped; lobes obliquely narrowly ovate to sub-orbicular, reddish, maroonish or purplish on inside. Corolline corona lobes free, fleshy, tri- or bi-segmented, central segment corniculate or ligulate and absent or present, lateral segments flap-like with apices rounded. Stamens sub-sessile from inner bases of coronal feet, anthers broadly hastate, whitish, acuminate. Interstaminal nectary lobes scoop-shaped, laterally fused with inner bases of coronal feet, arranged cone-like around styles. Style-head very broadly ovoid, dark coloured; translators narrowly elliptic and apically split, stalked. Follicles paired, narrowly ovoid or ellipsoidal. Seeds comose.

A genus of two species, endemic to Africa.

Corona lobes with three segments (a central corniculate or ligulate lobule and two flap-like lateral lobules); sepals ovate to elliptic with acute to acuminate apex . . . . **2.** *whitei*
Corona lobes with two segments (only flap-like lateral lobules present, and corniculate or ligulate central lobule absent); sepals broadly elliptic to orbicular with obtuse apex . . . . **1.** *ecornuta*

1. **Mondia ecornuta** (N.E. Br.) Bullock in Kew Bull. **15**: 203 (1961). —Venter in F.T.E.A., Apocynaceae (part 2): 143 (2012). Type: Kenya, Ribe, near Mombasa, v.1880, *Wakefield* s.n. (K holotype). FIGURE 7.3.**128B**.

Chlorocodon ecornutum N.E. Br. in Bull. Misc. Inform. Kew **1895**: 111 (1895).

Stems 5+ m long. Leaves thin-coriaceous; petiole 10–35 mm long; lamina 9–12(15) × 4–6(10) cm, elliptic to ovate, acuminate to cuspidate, sometimes recurved, base obtuse to cordate, bright green above, paler green beneath, secondary veins 8–10 per side. Inflorescences few-flowered, glabrous; peduncles 1–3 cm long; pedicels 1–4 cm long; bracts ± 2 mm long. Sepals 2–3 × 2–3 mm, broadly elliptic to orbicular, obtuse, glabrous, margin sparsely fimbriate. Corolla glabrous; tube ± 0.5 mm long; lobes 7–10 × 4–5 mm, inside papillose, purple, dark maroon, maroon or

---

[8] by H.J.T. Venter

brownish, outside glabrous, pale green to greenish-yellow, sometimes spotted green-maroon. Corona lobes bi-segmented, flap-like lobules 1–2 × 1–2 mm. Anthers 4–5 × 3–4 mm. Interstaminal nectary lobes with margins entire, agglutinated together into a hollow cone around styles and upper part of ovaries, small apertures left at inner base of coronal feet. Styles ± 0.5 mm long; style-head ± 3 × 3 mm. Follicles 10–12 × 2–2.5 cm, narrowly ellipsoidal.

Mozambique. N: Cabo Delgado Prov., Messalo [Misala] River, fl. i.1912, *Allen* 139 (K).

Also in Kenya and Tanzania. Found along edges of coastal scrub, woodland, dense forest or mangrove forest, on coral or limestone outcrops; also more inland; sea coast–700 m.

An extract from the roots is used as remedy for malaria and oxyuriasis; masserated root "skin" is given orally for bilharziasis.

2. **Mondia whitei** (Hook. f.) Skeels in Bull. Bur. Pl. Industr. U.S.D.A. **223**: 45 (1911). —Venter in F.T.E.A., Apocynaceae (part 2): 145–146 (2012). Type: South Africa, KwaZulu-Natal, Mfundisweni, v.1880, *White* s.n. (K holotype, K). FIGURE 7.3.**128A**.

    Chlorocodon whitei Hook.f. in Bot. Mag. **97**: t.5898 (1871).

    Periploca latifolia K. Schum. in Engler, Pflanzenw. Ost-Afrikas **C**: 321 (1895); in Bot. Jahrb. Syst. **23**: 232 (1896). Type: Burundi, *Stuhlmann* 1619 (B† holotype).

    Tacazzea amplifolia S. Moore in J. Bot. **50**: 337 (1912). Type: Angola, Cazengo, *Gossweiler* 616 (BM holotype, K, P).

Large liana. Roots aromatic. Stems up to 20 m long. Leaves herbaceous; petiole 2–7 cm long, glabrous to puberulent; lamina (6)12–20(28) × (3)6–14(20) cm, ovate, broadly ovate, elliptic, broadly elliptic or sub-orbicular, acuminate to cuspidate, sometimes recurved, base obtuse to cordate, glabrous to sparsely puberulent or sparsely puberulent on veins only, bright green above, paler green beneath, secondary veins 7–10 per side. Inflorescences 10–20-flowered, glabrous to puberulent; peduncles 2–4 cm long; pedicels 1–1.5 cm long; bracts 2–5 mm long. Sepals 2–3 × 1–2 mm, ovate to elliptic, acute to acuminate, outside glabrous to puberulent, margin fimbriate. Corolla glabrous; tube 2–3 mm long; lobes 9–11 × 4–6 mm, inside glabrous, violet, maroon, wine red, mauve red, outside glabrous, pale green. Corona lobes tri-segmented; lateral flaplike lobules 1–2 × 1–2 mm, green-yellow or creamy yellow; central corniculate or ligulate lobule 5–8 mm long, darker coloured. Anthers 4–5 × 3–4 mm. Interstaminal nectary lobes with margins dentate, not agglutinated. Styles 0.5–1 mm long; style-head ± 3 mm long, ± 3 mm wide, acute. Follicles 8–12 × 2–4 cm. Seeds 8–10 mm long, obliquely ovate, coma 2–2.5 cm long.

Zambia. W: Kitwe, fl. 8.i.1956, *Fanshawe* 2699 (K, NDO). C: cult. Protea Hill Farm 13 km SE of Lusaka, fl. 8.i.1996, *Bingham* 10762 (K). E: Sasare, fl. 8.xii.1958, *Robson* 868 (BM, BR, K, SRGH). **Zimbabwe**. N: Lomagundi, fl. 1.xi.1973, *Cannell* 574 (SRGH). W: Victoria Falls, fl. i.1906, *Allen* 259 (K). C: Harare [Salisbury], fl. 28.xi.1923, *Eyles* 3385 (K, SRGH). E: Chipinga Dist., Gungunyana Forest Reserve, fl. ii.1964, *Goldsmith* 5/64 (K, SRGH). S: Dist. Belingwe, Mount Buhwa, fr. 4.v.1973, *Biegel, Pope & Simon* 4297 (K, SRGH). **Malawi**. N: Rumphi Dist., Lura to Chiwe road, SE of Chilumba, fl. 8.i.1978, *Pawek* 13590 (K, MAL, MO, SRGH, UC). S: Zomba Dist., National Herbarium and Botanic Garden, fl. & fr. 17.ii.1992, *Goyder & Paton* 3665 (BR, K, MAL, MO, PRE). **Mozambique**. N: Niassa Dist., Meconta, fl. 18.i.1964, *Torre & Paiva* 10056 (LISC). MS: Manica e Sofala, Tsetserra, road to Mavita, fl. 30.xi.1966, *Müller* 511 (K, SRGH).

Also in Angola, Burundi, Cameroon, Central African Republic, D.R. Congo, Gabon, Ghana, Guinea, Guinea Bissau, Ivory Coast, Kenya, Liberia, Namibia, Nigeria, Senegal, Sierra Leone, South Africa, Sudan, Swaziland, Tanzania, Uganda. Found in a variety of moist habitats ranging from swamp forest, swampy shrubby grassland and riverine forest to disturbed forest; sea coast–1800 m.

Flowers smell most unpleasant. The ginger-scented roots of *Mondia whitei* are used to alleviate flatulence, to settle the stomach, as a tonic and as an aphrodisiac. Vernacular name: "citumbulo" (Malawi).

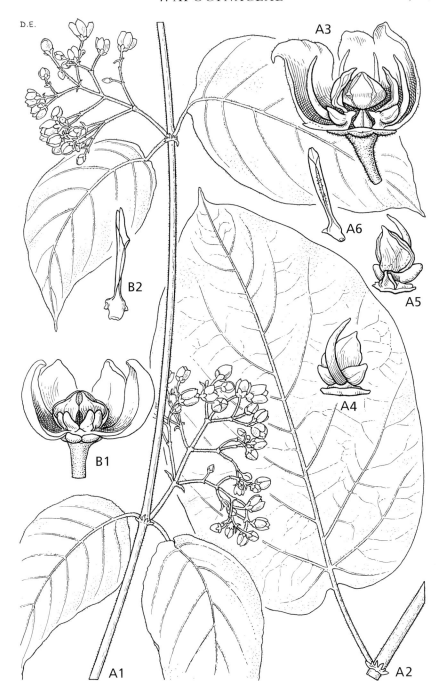

Fig. 7.3.**128**. A. —MONDIA WHITEI. A1, A2, habit (× ²/₃); A3, half-flower (× 3); A4, A5, stamen in dorsal and ventral view, showing attachment of corona lobe (× 3); A6, translator (× 8). B. — MONDIA ECORNUTA. B1, flower with one corona lobe removed to expose gynostegium (× 2); B2, translator (× 8). A1, A2 from *Drummond & Hemsley* 1432; A3–A6 from *Dawkins* 706; B1, B2 from *Faulkner* 558. Drawn by D. Erasmus. Reproduced from Flora of Tropical East Africa (2012).

37. **RAPHIONACME** Harv.[9]

Raphionacme Harv. in London J. Bot. **1**: 22 (1842). —Bentham in Bentham & Hooker, Gen. Pl. **2**(2): 745 (1876). —Brown in F.T.A. **4**(1): 268 (1902); in Fl. Cap. **4**(1): 532 (1907). —Venter in Hedberg & Edwards, Fl. Ethiopia Eritrea 4(1): 108–110 (2003); in Fl. Somalia **3**: 139–140 (2006); in S. African J. Bot. **75**: 292–350 (2009); in F.T.E.A., Apocynaceae (part 2): 146–157 (2012).

Apoxyanthera Hochst. in Flora **26**: 78 (1843).

Zuchellia Decne. in Candolle, Prodr. **8**: 492 (1844).

Chlorocyathus Oliv. in Hooker's Icon. Pl. **16**: t.1557 (1887).

Zaczatea Baill. in Bull. Mens. Soc. Linn. Paris **2**: 806 (1889).

Mafekingia Baill., Hist. Pl. **10**: 303 (1890).

Raphiacme K. Schum. in Bot. Jahrb. Syst. **17**: 117 (1893).

Pentagonanthus Bullock in Hooker's Icon. Pl. **36**: t.3583 (1962).

Geophytic, suffrutescent herbaceous prostrate, erect or climbing plants, rarely woody, usually with white latex. Taproot tuber turnip-shaped or cylindrical. Subterranean stems perennial, erect, one or few from crown of tuber. Aerial stems annual, seldom perennial, twining, erect or procumbent, branching dichotomous or lateral, glabrous or hairy, interpetiolar ridges with reddish dentate colleters. Leaves opposite, sessile or petiolate; blade broadly to narrowly ovate, elliptic, obovate or linear, herbaceous to leathery, glabrous to hairy, main veins prominent beneath, secondary veins arching, divaricate or patent. Inflorescences racemose, plumose or globose, few to many-flowered, terminal and/or pseudo-axillary, glabrous or hairy. Sepals with or without paired colleters at inner bases. Corolla with inside glabrous or rarely hairy, glabrous or hairy outside; tube campanulate to cylindrical with inner face vertically fluted (coronal ridges); lobes ovate, obovate or triangular, spreading to reflexed. Corona arising from corolla mouth, lobes opposite sepals; coronal feet inconspicuous or conspicuous, columnar, cone-shaped or knee-shaped; lobes free or seldom fused collar-like, entire or basally tri-segmented, filiform, corniculate, columnar or acicular, apices simple or variously incised, glabrous or hairy. Stamens arising from coronal feet, glabrous, filaments free; anthers ovate, oblong-ovate, triangular or hastate, dehiscence through full length slits or terminal half length slits; pollen in tetrads, single grains 8–16-porate. Nectaries pocket-like at base of corolla tube, deep green, fused to coronal feet. Styles basally free becoming one compound style, terete or rarely fluted, glabrous or hairy, terminally enlarged into style-head; style-head pentangular, broadly ovoid, oblong-ovoid or deltoid. Gynostegium with base in corolla mouth or elevated above corolla mouth. Follicles paired or single, erect or divaricate, narrowly ovoid to cylindrical. Seeds obliquely ovate with coma or ring of hairs.

With the exception of one species that is found in Arabia, all other 36 species of *Raphionacme* are endemic to Africa and widely distributed over the continent, the winter rainfall areas excluded. Most *Raphionacme* species are from grassland or savanna, but a few inhabit semi-desert, true desert, swamps or forests.

1. Corolla lobes with inner face green, yellowish-green, lemon-green, whitish-green or yellow, often with base or centre or tips or spots brown, reddish-brown, purple or magenta 2
– Corolla lobes with inner face white, pink, mauve, magenta, violet or blue 16
2. Corona lobes simple and ligulate, narrowly ovate or narrowly triangular . . . . **15.** *lanceolata*
– Corona lobes tri-segmented from their bases (central segment filiform, strap-shaped or acicular and lateral segments subulate, corniculate or globular) 3
3. Corona lobes with central segment filiform and lateral segments globular with apiculate apex . . . . **1.** *inconspicua*
– Corona lobes with central segment filiform, strap-shaped or acicular and lateral segments subulate or corniculate 4

---

[9] by H.J.T. Venter

4. Corolla lobes with a bright violet or magenta inverted "V" band in upper half . . . . **2.** *dyeri*
   − Corolla lobes without a violet or magenta inverted "V" band in upper half 5
5. Plants prostrate, procumbent or climbing. 6
   − Plants erect to spreading. 11
6. Plants prostrate or procumbent, corona reddish to purple . . . . **4.** *procumbens*
   − Plants climbing, corona green, pale green, pale yellow, whitish or when purple then leaves linear to very narrowly ovate 7
7. Leaves sessile, linear to very narrowly ovate and puberulous . . . . **5.** *longifolia*
   − Leaves petiolate, narrowly ovate, ovate, elliptic, narrowly obovate or obovate and pubescent, velutinous or tomentose. 8
8. Leaves pubescent, secondary veins 6–10 to a side, arching towards blade apex; gynostegium in corolla mouth; corolla lobes uniformly green or with random purple specks 9
   − Leaves velutinous or tomentose, secondary veins 10–24 to a side, divaricate to patent; corolla lobes with brownish, reddish or purplish markings at inner bases 10
9. Corolla lobes uniformly green; corona lobes with lateral segments fused pocket-like to corolla lobes and also fused to central segment, central segment filiform to subulate, not cleft . . . . **3.** *monteiroae*
   − Corolla lobes green with random purple specks; corona lobes not fused to corolla lobes, lateral segments free from central segment, central segment ensiform and deeply cleft. . . . . **6.** *sylvicola*
10. Suffrutescent herbaceous climber with stems up to 1 m long; leaves with 10–13 secondary veins to a side; corolla glabrous; corona lobes with central segment acicular to ensiform with apex attenuate to deeply cleft, lateral segments fused to central segment and acicular . . . . **7.** *welwitschii*
   − Perennial woody climber with stems 2 m and more long; leaves with 15–24 secondary veins to a side; corolla velutinous on outside; corona lobes with central segment filiform from dilated base and apex tortuous, lateral segments free from central segment, corniculate. . . . . **8.** *flanaganii*
11. Corolla lobes sulphur-yellow on inside; leaves glossy above . . . .**9.** *lucens*
   − Corolla lobes on inside green, white, whitish-green, yellowish, chrome-yellow; leaves never glossy 12
12. Corona lobes fused into an annulus by way of their lateral segments. . . . **10.** *utilis*
   − Corona lobes free, not fused into an annulus 13
13. Inflorescence densely globose 14
   − Inflorescence plumose or spike-like. 15
14. Corolla lobes on their inside, whitish green, creamy yellow or yellowish, sometimes tinged violet, brown or red, lobe tips violet, pink or red; leaf blade with secondary veins invisible or if visible then ± 5 veins on either side of midvein; follicles glabrous . . . .**11.** *globosa*
   − Corolla lobes on their inside green, brownish-green, maroon-green or violet-green with base brown to brownish-green; leaf blade with secondary veins conspicuous and 10–25 per side; follicles puberulous . . . . **12.** *galpinii*
15. Leaf blade hairy on both sides; corolla lobes on inside green or dull green to whitish-green with brown, pink, magenta or violet centre streaks or tips . . . .**13.** *madiensis*
   − Leaf blade glabrous above (main veins rarely somewhat hairy); corolla lobes on inside uniformly green to yellowish . . . . **14.** *velutina*
16. Corolla lobes with inner face white 17
   − Corolla lobes with inner face pink, mauve, magenta or violet 19

17. Inflorescence many-flowered, globose; corona lobes tri-segmented with central segment filiform and lateral segments subulate . . . . **11.** *globosa*
– Inflorescence few-flowered, not globose; corona lobes simple, filiform or narrowly ovate 18
18. Corona lobes filiform; leafblade linear-ovate, densely hirsute and glandular, secondary veins arching . . . . **21.** *chimanimaniana*
– Corona lobes narrowly ovate; leafblade elliptic to ovate, sparsely hispid with margin densely hispid, secondary veins divaricate to patent . . . . . . **12.** *longituba*
19. Staminal filament bases semi-ovoid with upper half filiform . . . . **16.** *michelii*
– Staminal filaments filiform, without any basal swelling 20
20. Inside of corolla lobes and base of corniculate corona lobes pubescent to tomentose . . . .**17.** *linearis*
– Inside of corolla lobes glabrous, corona lobes glabrous and filiform or subulate 21
21. Corolla tube pentangular-campanulate, outwardly 5-ridged and 5-pocketed towards base . . . . **18.** *grandiflora*
– Corolla tube campanulate or cylindrical, without any ridges or pockets 22
22. Corona lobes tri-segmented with central segment filiform and lateral segments corniculate or subulate . . . .**19.** *pulchella*
– Corona lobes simple and filiform, ovate, angular ovate, narrowly obovate or obtriangular 23
23 Corona lobes filiform with base columnar 24
– Corona lobes ovate, angular ovate, narrowly obovate or obtriangular 25
24. Leaves above glabrous to sparsely puberulous; style vertically partly grooved and hairy; corona lobes fused collar-like around the stamens, often with a small lobule upon the collar between corona lobes . . . . **20.** *splendens*
– Leaves above densely hirsute and glandular; style not grooved, glabrous; corona lobes free from one another with no collar formed . . . . **21.** *chimanimaniana*
25. Leaf blade sparsely hispid, but margins densely hispid, thickened and undulating; corolla lobes with inside white or white with a tinge of pink or violet; corona lobes green to creamy-yellow . . . . **22.** *longituba*
– Leaf blade glabrous or densely hirsute with margins glabrous or hirsute, thin and not undulating; corolla lobes with inside pink, maroon, mauve, whitish-violet or violet; corona lobes white to violet 26
26. Inflorescences compact, congested, of numerous flowers; corona lobe apices frilled or trisegmented; follicles usually paired and widely divergent, narrowly ovoid, less than 10 cm long . . . . **23.** *hirsuta*
– Inflorescences lax, raceme-like, of one or a few monochasia, each with few flowers; corona lobe apices bifurcate; follicles single, erect, cylindrical obovoid (spear-like), up to 30 cm long . . . .**24.** *palustris*

1. **Raphionacme inconspicua** H. Huber in Mitt. Bot. Staatssamml. München **12:** 73 (1955); in Merxmüller, Prodr. Fl. Sudwestafr. **113:** 6 (1967). Type: Namibia, Outjo, *Volk* 2718 (M holotype).

Erect herb. Tuber turnip-shaped. Aerial stems up to 16 cm × 1–3 mm, branching lateral. Leaves spreading, greyish-green, petiolate; petioles 5–6 mm long; blade (2.6)5.5–6.0 × (0.4)1.4–2.2 cm, oblanceolate to linear-oblanceolate, acute, base cuneate, pubescent to tomentose on both sides, margin entire, sometimes undulate, secondary veins arching and ± 7 to a side. Inflorescences terminal and axillary, pubescent to tomentose, hairs white, many-flowered, peduncles 5–10 mm long, pedicels 5–10 mm long; bracts ± 1 mm long, narrowly triangular, caudate, pubescent. Sepals 0.6–1.0 × 0.5–1 mm long, triangular, acute to apiculate, outside pubescent. Corolla tube 1.5–2 mm long, campanulate, outside puberulent, inside glabrous; lobes reflexed, 4.5–6 × 2 mm, ovate or triangular, obtuse to acute, outside puberulent, inside glabrous, green. Corona purplish, coronal feet tri-segmented, free; central segment 4–5 mm long, filiform, attenuate; lateral segments 1 × 1 mm, fleshy, globular, apiculate. Stamens from apices of coronal feet; filaments ± 1 mm long, base dilated, upper part filiform; anthers 1–2 mm long, narrowly ovate, whitish with connective purplish, full length slits. Ovaries ± 1 mm long; style ± 2 mm long, terete; style-head, ± 1.5 × 1 mm, ovoid, acute; translators ± 2 mm long, spoon elliptic. Gynostegium elevated above corolla mouth. Follicles and seeds unknown.

Botswana. N: 10 km S of Kazangula, fl. 11.xii.1991, *Venter et al.* 144 (BLFU).

Also from Angola and Namibia. In arid hot savanna of *Colophospermum* and *Adansonia*. *Raphionacme inconspicua* is mostly found in mountainous habitats amongst limestone or dolomite rocks in shade of trees and shrubs; 1000–1500 m.

2. **Raphionacme dyeri** Retief & Venter in S. African J. Bot. **2**: 326 (1983). Type: South Africa, Free State Province, Bloemfontein, near airport, *Rawlinson* s.n. in PRE-57731 (PRE holotype, BFLU).

A decumbent to spreading herb. Tuber ± 10 cm diam., ovoid. Aerial stems up to 15 cm × 2–3 mm, reddish-green, hirsute, branching dichotomous. Leaves spreading, mostly folded lengthwise, petiolate; petiole 3–10 mm long; blade 20–65 × 4–15 mm, ovate to narrowly ovate, obtuse to acute, mucronate, base cuneate to obtuse, above dull green glaucous, glabrous, below hirsute, secondary veins arching and 10–12 to a side. Inflorescences terminal or axillary, paniculate, monochasial branches 3–5-flowered, peduncles 2–10 mm long, pedicels 2–8 mm long; bracts 2–3 mm long, subulate. Sepals 1.5–2.0 × 1 mm, ovate, purplish-green, outside hirsute. Corolla tube 2.5–4.0 mm long, campanulate, outside purplish-green to green, hirsute, inside yellowish-green, glabrous; lobes spreading, 4–8 × 2.0–3.5 mm, oblong-ovate to narrowly ovate, apex obtuse to obliquely obtuse, outside green to purplish-green, hirsute, inside yellowish-green with bright purple or magenta inverted 'V'. Coronal feet free, lobes tri-partite; central segment 10–11 mm long, filiform; lateral segments 1.5 mm long, corniculate, fleshy. Stamens from inner base of coronal feet; filaments ± 1 mm long, filiform; anthers 1–2 mm long, narrowly ovate, acute, greenish white to mauve, full length slits. Ovaries ± 1 mm long; style 2 mm long, terete; style-head 1–2 × 1–2 mm, broadly ovoid, greenish white to mauve; translators ± 2.5 mm long, spoon broadly ovate. Gynostegium in corolla mouth. Follicles paired, horizontally divergent, 5–8 × 0.8–1.4 cm, narrowly ovoid. Seeds 8 × 3 mm, obliquely ovate; coma 2.0–2.5 cm long, coppery-white.

Zimbabwe. S: Beitbridge, 10 km along road to Chituripasi, fl. 9.i.1999, *Bruyns* 7775 (BOL).

Also in South Africa. In grassland or open savanna on sandy soils, red loam soils, as well as clayey soils; 1000–1400 m.

3. **Raphionacme monteiroae** (Oliv.) N.E. Br. in Fl. Cap. 4(1): 533 (1907). Type: Mozambique, Maputo (Delagoa Bay), 1882, *Monteiro* s.n. (K holotype).

    Chlorocyathus monteiroae Oliv. in Hooker's Icon. Pl. **16**: t.1557 (1887). —Schlechter in J. Bot. **4**: 314 (1896), not *Cryptolepis monteiroae* Oliv. in Hooker's Icon. Pl. **16**: t.1591 (1887). —Venter in F.T.E.A., Apocynaceae (part 2): 138 (2012).

    Raphionacme loandae Schltr. & Rendle in Hiern, Cat. Afr. Pl. **1**: 679 (1898). —Brown in F.T.A. 4(1): 275 (1902). Type: Angola, Loanda, near Boa Vista, iii.1858, *Welwitsch* 4274 (BM holotype, LISU).

A few-stemmed perennial climber. Tubers numerous from lateral roots, globoid to cylindrical-ovoid, 10–30 × 3–7 cm. Aerial stems woody, twining, slender, up to 8 m long, branching lateral, pubescent. Leaf petiole 1–10 mm long; blade narrowly ovate, ovate, elliptic, narrowly obovate, obovate or suborbicular, acute to obtuse, base cuneate to obtuse, 1–6 × 0.5–3 cm, semi-succulent, above dark green, glossy, sparsely pubescent, beneath pale green, pubescent, main and secondary veins visible, secondaries 6–8 per side. Inflorescences raceme-like, 2–5-flowered, pubescent; peduncles 0.5–2 cm long, pedicels 0.2–1 cm long, bracts ovate, pubescent, 1–3 mm long. Sepals ovate, acute, 2–4 mm long, 1–2 mm wide, outside pubescent. Corolla bright green to light green, outside pubescent; tube cylindrical-campanulate, 4–6 mm long; lobes ovate to narrowly ovate, obtuse to acute, 6–10 × 3–5 mm, spreading. Corona lobes trisegmented; lobe base obtriangular, laterally fused to corolla lobes, succulent; central segment filiform to subulate, 2–3 mm long, creamy white to white; lateral segments corniculate, ± 1 mm long, pinkish to purplish. Staminal filaments filiform, 1–2 mm long; anthers narrowly ovate, acute, 2–3 mm long. Style terete, 2–4 mm long; style-head ovoid, obtuse; translators ± 3 mm long, spoon broadly ovate, stalked. Follicles horizontally paired, very narrowly ovoid, tapering, 7–11 × 0.7 cm. Seeds narrowly ovate or rhomboid, 7 mm long, coma white, 3 cm long.

Zimbabwe. N: Binga, Mwenda Research Station, fl. 24.i.1966, *Magadza* 44 (SRGH). W: Kariba, Sinamwenda, fl. 25.ii.1966, *Jarman* 479 (SRGH). E: Nyanga Dist., at lake 3 km S of Regina Coeli Mission, fl. 10.ii.1997, *Brummitt & Pope* 19577 (K). S: Beitbridge Dist., Nuli Hills, fl. 20.iii.1967, *Rushworth* 463 (K, PRE, SRGH). **Mozambique**. N: Cabo Delgado, Pemba, *Groenendijk & Dungo* 585 (LMU, MO). T: Chiringa, fl. 11.iv.1972, *Macedo* 5179A (LMA, LISC). M: Maputo [Delagoa Bay], fl. 31.xii.1897, *Schlechter* 11965 (BM, BR, COI, G, K, NBG, WAG).

Also in Angola, Kenya, Namibia, South Africa and Tanzania, and probably in Botswana as well. Found in succulent thicket and dry thorn savanna; sea coast–800 m.

4. **Raphionacme procumbens** Schltr. in Bot. Jahrb. Syst. **20**(Beibl. 51): 11 (1895). Type: South Africa, Mpumalanga Province [Eastern Transvaal], Elandspruit Mountains, 5.xii.1893, *Schlechter* 3867 (B† holotype, Z lectotype, BOL, GRA, K, PRE, Z), lectotypified by Venter in S. African J. Bot. **75**: 333 (2009).

Prostrate to procumbent, many-stemmed herb. Tuber up to 15 cm diam., sub-spherical. Aerial stems annual, procumbent, up to 40 cm × 24 mm, pubescent to velutinous, brown, branching lateral. Leaves spreading, petiolate; petiole 3–13 mm long, velutinous; blade (3)5–6.5 × 1–3 cm, narrowly obovate to narrowly oblong-obovate, mucronate or acute, often recurved, base obtuse to tapering, coriaceous, above dark green, velutinous, beneath pale green, densely velutinous, secondary veins patent to divaricate and 13–22 to a side. Inflorescences axillary at apical 3–5 nodes, compact, globose, many-flowered fascicle of monochasia, densely velutinous, peduncles 3–7 mm long, pedicels 3–4 mm long; bracts 2–3 mm long, acicular. Sepals 2–3 × ± 1 mm, narrowly triangular to narrowly ovate, attenuate, outside densely pubescent. Corolla tube ± 1 mm long, campanulate, outside glabrous to sparsely pubescent, inside glabrous; lobes 3–4 × 2 mm, narrowly ovate to narrowly ovate-triangular, obtuse to acute, outside pubescent, inside papillose and green or greenish yellow, base maroon to violet. Corona reddish to purple, papillose; coronal feet dilated, fleshy, tri-segmented, free; central segment ± 2 mm long, subulate, apex simple or deeply bifid; lateral segments ± 1 mm long, subulate or triangular, apex 3–4-fid. Stamens from inner face of coronal feet; filament base ± 0.7 mm long, columnar, fleshy; upper part ± 0.5 mm long, filiform; anthers ± 2 mm long, narrowly ovate, acute, whitish, full length slits. Ovaries ± 0.5 mm long; style ± 2 mm long, terete; style-head 1.5–2 × 1 mm, ovoid, acute; translators 0.7–0.9 mm long, spoon elliptic. Gynostegium in corolla mouth. Follicles solitary, pendulous, 3.5–4.5 × 1.2–1.5 cm, ovoid, acute. Seeds 7–8 mm long, obliquely obovate, brown; coma 1–1.5 cm long, silvery white.

Zimbabwe. W: Matobo Dist., Broadleaf Farm, fl. i.1961, *Miller* 7653 (K, SRGH). C: Gwelo, fl. 15.iv.1905, *Gardner* 13 (K). E: Kukwanisa, 43 km from Mutare [Umtali] on road to Inyanga National Park, fl. 5.i.1967, *Biegel* 1639 (K, PRE, SRGH). **Mozambique**. M: Lebombo Mountains, near border with Swaziland c.20 km N of South African border, fl. 6.xii.2001, *Goyder* 5030 (K, LMU).

Also in South Africa (Limpopo, Mpumalanga and KwaZulu-Natal Provinces) and Swaziland. Associated with warmer, moister highlands and mountain ranges; 350–1900 m.

A component of grassland or the grass layer of savanna and may be quite common in places.

5. **Raphionacme longifolia** N.E. Br. in Bull. Misc. Inform. Kew **1895**: 110 (1895). —
Venter in F.T.E.A., Apocynaceae (part 2): 150 (2012). Type: Mozambique, Lower
Shire Valley, Moramballa, 18.i.1863, *Kirk* s.n. (K lectotype), lectotypified by Venter
in S. African J. Bot. **75**: 322 (2009).

 Raphionacme decolor Schltr. in Fries, Wiss. Ergebn. Schwed. Rhodesia-Kongo-Exped. 1911-
1912 **1**: 265 (1916). Type: Zambia, Kalambo, between Mbala and Kasanga, *Fries* 1396 (UPS
holotype).

 Erect, terminally twining, single or few-stemmed herb. Tuber up to 10 cm diam., ovoid to subglobose. Aerial stems up to 75 cm × 2–4 mm, puberulous, brownish, branching lateral. Leaves
spreading or ascending, sessile; blade (5)12–15(20) × (0.4)0.8–1(2.2) cm, linear to very narrowly
ovate, attenuate, base tapering, puberulous, often involute, above green, beneath pale green,
midvein whitish, secondary veins arching and (4)10–15 to a side. Inflorescences axillary and
terminal, fascicles 5–10-flowered, puberulous to pubescent, peduncles 2–10 mm long, pedicels
3–6 mm long; bracts 2–3 mm long, subulate. Sepals 1–2 × 1 mm, narrowly triangular to narrowly
ovate, attenuate, densely puberulous. Corolla outside densely puberulent, inside glabrous; tube
2–3 mm long, campanulate; lobes 4–6 × 2–3 mm, oblong-ovate to triangular, acute to attenuate,
inside green, green with violet base, green-brown, green-violet, maroon-brown or dull violet-
red. Corona whitish to violet; coronal feet ± 1 mm long, columnar, free, tri-segmented from
outer face; central segment 5–6 mm long, filiform, papillate; lateral segments ± 1 mm long,
subulate. Stamens from inner apex of coronal feet; filaments 1–2 mm long, filiform; anthers
1–2 mm long, oblong-ovate, attenuate, white, full length slits. Ovaries ± 1 mm long; style 1.5–2.0
mm long, terete, glabrous; style-head 1.5–2 × 1 mm, ovoid; translators 8–2.3 mm long, spoon
elliptic. Gynostegium in corolla mouth. Follicles paired, erect, 10 × 0.4 cm, cylindrical-ovoid.
Seeds unknown.

 Botswana. N: Chobe Dist., 57 km N of Pandamatenga, fl. 23.xii.1996, *Bruyns* 6962
(BOL). **Zambia**. C: Lazy J Ranch, 20 km SE of Lusaka, fl. 20.xii.1994, *Bingham &*
Truluck 10208 (K). E: Chipata, fl. 15.xii.1964, *Robinson* 6286 (SRGH). **Zimbabwe**. N:
Lomagundi Dist., Molly South Hill, fl. 14.xii.1964, *Wild & Drummond* 6678 (BR, K,
LISC, SRGH). W: Hwange, near Gwaai River, fl. 27.i.1967, *Wild* 7608 (SRGH). C: Harare
Dist., Rumane, fl. 17.xi.1945, *Greatrex* s.n. in GH13885 (K, SRGH). E: Chiredzi, 7 km
from Chipinda Pool, fl. 3.xii.1970, *Kelly* 271 (SRGH). **Malawi**. C: Dedza Dist., Sosola
Rest House, Mua, fl. 15.xii.1969, *Salubeni* 1436 (MAL, SRGH). S: E of Mangochi, fl.
22.xii.1997, *Bruyns* 7426 (BOL). **Mozambique**. N: 16 km N of Mandimba, fl. 2.xii.1961,
Leach 11373 (SRGH). Z: Mocuba Dist., Namagoa, fl. 10.i.1948, fr. 16.ii.1949, *Faulkner*
Kew 203 (BR, K).

 Also in Tanzania and Central African Republic. *Raphionacme longifolia* is a grassland
component of miombo and mopane savanna; 500–1600 m.

 The species flowers in summer, peaking in December. Vernacular name: "Mongwa"
(Zimbabwe). The tuber is eaten as food by local inhabitants.

6. **Raphionacme sylvicola** Venter & R.L. Verh. in Novon **10**: 170 (2000). Type: Zambia,
Kaputa Region, Nsumbu (Sumbu) National Forest, Nkamba Bay Valley Area,
6.xii.1993, *Merello et al.* 962 (MO holotype, K).

 Woody climber. Roots unknown. Tuber unknown. Aerial stems twining, bark pale brown,
pubescent, branching lateral. Leaves spreading, petiolate; petiole 2–3 mm long, slender,
pubescent, with a few reddish colleters above; blade (9)11–13 × 4.5–6.5 cm, ovate to elliptic,
acuminate, base cuneate, pubescent, green above, somewhat paler green beneath, herbaceous
and thin-textured, secondary veins arching and 7–10 to a side. Inflorescences axillary and subterminal, 5–8-flowered, pubescent; peduncles 15–35 mm long, slender; pedicels 8–12 mm long,
slender; bracts 2 mm long, narrowly ovate, membranous, margins fimbriate. Sepals 1.5 × 1 mm,
ovate to triangular, acuminate, membranous, pubescent. Corolla green with purple specks;
tube 3 mm long, funnel-shaped, glabrous; lobes 4–5 × 1.5–2 mm, oblong-ovate, outside sparsely
pubescent, inside glabrous, apex obtuse. Corona green, glabrous; coronal feet tri-segmented,

free; central segment 8 mm long, strap-shaped, deeply bifid; lateral segments 1 mm long, subulate. Stamens fused to inner face and lateral segments of coronal feet; filaments 1 mm long, filiform; anthers 1.5 mm long, ovate, attenuate, white, full length slits. Ovaries ± 1 mm long; style ± 2 mm long, terete; style-head ovoid, ± 1.5 mm long; translators ± 1 mm long, spoon broadly ovate. Gynostegium in corolla mouth. Follicles and seeds unknown.

Zambia. N: Kaputa Region, Nsumbu (Sumbu) National Forest, Nkamba Bay Valley, fl. 6.xii.1993, *Merello et al.* 962 (K, MO).

In moist forest of the Zambezian miombo woodland; approximately 900 m. This uncommon species flowers in December.

7. **Raphionacme welwitschii** Schltr. & Rendle in J. Bot. **34**: 97 (1896). —Venter in
    F.T.E.A., Apocynaceae (part 2): 155 (2012). Type: Angola, Ambaca, between Halo
    and Zamba, x.1856, *Welwitsch* 4234 (BM holotype, K). FIGURE 7.3.**129**.

    Zucchellia angolensis Decne. in Candolle, Prodr. **8**: 492 (1844), non *Raphionacme angolensis*
    (Baill.) N.E. Br. in F.T.A. **4**(1): 271 (1902). Type: Angola, anon. s.n. (P holotype).
    Raphionacme denticulata N.E. Br. in F.T.A. **4**(1): 275 (1902). Type: Malawi, *Mahon* s.n. (K
    holotype).
    Raphionacme verdickii De Wild. in Ann. Mus. Congo Belge, Bot. sér. 5 **1**: 182 (1904). Type:
    D.R. Congo, Katanga, Lufira, xi.1900, *Verdick* 283 (BR holotype).
    Chlorocyathus welwitschii (Schltr. & Rendle) Bullock in Publ. Cult. Diamang Angola **42**: 130
    (1959).

Suffrutescent herbaceous climber. Tuber hemi-spherical, up 20 cm diam.. Aerial stems annual, branching lateral, twining, up to 1 m × 3–5 mm, velutinous. Leaf petiole (0.7)1.5–2.5(4) cm long, velutinous; blade (4)5–11(21) × (1.5)3–7(9) cm, narrowly ovate, ovate, elliptic, narrowly obovate or obovate, acute to obtuse-mucronate, base cuneate to obtuse, herbaceous to coriaceous, dark green above, pale green beneath, velutinous to tomentose, lateral veins semi-divaricate. Inflorescences axillary, compact, 10–15-flowered fascicle of monochasia, velutinous; peduncles 3–15 mm long; pedicels 2–6 mm long; bracts 1–2 mm long, broadly ovate. Sepals 2–3 × 1–2 mm, broadly triangular, broadly ovate or ovate, acute to obtuse, outside velutinous, green tinged red. Corolla tube 2–4 mm long, campanulate, greenish, glabrous; lobes reflexed, 5–6(12) × 3–4 mm, ovate, acute to obtuse, greenish-yellow to green with brown, reddish-brown or purplish-brown markings near base of lobe, outside tomentose. Corona glabrous, pale yellow, pale green or green, lobes tri-segmented to tri-partite; central segment 2–5 mm long, filiform to acicular, apex simple or bifid; lateral segments acicular to claw-shaped. Staminal filaments 1–2 mm long, filiform with base dilated; anthers 2 mm long, narrowly ovate, acute to acuminate, greenish to white. Style 2–3 mm long, terete; style-head ± 2 mm long, ovoid, acute; translator 2 mm long, spoon broadly ovate. Gynostegium elevated above corolla mouth. Follicles solitary or rarely paired, 4–6 × 1–1.3 cm, narrowly ovoid, tapering; seeds 7 mm long, obliquely ovate, brown, coma 2.5 cm long, copper coloured.

Zambia. N: Mporokoso Dist., L. Chisi, Mweru-Wantipa, fl. 13.xii.1960, *Richards* 13679 (K). W: Mwinilunga Dist., Matonchi Farm, fl. 12.xi.1962, *Richards* 17287 (K, SRGH). C: Lusaka, Mt. Makulu, fl. 26.xii.1959, *White* 6022 (FHO, K). E: Luangwa Valley, Chindeni Hills, fl. 3.xii.2004, *Bingham* 12837 (K). **Zimbabwe**. E: Inyanga Dist., Honde Valley, fl. 16.xi.1960, *Wild* 5254 (K, SRGH). **Malawi**. N: Mzimba Dist., 12 km S of Eutini, W to Rukuru River, fl. 29.xii.1975, *Pawek* 10631 (K, MAL). S: Kasupe, Machinga, fr. 2.ii.1978, *Seyani* 783 (MAL, SRGH). **Mozambique**. N: Mocimboa – Mueda, fl. 11.xii.2003, *Luke et al.* 9987 (EA, K, LMA, MO, NHT).

Also in Angola, D.R. Congo and Tanzania. Found in Miombo woodland and grassland where soils vary from sandy to peaty to rocky; 400–1600 m.

8. **Raphionacme flanaganii** Schltr. in Bot. Jahrb. Syst. **18**(Beibl. 45): 2 (1894). —
    Venter in F.T.E.A., Apocynaceae (part 2): 149 (2012). Type: South Africa, Eastern
    Cape Province, Komgha, near Kabousie River, *Flanagan* 118 (B† holotype, PRE
    lectotype, GRA, NBG), lectotypified by Venter in S. African J. Bot. **75**: 306 (2009).

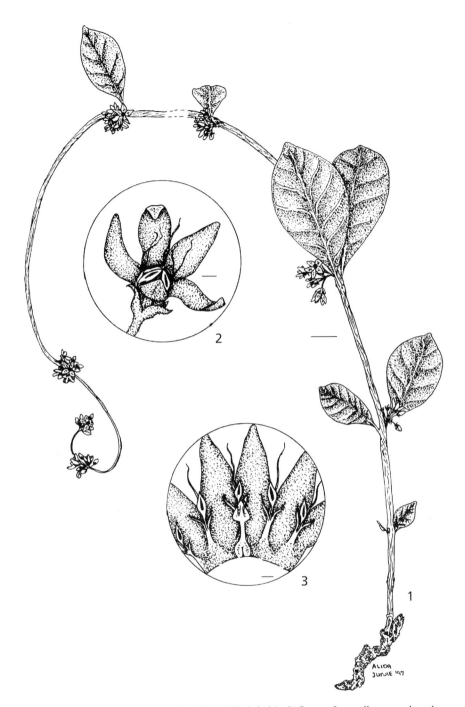

Fig. 7.3.**129**. RAPHIONACME WELWITSCHII. 1, habit; 2, flower; 3, corolla opened to show tri-segmented corona lobes, stamens and gynoecium. Scale bars: 1 = 10 mm, 2, 3 = 2 mm. 1, 2 from *Luke & Mbinda* 5978; 3 from *Drummond & Hemsley* 3836. Reproduced with permission from South African Journal of Botany (2009).

Raphionacme scandens N.E. Br. in Bull. Misc. Inform. Kew **1895**: 111 (1895). Type: South Africa, KwaZulu-Natal, Tugela, *Gerrard* 1312 (K holotype, BM, NH, W).

A perennial climber with one to few main stems. Tuber spindle-shaped, up to 30 cm diam.. Aerial stems up to at least 2 m long, branching lateral, velutinous. Leaf petiole 1–15 mm long; blade 3–11 × 0.5–5 cm, obovate, narrowly obovate or ovate, acute, obtuse or truncate-mucronate, base cuneate to obtuse, above velutinous to glabrous, beneath velutinous, primary vein prominent, secondary veins patent. Inflorescences sub-terminal, 7–30-flowered, velutinous; peduncles 5–25 mm long; pedicels 2–6 mm long; bracts 2–3 mm long, narrowly ovate. Sepals 2–4 × 1–2 mm, ovate, acute to obtuse, outside velutinous. Corolla green, outside velutinous, inside glabrous; tube campanulate, 2–3 mm long; lobes reflexed, 5–7 × 2–3 mm, oblong-ovate to ovate, obtuse to acute, yellow-green with reddish to purple triangular spot at base. Corona lobes tri-partite, white; central segment 4–10 mm long, filiform with dilated base, apex tortuose, papillose; lateral segments 2–3 mm long, corniculate. Staminal filaments filiform, ± 2 mm long; anthers ± 2 mm long, ovate, white. Style terete, ± 1.5 mm long; style-head ± 1.5 mm long, ovoid, obtuse; translators ± 2 mm long, spoon broadly ovate. Gynostegium in corolla mouth. Follicles paired, seldom solitary, 5–10 × ± 1 cm, cylindrical-ovoid, pendulous, puberulous. Seeds 6–7 mm long, coma 1–2 cm long.

Mozambique. M: Changalane, fl., *Gomez e Sousa* 3948 (LISC, PRE).

Also in Kenya and South Africa (KwaZulu-Natal and Eastern Cape Provinces). In thorn or other savanna, dry forest or streambank forest on soils varying from clay to sand; 1000–1800 m.

Vernacular name: "matamela" (Shisonga, Mozambique).

9. **Raphionacme lucens** Venter & R.L. Verh. in S. African J. Bot. **54**: 607 (1988). Type: South Africa, KwaZulu-Natal, between Mbazwana and Sibayi Lake, *Venter* 9086 (BFLU holotype, K, PRE).

Erect herb. Tuber 10–15 cm diam., flattened globose to spindle-shaped. Aerial stems 1 or 2, erect, up to 30 cm × 2–3 mm, puberulous, branching lateral. Leaves spreading, puberulous, petiolate; petiole 2–5 mm long; blade 4–4.5 × 2–2.5 cm, elliptic, sometimes ovate to obovate, obtuse-mucronate, base cuneate to obtuse, coriaceous, above glossy, green, beneath pale green, secondary veins divaricate and 7–10 to a side. Inflorescences terminal, compact, globose, 8–15-flowered, puberulous, peduncles 1–1.5 cm long, pedicels 0–3 mm long; bracts 2–3 mm long, subulate. Sepals 4–5 × 1 mm, linear-triangular, attenuate, puberulous, green. Corolla tube 2–3 mm long, campanulate, outside puberulous, citron-green, inside glabrous and sulphur-yellow with maroon ridges; lobes 4–5 × 2–2.5 mm, ovate, obtuse to acute, outside puberulous, inside citron-green with maroon base. Coronal feet hemi-globose, free, tri-segmented; central segment 4–5 mm long, filiform, pinkish to greenish, papillose, apex entire, tortuose; lateral segments 1–1.5 mm long, subulate, green. Stamens from inner base of coronal feet; filaments 2.5–3 mm long, filiform; anthers 1.5–2 mm long, narrowly ovate-triangular, attenuate, full length slits. Ovaries 0.5–1 mm long; style 2 mm long, terete, nectiferous, upper half green, lower maroon; style-head 2 × 1 mm, ovoid, apex green, acute; translators ± 2 mm long, spoon elliptic. Gynostegium in corolla mouth. Follicles erect, solitary or paired, 11–15 cm × 4–5 mm, very narrowly ovoid. Seeds obliquely ovate and concave, brown; coma 2.5–3 cm long, whitish.

Mozambique. M: Maputo, near Zitundo on road to Bela Vista, fr. 27.xi.2001, *Goyder* 5010 (K, LMU).

Also in South Africa (KwaZulu-Natal). *Raphionacme lucens* is found in grassland on sandy coastal flats where the climate is relatively moist and hot; sea coast–50 m.

10. **Raphionacme utilis** N.E. Br. & Stapf in Bull. Misc. Inform. Kew **1908**: 215 (1908). Type: Angola, Ecanda, *Compania de Mozambique* s.n. (K holotype).

Erect single to few-stemmed herb with leaves rosette-like on ground surface. Tuber 5–15 cm diam., broadly sub-spherical to spindle-shaped. Aerial stems up to 8 cm × 2–3 mm, brownish, glabrous or puberulous, branching lateral. Leaves spreading, sub-sessile to petiolate; petiole

0–10 mm long, glabrous, puberulous or pubescent; blade 5–9 × 3–4 cm, ovate, elliptic or obovate, obtuse, truncate or acute, base obtuse to cuneate, above dark green, glabrous, puberulous or pubescent, beneath purplish, glabrous or puberulous, margin wavy, secondary and tertiary veins visible, secondary veins arching and 5–10 to a side. Inflorescences mostly only terminal, compact to open sub-globose raceme of monochasia, many-flowered, glabrous, puberulous or pubescent, peduncles ± 5 mm long, pedicels ± 5 mm long; bracts ± 2 mm long, acicular. Sepals 2–3 mm long, very narrowly triangular to very narrowly ovate, attenuate, outside glabrous, puberulous or pubescent. Corolla outside glabrous, puberulous or pubescent, inside glabrous; tube 1.5–2 mm long, campanulate; lobes 5–6 × 2 mm, oblong-ovate to narrowly ovate, acute or obtuse, green to violet. Corona whitish, glabrous or papillose; coronal feet tri-segmented, central segment 5–6 mm long, filiform with apex tortuose, or linear with deeply cleft bifid apex; lateral segments of neighbour corona lobes fused completing the coronal annulus; compound lateral segments 1–2 × 1–2 mm, sub-squarish, oblong or sub-ovate, apices unevenly frayed or bilobed. Stamens fused to inner face of coronal feet; filaments ± 1 mm long, base fleshy, dilated, upper part filiform; anthers ± 1 mm long, oblong-ovate, attenuate, whitish, full length slits. Ovaries ± 1 mm long; style 2 mm long, terete, deep purple; style-head ± 1 mm long, triangular, bilobed, deep purple; translators ± 1 mm long, spoon elliptic. Gynostegium in corolla mouth. Follicles paired and horizontally divergent or single, 4 × 0.8 cm, narrowly ovoid, attenuate. Seeds 4–5 × 3–4 mm, broadly ovate, warty, brown; coma 12–15 mm long.

Zimbabwe. N: Mazoe, Umvukwes, Ruorka Ranch, fl. 21.xii.1952, *Wild* 3916 (K, SRGH). **Mozambique**. MS: cult. Zimbiti Ag. Exp. Station, fl. 6.ii.1909, *Sheppard* 308 (K).

Also in Angola, Cameroon and the D.R. Congo. Introduced from Angola to the Zimbiti Agricultural Experimental Station in Mozambique early in the previous century. In highland grassland at about 1500 m altitude, but it is also found in savanna.

Locally it can be very abundant, so much so, that in the beginning of the previous century wagon loads full of its tubers were collected to produce "Ecanda rubber" or "Bitinga rubber". Little is known about the habitat in which this plant grows. In Zimbabwe it was found in chrome rich soil.

11. **Raphionacme globosa** K. Schum. in Bot. Jahrb. Syst. **17**: 118 (1893). —Bullock in Kew Bull. **9**: 62 (1954). —Venter in F.T.E.A., Apocynaceae (part 2): 149–150 (2012). Type: Angola, Malange, xi.1879, *Mechow* 327 (B† holotype, Z lectotype, K), lectotypified by Venter in S. African J. Bot. **75**: 309 (2009).

   Raphionacme virgultorum S. Moore in J. Bot. **50**: 338 (1912). Type: Angola, Kubango, near Fort Princeza Amelia, *Gossweiler* 2267 (BM lectotype), lectotypified by Venter in S. African J. Bot. **75**: 309 (2009).

Erect, single to few-stemmed herb. Tuber up to 10 cm diam., conical to spherical. Aerial stems up to 0.5 m × 1–5 mm, glabrous, sparsely hirsute or scabrous, branching lateral. Leaves ascending to spreading, petiole 2–5 mm long; blade 5–19 × 0.2–2 cm, linear to very narrowly ovate to less often narrowly ovate, acute to attenuate, base cuneate, glabrous with midrib and margin hirsute or scabrous, or sparsely hirsute or scabrous all over, above dark green and sometimes densely dotted, beneath pale green and mostly densely dotted, margin reddish, secondary veins invisible or visible and arching nearly parallel to midrib, ± 5 to a side. Inflorescences terminal, rarely also subterminal, many-flowered globose pseudo-raceme of monochasia, hairy as on stems but denser, peduncles 1–4.5 cm long, pedicels ± 2 mm long; bracts 2–3 mm long, subulate. Sepals 2–3 × 0.8–1 mm, subulate to narrowly triangular, attenuate, outside hairy as on stem. Corolla outside hairy as on stem, inside glabrous; tube 2–3 mm long, campanulate; lobes 3–4 × 1–2 mm, ovate to triangular, bluntly acute, insides white, whitish green, creamy or yellowish, sometimes tinged violet, brown or red, tips purple, pink or red. Corona pale green tinged purple, coronal feet free, lobes tri-segmented; central segment 2–3 mm long, filiform, papillose, apex twirled; lateral segments ± 1.5 mm long, subulate, glabrous. Stamens from inner base of coronal feet; filaments ± 0.2 mm long, linear; anthers ± 1 mm long, narrowly angular-ovate, acuminate, white with reddish connective, full length slits. Ovaries 1–1.5 mm long; style 1–2 mm long, terete, reddish; style-head 1–2 × 0.5 mm, ovoid, obtuse; translators ± 1.6 mm long, spoon oblong-ovate. Gynostegium in corolla mouth. Follicles solitary, erect, ± 7.5 × 0.7 cm, cylindrical-ovoid tapering into a blunt tip, glabrous, smooth. Seeds ± 5 × 2 mm, obliquely obovate; coma 2.0–2.5 cm long, whitish.

Zambia. B: Barotseland, Balovale, fr. i.1954, *Gilges* 359 (K, PRE, SRGH). N: 15 km E of Kasama, fl. 20.xii.1961, *Robinson* 4736 (K, SRGH). W: Chingola, fl. 22.xi.1955, *Fanshawe* 2614 (K, NDO). C: Luangwa Valley, Kapampa River, fl. 7.i.1966, *Astle* 4291 (SRGH). E: Eastern Province, Sasare, fl. 8.xii.1958, *Robson* 866 (K, SRGH). **Malawi**. N: Chipita Dist., Kaseye Mission, fl. 4.i.1974, *Pawek* 7756 (K, MO, SRGH, UC).

Also in Angola, D.R. Congo and Tanzania. A highland species at altitudes ranging from 800 –1800 m in savanna, but also in grassland and forest. Occurs in a variety of habitats ranging from rocky shallow soils to deeper loamy soils.

Raphionacme globosa and *R. galpinii* are very similar in morphology and may be confused with one another, however there are a few distinctive characteristics, as indicated in the key, that separate the two taxa. *Raphionacme globosa* is found north of 15°S and *R. galpinii* south of 22°S. *Raphionacme globosa* flowers from November to February, peaking in December.

12. **Raphionacme galpinii** Schltr. in Bot. Jahrb. Syst. **18**(Beibl. 45): 14 (1894); in Bot. Jahrb. Syst. **20**(Beibl. 51): 10 (1895). Type: South Africa, Mpumalanga Province [Transvaal], Barberton, Saddleback Mountain, x.1889, *Galpin* 613 (B† holotype, PRE lectotype, NH, SAM), lectotypified by Venter in S. African J. Bot. **69**: 212 (2003).

    Raphionacme elata N.E. Br. in Fl. Cap. 4(1): 535 (1907). Type: South Africa, Northwest Province [Transvaal], Rustenburg, 14.xii.1903, *Pegler* 1054 (K lectotype, BOL, GRA, NH, PRE, SAM), lectotypified by Venter in S. African J. Bot. **69**: 212 (2003), as "holotype".

    Raphionacme macrorrhiza Schltr. in Bot. Jahrb. Syst. **20**(Beibl. 51): 10 (1895). Type: South Africa, KwaZulu-Natal Province, Durban, Clairmont, viii.1893, *Schlechter* 3084 (PRE lectotype, B†, BOL), lectotypified by Venter in S. African J. Bot. **75**: 307 (2009).

Erect, single to few-stemmed herb. Tuber up to 20 cm diam., conical to spherical. Aerial stems up to 1 m × 1–8 mm, densely velutinous, pubescent, puberulous or glabrous, branching lateral. Leaves ascending, petiolate; petiole 3–15 mm long; blade 4–16 × 0.5–3 cm, broadly to narrowly ovate, broadly to narrowly obovate, elliptic to narrowly elliptic or linear, obtuse, acute or acuminate, base cuneate to obtuse, dark olive green above, pale green below, indumentum as on stem, often more hairy beneath than above, secondary veins arching, seldom divaricate, 7–12 to a side. Inflorescences terminal or subterminal, many-flowered, globose pseudo-raceme of monochasiums; indumentum as on stems, peduncles 5–20 mm long, pedicels 2–6 mm long; bracts 2–10 mm long, subulate. Sepals 2–5 × 1–2 mm, subulate, apex attenuate, pubescent outside. Corolla pubescent outside, inside glabrous; tube 3–5 mm long, campanulate, green; lobes 3–5 × 2–3 mm, narrowly ovate to ovate, obtuse, spreading, inside green or green with base brown to brownish-green. Corona green, greenish-maroon or greenish-violet, coronal feet free, lobes tri-segmented; central segment 4–5 mm long, filiform; lateral segments 2–3 mm long, filiform, subulate or triangular. Stamens from inner base of coronal feet; filaments 1–2 mm long, filiform; anthers 1–2 mm long, ovate to narrowly ovate, full length slits, acute. Ovaries 1–2 mm long; style 1–2 mm long, terete; style-head 1–2 mm long, ovoid; translators ± 1.2 mm long, spoon elliptic. Gynostegium in corolla mouth. Follicles 1 or 2, erect, 5–22 × 0.4–1.0 cm, cylindrical-ovoid to narrowly ovoid, pubescent to glabrous. Seeds 5–6 × 2 mm, oblong-ovate; coma 1.3–1.5 cm long, white.

Mozambique. M: *Raphionacme galpinii* is common on the Lebombo Mountains which form the border between South Africa, Swaziland and southern Mozambique, and thus most probably also present on the Mozambique side of the border.

Also in South Africa (Provinces of Eastern Cape, Free State, Gauteng, KwaZulu-Natal, Limpopo, Mpumalanga and North-West) and Swaziland. A component of grassland and savanna, in a variety of habitats ranging from sub-tropical, fairly humid coastal flats and hills to inland plateaux and mountain sides with less humid and colder winter climate; sea coast–1800 m.

13. **Raphionacme madiensis** S. Moore in J. Bot. **46**: 293 (1908). —Venter in F.T.E.A., Apocynaceae (part 2): 153 (2012). Type: Uganda, Madi, near Nimuli, 17.v.1907, *Bagshawe* 1611 (BM holotype).

   Raphionacme wilczekiana R. Germ. in Bull. Jard. Bot. État. Bruxelles **22**: 74 (1952). Type: D.R. Congo, Plain of Ruzizi, Kabunambo to River Ruzizi, *Germain* 5567 (BR holotype, SRGH).

Erect to spreading, profusely lactiferous herb. Tuber up to 16 cm diam., sub-spherical. Aerial stems 1–few per plant, erect to spreading, 10–16 cm × 3–5 mm, puberulous to scaberulous, branching lateral. Leaves spreading, petiolate; petiole 5–20 mm long, puberulent; blade (4.5)7.5–14.5 × (1.4)2.5–4.5 cm, ovate, elliptic, obovate, narrowly ovate, narrowly elliptic or narrowly obovate, acute, base tapering, wrinkled, above green, puberulent, veins pale and densely puberulent, below pale green, puberulent and veins densely puberulent, margin wavy, secondary veins divaricate or arching and 7–12 to a side. Inflorescences terminal and axillary, plumose, compact, many-flowered, densely puberulous to scaberulous, peduncles 5–10 mm long, pedicels 5–7 mm long; bracts 3–5 mm long, subulate. Sepals 2–3 × 1 mm, very narrowly triangular or ovate, attenuate. Corolla tube ± 2 mm long, campanulate, outside puberulous, inside glabrous; lobes 5–7 × 1.5–2 mm, oblong-ovate, oblong-triangular, narrowly ovate or narrowly triangular, acute, outside puberulent, green, inside glabrous, green or dull green to whitish with brown, pink, magenta or violet centre streaks or tips or margins. Coronal feet ± 1 mm long, columnar, free, laterally fused to inner base of corolla lobes, apically tri-segmented; central segment 5–7 mm long, linear to filiform, papillose, lateral segments 2–3 mm long, filiform, glabrous or papillose, apex bifid or not, tortuous. Stamens borne at inner base of coronal feet; filaments ± 1 mm long, linear, dilating towards apex; anthers ± 2 mm long, ovate, white, full length slits, apices fused together over style-head. Ovaries ± 1 mm long; style ± 1 mm long, terete; style-head ± 2 mm long, ovoid, apex acute; translators ± 1.5 mm long, spoon angular ovate. Gynostegium in corolla mouth. Follicles solitary, erect, 4.5–6 × 1.2–1.5 cm, ovoid, tapering. Seeds unknown.

Zambia. N: Mbala [Abercorn] – Kambole, fl. 31.xii.1949, *Bullock* 2156 (K). **Zimbabwe**. N: Umvukwes, Ruorka Ranch, 15.xii.1952, *Wild* 3916 (MO). E: Mutare, Vumba Mountains, fl. xii.1937, *Obermeyer* 2123 (PRE).

Also in Angola, Burundi, D.R. Congo, Kenya, Rwanda, Tanzania and Uganda. In a variety of habitats from open savanna and grassland to rocky outcrops and shallow rock basins on limestone silt, grey clay, grey sandy clay-loam and orange-red loam; 700–2050 m.

Flowers from September to January south of the equator, from March to June north of the equator. The tuber is said to be edible.

14. **Raphionacme velutina** Schltr. in Bot. Jahrb. Syst. **20**(Beibl. 51): 12 (1895). Type: South Africa, Heidelberg, 20.x.1893, *Schlechter* 3509 (B† holotype, Z lectotype, BOL, SAM), lectotypified by Venter in S. African J. Bot. **69**: 213 (2003).

   Raphionacme burkei N.E. Br. in Fl. Cap. **4**(1): 537 (1907). Type: South Africa, Magaliesberg, *Burke* 64 (K lectotype, PRE, TCD), lectotypified by Venter in S. African J. Bot. **69**: 213 (2003).
   Raphionacme dinteri Schinz in Vierteljahrsschr. Naturf. Ges. Zürich **55**: 245 (1910). Type: Namibia, 10 km E of Orumbo, *Dinter* 1326 (Z lectotype), lectotypified by Venter in S. African J. Bot. **69**: 213 (2003).
   Brachystelma viridiflorum Turrill in Bull. Misc. Inform. Kew **1924**: 259 (1924). Type: "Cult. Kew, 6.vi.1923, received from Pretoria" (K holotype).

Spreading, many-stemmed herb. Tuber up to 25(45) × 20 cm, spindle-shaped. Aerial stems up to 40 cm × 2–5 mm, velutinous in younger stems, sparsely velutinous in older, purplish, branching lateral. Leaves spreading, sessile or petiolate; petiole up to 5 mm long, above with reddish colleters; blade usually folded lengthwise, (1)3–5(7) × 0.2–2 cm, very narrowly ovate, narrowly elliptic to narrowly obovate, ovate, elliptic or obovate or sub-orbicular, acute or obtuse-mucronate, base cuneate; above glabrous with midvein glabrous or sparsely puberulous, densely dotted; beneath densely velutinous, midvein purplish, secondary veins arching and 5–10 to a side. Inflorescences axillary, numerous along stem, compact pseudo-heads of ± 10 flowers, densely velutinous, peduncles 1–3 mm long, pedicels 1–2 mm long; bracts 1–2 mm long, subulate. Sepals 2 × 0.5 mm, subulate to very narrowly triangular, attenuate, velutinous. Corolla velutinous outside; tube 2–3 mm long, campanulate; lobes spreading, 3–4 × 2 mm, oblong-ovate to ovate, obtuse, chrome-yellow to yellow-green to green with base purplish. Coronal feet fleshy, tri-segmented, free; central segment 2–3 mm long, filiform, papillose or glabrous, sparsely pubescent at base, purple, attenuate; lateral segments 1–2 mm long, subulate, whitish, yellowish or greenish. Stamens from inner face of coronal feet; anthers sub-sessile, 2–4 mm long, narrowly ovate, mucronate, white, connective purple, full length slits. Ovaries 1 mm long, style 1–2 mm long, terete, style-head ± 0.75 mm long, narrowly ovoid, obtuse; translators ± 0.7 mm long, spoon elliptic. Gynostegium in corolla mouth. Follicles solitary, 5–9 × 0.5 cm, very narrowly obovoid, tapering, curved. Seeds 10 × 1 mm, obliquely oblong-ovate, brown.

Botswana. N: 21 km W of Nokaneng, fr. 11.iii.1965, *Wild & Drummond* 6870 (SRGH). SW: Ghanzi & Kgalagadi Dist., 4 km NE of Tshobokwane bore hole, fl. 6.xi.1979, *Skarpe* 341 (K). SE: railway crossing 5 km N of Gaborone, fl. 12.x.1977, *Hansen* 3222 (C, GAB, K, PRE). **Zambia**. S: Machili, fr. 16.xii.1960, *Fanshawe* 5977 (K, NDO). **Zimbabwe**. W: Hwange [Wankie] National Park, fl. 26.xi.1968, *Rushworth* 1298 (K, SRGH). C: Gweru, fl. 27.x.1965, *Biegel* 478 (SRGH). S: Chiredzi, 6.5 km from Chipinda Pools, fl. 3.xii.1970, *Kelly* 271 (SRGH).

Also in Angola, Namibia and South Africa (Provinces of Limpopo, North West and Northen Cape); 400–1700 m.

This is one of the more commonly encountered species in the genus and is found in grassland or the grassland component of savanna, usually on red sandy soil, but also on loamy, stony or calcrete soils. The species flowers during late spring to early summer, peaking in November. The tuber is used as source of water and food.

15. **Raphionacme lanceolata** Schinz in Verh. Bot. Vereins. Prov. Brandenburg **30**: 263 (1888). —Huber in Merxmüller, Prodr. Fl. Sudwestafr. **113**: 7 (1967). Type: Namibia, Ovamboland, Omandongo, *Schinz* 167 (Z lectotype, K, Z), lectotypified by Venter in S. African J. Bot. **69**: 212 (2003).

Raphionacme lanceolata var. *latifolia* N.E. Br. in F.T.A. 4(1): 274 (1902). Type: Botswana, near Chukutsa Salt-pan, 1899, *Lugard* 260 (K lectotype), lectotypified Venter in S. African J. Bot. **69**: 212 (2003).

Spreading, seldom erect herb. Tuber up to 35 cm diam., sub-spherical. Aerial stems up to 0.5(1.0) m × 2–5 mm, densely to sparsely pubescent or puberulous, sometimes velutinous, branching dichotomous, opposite or lateral. Leaves spreading, petiolate; petiole 2–10 mm long, sometimes with reddish colleters above; blade (1)3–9 × (0.5)1–4.5 cm, narrowly ovate to ovate or elliptic to obovate, acute to obtuse-mucronate, base obtuse to cuneate, indumentum above and beneath as on stem, secondary veins divaricate or arching and 7–9 to a side. Inflorescences axillary, pseudo-panicle of monochasiums, 5–10(15)-flowered, indumentum as on stems, peduncle 5–20 mm long, pedicel 5–10 mm long; bracts 2 mm long, subulate to narrowly ovate. Sepals 2 × 1 mm, ovate, acuminate, pubescent outside. Corolla tube 2–4 mm long, campanulate, sparsely pubescent outside; lobes spreading, 4–6 × 2 mm, oblong-ovate to ovate, obtuse, pubescent outside, inside glabrous and mostly green, sometimes mauve, bluish or greenish brown with mauve tips. Corona green to brownish or purplish-green, papillose; coronal feet semi-columnar, free; lobes 4–8 × 2 mm, simple, ligulate or narrowly ovate to narrowly triangular, deeply bifid. Stamens from inner apex of coronal feet; filaments ± 0.2–0.5 mm long, filiform; anthers 2.0–2.5 mm long, narrowly ovate, acuminate, full length slits. Ovaries ± 1 mm long, style 2–3 mm long, terete, glabrous; style-head 2–2.5 mm long, ovoid, acute; translators ± 1 mm long, spoon elliptic. Gynostegium in corolla mouth. Follicles solitary, rarely paired, erect, 5.5–11.5 × 0.5–1. cm, narrowly ovoid with tapering apex, puberulous. Seeds brownish, 5–10 mm long; coma 1.2–1.5 cm long.

Botswana. N: Northern Dist., Aha Hills, fr. 23.iv.1980, *P.A. Smith* 3386 (K, PRE, PSUB, SRGH). SW: Farm Dekar, fl. 20.i.1970, *Brown* 7972 (K). **Zambia**. B: Machili, fl. 17.xi.1960, *Fanshawe* 5901 (K); N/E: Mpika, Luangwa Valley Game Reserve, fl.

13.xii.1966, *Prince* 24 (SRGH). S: 112 km W of Livingstone, fl. i.1973, *Williamson* 2256 (SRGH). **Zimbabwe**. N: Shamwa, Chipoli, fl. 4.i.1981, *Ellert* s.n. in SRGH-266940 (SRGH); Gokwe, *Jacobson* 3514 (PRE, SRGH). W: Hwange National Park, fl. 26.xi.1968, *Rushworth* 1580 (SRGH). C: Harare, fl. 30.x.1945, *Wild* s.n. in GH13611 (K, SRGH). E: Mutare, Sabi Drift, fl. 1.xii.1954, *Wild* 4660 (BR, PRE, SRGH).

Also in Angola and Namibia. In grassland of mopane or thorn savanna; 500–1500 m.

Although mostly found in sandy habitats of flatlands, it also grows in stony granitic or quarts rich gravelly soils of hilly country. Climatically the environment is arid, with extremely hot summers and rains falling late in the season, the winters range from fairly warm to frosty at night. Flowering in late summer, peaking in February. The tuber is a source of water to local inhabitants. Vernacular name: 'Zurukaka' (Ovambo), 'Etundo' (Herero).

16. **Raphionacme michelii** De Wild. in Ann. Mus. Congo Belge, Bot. sér. 5, **1**: 181 (1904). —Venter in F.T.E.A., Apocynaceae (part 2): 153–154 (2012). Type: D.R. Congo, Kimbele, 8.x.1902, *Cabra & Michel* 47 (BR lectotype), lectotypified by Venter in S. African J. Bot. **75**: 327 (2009).

Raphionacme gossweileri S. Moore in J. Bot. **46**: 294 (1908). Type: Angola, Kuiriri, east of Kossuogo, x.1906, *Gossweiler* 3273 (BM holotype, K).

Erect or spreading herb. Tuber up to 10 cm diam., sub-spherical. Aerial stems 1–few, 2.5–8 cm × ± 2 mm, puberulent to pubescent, branching lateral. Leaves spreading to ascending, sub-sessile to petiolate; petiole 0–10 mm long, sometimes stem-clasping, puberulent to pubescent; blade 3.5–6 × 0.7–2 cm, narrowly obovate to obovate, acute, base tapering, above glabrous or pubescent, spotted, beneath puberulent to pubescent and denser on veins, margin wavy, secondary veins arching to divaricate and 7–8 a side. Inflorescences terminal, open, few-flowered monochasia, puberulent to pubescent, peduncles 2–17 mm long, pedicels 7–18 mm long; bracts 2–4 mm long, subulate. Sepals 2–3 × 1 mm, narrowly triangular to narrowly ovate, attenuate, outside puberulent to pubescent. Corolla tube 3–5 mm long, campanulate, outside glabrous to pubescent, inside glabrous; lobes spreading, 12–20 × 3–5 mm, oblong-ovate, obtuse to sub-acute, outside glabrous to pubescent, inside mauve. Coronal feet fleshy; lobes from outer apex of coronal feet, 7–20 mm long, base dilated becoming filiform, papillose or glabrous, apex tortuous. Stamens from inner apex of coronal feet; filament base 2–3 mm long, semi-ovoid, fleshy; upper part 2–4 mm long, filiform; anthers 3–4 × 2 mm, oblong-ovate, attenuate, whitish, full length slits. Ovaries 1–2 mm long; style 3–4 mm long, terete; style-head 3–4 × 1.5–2 mm, ovoid, acute; translators ± 4 mm long, spoon elliptic. Gynostegium elevated above corolla mouth. Follicles and seeds unknown.

Zambia. W: Kafue National Park, Shakalona Dambo, fl. 5.xii.1962, *Simpson* 16/19 (SRGH).

Also in Angola, D.R. Congo and Kenya. In savanna and on rocky river banks; 1100–1290 m.

Flowering in September and October.

17. **Raphionacme linearis** K. Schum. in Bot. Jahrb. Syst. **17**: 117 (1893). Type: Angola, Malange, *Mechow* 359 (Z lectotype, B†), lectotypified by Venter in S. African J. Bot. **75**: 321 (2009).

Raphionacme linearis var. *puberula* K. Schum. in Bot. Jahrb. Syst. **17**: 118 (1893). Type: Angola, Malange, *Mechow* 359, in part (Z lectotype, B†), lectotypified by Venter in S. African J. Bot. **75**: 321 (2009).

Raphionacme linearis var. *glabra* K. Schum. in Bot. Jahrb. Syst. **17**: 118 (1893). Type: Angola, Malange, *Mechow* 359, in part (Z lectotype, B†), lectotypified by Venter in S. African J. Bot. **75**: 321 (2009).

Erect, single or few-stemmed herb. Tuber cylindrical. Aerial stems up to 0.9 m × 3–5 mm, brownish, puberulent to pubescent, branching lateral. Leaves ascending, sessile; blade 5–10 × 0.3–0.5 cm, linear to linear-ovate, acute to attenuate, above dull green, puberulous with midrib deeply sunk, below glabrous, pale green with midrib pubescent, margin revolute, secondary veins invisible. Inflorescences terminal and sub-terminal, few-flowered monochasia, puberulent to pubescent; peduncle 0–4 cm long, dull purple; pedicels 1–4 cm long, dull purple; bracts 3–4 mm long, acicular. Sepals 5–10 × 1–2 mm, very narrowly triangular, attenuate, pubescent. Corolla outside sparsely pubescent; tube 2–3 mm long, broadly campanulate, outside mauve; lobes reflexed, 2–5 × (0.4)0.7–1.1 cm, obliquely linear to obliquely linear-obovate, obliquely obtuse, inside violet-blue to violet, mottled, base inside ridged, ridge pubescent to tomentose and papillose. Coronal feet free; lobes 7–10 mm long, corniculate, violet-blue to violet, glabrous with base pubescent to tomentose. Stamens from inner base of coronal feet; filaments ± 1 mm long, filiform; anthers 5–6 × 2 mm, oblong-ovate, attenuate, white, full length slits, with two basal auricles. Ovaries ± 1 mm long; style 4–5 mm long, terete; style-head 4–5 × 2 mm, oblong-ovoid, obtuse. Gynostegium in corolla mouth. Follicles 15 × 0.5 cm, very narrowly ovoid. Seeds comose.

Zambia. N: Kawambwa Dist., Mbereshi Dambo on road from Kawambwa to Nchelenge, fl. 7.xii.2002, *Bingham* 12581 (K). In swampy habitat or moist grassland where it may be common; 1000–2000 m.

Flowers sweet scented.

18. **Raphionacme grandiflora** N.E. Br. in Bull. Misc. Inform. Kew **1895**: 111 (1895). —Schumann in Engler & Prantl, Nat. Pflanzenfam. **4**(2): 221 (1895). —Venter in F.T.E.A., Apocynaceae (part 2): 150 (2012). Type: Tanzania, Niomkolo, 1890, *Carson* 5 (K lectotype), lectotypified by Bullock in Kew Bull. **8**: 353 (1953). FIGURE 7.3.**130**.

   Raphionacme grandiflora subsp. *glabrescens* Bullock in Kew Bull. **8**: 353 (1953). Type: Mozambique, Moebede, Lugela, Mocuba, Quelimane Dist., 1944, *Faulkner* 7 (K holotype, PRE).

   Pentagonanthus grandiflorus (N.E. Br.) Bullock in Hooker's Icon. Pl. **36**: t.3583 (1962).

   Pentagonanthus grandiflorus subsp. *glabrescens* (Bullock) Bullock in Hooker's Icon. Pl. **36**: t.3583 (1962).

Erect, single or few-stemmed herb. Tuber up to 7.5 cm diam., conical to spherical. Aerial stems up to 80 cm × 2–5 mm, sparsely pubescent or sparsely velutinous to pubescent or velutinous, branching lateral. Leaves spreading, rarely ascending, sessile to petiolate; petiole 0–2.5 cm long, with reddish axillary colleters; blade (2)10–21 × (0.3)1–6 cm, obovate, ovate, narrowly obovate, narrowly ovate to linear, often progressively narrowing from base to apex of stem, acute to attenuate, base cuneate, above dark green, glabrous to pubescent or velutinous and often densely dotted, beneath pale green and glabrous with veins pubescent or velutinous or blade hairy all over, secondary veins arching parallel to midrib and 8–10 to a side. Inflorescences terminal and from upper 2–3 nodes, pseudo-umbel of 1–3 flowers, hairy as on stem, peduncles rarely present, pedicels 1.5–4.5 cm long; bracts 1–5 mm long, subulate to narrowly triangular. Sepals 5–6 × ± 2 mm, narrowly ovate, attenuate hirsute, margin fimbriate. Corolla glabrous; tube 5–9 mm long, pentangular-campanulate, basally 5-saccate; lobes 1.5–2.0 × 0.8–1.0 cm, ovate to triangular, obtuse to acute, inside deep violet, deep violet-blue or deep red, sometimes with green tips, spreading. Corona white, whitish flecked violet, violet or greenish, coronal feet free, fused to corolla lobe bases; lobes 4–5 × 4–5 mm, rectangular, fleshy, glabrous, apex tri-segmented, central segment 5–7 mm long, filiform, lateral segments 1–2 mm long, corniculate to subulate. Stamens from inner base of coronal feet; filaments 1–2 mm long, filiform; anthers 7–8 mm long, narrowly ovate-hastate, acute, creamy-white with connective purplish, upper half with slits, lower of two thickened horn-like callosities. Ovaries ± 1 mm long; style 5–6 mm long, terete; style-head 5–9 × 4–8 mm, broadly oblong-ovoid, obtuse; translators ± 6 mm long, spoon broadly ovate. Gynostegium base in corolla mouth. Follicles paired, erect, 15–18 × 0.8–1.0 cm, linear-ovoid, attenuate, glabrous. Seeds 8–10 × 3 mm, narrowly ovate; coma 3.0 cm long, whitish.

Zambia. N: Lake Tanganyika, Mbulu Island, fl. 29.xii.1959, *Richards* 185 (BR, K). E: Petauke, fl. 17.xii.1958, *Robson* 971 (K). S: Mazabuka Dist., 3 km from Chirundu Bridge on Lusaka road, fl. 1.ii.1958, *Drummond* 5430 (BR, K, PRE, SRGH). **Zimbabwe**. N: Umvukwes, Umsengedzi Farm, fl. 22.xii.1952, *Wild* 3983 (K, SRGH). **Malawi**. N:

Fig. 7.3.**130**. RAPHIONACME GRANDIFLORA. 1, leafy shoot with tuber (× ¹/₃); 2, stem with leaves and flowers (× 1); 3, stem with leaves and fruit (× ¹/₃); 4, sepal (× 6); 5, section of flower to show corolla lobe and corona lobe (× 1); 6, corona lobe from outside (× 3); 7, pistil (× 6); 8–10, translator from outside, inside and laterally (× 6); 11, seed (× 1.5). Drawn by M. Stones. Reproduced with permission from Hooker's Icones Plantarum (1962).

Karonga Dist., Thulwe Hills, fl. 31.i.1992, *Goyder et al.* 3597 (BR, K, MAL). S: Zomba, Lake Chirwa, fl. 28.i.1959, *Robson* 1332 (BM, K, PRE, SRGH, WAG). **Mozambique.** N: North of Nampula, fl.fr. 26.ii.1937, *Torre* 1356 (COI). Z: Quelimane Dist., Muobede, fl. 1944, *Faulkner* 7 (K, PRE).

Also in Rwanda and Tanzania. In grassland and miombo woodland with *Brachystegia*, *Uapaca* and *Pterocarpus* species; 500–1750 m altitude.

Habitat ranges from alluvial to sandy to laterite soils, often stony and among rocks. Flowers from November to February.

19. **Raphionacme pulchella** Venter & R.L. Verh. in S. African J. Bot. **54**: 72 (1988). Type: Mozambique, Manica e Sofala Dist., Makurupini Forest, *Williams* RSES-89 (SRGH holotype).

Erect herb. Roots and tuber unknown. Aerial stems up to 25 cm × 1–2 mm, sparsely pubescent, branching lateral. Leaves ascending, sessile; blade 5–10 × 0.3–0.4 cm, linear, acute, base cuneate, sparsely pubescent all over, or sparsely pubescent above and nearly glabrous beneath, secondary veins arching and ± 8 to a side. Inflorescences axillary from upper nodes, lax, 1–4 monochasial branches, each branch with up to 10 flowers, densely pubescent, peduncles 8–15 mm long, pedicels 3–5 mm long; bracts 2 mm long, narrowly ovate to subulate. Sepals 2–4 × 1 mm, ovate to narrowly ovate, attenuate, pubescent, maroon. Corolla tube 1–2 mm long, broadly campanulate, outside nearly glabrous, inside glabrous; lobes 3–4 × 1.5–2 mm, ovate to broadly ovate, acute, violet, glabrous. Coronal feet fleshy, dilated, tri-segmented, free, glabrous; central segment 3–4 mm long, base dilated becoming filiform, apex tortuous; lateral segments 0.5–1.5 mm long, corniculate or subulate. Stamens arising from inner apex of coronal feet; filaments 0.5–0.7 mm long, filament base fleshy becoming filiform; anthers 1–2 mm long, ovate, full length slits, apex acuminate. Ovaries 0.5 mm long; style 1–1.5 mm long, terete; style-head 1–2 × 1–1.5 mm, ovoid to rhomboid; translators 0.6–0.8 mm long, spoon elliptic. Gynostegium in corolla mouth. Follicles and seeds unknown.

Zimbabwe. N: 48 km NE of Manguila, Mukamba Farm, fl. 13.xii.1975, *Lancaster* 140 (SRGH). E: Mutare, near confluence of Haroni and Lusitu Rivers, fl. 5.xii.1965, *Mare* 57 (SRGH). **Mozambique.** MS: Manica Dist., Makurupini Forest, fl. 15.i.1965, *Williams* 89 (SRGH).

The distribution of *Raphionacme pulchella* is poorly known and disjunct. It was collected at between 1000 and 2000 m altitude in the high mountain range running from Manicaland on the border with Mozambique to north-central Zimbabwe in the Harare Region. In open grassland in *Brachystegia* woodland.

20. **Raphionacme splendens** Schltr. in J. Bot. **33**: 301 (1895). —Venter in F.T.E.A., Apocynaceae (part 2): 154 (2012). Type: Uganda, Ruwenzori, *Scott-Elliot* s.n. (BM holotype, K).

Raphionacme excisa Schltr. in J. Bot. **33**: 301 (1895). —Venter in Hedberg & Edwards, Fl. Ethiopia Eritrea **4**(1): 109 (2003). Type: Uganda, Ruwenzori, *Scott-Elliot* s.n. (BM holotype, K).

Raphionacme bagshawei S. Moore in J. Bot. **45**: 50 (1907). Type: Uganda, Unyoro, above Kibero on Lake Albert, 19.ii.1906, *Bagshawe* 910 (BM holotype).

Subsp. **splendens**.

Erect, 1–3-stemmed herb. Tuber 8–20 × 2–6 cm, narrowly ovoid, ovoid, deltoid, very narrowly deltoid or obconical. Aerial stems erect, 20–50 cm × 2–4 mm, reddish-green, sparsely puberulous, sparsely scabrous or pubescent, branching lateral. Leaves often only develop after flowering, ascending, petiolate; petioles 0–3 mm long; blade 2.5–12 × 0.5–3.5 cm, linear, linear-ovate, narrowly ovate, ovate, narrowly obovate, obovate, elliptic, broadly ovate or broadly elliptic, acute, attenuate, acuminate or obtuse-mucronate, base cuneate to obtuse, above glabrous to sparsely puberulous or sparsely scaberulous, beneath sparsely puberulous, sparsely scaberulous, densely pubescent or scabrous, margin often wavy, secondary veins arching to divaricate and 4–15 to a side. Inflorescences terminal and axillary, puberulous to pubescent, lax, racemose, of 1–few monochasia, each monochasium 1–10-flowered; peduncles 1–15 mm long, reddish; pedicels 5–30 mm long, reddish; bracts 1–2 mm long, ovate to narrowly triangular. Sepals 2–4 × 1–2 mm, ovate to triangular, acute, outside sparsely puberulous to pubescent, reddish-green. Corolla tube 2–4 mm long, campanulate or narrowly urceolate to broadly urceolate, outside glabrous to sparsely pubescent or puberulous, greenish, inside glabrous; lobes reflexed, (5)10–20(30) × 2–4(9) mm, narrowly ovate, narrowly elliptic, narrowly obovate or narrowly triangular, obtuse to acute, outside brownish-pink, dark brown or greenish, sparsely pubescent or hirsute, inside whitish pink, pink, mauve or purple to bluish. Coronal feet columnar; columns erect, exserted, 1–2 mm long, basally fused into an annulus with fleshy intercolumnar lobules of ± 1 × 1 mm; lobes from outer apices of coronal columns, (4)7–12(20) mm long, filiform, glabrous, purple to white, apex entire, tortuous. Stamens from inner apex of coronal columns; filaments 2–5 mm long, semi-erect, purple; anthers 2–4(6) mm long, ovate to narrowly ovate, acute, greenish. Ovaries 1–1.5 mm long; style 4–8 mm long, terete; central part of style swollen cone- or oblong-like and vertically grooved with fine hairs; style-head (2)4–6(8) mm long, oblong-ovoid, acute, greenish; translators 2–7 mm long, spoon broadly ovate. Gynostegium elevated above corolla mouth. Follicles single, rarely paired, erect, 5–11 × 0.4–1.5 cm, very narrowly ovoid, glabrous, puberulous or hirsute. Seeds 5–8 mm long, narrowly ovate, brown; coma 2.5–3 cm long, silvery to coppery coloured.

Only subsp. *splendens* is known from the F.Z. area; subsp. *bingeri* (A. Chev.) Venter is found in West Africa north of the equator.

Zambia. N: Kangiri, NW of Mweru-wa-Ntipa, fl. 7.viii.1962, *Tyrer* 361 (BM, K). C: 16 km W of Luangwa River Bridge, fl. 6.ix.1947, *Greenway & Brenan* 8060 (K). **Zimbabwe**. W: Sebungwe, Nyabini, fl. ix.1955, *Davies* 1540 (SRGH). **Malawi**. N: Nyungwe, fl. 17.ix.1930, *Migeod* 927 (BM). **Mozambique**. N: Cabo Delgado, Palma – Pundanhar road, c.4.5 km W of junction with road to Nhica do Rovuma, fl. 8.xi.2009, *Goyder et al.* 6025 (K, LMA, LMU, P). Z: entre o Gilé e o rio Mulela, fl. 10.x.1949, *Barbosa & Carvalho* 4366 (K, LMA). MS: Entrance to Gorongosa Game Reserve, fl. 23.ix.1961, *Methuen* 237 (K).

Raphionacme splendens is the most widespread species of *Raphionacme*, ranging from Senegal in the west to Tanzania in the east, and from Sudan in the north to Mozambique in the south. As could be expected from a species with such an enormous distribution range, *R. splendens* inhabits a variety of vegetation types and habitats from coast to 1500 m altitude, but these all seem to have in common a mesic to moist environment in the growing season. The vegetation types range from scrub forest to coastal and inland open or dense savanna to pure grassland. The habitat could be shallow soil on granite domes or other rock faces, sandy acidic soil, heavy black clay soils or light sandy loam soils which all may be seasonally inundated. This species, furthermore, seems to be pyrophytic. Although flowering plants of *R. splendens* may be found throughout the year, south of the equator its flowering peaks in August to October.

A decoction of leaves is used for conjunctivitis. The raw edible tuber supposedly tastes like sweet patato.

21. **Raphionacme chimanimaniana** Venter & R.L. Verh. in S. African J. Bot. **54**: 380 (1988). Type: Zimbabwe, Mandidzudzure Dist., Chimanimani Mountains, 14.xi.1967, *Mavi* 634 (SRGH holotype, K).

Erect herb. Tuber unknown. Aerial stems up to 7 cm × 2 mm, densely hirsute and glandular, branching lateral. Leaves ascending, sessile; blade 70–90 × 2–4 mm, linear-ovate, attenuate, base cuneate, hirsute and glandular above, sparsely hirsute below, margin and midrib hirsute, secondary veins arching and ± 10 to a side. Inflorescences lax, terminal and axillary, monochasial branches 2–3-flowered, densely glandular-hirsute; peduncles 4 mm long; pedicels 9 mm long; bracts 3 mm long, subulate, reddish. Sepals 2 × 1 mm, narrowly triangular, attenuate, densely glandular-hirsute. Corolla tube 3–4 mm long, campanulate, outside hirsute, inside glabrous; lobes 10–12 × 2.5–3 mm, obovate to ovate, apex acute, spreading, outside hirsute, inside glabrous, magenta to white. Coronal feet 2 mm long, columnar, free; lobes 7–10 mm long, filiform, glabrous, tortuous. Stamens from inner apex of coronal feet; filaments 2.0–2.5 mm long, filiform; anthers 4–5 mm long, narrowly triangular, acute, full length slits. Ovaries 1 mm long; style 6 mm long, terete, style-head 3–4 mm long, broadly oblong ovoid, apex acute, simple or bifid; translators 3.5–4.0 mm long, spoon elliptic. Gynostegium elevated above corolla mouth. Follicles paired, erect, cylindrical-ellipsoid, 4.5 × 0.3 cm. Seeds unknown.

Zimbabwe. E: Chimanimani Mountains, Musapa Gap, fl. 20.xii.1957, *Phipps* 839 (K, SRGH).

Only two collections are known, both from the Chimanimani Mountains at ± 1500 m. In grassland above the level of the *Brachystegia tamarindoides* woodland on the northern slopes of the mountains, on granite or quartzite.

22. **Raphionacme longituba** E.A. Bruce in Bull. Misc. Inform. Kew **1937**: 419 (1938). —Venter in F.T.E.A., Apocynaceae (part 2): 152 (2012). Type: Tanzania, Kakoma, south of Tabora, 13.i.1936, *H.M. Lloyd* 45 (K holotype, EA).

   Raphionacme ernstiana Meve in Bradleya **18**: 71 (2000). Type: Tanzania, Singida Prov., 51 km SW of Itigi on road to Rungwa, 10.i.1998, *E. & M Specks* 1038 (K holotype, UBT).

Spreading herb. Tuber up to 25 cm diam., spindle-shaped. Aerial stems spreading to decumbent, 15–20 cm × 4–5 mm, hispid, branching dichotomous. Leaves spreading, petiolate; petiole 1–3 mm long, hispid, above with reddish colleters; blade 4.5–6.5 × 2–3.5 cm, elliptic to ovate, obtuse-mucronate, base obtuse, often succulent, sparsely hispid, dotted, margin thickened, undulating and densely hispid; secondary veins divaricate to patent and 16–22 to a side. Inflorescences axillary, hispid, 3–4-flowered, peduncle 10–30 mm long, pedicels 5–6 mm long; bracts 3–7 mm long, subulate. Sepals 4–6 × 1–2 mm, very narrowly triangular, attenuate, outside hispid. Corolla with outside green tinged maroon or purplish to brown, hispid, inside white or white tinged pink to purplish, glabrous; tube 7–10 mm long, cylindrical-campanulate; lobes spreading, 9–11 × 3–4 mm, oblong to narrowly ovate, obtuse to acute. Corona conniving dome-like over gynostegium; coronal feet free; lobes 5–6 × 2–3 mm, narrowly ovate, often concave, acuminate or bifid, fleshy, green to creamy-yellow. Stamens from inner base of coronal feet; filaments ± 1 mm long, linear; anthers 2–3 mm long, oblong-ovate to narrowly ovate, acuminate, slits in upper 3/4. Ovaries ± 1 mm long; styles 7–8 mm long, terete, apically fused; style-head ± 2 × 1 mm, ovoid, acute; translators ±2 mm long, spoon elliptic. Gynostegium in corolla mouth. Follicles solitary, erect, 17–18 × 1–1.5 cm, very narrowly ovoid, tapering. Seeds ± 9 × 3 mm, coma ± 3 cm long.

Zambia. N: Isoka Dist., 18 km from Tunduma on road to Mbala, fl. 10.i.1975, *Brummitt & Polhill* 13697 (BR, K). C: Kabwe Dist., Mpunde, fl. 7.xii.1972, *Kornas* 2768 (K). E: Lutembe River just below Great East Road Bridge, fl. 9.i.1959, *Robson* 1110 (K, PRE, SRGH). S: Mazabuka, fl. 7.xii.1931, *Trapnell* in CRS540 (K). **Malawi**. N: Mzimba Dist., 12 km S of Eutini, fl. 31.xi.1976, *Pawek* 10788 (K, MAL, MO, SRGH, UC).

Also in Tanzania. *Raphionacme longituba* is a highland species in grassland and savanna such as *Brachystegia* woodland, sometimes common; 1500–1750 m.

The habitat varies from sandy to gritty to rocky soils, which may be deep or shallow. The habitat may even be temporarily marshy. Flowering occurs from November to January, but peaks in December.

23. **Raphionacme hirsuta** (E. Mey.) R.A. Dyer in Fl. Pl. South Africa **22**: t.853 (1942). Types: South Africa, Eastern Cape Province, *Drège* s.n. (LUB† holotype); Kei River, xi.1889, *Flanagan* 394 (PRE neotype, GRA), designated by Venter in S. African J. Bot. **69**: 212 (2003).

   Brachystelma hirsutum E. Mey., Comm. Pl. Afr. Austr. **2**: 197 (1838).
   Raphionacme divaricata Harv. in London J. Bot. **1**: 23 (1842). —Brown in Fl. Cap. **4**(1): 553. Type: South Africa, KwaZulu-Natal Province, Durban, *Owen* s.n. (K holotype, TCD).

Apoxyanthera pubescens Hochst. in Flora **26**: 78 (1843). Type: South Africa, KwaZulu-Natal Province, Pietermaritzburg, *Krauss* 106b (TUB holotype).

Raphionacme pubescens (Hochst.) Hochst. in Flora **27**: 827 (1844).

Raphionacme obovata Turcz. in Bull. Soc. Imp. Naturalistes Moscou **21**: 250 (1848). Type: South Africa, Eastern Cape Province, Uitenhage, Olifanthoek, xi.1833, *Ecklon & Zeyher* 64 (S lectotype, G, LD, P, TCD), lectotypified by Venter in S. African J. Bot. **75**: 312 (2009).

Raphionacme purpurea Harv., Thes. Cap. **1**: 41, pl.66 (1859). Type: South Africa, KwaZulu-Natal Province, Pietermaritzburg, Field's Hill, ix.1858, *Sanderson* 84 (TCD holotype, K).

Mafekingia parquetiana Baill., Hist. Pl. **10**: 303 (1890). Type: South Africa, North West Province, Mafeking, i.1887, *Duparquet* 367 (P lectotype), lectotypified by Venter in S. African J. Bot. **75**: 313 (2009).

Raphionacme divaricata var. *glabra* N.E. Br. in Fl. Cap. **4**(1): 539 (1907). Type: South Africa, *Zeyher* 1140 (K holotype, BM).

Raphionacme hirsuta var. *glabra* (N.E. Br.) R.A. Dyer in Fl. Pl. South Africa **22**: t.853 (1942).

Spreading herb. Tuber 15–20 × 7–15 cm, spindle-shaped to sub-spherical. Aerial stems few to many, up to 30 cm × 2–5 mm, pubescent, dichotomously branched. Leaves spreading, petiolate; petiole 2–5 mm long, with axillary colleters; blade 1–5 × 0.5–4 cm, ovate to broadly ovate, elliptic to broadly elliptic, obovate to broadly obovate or sub-orbicular, acute, cuspidate or obtuse-mucronate, base cuneate to obtuse, green with margin often reddish, glabrous to densely hirsute below and above, secondary veins divaricate to patent or rarely arching and 6–10(–16) to a side. Inflorescences terminal or/and axillary, compact, branched, congested with numerous flowers, sparsely hirsute to glabrous, peduncles 0.2–4 cm long, pedicels 0.3–1 cm long; bracts 2–3 mm long, ovate to subulate. Sepals 2–5 × 1 mm, subulate, acuminate, green to purplish-green, sparsely to densely hirsute outside. Corolla pink, maroon, mauve or violet, outside sparsely hirsute, inside glabrous; tube 3–5 mm long, campanulate; lobes 5–8 × 2–3 mm, ovate to narrowly ovate, obtuse, spreading. Corona white to violet; coronal feet free; lobes 4–6 mm long, narrowly obovate, ovate or obtriangular, apex frilled or tri-segmented with central segment a single or bifid subulate tooth; lateral segments broad and one or more toothed. Stamens from inner base of coronal feet; filaments 1–2 mm long, filiform; anthers 2–3 mm long, ovate, attenuate, white with pink or mauve connectives, full length slits. Ovaries 1–2 mm long; style terete, 2 mm long; style-head ovoid, 1–2 mm long, acute; translators 0.8–1.5 mm long, spoon ovate. Gynostegium in corolla mouth. Follicles solitary or paired, erect to spreading, 4.5–11 × 0.6–0.8 cm, narrowly ovoid, attenuate, pubescent. Seeds 7–8 × 3 mm, obliquely ovate, ridged, smooth, coma 1–1.5 cm long, white.

Botswana. SE: Probably south-eastern region on the border with South Africa. **Mozambique**. M: Probably across the border from Swaziland and South Africa (KwaZulu-Natal Province).

Also in Lesotho, South Africa (all provinces, except the Western Cape) and Swaziland. A species with wide ecological amplitude in habitats ranging from the eastern moister and drier subtropical, frost free, coastal or lowland grassland and savanna to inland valley, highland and mountain frost prone grassland to western arid savanna and grassland; coastal area–1800 m. The species exhibits a variety of forms with regard to leaf shape, venation and vesture.

24. **Raphionacme palustris** Venter & R.L. Verh. in S. African J. Bot. **52**: 149 (1986). Type: South Africa, KwaZulu-Natal, Karkloof Range, Farm Bennie, *Venter* 9005 (BLFU holotype, K, PRE).

Erect, glabrous herb of up to 0.5 m high. Tuber(s) 1–few, 10–25 × 10–20 cm, sub-spherical, ovoid or cylindrical. Aerial stems up to 50 cm × 5 mm, erect or sub-erect, becoming violet with age, branching dichotomous or opposite. Leaves petiolate; petiole 2–10 mm long, violet; blade 5–6 × 2–3 cm, oblong-ovate, obtuse-mucronate, base cuneate to obtuse, above dark green, beneath green, margin violet, veins translucent, secondary veins divaricate and 10–19 to a side. Inflorescences terminal or axillary, violet, solitary monochasium or plumose-like from a few monochasia, each monochasium few-flowered, peduncles 10–20 mm long, pedicels 5–10 mm long; bracts 3 mm long, very narrowly ovate. Sepals 3 × 1 mm, narrowly triangular, brownish-violet, attenuate. Corolla tube 3–4 mm long, campanulate, glabrous, outside pale greenish-violet, inside whitish-violet; lobes spreading, 8–10 × 5 mm, obovate to oblong-obovate, obtuse, outside pale greenish-violet, inside whitish-violet. Corona whitish-violet; coronal feet fleshy, apically lobed, free; lobes 1–2 mm long, angular-ovate, arched over style-head, apex bifurcate. Stamens arising from inner base of coronal feet; filaments 1–2 mm long, filiform, violet; anthers ± 2 mm long, narrowly angular-ovate, apex violet and acute, lemon-yellow, full length slits. Ovaries 1–2 mm long, style 1–3 mm long, terete; style-head 2 × 3 mm, broadly rhomboid, pale violet; translators 1.2–2.3 mm long, spoon oblong-ovate. Gynostegium in corolla mouth. Follicles erect, normally solitary, 18–27 × 0.8–1 cm, very narrowly cylindrical obovoid, acuminate. Seeds 8–10 mm long, obliquely narrowly oblong, pale brown, smooth, margin thickened; coma 4–5 cm long, whitish.

Zimbabwe. C: Kadoma, fl. 7.xii.1924, *Hoffe* 9 (PRE).

Probably also in Mozambique. Also in South Africa (KwaZulu-Natal) and probably to be found in Swaziland. Found in sub-tropical swamps, moist grassland or heathland and flowers from September to October; sea coast–1500 m.

38. SCHLECHTERELLA K. Schum.[10]

Schlechterella K. Schum. in Engler & Prantl, Nat. Pflanzenfam., Index: 462 (1899). —Venter & Verhoeven in S. African J. Bot. **64**: 350–355 (1998). —Venter in Fl. Ethiopia Erithrea **4**: 110–112 (2003); in Fl. Somalia **3**: 140 (2006).

Pleurostelma Schltr. in J. Bot. **33**: 303, t.351 (1895), illegitimate name, non Baill. (1890).

Triodoglossum Bullock in Hooker's Icon. Pl. **36**: t.3584 (1962).

Perennial suffrutescent climber with white latex. Tubers turnip-shaped or cylindrical. Stems twining. Leaves opposite or fascicled, linear, narrowly ovate or obovate; interpetiolar ridges with dentate colleters. Inflorescences terminal and axillary. Sepals acicular, ovate or triangular. Corolla tube campanulate to shallowly campanulate; lobes oblong-ovate or obovate. Corona inserted in corolla mouth, coronal feet fused with staminal filaments and nectaries forming an annulus, lobes ligulate or cylindrical, 2-, 3-, or 5-fid. Stamens inserted at inner base of coronal feet; filaments filiform; anthers ovate to ovate-hastiform, base with white calosities, pollen in pollinia. Interstaminal nectaries lobular, fused laterally with coronal feet. Style-head broadly ovoid. Gynostegium in corolla mouth. Follicles paired, widely divergent, cylindrical-ovoid.

An African genus of two species found from Ethiopia and Somalia in the north-east to Mozambique in the south-east.

Schlechterella africana (Schltr.) K. Schum. in Engler & Prantl, Nat. Pflanzenfam. Nachtr. **2**: 60 (1900). —Venter in F.T.E.A., Apocynaceae (part 2): 159 (2012). Type: Kenya, Machakos Dist., Ngomeni, "Ruwenzori Expedition 1893–1894", 20.xi.1893, *Scott Elliott* 6175 (BM holotype, K). FIGURE 7.3.**131**.

Pleurostelma africanum Schltr. in J. Bot. **33**: 303, t.351 (1895).

Tacazzea africana (Schltr.) N.E. Br. in F.T.A. **4**(1): 261 (1902).

Woody, glabrous climber. Root tuber cylindrical. Aerial stems up to 1.5 m × 5 mm, lateral shoots sometimes stunted. Leaves opposite on normal shoots, fascicled in verticillate clusters on stunted shoots, sessile to shortly petiolate; blade 4.5–8.5 × 0.2–1 cm, linear to very narrowly ovate, obtuse to attenuate, base cuneate to rounded, margin revolute and undulate. Inflorescences lax, frail, with few-flowered monochasial branches; peduncles 1–2 cm long, pedicels 3–5 mm long, bracts ± 1 mm long, triangular. Sepals ovate-triangular, acute, 1 × 0.5 mm. Corolla tube shallowly campanulate, 0.5–1.0 mm long; lobes linear to oblong-ovate, acute, slightly reflexed, 6–11 × 2–4 mm, pale greenish outside, inside creamy yellow, white flushed mauve-pink to mauve-

---

[10] by H.J.T. Venter

Fig. 7.3.**131**. SCHLECHTERELLA AFRICANA. 1, flowering stem ($\times \,^2/_3$); 2, stems showing fascicles of leaves on stunted lateral shoots ($\times \,^2/_3$); 3, fruiting specimen ($\times \,^2/_3$); 4, partial inflorescence ($\times 2$); 5, centre of flower showing corona and gynostegium ($\times 8$); 6, half-flower, ($\times 8$); 7, stamens ($\times 12$); 8, translator ($\times 12$). 1, 4–8 from *Newbould* 3242; 2, 3 from *Greenway* 9546. Drawn by D. Erasmus. Reproduced from Flora of Tropical East Africa (2012).

purple or pale mauve. Corona creamy yellow, annulus ± 0.5 mm high, lobes cylindrical, 2–3 mm long; terminal part 2- or 3-fid, inner segment filiform, longest, 2–4 mm long. Ovaries ± 0.5 mm long, style ± 1 mm long, style-head 0.5 × 0.3 mm; translators ± 0.8 mm long, elliptic. Follicles cylindrical-ovoid, blunt-acute, 8–9 cm × 5 mm, reddish brown. Seed unknown.

Mozambique. Z: 11 km from Namina, fl. 24.vii.1962, *Leach & Schelpe* 1441 (SRGH). Also in Ethiopia, Kenya, Somalia and Tanzania. In woodland or dry scrub, usually on sandy red soil, but also on limestone and basement complex; 300–1400 m.

Subfam. **SECAMONOIDEAE**

The smallest of the subfamilies of Apocynaceae. This group is most diverse in Madagascar, both in numbers of species and genera represented. Approximately 7 genera are recognised, with c.150 species. The single genus *Secamone* occurs in continental Africa, and just 10 species are to be found within the Flora area.

39. **SECAMONE** R. Br.[11]

Secamone R. Br., Prodr. Fl. Nov. Holland.: 464 (1810); in Mem. Wern. Nat. Hist. Soc. **1**: 55 (1811). —Bentham in Bentham & Hooker, Gen. Pl. **2**(2): 746 (1876). —Goyder in Kew Bull. **47**: 437–474 (1992). —Klackenberg in Adansonia, sér. 3, **23**: 317–335 (2001).
Rhynchostigma Benth. in Bentham & Hooker, Gen. Pl. **2**(2): 771 (1876); in Hooker's Icon. Pl. **12**: 77 (1876).

Lianas and twining shrubs with white latex. Young shoots and inflorescences glabrous or puberulent. Leaves opposite, petiolate. Inflorescences terminal or axillary (extra-axillary is the norm in Asclepiadoideae), cymose or composed of an aggregation of cymes. Calyx of 5 ± free, triangular to ovate or suborbicular lobes with ciliate margins. Corolla rotate or campanulate, 1–8 mm long, glabrous or hairy on the inner surface, 5-lobed to halfway or beyond; corolline corona present or occasionally absent, formed of 5, usually paired, nonvascularised fleshy ridges running from the base of the corolla tube to the corolla lobe sinus, sometimes extending up the inner face of the corolla lobes or forming a pocket obscuring the sinus. Gynostegial corona staminal, of 5 laterally or more rarely dorsally compressed or subulate lobes. Anthers with entire or fimbriate appendages. Pollinaria minute, with 4 pollinia attached directly or by extremely short, flattened caudicles to the pale, porous corpusculum. Stylar head appendage terete, lobed or dilated, exserted beyond the top of the staminal column or not. Follicles paired but sometimes only one developing, smooth or shallowly striate. Seeds ovate, compressed, with a coma of white or ivory hairs.

A genus of c.90 species distributed across the paleotropics, with its centre of diversity in Madagascar, and with 21 species in continental Africa.

1. Staminal corona lobes dorsally compressed . . . .**10.** *brevipes*
   – Staminal corona lobes laterally compressed or subulate 2
2. Corolla lobes densely pubescent above . . . .**4.** *alpini*
   – Corolla lobes glabrous above, or with few scattered hairs only 3
3. Apical portion of stylar head obconic, clavate or with 2 divergent lobes. 4
   – Apical portion of stylar head not dilated 9
4. Mature leaves rusty-pubescent below 5
   – Mature leaves glabrous or with few scattered hairs only 7

[11] by D.J. Goyder

5. Corolla lobes ± as long as the tube; lower epidermis without tuberculate papillae in addition to the indumentum (viewed under at least × 25) .... **8.** *erythradenia*
– Corolla lobes at least twice as long as the tube; lower leaf epidermis with minute but dense tuberculate papillae in addition to the indumentum (viewed under at least × 25) 6
6. Corolla 1.5–3 mm long, ovoid or subglobose in bud .... **7.** *stuhlmannii*
– Corolla 3.5–5 mm long, conical in bud .... **9.** *dewevrei*
7. Corolline corona formed of 5 membranous pockets obscuring corolla lobe sinus; corolla tube (including coronal pockets) at least 1.2 mm long..... **6.** *delagoensis*
– Corolline corona pockets absent; corolla tube less than 1 mm long.......... 8
8. Inflorescences lax, 2–3 times dichotomously branched; lower leaf epidermis minutely but densely tuberculate papillate (viewed under at least × 25) . **1.** *retusa*
– Inflorescences congested, with few branches; lower leaf epidermis without tuberculate papillae (viewed under at least × 25) .... **2.** *punctulata*
9. Corolla tube at least 1 mm long; corolline corona of 5 membranous pockets obscuring the corolla lobe sinus............................... **6.** *delagoensis*
– Corolla tube c.0.5 mm long; corolline corona pockets absent 10
10. Staminal corona lobes falcate, half to as long as the staminal column; leaves mostly ovate to suborbicular, rarely lanceolate .... **3.** *parvifolia*
– Staminal corona lobes triangular or quadrate, about half length of staminal column or less; leaves mostly linear to lanceolate elliptic........... **5.** *filiformis*

1. **Secamone retusa** N.E. Br. in Bull. Misc. Inform., Kew **1895**: 248 (1895); in F.T.A. 4(1): 279 (1902). —Goyder & Harris in F.T.E.A., Apocynaceae (part 2): 169 (2012). Type: Tanzania, Zanzibar, *Kirk* s.n. (K holotype).

Woody climber, stems twining, glabrous except for a minute pubescence at the joints of the inflorescence and a few scattered hairs on undersides of leaves. Leaves with petiole (1)2–4.5(7) mm long; lamina 2.7–6.5(7.7) × 0.8–2.8 cm, oblong or obovate, apex obtuse or rounded, occasionally retuse, apiculate, base rounded to cuneate, the margins commonly inrolled, veins prominent underneath, coriaceous, lower surface paler than the upper, lower epidermis minutely but densely tuberculate-papillate. Inflorescence terminal and axillary, lax, c.15–30(40) mm long, 2–3 times dichotomously branched, the branches slender, spreading, often at c.90°; pedicels slender, 3–6(8) mm long. Calyx lobes 0.5–1 × 0.4–0.7 mm, broadly ovate or oblong, obtuse, margins ciliate. Corolla 2–3 mm long, yellow or greenish white, drying orange-brown with paler margins, united for about ¼ of its length; tube c.0.5 mm long, campanulate; lobes 1.5–2.5 × c.1 mm oblong, obtuse or rounded at the apex. Corolline corona of 5 pairs of fleshy ridges extending from the base of the corolla tube to either side of the corolla lobe sinus. Staminal corona lobes minute, to about half the height of the staminal column, deltoid-subulate, spreading or occasionally erect and slightly incurved. Stylar head appendage exserted c.0.3–0.4 mm from the staminal column, broadly obconic and shortly bilobed, the lobes truncate or occasionally rounded. Follicles c.75 × 7 mm, fawn, striate.

Mozambique. N: Cabo Delgado Province, Palma Dist., track between Lake Nompuid/ Mikulumu and the Rovuma River, fl. 15.xi.2009, *Clarke* 121 (K, LMA, LMU, P).
Also in Kenya and Tanzania. Lowland and coastal forest on sandy soils; 0–300 m.

2. **Secamone punctulata** Decne. in Candolle, Prodr. **8**: 502 (1844). —Brown in F.T.A. 4(1): 284 (1902). —Goyder & Harris in F.T.E.A., Apocynaceae (part 2): 170 (2012). Type: Tanzania, Pemba, *Bojer* s.n. (P holotype, K).
Secamone micrandra K. Schum. in Bot. Jahrb. Syst. **17**: 142 (1893). —Brown in F.T.A. 4(1): 281 (1902). —White, Forest Fl. N. Rhodesia: 360 (1962). —Fanshawe, Checkl. Woody Pl. Zambia: 39 (1973). Type: Angola, Cuanza Norte, Golungo Alto, Sobato Quilombo, *Welwitsch* 5942 (K lectotype, COI, LISU), lectotype designated here.

Secamone stenophylla K. Schum. in Engler, Pflanzenw. Ost-Afrikas **C**: 325 (1895). Type: Tanzania, Mascheua, *Holst* 3510 (B† holotype, K lectotype, M, W), lectotype designated here.

Secamone sansibariensis K. Schum. in Engler, Pflanzenw. Ost-Afrikas **C**: 325 (1895). — Brown in F.T.A. **4**(1): 282 (1902). Type: Tanzania, Zanzibar, 10.x.1889, *Stuhlmann* 490 (HBG502785 lectotype, B†, HBG), lectotype designated here. Paralectotypes: Tanzania, Zanzibar, *Stuhlmann I* 533 (B†, HBG, K), 572 (B†, HBG) & 772 (B†, HBG).

Secamone punctulata var. *stenophylla* (K. Schum.) N.E. Br. in F.T.A. **4**(1): 284 (1902). — White, Forest Fl. N. Rhodesia: 360 (1962).

Thin wiry climber or small shrub, glabrous or sparsely rusty-puberulent on young shoots, the inflorescence sparsely to densely covered with spreading red hairs. Leaves shortly petiolate; lamina extremely variable in size and shape, but in the Flora area 1.5–4 × 0.5–1.2 cm, elliptic to lanceolate, apex acute to rounded, apiculate, base rounded or obtuse, glabrous or glabrescent, lower epidermis smooth or finely granulate, veins commonly indistinct. Inflorescences numerous, 5–20 × 2–25 mm, terminal and axillary, the cymes few-flowered and subumbelliform on short peduncles or larger and more diffuse, sometimes on leafless shoots, flowers ovoid in bud, sweetly scented. Peduncles (1)4–15 mm long; pedicels 0.5–2(5) mm long, glabrous or sparsely pubescent. Calyx lobes 0.5–1 × 0.5–1 mm, ovate or suborbicular, the apex obtuse or rounded, margins ciliate. Corolla 1–2(3) mm long, orange or yellow, lobed to c.2/3; tube 0.3–0.6(0.8) mm long, broadly campanulate; lobes 1.2–2 × 0.6–0.8 mm, oblong, obtuse. Corolline corona of 5 pairs of fleshy ridges extending from the base of the corolla tube to either side of the corolla lobe sinus. Staminal corona lobes erect, green or white, reaching to just below the top of the staminal column or variously reduced to an upward or outward pointing peg about 1/2 the height of the column. Stylar head appendage exserted c.0.2–0.4 mm from the staminal column, obconic or clavate at the apex. Follicles widely divergent, (3.5)4.5–5.5 × 0.3–0.5 cm, slender, attenuate, striate, olive-green.

Zambia. W: Kitwe, fl. 27.ii.1955, *Fanshawe* 2106 (BR, K, NDO, SRGH). S: Kafwala, Kafue National Park, 17.x.2004, *Bingham* 12799 (K).

Also known from Ivory Coast, Ghana, D.R. Congo, Angola and eastern Africa from Ethiopia and Somalia to Tanzania. Riverine forest; 1000–1300 m.

3. **Secamone parvifolia** (Oliv.) Bullock in Kew Bull. **9**: 368 (1954). —White, Forest Fl. N. Rhodesia: 360 (1962). —Binns, First Checkl. Herb. Fl. Malawi: 22 (1968). —Ross, Fl. Natal: 290 (1972). —Compton, Fl. Swaziland: 460 (1976). —Goyder & Harris in F.T.E.A., Apocynaceae (part 2): 171 (2012). Type: Tanzania, Kilimanjaro, *Johnson* s.n. (K holotype). FIGURE 7.3.**132**.

Gymnema parvifolium Oliv. in Trans. Linn. Soc., Bot. **2**: 342 (1887).

Secamone schweinfurthii K. Schum. in Bot. Jahrb. Syst. **17**: 143 (1893). —Engler, Pflanzenw. Ost-Afrikas **C**: 325 (1895). —Brown in F.T.A. **4**(1): 284 (1902). Type: Sudan, Kulongo, Bongo, *Schweinfurth* 2232 (B† holotype, K).

Secamone zambesiaca Schltr. in J. Bot. **33**: 303 (1895). —Brown in F.T.A. **4**(1): 285 (1902). —Binns, First Checkl. Herb. Fl. Malawi: 22 (1968). Type: Malawi, Chiromo, Shire River, *Scott Elliot* 2803 (K holotype).

Secamone kirkii N.E. Br. in Bull. Misc. Inform. Kew **1895**: 248 (1895); in F.T.A. **4**(1): 285 (1902). Type: Tanzania, Zanzibar, *Kirk* s.n. (K holotype).

Secamone emetica (Retz.) R. Br. var. *glabra* K. Schum. in Engler, Pflanzenw. Ost-Afrikas **C**: 324 (1895). Type: Tanzania, Maschaua, *Holst* 3555 (B† holotype, K).

Secamone usambarica N.E. Br. in F.T.A. **4**(1): 281 (1902). Type: Tanzania, Maschaua, *Holst* 3555 (K holotype, B†).

Secamone mombasica N.E. Br. in F.T.A. **4**(1): 284 (1902). Type: Kenya, near Mombasa, *Hildebrandt* 1979 (K holotype, B†, W).

Secamone zambesiaca var. *parvifolia* N.E. Br. in Fl. Cap. **4**(1): 544 (1907). Type: Mozambique, Maputo [Lourenço Marques, Delagoa Bay], 9.xii.1897, *Schlechter* 11669 (K lectotype, B†, BOL), lectotype designated by Goyder in Kew Bull. **47**: 451 (1992) as holotype.

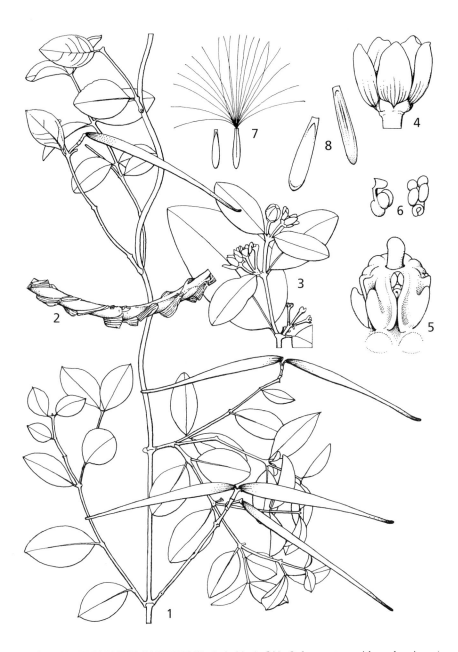

Fig. 7.3.**132**. SECAMONE PARVIFOLIA. 1, habit (× ²/₃); 2, lower stem with corky wings (× ½); 3, flowering shoot (× 2); 4 flower (× 8); 5, gynostegium with staminal corona lobes and emergent stylar head appendages (× 22); 6, pollinaria (× 34); 7, seeds, with coma (× 1); 8, seeds without coma (× 2). 1, 7 from *Gilbert & Phillips* 8296; 2 from *Gilbert et al.* 7821; 3–6 from *Gilbert & Thulin* 171. Drawn by E. Papadopoulos. Reproduced with permission from Flora of Ethiopia and Eritrea (2003).

Woody climber, old stems with two spiralling, corky wings alternating at the nodes; young shoots puberulent, the hairs white or red, ± adpressed. Leaves with petiole 1–4(6) mm long; lamina (0.8)1.1–5.6(6.4) × 0.4–3.0(4.6) cm, lanceolate to broadly ovate or suborbicular, apex acute to acuminate or obtuse, sometimes apiculate, the base rounded or rarely subcordate, usually thin, lower epidermis smooth or minutely but densely tuberculate-papillate, veins prominent, the margins smooth or crispate. Flowers 1 to many in simple or branched cymose, terminal or axillary cymose inflorescences, scented, commonly on short lateral shoots, the inflorescences lax or condensed, 5–30 × 10–30 mm. Peduncles pubescent, 1–8(15) mm long; pedicels 1–6 mm long, glabrous or with a line of hairs. Calyx lobes 0.5–1 × 0.5–1 mm, oblong, broadly ovate or suborbicular, obtuse or rounded, ciliate. Corolla 2–2.5 mm long, white, cream or yellow, divided for $^4/_5$–$^3/_4$ of its length; tube c.0.5 mm long, campanulate; lobes 1.5–2 × c.1 mm, ovate or oblong, obtuse or rounded at the apex. Corolline corona of 5 pairs of fleshy ridges extending from the base of the corolla tube to either side of the corolla lobe sinus. Staminal corona lobes from ½ to as long as the staminal column, laterally compressed, falcate. Anthers sometimes with white fimbriate appendages surrounding the stylar head appendage. Stylar head appendage barely exserted to exserted for c.1 mm from the column, entire to deeply bifid, but not spreading or strongly dilated. Follicles widely divergent, 4.5–10 × 0.5–0.7 cm, the upper face flattened, tapering gradually to a long point. Seeds reddish brown, c.5–11 × 1–1.5 mm, channelled down one face.

Botswana. SE: Kweneng Dist., Gabane Hills, fl. 5.xi.1978, *Hansen* 3529 (BM, C, GAB, K, PRE, SRGH, WAG). **Zimbabwe**. E: Chimanimani (Melsetter) Dist., Umvumvumu R., fl. 20.xii.1947, *Chase* 464 (K, SRGH). S: Nberengwa (Belingwe) Dist., N bank of Ngesi R., 6 km N of Bukwa Mt, fl. 1.xi.1973, *Pope et al.* 1153 (K, SRGH). **Malawi**. S: Chikwawa Dist., Lengwe Game Reserve, 1 km from camp, fr. 6.iii.1970, *Brummitt* 8928 (K, SRGH). **Mozambique**. N: Faro de Cabo Delgado c.20 km NE of Palma, fl. 4.xii.2008, *Goyder et al.* 5079 (K, LMA, LMU, P). Z: 20 km NE of Mopeia Velha on road to Quelimane, fl. 7.xii.1971, *Müller & Pope* 1937 (K, SRGH). GI: Sabie River, Massengena, fr. vii.1932, *Smuts* P371 (K, PRE). M: Foot of Lebombo mountains c.20 km N of South African border, fl. 6.xii.2001, *Goyder* 5029 (K, LMU).

Widely distributed in eastern Africa from Sudan and Ethiopia to South Africa (Mpumalanga and KwaZulu-Natal). Thickets, mixed deciduous woodland, coastal and riverine forest; 0–1500 m.

A single collection from Northern Zambia, *Richards* 20511, resembles *Secamone parvifolia* in most characters, but the stylar head apex is strongly dilated as in *S. punctulata*.

4. **Secamone alpini** Schult. in Roemer & Schultes, Syst. Veg., ed. 15bis **6**: 125 (1820). —Brown in F.T.A. **4**(1): 279 (1902); in Fl .Cap. **4**(1): 544 (1907). —Binns, First Checkl. Herb. Fl. Malawi: 22 (1968). —Ross, Fl. Natal: 290 (1972). —Compton, Fl. Swaziland: 460 (1976). —Goyder & Harris in F.T.E.A., Apocynaceae (part 2): 173 (2012). Type: '*Periploca secamone*' (LINN 307.2), lectotypified by Goyder & Singh in Taxon **40**: 630 (1991).

    Periploca secamone L., Mant. Pl. Altera: 216 (1771), excl. Alp. Aegypt. t.134. —Thunberg, Prodr. Pl. Cap.: 47 (1794).

    Secamone thunbergii E. Mey., Comm. Pl. Afr. Austr.: 224 (1838). Type: South Africa, Cape, 'in sylvis Houtniquas', Hb. *Thunberg* 6222 (UPS lectotype), designated by Goyder & Singh in Taxon **40**: 630 (1991).

    Secamone thunbergii var. *retusa* E. Mey., Comm. Pl. Afr. Austr.: 224 (1838). Type: South Africa, near Glenfilling, *Drège* 3474 (B† holotype, K).

Robust liane or shrub, minutely rusty puberulent on young shoots and inflorescence. Leaves with petiole 1–8.5(11.5) mm long; lamina (1.6)2.5–7.4(9.4) × 0.6–2.3(3.4) cm, obovate, oblong, lanceolate- or ovate-oblong or occasionally elliptic, the apex acute, obtuse or rounded, occasionally retuse, often apiculate, base rounded or subcuneate, young leaves with scattered hairs on both surfaces, mature leaves coriaceous, glabrescent, lower epidermis smooth or minutely but densely tuberculate-papillate. Cymes arranged in axillary or terminal, pyramidal or corymbose panicles, rusty-puberulent except for the corolla, the inflorescence 10–55 × 10–45 mm, flowers scented with a faint musty odour. Peduncles 4–12 mm to first inflorescence branch, subsequent internodes of similar length; pedicels 1–4 mm long. Calyx lobes 0.5–1 × 0.5–1 mm, broadly ovate, obtuse or subacute. Corolla 1.5–3 mm long, greenish-yellow or white, rotate, lobed almost to the base, the tube 0.5–1 mm long, lobes 1–2 × 0.8–1 mm, ovate convex, glabrous on the outer face, pubescent with spreading white hairs inside except for the glabrous margin. Corolline corona absent. Staminal corona lobes equalling or usually exceeding the staminal column, subulate, erect or incurved over the tips of the anthers. Stylar head appendage shorter than the anthers or exserted for up to c.0.2 mm, the apical part truncate, angled but not 2-lobed. Follicles widely divergent, 4.5–8(10) × 0.3–0.8 cm, tapering to a long point, dark greenish brown, striate. Seeds (4)8–12 × 1–2 mm, linear-lanceolate, channelled down one face, somewhat flattened, one face concave, the other convex dark reddish brown, glabrous.

Zambia. E: Nyika Plateau, Chowo Forest, fl. xii.1980, *Dowsett-Lemaire* 99 (K). **Zimbabwe**. W: Matobo Dist., fl. xii.1959, *Miller* 6095 (K, SRGH). C: Wedza Mt, fr. 14.v.1964, *Wild* 6562 (BR, K, LISC, SRGH). E: Mutare (Umtali) Dist., mountain ravine S of Mutare, fl. 25.x.1953, *Chase* 5117 (BM, BR, LISC, K, P, SRGH). S: Nberengwa (Belingwe) Dist., Bukwa Mt, upper W slope, fl. 30.x.1973, *Pope et al.* 1129 (K, SRGH). **Malawi**. N: Chitipa Dist., Jembya Forest Reserve 18 km SSE of Chisenga, fl. 15.xii.1988, *Thompson & Rawlins* 5572 CM, K). C: Dedza Mt Forest, fl. 7.xi.1968, *Salubeni* 871 (K, LISC, MAL, SRGH). S: Mt Mulanje, Lichenya Plateau, Nessa Path, fl. 15.xii.1988, *Chapman & Chapman* 9436 (K, MO). **Mozambique**. Z: Mabu Mountain, fl. 21.x.2008, *Harris et al.* 637 (K, LMA, LMU, MAL). MS: Tsetsera, near road to Mavita, fl. 30.xi.1966, *Müller* 519 (K, LISC, SRGH).

Widely distributed in eastern and southern Africa from Kenya and Uganda to the Western and Eastern Cape provinces of South Africa. Montane forest; 1000–2000 m.

5. **Secamone filiformis** (L.f.) J.H. Ross in Bothalia **11**: 277 (1974). Type: South Africa, Cape Province, Hb. *Thunberg* 5606 (UPS holotype).

Celastrus filiformis L.f., Suppl. Pl.: 153 (1782).

Astephanus frutescens E. Mey., Comm. Pl. Afr. Austr.: 223 (1838). Types: South Africa, Cape Province, near Kat R., *Drège* s.n. (K syntype); near Zwartkop R., *Drège* 2230 (K syntype); Beans R., Zuurberg Range, *Drège* s.n. (K syntype).

Secamone frutescens (E. Mey.) Decne. in Candolle, Prodr. **8**: 501 (1844). —Brown in F.T.A. **4**(1): 283 (1902); in Fl.Cap. **4**(1): 546 (1907). —Ross, Fl. Natal: 290 (1972). —Compton, Fl. Swaziland: 460 (1976).

Woody climber, inflorescence and young shoots glabrous or pubescent. Leaves shortly petiolate, the petiole 0.5–2(3) mm long; lamina 0.9–3.1(3.7) × 0.2–0.7(0.9) cm, linear or oblong to lanceolate or elliptic, the apex acute, obtuse or rounded, the base rounded to subacute, glabrous or with scattered reddish hairs on the petiole and the under-surface of the leaf, the underside pale silvery grey, lower epidermis ± smooth or minutely but densely tuberculate-papillate. Cymes few-flowered, contracted onto short terminal or axillary peduncles, the peduncles 0.5–2(4) mm long. Pedicels 1.5–3 mm long. Calyx lobes 0.3–0.5 × 0.3–0.5 mm, ovate to suborbicular, obtuse, margins ciliate. Corolla 1.5–2 mm long; tube c.0.5 mm long, campanulate; lobes 1–1.6 × 0.5–1 mm, ovate-oblong, obtuse; corolline corona of 5 weak ridges linking the corolla lobe sinus to the base of the tube. Staminal corona lobes minute, laterally compressed and adnate to the staminal column, quadrate, about half the height of the column. Apical portion of stylar head exserted c.0.2 mm from the staminal column, cushion-shaped, often with a shallow groove across the top. Follicles slender, 4–6 × c.0.3 cm, brown. Seeds c.10 × 1.5 mm, brown.

Zimbabwe. E: Chipinga Dist., above Kondo, Sabi Valley, fl. & fr. 16.ii.1960, *Goodier* 917 (K, SRGH). **Mozambique**. GI: Inhambane, between Zavala and Inharrime, fr. 3.iv.1959, *Barbosa & Lemos* 8481 (K, LMA, SRGH). M: Maputo Elephant Reserve, c.5 km along track W of Ponta Milibangalala, fl. 30.xi.2001, *Goyder* 5016 (K, LMU).

Also known from Swaziland and eastern South Africa. Dry coastal or montane forest; 40–500 m.

6. **Secamone delagoensis** Schltr. in Bot. Jahrb. Syst. **38**: 35 (1905). —Brown in Fl. Cap. **4**(1): 546 (1907). —Ross, Fl. Natal: 290 (1972). Type: Mozambique, Maputo [Lourenço Marques], 150 ft., 6.xii.1897, *Schlechter* 11625 (K lectotype, B†, BOL, BR, E, G, GRA, HBG, L, MO, P, PH, PRE, S, STU, W, WAG), lectotype designated here.

Slender woody climber, glabrous in all parts. Leaves shortly petiolate, the petiole c.0.5 mm long, lamina 1–2.5(2.9) × 0.1–0.3 cm, linear to oblong, apex acute or obtuse, apiculate, the base rounded or subcuneate, lower epidermis smooth. Cymes terminal and axillary, 1–4-flowered. Peduncle 2–4(12) mm long. Pedicels 2–3(5) mm long. Calyx lobes c.1 × 1 mm, broadly ovate to suborbicular, rounded, the margins minutely ciliate. Corolla 2.5–3.5 mm long, white, ovoid-conical in bud; tube 1.2–1.8 mm long excluding the coronal pockets, tubular-campanulate; lobes 1.2–2 × 1–1.2 mm, oblong, rounded or obtuse, held ± erect; corolline corona of 5 pockets on the inner surface of the corolla obscuring the corolla lobe sinus for 0.2–0.4 mm above the mouth of the corolla tube, each pocket formed of 2 fleshy auricles. Staminal corona lobes laterally compressed forming flanges from the lower half of the staminal column. Apical portion of stylar head exserted for up to 1 mm from the staminal column, cylindrical or ovoid-clavate. Fruits not seen.

Mozambique. GI: Panda Dist., Chichococha, fl. 1.ii.2019, *Osborne et al.* 1600 (K, LMA, LMU). M: Licuati Forest Reserve 20 km W of Bela Vista, fl. 3.xii.2001, *Goyder* 5023 (K, LMU).

Also occurs in the Maputaland area of northern KwaZulu-Natal. Sand forest; 50–100 m.

7. **Secamone stuhlmannii** K. Schum. in Engler, Pflanzenw. Ost-Afrikas **C**: 325 (1895). —Brown in F.T.A. **4**(1): 283 (1902). —White, Forest Fl. N. Rhodesia: 360 (1962). —Binns, First Checkl. Herb. Fl. Malawi: 22 (1968). —Fanshawe, Checkl. Woody Pl. Zambia: 39 (1973). —Goyder & Harris in F.T.E.A., Apocynaceae (part 2): 174 (2012). Type: Tanzania, Karagwe, Kafuro, 19.iii.1891, *Stuhlmann* 1894 (B† holotype, K).

    Secamone floribunda N.E. Br. in F.T.A. **4**(1): 282 (1902). Type: Tanzania, Usmawo, Kageyi, *Fischer* 396 (K holotype, B†).

    Secamone phillyreoides S. Moore in J. Linn. Soc., Bot. **37**: 182 (1905). Type: Uganda, Mulema, South Ankole, *Bagshawe* 283 (BM holotype).

    Secamone rariflora S. Moore in J. Linn. Soc., Bot. **37**: 183 (1905). Type: Uganda, Buvuma Is., *Bagshawe* 646 (BM holotype).

Woody climber, the whole plant rusty pubescent with spreading hairs, particularly dense on the inflorescence, young stems, petioles and the underside of the leaves. Leaves petiolate, the petioles 0.5–4.5 mm long; lamina (1)1.3–4(6.5) × (0.2)0.5–2(2.4) cm, oblong or ovate to lanceolate-elliptic, apex acute or occasionally rounded, the base rounded, obtuse or acute, lower epidermis pale, with minute but dense tuberculate papillae in addition to the reddish indumentum. Cymes terminal and axillary, up to 6-flowered. Peduncles 1–12 mm long. Pedicels 1–3(8) mm long. Flowers sweetly scented, ovoid or subglobose in bud. Calyx lobes 0.5–1.5 × c.0.5 mm, narrowly to broadly ovate, acute or obtuse, pubescent, the margins ciliate, paler than the rest of the calyx. Corolla 1.5–3 mm long; tube 0.2–0.4 mm long, salveriform; lobes 1.5–2.5 × c.1 mm, ovate-oblong, obtuse, yellow when fresh, drying reddish brown with a paler margin, glabrous or with scattered white hairs; corolline corona ridges minute, forming a small pocket at the corolla lobe sinus. Staminal corona lobes attached near the base of the staminal column, 1/3 to as long as the column, laterally compressed and broadly triangular or falcate with an incurved tip. Apical portion of stylar head exserted for c.1 mm from the top of the staminal column, obconic or clavate, deeply 2-lobed or subentire. Follicles 6.5–9.5 × 0.7–0.9 cm, tapering gradually to a drawn-out point, silvery brown or olive green, striate, puberulent. Seeds 7–10 × 1–1.5 mm, reddish brown, channelled down one face.

Subsp. **whytei** (N.E. Br.) Goyder & T. Harris in Kew Bull. **62**: 282 (2007). Type: Malawi,
Mt Malosa, *Whyte* (K holotype, B†).

Secamone whytei N.E. Br. in Bull. Misc. Inform., Kew **1898**: 308 (1898); in F.T.A. **4**(1): 283
(1902). —Topham, Checkl. Forest Trees Shrubs Nyasaland Prot.: 31 (1958).

Malawi. S: Mt Malosa, fl. xi/xii.1896, *Whyte* s.n. (K).
Known only from the type which was collected between 1200 and 1800 m.

8. **Secamone erythradenia** K. Schum. in Bot. Jahrb. Syst. **17**: 141 (1893). —Brown in
F.T.A. **4**(1): 278 (1902); in Fl. Cap. **4**(1): 546 (1907). —Fanshawe, Checkl. Woody
Pl. Zambia: 39 (1973). Type: Angola, Huilla, wooded parts of Morro de Lopollo,
Welwitsch 5941 (K lectotype, C, COI, LISU), lectotype designated here.

Woody climber with twining stems. Young shoots, petioles, midribs and inflorescences
pubescent, the hairs reddish, spreading. Leaves shortly petiolate; petiole 0.7–2.2 mm long; lamina
0.9–3.1 × 0.5–1.4 cm, ovate or ovate-lanceolate, apex acute or occasionally obtuse or rounded,
generally apiculate, the base rounded or slightly cordate, lower epidermis smooth, pubescent.
Flowers solitary or 2–5 in lax, axillary cymes. Inflorescences 1–2 cm long. Pedicels 2–5 mm long.
Calyx lobes 1.5–2 × 0.7–1 mm, ovate, obtuse. Corolla 3.5–5 mm long, white; tube 2–2.5 mm long,
campanulate; lobes 2–3 × c.1 mm, oblong, obtuse; corolline corona of 5 pairs of fleshy ridges
stretching from the base of the corolla tube to the corolla lobe sinus, widest about halfway up the
tube. Staminal corona lobes longer than the staminal column, flattened dorsally or triangular in
section, free in the upper half and incurved over the stamens, the tip entire or shallowly 3-lobed.
Apical portion of stylar head exserted for up to c.1 mm from the staminal column, ovoid-clavate,
bifid, occasionally partially enclosed by the staminal corona lobes. Follicles 6.5–7.5 × 0.7–0.8 cm,
pubescent towards the base, shallowly striate, fawn to brown, reflexed.

Zambia. N: Kawambwa, fl. 16.xi.1957, *Fanshawe* 4076 (K, NDO). W: Kaoma – Lukulu
road, fl. & fr. 3.iii.1996, *Luwiika et al.* 257 (K, MO). C: Mt Makulu near Chilanga,
fl. 1.v.1958, *Angus* 1808 (BR, FHO, K, SRGH). S: Mazabuka Dist., Siamambo Forest
Reserve, 17.i.1960, *White* 6308 (K, FHO).
Also known from D.R. Congo and Angola. Mushitu forest and dry evergreen
thickets; 1000–1300 m.

9. **Secamone dewevrei** De Wild. in Ann. Mus. Congo Belge, Bot., sér. 5, **1**: 191 (1904).
Type: Zaire, Mbandaka (Coquilhatville), 13.i.1866, *Dewèvre* 602 (BR holotype).
Secamone gabonensis P.T. Li in J. S. China Agric. Univ. **15**: 63 (1994). Type: Gabon, *Le Testu*
6483 (MO holotype, BM, BR, MA, P, WAG).

Slender woody twiner; young shoots, leaves and inflorescences densely pubescent with stiff,
spreading, reddish hairs. Leaves petiolate; petiole 1–4 mm long; lamina 1.2–4.5 × (0.6)1–1.7
cm, ovate-lanceolate, elliptic or oblong-elliptic, apex acute, base rounded to acute, margins
commonly slightly revolute in dried material, adaxial surface dark green and somewhat leathery,
glabrescent, midrib and occasionally the principle lateral veins in channels in the leaf surface,
lower epidermis minutely but densely tuberculate-papillate, sparsely pubescent with stiff, erect
hairs, midrib and lateral veins prominent, densely pubescent. Inflorescence cymose, irregularly
branched, the branches spreading commonly at right-angles, sparsely to densely pubescent
throughout except for the glabrous corolla. Peduncles (1)2–10 mm long, slender. Pedicels 2–9
mm long, slender and often somewhat flexuous. Calyx lobes 1–2 × 0.5–1 mm, narrowly triangular
to ovate, apex acute or obtuse. Corolla 3.5–5 mm long; tube 0.5–1 mm long, campanulate; lobes
3–4 × 0.5–1 mm, oblong, apex rounded or emarginate; corolline corona of 5 minute pockets at
the corolla lobe sinuses. Staminal corona lobes laterally compressed, 2/3 to as long as the column,
triangular or subulate, spreading or suberect. Apical portion of stylar head exserted c.0.4–0.8
mm from the staminal column, obconic and weakly to strongly 2-lobed. Follicles 5–6 × 0.3–0.7
cm, pubescent at least towards the base and around the suture.

Subsp. **elliptica** Goyder in Kew Bull. **47**: 462 (1992). Type: Zambia, Mwinilunga Dist., 4.x. 1952, *White* 3452 (K holotype, BR, EA, FHO, NY).

Leaf lamina elliptic or oblong-elliptic, apex acute or rounded, apiculate, the base rounded to acute. Inflorescence densely pubescent. Corolla 3–4 mm long, greenish yellow; tube 0.5–1 mm long; lobes c.3 × 1 mm. Staminal corona lobes about ⅔ as long as column, triangular, spreading or suberect.

Zambia. W: Mwinilunga, fl. 22.ix.1955, *Holmes* 1219 (K, NDO).

Also in Angola. Dry evergreen forest or *Cryptosepalum* woodland on Kalahari sands; 1200 m.

10. **Secamone brevipes** (Benth.) Klack. in Adansonia, sér. 3 **23**: 326 (2001). Type: Cameroon, Nun River, ix.1860, *Mann* 484 (K holotype).

   Rhynchostigma brevipes Benth. in Hooker's Icon. Pl. **12**: 78, t.1189 (1876).
   Rhynchostigma lujaei De Wild. & T. Durand in Bull. Soc. Roy. Bot. Belgique, Compt. Rend. **38**: 208 (1900). Type: D.R. Congo, Stanley Port, Sabuka, 24.ix.1898, *Luja* 52 (BR holotype).
   Toxocarpus brevipes (Benth.) N.E. Br. in F.T.A. **4**(1): 287 (1902).
   Toxocarpus parviflorus (Benth.) N.E. Br. in F.T.A. **4**(1): 287 (1902).
   Toxocarpus lujaei (De Wild. & T. Durand) De Wild. in Ann. Mus. Congo Belge, Bot., sér. 5 **1**: 191 (1904).

Woody liana; young stems reddish, pubescent; the hairs reddish brown, spreading. Leaf pairs widely dispersed along twining branches; petiole 3–7 mm, robust, densely pubescent; lamina coriaceous, 33–50(95) × 13–25(32) mm, elliptic to oblong or obovate, apex obtuse or rounded, apiculate, or retuse, base rounded to cuneate, lower surface paler than the upper, lamina below sparsely pubescent, becoming glabrous with age, margin and midrib densely pubescent when young, midrib recessed into a channel on the upper surface, prominent below, lateral veins conspicuous below, slightly ascending, parallel, anastamosing into an arced, submarginal vein. Cymes axillary, densely rusty-pubescent, 5–7-flowered, shorter than the leaves; peduncles 1–8 mm; pedicels slender, 6–10 mm, densely covered with spreading, reddish hairs. Flowers sweetly scented; sepals 2.5–3 × 1.5 mm, ovate, acute, with adpressed reddish hairs at least on the midrib, margins ciliate. Corolla white, 5–6 mm, long, lobed nearly to the base and contorted in bud, the lobes spreading, c.4 × 1.5 mm, oblong, glabrous except for a tuft of white hairs at the base of each lobe. Corona lobes dorsally flattened and adpressed onto the staminal column, slightly exceeding the column or not, narrowly oblong with a truncate, shallowly toothed apex, the base slightly dilated. Staminal column c.1.5 mm high; pollinia c.0.25 mm. Stylar head appendage spirally grooved, exserted above the stamens for 1–2 mm, cylindrical or ovoid-clavate. Follicles not seen.

Zambia. W: Kalene Hill, Mwinilunga, fl. 18.v.1969, *Mutimushi* 3293 (K, NDO).

Also in Angola, D.R. Congo, Cameroon and Nigeria. Scrambling over trees and shrubs in mushitu forests; c.1200–1400 m.

Subfam. **ASCLEPIADOIDEAE**

This is the largest subfamily in the expanded Apocynaceae and is of Old World origin, but with one additional major radiation in New World tropics. The centre of diversity is tropical and subtropical Africa. The subfamily consists of four tribes: Fockeeae, Marsdenieae, Ceropegieae and Asclepiadeae, all of which are represented in the Flora area. The arrangement of the account presented here reflects the evolutionary relationships elucidated in major molecular surveys across the subfamily in recent years as summarised by Goyder in Ghazanfar & Beentje, *Taxonomy and ecology of African plants, their conservation and sustainable use*: 205–214 (2006). This in turn was based on the global overviews of Rapini *et al.* in Taxon **52**: 33–50 (2003) and Liede-Schumann *et al.* in Syst. Bot. **30**: 184–195 (2005), and the many supporting studies of individual tribes and subtribes cited in these works, or published since.

Approximately 172 genera and 3000 species in tropical and subtropical regions of the world. 39 genera and 281 species occur within the Flora region.

Spot Characters.

Stem succulence — Succulent stems (frequently but not always associated with reduced leaves) are characteristic of one subtribe, and can occur in species of a handful of other genera:

1) The stapeliads – Ceropegieae subtribe Stapeliinae (genera 49–57) (*Australluma, Duvalia, Hoodia, Huernia, Orbea, Piaranthus, Stapelia, Tavaresia, Tridentea*).

2) Some species of *Ceropegia* (genus 48).

3) *Schizostephanus alatus* (genus 74) and some species of *Cynanchum* (genus 76).

4) *Fockea multiflora* (genus 40) – massive swollen stem tapering into the basal tuber.

Erect herbaceous habit — Perennial herbs with annually produced erect shoots from a tuberous or woody underground rootstock:

1) Characteristic of many genera in the *Asclepias* radiation (genera 61–73) (*Asclepias, Aspidoglossum, Glossostelma, Gomphocarpus, Kanahia, Margaretta, Pachycarpus, Parapodium, Periglossum, Schizoglossum, Stathmostelma, Stenostelma, Xysmalobium*).

2) This growth form also occurs sporadically in *Ceropegia* (genus 48) and *Vincetoxicum caffrum* (genus 77) is a pyrophyte with erect or sprawling stems.

3) Check also *Raphionacme* (genus 37) in subfam. Periplocoideae.

Leaves with toothed or dissected margins — Families in the Gentianales are normally characterised by opposite leaves with entire margins. There are exception however:

1) Some species of *Ceropegia* (genus 48) have irregularly dentate leaves.

2) *Emicocarpus fissifolius* (genus 58) has highly dissected leaves.

Latex colour — Although it is often incorrectly assumed that all Apocynaceae have white or milky latex, certain taxonomic groups within the Asclepiadoideae (in particular) are exceptional in having clear, or in the more succulent genera, slightly cloudy, latex. Very often, if the collection data on the label omits to mention latex, it is because white exudate was not observed. This may be an indication that the latex was clear, and that it was not recognised as latex by the collector. If your plant has clear (or cloudy) latex, it might be:

1) Tribe Ceropegieae (genera 44–57) – *Ceropegia* and allies (*Ceropegia, Neoschumannia, Orthanthera, Riocreuxia, Sisyranthus*); and the succulent stapeliads.

2) *Schizostephanus* (genus 74) and *Vincetoxicum* (genus 77).

3) *Marsdenia* – some species apparently may have clear latex.

All other genera have white or milky latex, frequently in copious amounts.

Pollinarium structure, and orientation/attachment of pollinia to the corpusculum. FIGURE 7.3.**133**. — Although this requires careful observation under a stereo dissecting microscope, pollinarium structure provides the most reliable way of subdividing the subfamily into major tribal groups. Because these are not characters readily observable in the field, I have used them sparingly in constructing the artifical key to genera which follows, but in the herbarium, they give you a powerful diagnostic tool to ascertain the affinities of the plant under investigation. The Fockeeae (genera 40, 41: *Cibirhiza* and *Fockea*) have peculiar pollinaria that do not fit comfortably into the broad patterns outlined below – the pollen masses are attached directly to a little-differentiated translator. All other groups within the subfamily have a well-structured corpusculum with a longitudinal ventral groove, and arms (in various configurations) linking the corpusculum to the two pollinia. Look for the following:

1) Translator arm attached at the base of the pollinium so that the pollen masses are held erect in the anther cells – this takes you to tribes Marsdenieae (genera 42–43) and Ceropegieae (genera 44–57).

If there is a well defined translucent germination zone on the inner margin of the pollinia, this indicates Ceropegieae (genera 44–57).

If not, this indicates Marsdenieae (genera 42–43).

2) Translator arm attached to the top of the pollinium, so the pollen masses are pendant within the anther cells – this takes you to tribe Asclepiadeae (genera 58–78). But note that one genus in the Flora region (*Vincetoxicum*) has such small flowers with the vertical scale so reduced that the anther sacs may be displaced laterally, the pollinia are minute, and the point of attachment is variable in position.

A translucent germination zone on the pollinia in this group within the Flora area is indicative of *Aspidoglossum* (genus 65).

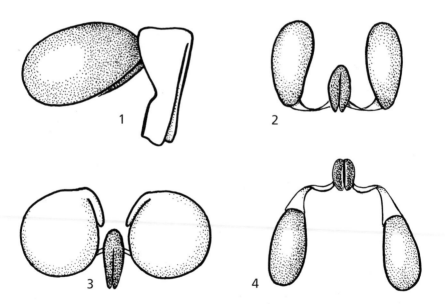

Fig. 7.3.**133**. Basic pollinarium types found in the ASCLEPIADOIDEAE. 1. —FOCKEEAE: two pollinaria attached directly to corpusculum. 2. —MARSDENIEAE: two erect (basifixed) pollinia attached indirectly to corpusculum. 3. —CEROPEGIEAE: two erect (basi- or medi-fixed) pollinaria with translucent germination zones on inner margins, attached indirectly to corpusculum. 4. —ASCLEPIADEAE: two pendant (apically attached) pollinia attached indirectly to corpusculum. Drawn by Margaret Tebbs. Reproduced from Flora of Tropical East Africa (2012).

1. Flowers emerging directly from the ground; stems and leaves absent at time of flowering. . . . . **76. Cynanchum praecox**
 – Not as above. 2
2. Leaves absent or reduced to scales; stems succulent. 3
 – Leaves well developed; stems succulent or not 14
3. Corolla tube well-developed, much longer than wide, usually with a distinct basal swelling; corolla lobes usually shorter than the tube, remaining attached at the tips to form a lantern or cage-like structure . . . . **48. Ceropegia**
 – Corolla not as above; if tube longer than lobes, then stems with prominent angles or rows of tubercles. 4
4. Stems smooth or striate, lacking prominent angles or rows of tubercles, frequently scrambling or twining; latex white . . . . **76. Cynanchum**
 – Stems with prominent angles or rows of tubercles, prostrate to erect, but never scrambling or twining; latex clear or at most cloudy. 5
5. Gynostegial corona whorls separated from each other vertically; corolla frequently with a clearly defined raised annulus (corolline corona). 6
 – Gynostegial corona whorls not separated from each other vertically, and partially or wholly fused to each other; corolla lacking a clearly defined raised annulus (corolline corona). 7
6. Leaf rudiments with small stipular denticles; gynostegial corona raised from base of corolla tube on a short stipe, the outer corona resting on corolline annulus. . . . . **51. Duvalia**
 – Leaf rudiments lacking stipular denticles; gynostegial corona sessile on corolla tube, if corolla with conspicuous annulus then this surrounds rather than raises the gynostegium and corona . . . . **54. Huernia**
7. Stem tubercles in 6 or more lines or angles. 8
 – Stem tubercles in 4 or 5 lines or angles 9
8. Inflorescences generally solitary, near base of stem; corolla tubular **56. Tavaresia**
 – Inflorescences several, towards apex of stem; corolla rotate (at least in the F.Z. region) . . . . **49. Hoodia**
9. Stems, pedicels, sepals and exterior of corolla at least minutely pubescent, but stem can be subglabrous; leaf rudiments erect, deciduous . . . . **50. Stapelia**
 – Stems, pedicels, sepals and exterior of corolla glabrous; leaf rudiments spreading or absent. 10
10. Inflorescences arising from upper half of stem, usually several per stem 11
 – Inflorescences arising on lower half of stem, generally solitary 13
11. Outer gynostegial corona reduced or absent; deep nectar chamber present beneath anther wings . . . . **53. Piaranthus**
 – Outer gynostegial corona well developed; nectar chamber shallow or absent . 12
12. Leaf rudiments constricted slightly at the base; southern Zimbabwe and Mozambique . . . . **57. Australluma**
 – Leaf rudiments not constricted at the base, tapering continuosly into stem tubercle . . . . **55. Orbea**
13. Leaf rudiments constricted slightly at the base; corolla usually covered with multicellular papillae; SW Botswana . . . . **52. Tridentea**
 – Leaf rudiments not constricted at the base, tapering continuosly into stem tubercle; corolla smooth to deeply rugulose but rarely with multicellular papillae . . . .**55. Orbea**
14. Corolla tube well developed, generally much longer than wide, usually with a distinct basal swelling; corolla lobes mostly shorter than the tube and remaining attached at the tips to form a lantern or cage-like structure (very rarely free at the tips and rigidly spreading); latex clear. 15
 – Corolla not as above; latex white or clear 16

15. Inflorescences generally paniculate, lax; corolla mostly (in Flora region) orange or yellowish; corolla and corona glabrous throughout; staminal corona lobes often rudimentary and exceeded by the anthers; rootstock woody, with fusiform roots . . . . **46. Riocreuxia**
– Inflorescences generally umbelliform, more compact; corolla of various colours, but rarely yellow or orange; corolla and corona glabrous or pubescent; staminal corona lobes generally exceeding the anthers; rootstock generally not woody, roots fibrous, fusiform, tuberous or rhizomatous . . . .**48. Ceropegia**
16. Prostrate herb with at least the lower leaves deeply dissected; follicles single-seeded, indehiscent; southern Mozambique . . . .**58. Emicocarpus**
– Not as above; all leaves with entire margins. 17
17. Shrubs to 5 m tall, larger plants arborescent with corky stems; leaves glaucous, semi-succulent, mostly obovate, to 26 x 15 cm; corona lobes laterally compressed, with a short, upturned basal spur . . . .**60. Calotropis**
– Erect perennials, twiners or scramblers, if shrubs then no more than 2 m tall; leaves much smaller than above; corona lobes lacking an upturned basal spur 18
18. Herbs, frequently with erect annual stems arising from a perennial rootstock; or tufted shrubby herbs woody at the base only, with ascending or erect branches, and linear or narrowly lanceolate leaves; stems never twining. 19
– Large woody climbers, lianescent shrubs or slender vines; stems mostly scrambling or twining distally, if not then leaves broad 37
19. Latex clear (often not mentioned in label data); pollinia erect, attached basally to translator arms, often ± exposed on top of stylar head 20
– Latex white; pollinia pendant, attachment to translator arms apical, generally concealed behind anthers. 22
20. Gynostegial corona in two series. . . . .**48. Ceropegia**
– Corona apparently of one series (outer series so reduced as to be insignificant) 21
21. Gynostegial corona lobes dorsiventrally flattened; corolla small, to 5 mm long, generally white to yellowish but sometimes marked with purple . . **47. Sisyranthus**
– Gynostegial corona lobes subglobose; corolla generally larger, 5–14 mm long, deep reddish purple within. . . . . **77. Vincetoxicum caffrum**
22. Staminal corona lobes absent; corona adnate to corolla tube basally with free upper margins, the lobes contiguous and forming a pentagonal disc around the gynostegium; SE Botswana . . . . **71. Parapodium**
– Staminal corona lobes present; disk-like corona absent 23
23. Corona lobes petaloid, larger and more showy than the corolla. . . **67. Margaretta**
– Corona lobes not petaloid. 24
24. Inflorescence axis indeterminate; rheophyte in seasonal watercourses. . . . . **61. Kanahia**
– Inflorescence umbelliform, sometimes subglobose, the axis determinate; found in various habitats including seasonally waterlogged grasslands or banks of rivers, but not the river bed 25
25. Short-lived perennial subshrubs or pyrophytic herbs; rootstock a conventional taproot, sometimes becoming woody (especially in pyrophytes), but not swollen to form a napiform tuber 26
– Pyrophytic herbs; rootstock a vertical napiform tuber and/or with fleshy fusiform lateral roots 27
26. Corona laterally compressed, with a central cavity; leaves glabrous or with soft indumentum . . . . **69. Gomphocarpus**
– Corona solid, fleshy, without a central cavity; leaves with stiff, ± hispid indumentum . . . . **72. Xysmalobium undulatum**

27. Inflorescences sessile and fasciculate; pollinia usually with transparent germination zone at point of attachment to translator arms . . . .**65. Aspidoglossum**
   – Inflorescences pedunculate (rarely sessile); pollinia lacking transparent germination zone (except in *Schizoglossum barbatum*) 28
28. Leaves with an indumentum of stiff hairs, feeling scabrid to the touch; lamina generally broad with prominent secondary veins and a truncate to cordate base; follicles generally ornamented with longitudinal wings or soft spine-like processes (but smooth and inflated in *Pachycarpus lineolatus* and *P. bisacculatus*) 29
   – Leaves glabrous or pubescent, if stiff scabrid hairs present, then confined to the margins; lamina narrow to broad, secondary veins often obscure, base frequently cuneate, occasionally truncate or cordate; follicles generally smooth, rarely with weak ornamentation of short filiform processes 30
29. Corona lobes solid, fleshy; corolla usually bearded within towards the apex; follicles covered with soft filiform processes . . . . . . **72. Xysmalobium undulatum**
   – Corona lobes generally dorsiventrally flattened, commonly with lobes or fleshy wings arising near the base of the ventral face; corolla not conspicuously bearded; follicles winged or smooth, lacking filiform processes . . . . **70. Pachycarpus**
30. Corona lobes solid, fleshy, occasionally with teeth or small processes apically . .31
   – Corona lobes dorsiventrally flattened or cucullate, with or without teeth or other ornamentation within the cavity of the lobe, sometimes somewhat reduced . . 32
31. Rootstock with a short vertical axis from which fusiform lateral tuberous roots arise; leaves semisucculent, venation obscure . . . . **73. Glossostelma**
   – Rootstock a vertical napiform tuber; leaves membranous, venation mostly easily observed . . . . **72. Xysmalobium**
32. Corona lobes dorsiventrally flattened 33
   – Corona lobes cucullate 36
33. Apex of corona extended into a subulate or attenuate tooth or horn 34
   – Apex of corona rounded, subacute or variously toothed, but not extended into a single point. 35
34. Corolla lobes reflexed from base; plant from Mt Mulanje . . . . **64. Schizoglossum**
   – Corolla lobes pouched in lower half, convex and reflexed above . . . .
   . . . . **63. Stenostelma**
35. Gynostegium barrel-shaped, tapering both basally and apically; pollinaria with slender sigmoid curved translator arms at least twice as long as the pollinia . . . .
   . . . . **62. Periglossum**
   – Not as above. . . . . **73. Glossostelma carsonii**
36. Pollinaria with differentially winged and contorted translator arms: proximal portion broad and membranous, distal portion filiform and pendulous; corolla and corona generally brightly coloured (red, orange, yellow), occasionally white with purple markings; flowers held erect . . . . **68. Stathmostelma**
   – Pollinaria not as above; corolla and corona dull, if brightly coloured then corona with prominent tooth arising from the cavity (*Asclepias curassavica*), or flowers nodding or held laterally. . . . .**66. Asclepias**
37. Gynostegial corona in three series – downward pointing staminal lobes partially fused to form a skirt which obscures the gynostegial stipe; a whorl of spreading or erect interstaminal lobes; and a further whorl of 5 erect dorsi-ventrally flattened staminal lobes with swollen bases; latex clear; slender twiner with a relict wet forest distribution . . . . **45. Neoschumannia**
   – Gynostegial corona not in three series, or if so, not in the configuration described above, and plant with white latex. 38

38. Twining herbs or slender to woody lianas arising from large subterranean or partially exposed tubers; gynostegial coronas in several series, one of which is tubular; pollinia attached directly to the corpusculum. 39
– Plants not arising from large subterranean or partially exposed tubers; gynostegial coronas, if present, in one or at most two series; pollinia attached indirectly to corpusculum 40
39. Corolla lobes linear, spreading; fused portion of corolla not extending beyond corona; tubular part of the corona longer than the gynostegium . . . . **40. Fockea**
– Corolla lobes broadly ovate or triangular; fused portion of corolla extending as a flat limb beyond the corona; tubular part of corona shorter than the gynostegium. . . . .**41. Cibirhiza**
40. Corolline corona present in corolla lobe sinuses; gynostegial corona minute or absent . . . .**42. Marsdenia sylvestris**
– Corolline corona absent, or forming an annulus around the base of the gynostegium; gynostegial corona mostly well developed (but reduced to scales in *Vincetoxicum cernuum*) 41
41. Pollinia erect in the anther cells, translator arms attached to the base of the pollinia; corolla mostly with well-developed tube, at least at the base (but rotate to broadly campanulate in *Marsdenia exellii*) 42
– Pollinia pendent in anther cells, translator arms attached to the apex of the pollinia (but note this character unreliable in some *Vincetoxicum*); corolla rotate or broadly campanulate, united portion short or not clearly tubular (except in *Pergularia*). 44
42. Corolla tube at least 7 mm long; corona lobes arrow-head shaped, adnate to the staminal column except for the free wing-like margins; pollinia with translucent germination zone at apex. . . . .**44. Orthanthera**
– Corolla tube rarely longer than 5 mm and then with the mouth of the tube obscured by inward-pointing hairs; corona lobes ovoid or flattened, attached to the column at the base only; pollinia without translucent germination zone. . 43
43. Staminal corona lobes with a ligule on the inner face . . . . **43. Telosma**
– Staminal corona lobes lacking a ligule on the inner face. . . . .**42. Marsdenia**
44. Staminal corona lobes semisagittate, with a free subulate apex arched over the head of the column and a basal projection; corolla lobes with bearded margins; rampant scramblers with copious white latex . . . .**59. Pergularia**
– Staminal corona lobes not as above; corolla lobes without bearded margins. . 45
45. Stems thick and fleshy; latex clear or yellowish, not milky; corona united into a fluted tube surrounding and obscuring the gynostegium, tube topped with 10 free erect lobes. . . . .**74. Schizostephanus**
– Plant not as above, stems not succulent; if corona tubular, surrounding the gynostegium, then latex white 46
46. Stylar head appendage long-rostrate, well-exserted from staminal column; gynostegial corona vestigial; latex clear . . . .**77. Vincetoxicum cernuum**
– Stylar head appendage not strongly exserted from staminal column, or if so, then corona lobes well-developed. 47
47. Latex white, generally copious 48
– Latex clear, often sparse . . . .**77. Vincetoxicum**
48. Staminal corona lobes fused at the base only, or if more highly fused, then with clearly visible "sutures" along the lines of fusion. . . . .**75. Pentarrhinum**
– Staminal corona lobes fused partially or for much of their length, without visible "sutures" along the lines of fusion . . . . **76. Cynanchum**

40. FOCKEA Endl.[12]

Fockea Endl. in Endlicher & Fenzl, Nov. Stirp. Dec.: 17 (1839). —Court in Asklepios **40**: 67–74 (1987). —Bruyns & Klak in Ann. Missouri Bot. Gard. **93**: 535–564 (2006).

Chymocormus Harv. in London J. Bot. **1**: 23 (1842).

Massive lianas, or erect to twining herbs arising from subterranean or partially exposed tuber; latex white. Leaves opposite. Inflorescences extra-axillary, umbelliform. Corolla rotate, fused portion not extending beyond corona; lobes linear, spreading. Corolline corona absent. Gynostegial corona forming a tube obscuring the gynostegium with several series of erect or spreading lobes at the mouth of the tube, and an inner series of 5 tall erect lobes arising from the inner face of the corona tube. Gynostegium with connivent, erect, inflated and somewhat translucent anther appendages filling the mouth of the corona tube and completely obscuring the much shorter stylar head. Pollinarium with two erect, flattened, clear or translucent pollinia attached directly to the pale brown corpusculum; corpusculum long and narrow, narrower than the pollinia; pollinia consisting of tetrads which are not enclosed by a wall. Follicles pendulous, single by abortion, fusiform and narrowing into a slender beak; seeds flattened, ovate with a narrow marginal rim, with coma at one end only (extending around margin in one species outside the Flora region).

Six species restricted to southern or south tropical Africa, but with two more widely distributed species with disjunct distributions extending into Kenya and Tanzania.

Stems slender but sometimes woody at base, arising from a basal tuber, at most to 1–2 m long; leaves less than 2.5 cm wide, glabrous or minutely pubescent beneath; inflorescences sessile, few-flowered; corolla twisted in bud, narrowly conical. . . .
. . . . **1.** *angustifolia*
Stems massive and swollen towards base, but not clearly differentiated into stem and basal tuber, often scrambling to 10 m or more; leaves generally more than 2.5 cm wide, tomentose beneath especially when young; inflorescences pedunculate with few to many flowers; corolla not twisted in bud, ovoid . . . . **2.** *multiflora*

1. **Fockea angustifolia** K. Schum. in Bot. Jahrb. Syst. **17**: 146 (1893). —Goyder in F.T.E.A., Apocynaceae (part 2): 186 (2012). Type: South Africa, Cape, Griqualand West, ii.1886, *Marloth* 1008 (B† holotype, M).

   Fockea sessiliflora Schltr. in Bot. Jahrb. Syst. **20**(Beibl. 51): 44 (1895). —Brown in F.T.A. **4**(1): 429 (1903). Type: South Africa, Limpopo Province, near Klippdam, 14.ii.1894, *Schlechter* 4493 (B† holotype, drawing at W lectotype), designated by Bruyns & Klak in Ann. Missouri Bot. Gard. **93**: 548 (2006).

   Fockea lugardii N.E. Br. in F.T.A. **4**(1): 429 (1903). Type: Botswana, Ngamiland, Kwebe Hills, *Lugard* 299 (K holotype).

   Fockea dammarana Schltr. in Bot. Jahrb. Syst. **38**: 56 (1905). Type: Namibia, Damaraland, 1879, *Een* s.n. (BM holotype).

   Fockea tugelensis N.E.Br. in Fl. Cap. **4**(1): 778 (1908). Type: South Africa, KwaZulu-Natal, Tugela, *Gerrard & McKen* 1310 (K holotype, BOL, TCD).

   Fockea mildbraedii Schltr. in Mildbraed, Wiss. Erg. Deut. Zentr.-Afr. Exped., Bot. **2**: 545 (1913). Type: Tanzania, Lembeni, 4.ix.1910, *Winkler* 3803 (K lectotype, B†, WRSL), designated by Goyder in F.T.E.A., Apocynaceae (part 2): 187 (2012).

   Fockea monroi S. Moore in J. Bot. **52**: 149 (1914). Type: Zimbabwe, Chimanimani Dist., Victoria, *Monro* 828 (BM lectotype, BOL, SRGH), designated by Goyder in F.T.E.A., Apocynaceae (part 2): 187 (2012).

   Cynanchum omissum Bullock in Kew Bull. **10**: 623 (1956). Type: Kenya, Kwale Dist., between Samburu and Mackinnon road, viii.1953, *Drummond & Hemsley* 4045 (K holotype, EA).

---

[12] by D.J. Goyder

Slender woody twiner to 1.5(2) m from a large napiform tuber, latex white; stems often reddish, minutely pubescent. Leaves opposite, petiole 1–3 mm long, minutely pubescent; lamina 2–8 × 0.1–2.5 cm, narrowly linear to oblong, elliptic or ovate, apex rounded to subacute or acuminate, base cuneate, glabrous or minutely pubescent beneath. Inflorescences extra-axillary, sessile but sometimes on short lateral shoots, 1–6 flowers open at one time; pedicels 0–1(2) mm long, minutely but densely pubescent. Sepals c.1 mm long, triangular, minutely pubescent. Corolla narrowly conical and strongly contorted in bud, divided ± to the base, lobes 3–6 × 0.5–1 mm, linear, green and densely white-pubescent abaxially, adaxially green to orange or brown, sparsely pubescent or papillose. Corona white, tubular portion 2–3(6) mm long and 1–1.5 mm in diam., urceolate or cylindrical, apical lobes 0.5–1 mm long; lobes from inner whorl of corona filiform, projecting from mouth of tube for up to 1.5 mm. Follicles 7–20 × 0.8–1.2 cm, smooth, grey-green often with purplish bands; seeds 8–10 × 4–6 mm, yellow-brown.

Botswana. N: 81 km W of Kuke gate, fl. 14.iii.1980, *P.A. Smith* 3167 (K, PRE, PSUB, SRGH). SW: Ghanzi, farm 48, fl. 9.iv.1969, *de Hoogh* 233 (K, SRGH). SE: Content Farm Gabarone, *Kelaole* A69 (SRGH). **Zambia**. S: Gwembe Dist., near Ntoboute village, *Scudder* s.n. (SRGH). **Zimbabwe**. S: Bikita Dist., W bank, Devuli River Bridge, fl. 15.iv.1963, *Chase* 7979 (K, SRGH).

Also known from SW Angola, Namibia, South Africa, Swaziland, Kenya and Tanzania. Mopane woodland or scrub, sandy or stony soil, sometimes over calcrete; 450–1150 m.

2. **Fockea multiflora** K. Schum. in Bot. Jahrb. Syst. **17**: 145 (1893). —Brown in F.T.A. 4(1): 428 (1903). —Goyder in F.T.E.A., Apocynaceae (part 2): 187 (2012). Type: Tanzania, French mission, Ussambiro, 22.x.1890, *Stuhlmann* 848 (B† holotype, K fragment). FIGURE 7.3.**134**.

Fockea schinzii N.E. Br. in Bull. Misc. Inform., Kew **1895**: 259 (1895); in F.T.A. 4(1): 428 (1903). Type: Angola, *Welwitsch* 4194 (K lectotype, LISU, P), designated by Bruyns & Klak in Ann. Missouri Bot. Gard. **93**: 557 (2006).

Robust twiner to 15 m, latex white; stems massively swollen and up to c.30 cm across towards the base (distinction between stem and basal tuber rapidly obscured with age), semi-succulent with shiny grey or reddish bark, young shoots densely pubescent to tomentose. Leaves opposite, petiole 0.5–4 cm, densely pubescent; lamina to c.15 × 8 cm, obovate to elliptic or suborbicular, apex rounded to subacute, base cuneate or rounded, tomentose beneath, at least in younger leaves, adaxial face pubescent. Inflorescences extra-axillary, forming simple or branched umbelliform clusters of up to c.30 flowers; peduncles 5–15(30) mm long, tomentose; pedicels 6–13 mm, tomentose or densely pubescent. Sepals c.1 mm long, triangular, tomentose or densely pubescent. Corolla not twisted in bud, ovoid; tube c.1 mm long; lobes 6–10 × c.2 mm, oblong, bright green or brown, glabrous to sparsely pubescent. Corona white, tubular portion 2–3 mm long and 1.5–2.5 mm in diam., urceolate or cylindrical, apical lobes 0.5–2 mm long; lobes from inner whorl of corona filiform, projecting from mouth of tube for c.1 mm. Follicles 10–22 × 1.5–3 cm, broadly fusiform and tapering to a long beak, smooth, purplish; seeds 8–10 × 7–8 mm.

Caprivi. Cemetery at Andara Mission Station, 26.i.1956, *de Winter & Wiss* 4430 (K, M, PRE, WIND). **Botswana**. N: S bank of river at Ngoma pont road, fl. 8.x.1954, *Story* 4809 (K, PRE). **Zambia**. N: Kaputa Dist., Lake Chishi, fl. 13.ix.1958, *Fanshawe* 4820 (K, NDO). S: Mazabuka Dist., Gwembe Valley, fl. & fr. 28.ix.1955, *Bainbridge* 139/55 (K, FHO). **Zimbabwe**. N: Kariba, fl. x.1959, *Goldsmith* 44/59 (K, SRGH). W: Near Hwange (Wankie), *Levy* 1123 (K, SRGH). **Mozambique**. T: Mazoe River, 16 km from Zimbabwe border, fl. 22.ix.1948, *Wild* 2584 (K, SRGH).

This species has been recorded from Liwonde National Park in southern Malawi (www.inaturalist.org/observations/8916335). Also found in southern Angola, northern Namibia and NE Tanzania. Mopane and Itigi thickets; 400–1200 m.

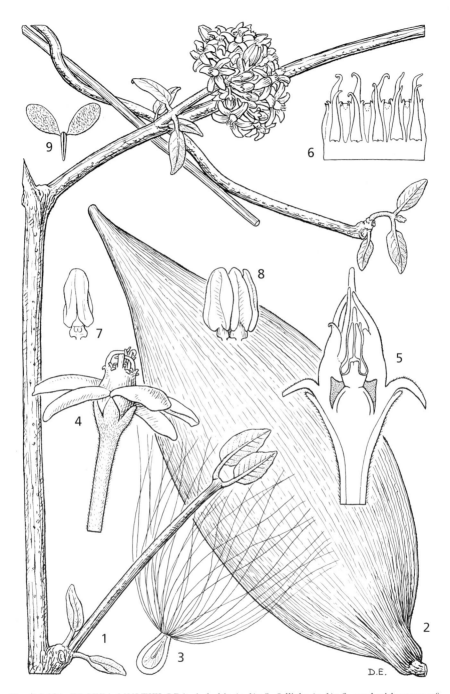

Fig. 7.3.**134**. FOCKEA MULTIFLORA. 1, habit (× 1); 2, follicle (× 1); 3, seed with coma of hairs (× 1); 4, flower (× 4); 5, half-flower (× 8); 6, tubular gynostegial corona, opened out and viewed from within (× 4 2/3); 7, 8, anthers, showing the large inflated anther appendages (× 8); 9, pollinarium (× 1). 1 from *Greenway* 9056; 2, 3 from *Bainbridge* 139/55; 4–9 from *Bullock* 1186. Drawn by D. Erasmus. Reproduced from Flora of Tropical East Africa (2012).

41. **CIBIRHIZA** Bruyns[13]

Cibirhiza Bruyns in Notes Roy. Bot. Gard. Edinburgh **45**: 51 (1988).

Slender woody twiners arising from a large tuberous rootstock; latex white. Leaves opposite. Inflorescences extra-axillary, umbelliform. Corolla rotate, fused portion extending as a flat limb beyond the corona; lobes broadly ovate or triangular. Corolline corona absent. Gynostegial corona in several series; tubular element shorter than the gynostegium; inner elements strongly lobed or divided. Anther appendages ± obsolete. Pollinarium with two pollinia attached directly to the pale brown corpusculum; corpusculum broad, both in relation to its length and in relation to the pollinia; pollinia consisting of single pollen grains enclosed by a wall. Follicles single by abortion.

3 species, one from Dhofar in western Oman, one from the Ogaden in eastern Ethiopia, and the third from Tanzania and Zambia.

Cibirhiza albersiana H. Kunze, Meve & Liede in Taxon **43**: 368 (1994). —Goyder in F.T.E.A., Apocynaceae (part 2): 189 (2012). Type: Zambia, Mazabuka Dist., 30 km N of Pemba near Milimo village, 12.ii.1960, *White* 6969 (K holotype, FHO). FIGURE 7.3.**135**.

Slender woody twiner to c.5 m from a large napiform tuber, latex white; stems minutely pubescent. Leaves opposite, petiole 0.5–5 cm long, minutely pubescent; lamina 2.5–13 × 1–9 cm, broadly ovate to oblong or elliptic, sometimes variable even on the same shoot, apex subacute to acuminate, base rounded to cuneate, minutely pubescent at least on the veins. Inflorescences extra-axillary, subsessile, forming a dense sub-umbelliform cluster of flowers, minutely pubescent; pedicels 3–5 mm long. Sepals c.1 mm long, ovate, minutely pubescent. Corolla rotate, yellow or green with brownish spots, c.10 mm in diam. and lobed for half its length; lobes 2–3 mm long, broadly ovate or triangular, apex rounded, glabrous. Gynostegial corona in several series; a low outer collar, and free erect bi- or tri-partite staminal lobes, the inner-most terminating in a long drawn-out filiform projection 1 mm or more in length. Anther wings c.0.3 mm long. Pollinaria with corpusculum c.0.2 mm long; translator arms absent; pollinia c.0.2 mm long, narrowly oblong. Stylar head flat, not projecting beyond anthers. Follicles and seeds unknown.

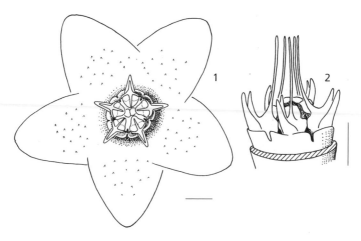

Fig. 7.3.**135**. CIBIRHIZA ALBERSIANA. 1, flower; 2, gynostegium and corona. Scale bars = 1 mm. Reproduced with permission from Taxon (1994).

---

[13] by D.J. Goyder

Zambia. N: Mbala (Abercorn), Crocodile Islands, Lake Tanganyika, fl. 9.ii.1964, *Richards* 18979 (K). S: Choma Dist., Maamba, fl. 10.ii.1978, *Chisumpa* 473 (K, NDO). Also known from very few localities in central and NE Tanzania. Dry deciduous forest, evergreen riparian thicket or mopane woodland; 600–1200 m.

42. MARSDENIA R. Br.[14]

Marsdenia R. Br., Prodr. Fl. Nov. Holland.: 460 (1810), conserved name. —Omlor, Gen. Revis. Marsdenieae (1998). —Forster in Austral. Syst. Bot. **8**: 703–933 (1995).

Gymnema R. Br., Prodr. Fl. Nov. Holland.: 461 (1810).

Dregea E. Mey., Comm. Pl. Afr. Austr.: 199 (1838), conserved name.

Pterophora Harv., Gen. S. Afr. Pl.: 223 (1838).

Pterygocarpus Hochst. in Flora **26**: 78 (1843).

Dregea sect. *Pterygocarpus* (Hochst.) K. Schum. in Engler & Prantl, Nat. Pflanzenfam. 4(2): 293 (1895).

Traunia K. Schum., Notizbl. Königl. Bot. Gart. Berlin **1**: 23 (1895); in Engler & Prantl, Nat. Pflanzenfam. 4(2): 287 (1895).

Dalzielia Turrill in Hooker's Icon. Pl. **31**: t.3061 (1916).

Dregea subgen. *Traunia* (K. Schum.) Bullock in Kew Bull. **11**: 516 (1957).

Gongronema sensu Bullock in Kew Bull. **15**: 197 (1961) pro parte, non (Endl.) Decne. (1844).

Woody or wiry twiners, occasionally scandent shrubs; latex white or clear. Leaves opposite, petiolate, glabrous or more usually pubescent. Inflorescences extra-axillary or axillary; sessile or pedunculate, umbelliform or irregularly branched. Corolla somewhat fleshy, mostly tubular but occasionally rotate; lobes frequently imbricate and contorted, but sometimes subvalvate. Corolline corona generally absent, if present then either forming longitudinal ridges down the corolla tube (*M. sylvestris*), or forming an annulus around the gynostegium (*M. exellii*). Gynostegial corona of 5 ovoid or flattened fleshy staminal lobes, attached to anthers basally, shorter or longer than anthers, apex free; rarely vestigial, or absent. Pollinia erect, club-shaped or rounded, basally attached to translator arms; corpusculum narrowly ovate to subcylindrical. Apex of stylar head flat, domed or rostrate, exserted beyond the anther appendages or concealed by them. Follicles single or paired, thick and often somewhat woody, smooth, weakly ribbed or with conspicuous longitudinal wings or surface entirely obscured by the strongly contorted, cristate wings; seeds flattened, with a broad or narrow marginal wing; with a coma of hairs at one end.

The delimitation of *Marsdenia* adopted here largely follows that of Omlor's generic revision of the Marsdenieae (1998), but with the addition of *Gymnema*, formerly separated on the position of its corona, which is on the corolla rather than on the gynostegium. The loss (or gain) of staminal coronas may occur more frequently than earlier thought, and has been demonstrated in sister species in the distantly related Andean genus *Philibertia* Kunth (Goyder in Kew Bull. **59**: 415–451, 2004). In *Philibertia*, taxa lacking a staminal corona frequently compensate by the possession of swellings on the corolla which perform the same function of restricting pollinator access to five regions of the tube and gynostegium. The situation in African *Marsdenia* parallels this exactly, with two morphologically similar species distinguished florally by the position of the corona – the predominantly tropical species, *M. sylvestris*, formerly *Gymnema sylvestre*, has a corona derived from the corolla, while in the southern African species, *M. dregea*, the corona is gynostegial. A similar generic concept was adopted by Forster (1995) for Asian species. Fruit ornamentation, while striking in some African *Marsdenias*, appears to behave in a similar way, with some species possessing prominent longitudinal wings, while apparently closely related species lack them.

---

[14] by D.J. Goyder

Spot characters: fruiting material [fruit not known in *M. chirindensis* and *M. exellii*]. FIGURE 7.3.**136**.

Follicles with 4 well-developed, longitudinal wings – *M. faulknerae, M. dregea.*

Follicle surface ± completely obscured by the many cristate wings – *M. abyssinica.*

Follicles lacking wings
  surface smooth – *M. angolensis, M. cynanchoides, M. gazensis, M. latifolia, M. sylvestris.*
  surface with longitudinal wrinkles at least when dry – *M. macrantha.*

1. Apex of stylar head extending into a rostrate appendage exserted from both staminal column and the mouth of the corolla tube . . . . **9.** *faulknerae*
– Apex of stylar head flat or domed, never long-rostrate 2
2. Gynostegium fully exposed on the base of the corolla, or only partially obscured by a short corolla tube 3
– Gynostegium (with the exception of the domed stylar head of some species) obscured by the corolla tube 7
3. Inflorescences divaricately branched with flowers scattered along the axes; flowers small – corolla lobes 1.5–2.5 mm long; pedicels 2–5 mm long . . . . **3.** *latifolia*
– Inflorescences not as above, mostly umbelliform; flowers larger – corolla lobes 3–9 mm long; pedicels 4–20 mm long 4
4. Corolla lobes glabrous adaxially, or with a tuft of hairs at the base 5
– Corolla lobes pubescent adaxially 6
5. United portion of corolla 5–7 mm across; pubescent cup or annulus surrounding the base of the gynostegium; corona lobes ovoid, not expanded into a flattened tongue apically. . . . . **11.** *exellii*
– Corolla united only for c.1 mm at base; additional cup or annulus absent; corona lobes laterally compressed below, expanded into a flattened tongue apically . . .
. . . . **6.** *gazensis*
6. Sepals foliaceous; corolla lobes 7–9 mm long, ± uniformly pubescent adaxially; surface of follicles not obscured by wings or other ornamentation, finely longitudinally wrinkled when dry. . . . . **7.** *macrantha*
– Sepals not foliaceous; corolla lobes 4–5 mm long, indumentum mostly towards apex and margins adaxially, centre of the lobe glabrous; surface of follicles entirely or mostly obscured by highly cristate convoluted wings . . . . **10.** *abyssinica*
7. Corona corolline, of five fleshy, hirsute ridges running from the corolla lobe sinus down the length of the corolla tube, frequently bifid and projecting somewhat at the mouth of the corolla tube. . . . . **1.** *sylvestris*
– Corona gynostegial, of 5 lobes adnate to the back of the stamens 8
8. Leaves deeply cordate at the base; upper and lower surfaces pubescent; infloresences pedunculate . . . .**2.** *angolensis*
– Leaves cuneate to rounded at the base; upper surface glabrous or sparsely pubescent, lower surface glabrous or subglabrous; inflorescences sessile or pedunculate. . . . 9
9. Leaves ovate, lamina of mature leaves less than 5 cm long. . . . .**8.** *dregea*
– Leaves elliptic to oblong, lamina of mature leaves at least 5 cm long 10
10. Inflorescences sessile or subsessile; coastal dry forest in central and northern Mozambique or neighbouring Zimbabwe. . . . .**4.** *cynanchoides*
– Inflorescences long-pedunculate; middle-altitude rainforest (Chirinda) . . . .
. . . . **5.** *chirindensis*

Note: *Marsdenia crinita* Oliv. has been reported from Zimbabwe, but the record, a specimen at Kew (*Craster* 210, with no additional locality information), appears to be the result of a label mix-up. The nearest confirmed localities for this species are from NW Angola.

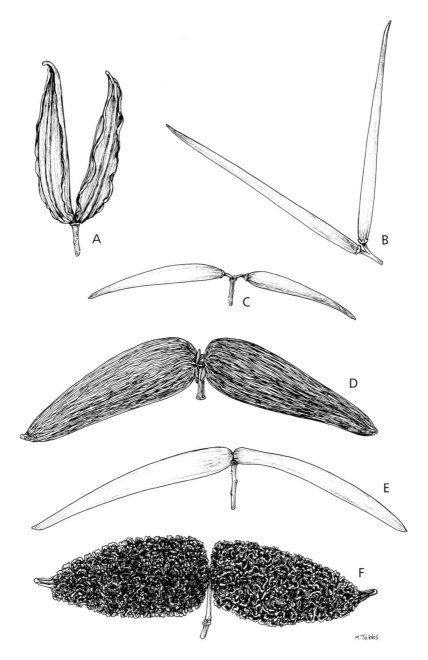

Fig. 7.3.**136**. MARSDENIA fruits. A. —MARSDENIA FAULKNERAE, paired follicles, from *Faulkner* 1019. B. —MARSDENIA ANGOLENSIS, only one follicle developing in each flower, from *Chandler* 1045. C. —MARSDENIA SYLVESTRIS, only one follicle developing in each flower, from *Greenway & Kanuri* s.n., iv.1970. D. —MARSDENIA MACRANTHA, paired follicles, from *Crawford* 101. E. —MARSDENIA LATIFOLIA, paired follicles, from *Dhetchuvi* 873. F. —MARSDENIA ABYSSINICA, paired follicles, from *Bullock* 3050. Drawn by Margaret Tebbs. Reproduced and adapted from Flora of Tropical East Africa (2012).

1. **Marsdenia sylvestris** (Retz.) P.I. Forst. in Austral. Syst. Bot. **8**: 694 (1995). —Goyder in F.T.E.A., Apocynaceae (part 2): 194 (2012). Type: *Koenig* 6733 (LD lectotype), designated by Forster in Austrobaileya **3**: 126 (1989). FIGURE 7.3.**136C**.

> *Periploca sylvestris* Retz., Observ. Bot. **2**: 15 (1791).
> *Gymnema sylvestre* (Retz.) A. Rees, Cycl. **17**: [188] no. 4 (1811). —Brown in F.T.A. **4**(1): 413 (1903).

Scandent shrub or woody twiner with white latex; stems to c.5 m long, young shoots villous with white or tawny hairs, older stems lenticellate, to c.2.5 cm in diam. at base, arising from a long taproot. Leaves opposite, petiole 0.5–1.5(2) cm long; lamina 3–7 × 1.5–5 cm, broadly ovate, obovate, elliptic or suborbicular, apex acute, shortly attenuate or narrowing abruptly into a short acumen, base rounded or very weakly cordate, pubescent on both surfaces. Inflorescences extra-axillary, single or paired, much shorter than the adjacent leaves, forming dense subumbelliform clusters of flowers; peduncles (0.3)0.5–1.5 cm long, densely pubescent; pedicels 2–5 mm long, pubescent. Sepals 1–2 × 0.5–1 mm, lanceolate to broadly ovate or orbicular, apex rounded, pubescent with ciliate margins. Corolla yellow, orange or cream, glabrous except on the corona; tube 1–1.5 mm long, ± completely enclosing the gynostegium; lobes spreading, c.1.5 × 1 mm, oblong or triangular with a rounded or somewhat truncate apex, margins ciliate. Corolline corona forming five fleshy ridges running from the corolla lobe sinus down the length of the corolla tube, pubescent along the lateral faces, frequently bifid and projecting somewhat at the mouth of the corolla tube. Staminal corona absent. Anther wings c.0.8 mm long. Pollinaria minute, corpusculum 0.1 mm long, ovoid; translator arms slender, c.0.1 mm long; pollinia c.0.1 mm long, ovoid. Stylar head projecting for c.1 mm beyond anthers, ovoid-conical. Follicles single, 5–9 × c.0.7 cm, slender, fusiform, tapering gently into a long attenuate tip, smooth, glabrous. Seeds flattened, c.9 × 5 mm, oblong with a narrow margin; coma c.3.5 cm long.

Caprivi. Katima Mulilo area, fl. 24.xii.1958, *Killick & Leistner* 3066 (K, PRE). **Botswana**. N: Selinda Reserve, near Wild Dog Pan, fl. 9.iv.2005, *A. & R. Heath* 1027 (GAB, K, PSUB). SE: Mochudi Hills, fl. i.1955, *Reyneke* 212 (K, PRE). **Zambia**. B: bank of R. Zambezi at Sesheke, fl. 27.xii.1952, *Angus* 1049A (FHO, K). N: Mbala (Abercorn) Dist., Lufubu River, Iyendwe valley, fl. 8.xii.1959, *Richards* 11902 (K). W: Kitwe, fl. 22.v.1958, *Fanshawe* 4280 (K, NDO). S: Island in R. Zambezi near Katombora, fr. 24.viii.1947, *Brenan* 7747 (FHO, K). **Zimbabwe**. N: Mazoe Dist., Glendale / Chiweshe, fr. 1.ii.1981, *Carr* s.n. (K, SRGH). W: Matobo Dist., Bessner Kobila Farm, fl. ii.1954, *Miller* 2192 (K, SRGH). C: Sebakwe R., fl. xii.1948, *Hodgson* H/39/48 (K, SRGH). S: Beitbridge, Umzingwane River, fl. 8.ii.2000, *Timberlake & Cunliffe* 4491 (K). **Malawi**. N: Nkhata Bay Dist., Mkhuluti Hills, Khuyu Road, fr. 23.viii.1984, *Salubeni et al.* 3860 (K, MAL). C: Lilongwe Dist., Nature Sanctuary Zone A, fl. 28.i.1985, *Patel & Banda* 1988 (K, MAL). **Mozambique**. T: Marueira, 4 km from Bucha towards R. Zambezi, fr. 11.iii.1972, *Macêdo* 5055 (K, LMA). MS: Xiluvu Hills S of Beira, fl. 2.iii.1962, *Pole-Evans* 6317 (K). GI: Between Zandamela and Zavala, fl. 3.iv.1959, *Barbosa & Lemos* 8479 (K, LMA). M: Majajane forest, Salamanga, fr. 17.iv.1984, *Groenendijk & Dungo* 1327 (K, LMU).

Widely distributed across Old World tropics - in Africa from the northeastern provinces of South Africa northwards to Senegal and Mauritania in the west and Ethiopia in the east, then via the Arabian Peninsula to the Indian subcontinent. Riverine forest margins, thicket and dry bushland, frequently on sand or stony soil; 100–1200 m.

Easily confused with the more southerly *Marsdenia dregea*, if can be distinguised by the pubescent rather than glabrous leaves, the smooth, unwinged follicles, and the position of the corona lobes which arise from the corolla lobe sinus rather than the gynostegium.

2. **Marsdenia angolensis** N.E. Br. in Bull. Misc. Inform., Kew **1895**: 258 (1895); in F.T.A. **4**(1): 423 (1903). —Goyder in F.T.E.A., Apocynaceae (part 2): 195 (2012). Type: Angola, *Welwitsch* 4245 (K lectotype, LISU), designated by Bullock in Kew Bull. **9**: 367 (1954). FIGURE 7.3.**136B**.

    Marsdenia gondarensis Chiov. in Ann. Bot. (Rome) **9**: 80 (1911). Type: Ethiopia, Gondar, Dembia, *Chiovenda* 1741 (FT holotype).

    Gongronema angolense (N.E. Br.) Bullock in Kew Bull. **15**: 199 (1961). —Goyder in F.T.E.A., Apocynaceae (part 2): 194 (2012).

Herbaceous or woody scrambler with white latex, frequently 3–4 m high, but in suitable habitats reaching the forest canopy; stems densely pubescent with spreading ivory or yellowish hairs to 1 mm long. Leaves with petiole 2–6 cm long, densely pubescent; lamina 5–12 × 3–7 cm, broadly ovate, apex acuminate, base deeply cordate, the auricles separated by a narrow sinus, both surfaces pubescent with spreading hairs, particularly on the veins. Inflorescences extra-axillary, ± as long as the petioles of the adjacent leaf pair, with 2–3 principal branches terminated by subumbelliform clusters of flowers, pubescent; bracts to c.6 mm long, linear, foliose, densely pubescent; pedicels 4–10 mm long. Sepals c.1.5–2 mm long, mostly ovate to suborbicular with a rounded apex, but occasionally lanceolate and acute, densely pubescent. Flower buds ovoid, the corolla lobes overlapping to the right. Corolla cream or yellowish green, united into a broadly cylindrical tube for about half its length ± enclosing the gynostegium, tube 1.5–2.5 mm long, pubescent outside, inner face with pubescent patches towards the base, glabrous or minutely papillose above; lobes spreading, oblong with a rounded apex, 1.5–2 mm long, 1.5 mm wide, pubescent outside, glabrous within. Coronal corona absent. Staminal corona lobes c.2 mm long, fleshy, quadrate to triangular in section below in lower half, adnate to the back of the anthers for 1 mm, the upper half of the lobe free, tapering into a dorsiventrally flattened tongue arched over the stylar head. Anther wings c.1 mm long, curved gently away from the column towards the base and almost completely obscured by the corona lobes. Corpusculum c.0.2 × 0.2 mm, rhomboid; translator arms c.0.1 mm long; pollinia c.0.2 × 0.1 mm, oblong-ovate in outline, somewhat flattened. Stylar head extending above the anthers for c.1 mm, domed. Follicles occurring singly or paired at an angle of c.60°, narrowly cylindrical and tapering to a slightly displaced tip, 8–12 × 0.5 cm, densely pubescent with short spreading hairs, at least when young. Seeds flattened, c.7 × 2 mm, oblong with a narrow yellowish margin; coma c.2 cm long.

Zambia. B: Masese, fl. 14.i.1961, *Fanshawe* 6133 (K, NDO). **Zimbabwe**. C: near Harare (Salisbury), fl. 13.xii.1936, *Eyles* 8828 (K). E: SE slope of Mutarampanda area of Inyamatshira mountain range, fl. 16.i.1964, *Chase* 8105 (K, SRGH).

Widespread in wetter regions of tropical Africa from West Africa to Ethiopia, and Angola to Zimbabwe. Scrambling over margins of evergreen forest or scrub; 1000–1600 m.

3. **Marsdenia latifolia** (Benth.) K. Schum., Just's Bot. Jahresb. **26**(1): 372 (1900). —Goyder in F.T.E.A., Apocynaceae (part 2): 196 (2012). Type: São Tomé [St. Thomas], *Don* s.n. (K holotype). FIGURE 7.3.**136E**.

    Gongronema latifolium Benth. in Hooker, Niger Fl.: 456 (1849). —Bullock in F.W.T.A. ed. 2, **2**: 98 (1963).

    Marsdenia leonensis Benth. in Hooker, Niger Fl.: 455 (1849). —Brown in F.T.A. **4**(1): 424 (1903). Type: Sierra Leone, *Vogel* s.n. (K holotype).

    Marsdenia glabriflora Benth. in Hooker, Niger Fl.: 455 (1849). —Brown in F.T.A. **4**(1): 424 (1903). Type: Sierra Leone, *Vogel* s.n. (K holotype).

    Marsdenia racemosa K. Schum. in Bot. Jahrb. Syst. **17**: 147 (1893). Type: Angola, Lunda, Lulua plain, *Pogge* 1249 (B† holotype, K).

    Marsdenia profusa N.E. Br. in Bull. Misc. Inform. Kew **1895**: 258 (1895); in F.T.A. **4**(1): 425 (1903). Type: Niger, Brass, *Barter* 16 (K holotype).

    Marsdenia glabriflora var. *orbicularis* N.E. Br. in F.T.A. **4**(1): 424 (1903). Type: Nigeria, Bonny River, x.1860, *Mann* s.n. (K holotype).

Slender to robust woody scrambler to c.4 m high with white latex; stems sparsely to densely pubescent with spreading white or ivory hairs c.0.5 mm long. Leaves with petiole 1–6 cm long, sparsely to densely pubescent; lamina 6–12 × 3–7 cm, ovate to ovate-oblong, apex acute to strongly acuminate, subtruncate to shallowly cordate at the base, the auricles separated by a broad sinus, both surfaces pubescent with spreading hairs, particularly on the veins. Inflorescences extra-axillary, initially ± as long as the petioles of the adjacent leaf pair and subumbelliform, but branches developing rapidly into long axes with flowers scattered along their lengths, pubescent; bracts 1–3 mm long, linear, densely pubescent; pedicels 2–5 mm long. Sepals 1(2) mm long, ovate to suborbicular with a rounded apex, densely pubescent. Flower buds ovoid, the corolla lobes overlapping to the right. Corolla cream or yellowish green, united into a tube for no more than a third of its length and only partially enclosing the gynostegium, tube 0.5–1 mm long, glabrous outside, inner face entirely glabrous to minutely papillose, with or without tufts of long inward-pointing hairs in the throat; lobes spreading, oblong with a rounded apex, 1.5–2.5 mm long, 1–1.5 mm wide, glabrous outside, minutely papillose within but with entirely glabrous margins. Corolline corona absent. Staminal corona very variable, lobes mostly well-developed, c.1–1.5 mm long, fleshy, quadrate to triangular in section and adnate to the back of the anthers in the lower half, the upper half free and arching over the stylar head, but sometimes reduced to a vestigial peg. Anther wings c.0.5–0.7 mm long, oblong, not obscured by the corona lobes. Corpusculum c.0.2–0.25 × 0.05 mm, subcylindrical; translator arms 0.1–0.2 mm long; pollinia c.0.2 × 0.15 mm, ovate to suborbicular in outline, somewhat flattened. Stylar head extending above the anthers for 0.5–1 mm, conical or domed. Follicles occurring singly or in pairs, lanceolate-subcylindrical and tapering to a rounded apex, 4–8 × 1–1.5 cm, glabrescent. Seeds not seen.

Zambia. B: Kataba, fl. 12.xii.1960, *Fanshawe* 5972 (K, NDO). N: Lake Chishi, fl. 10.x.1958, *Fanshawe* 4906 (K, NDO). S: Namwala Dist., Lubanga thicket, c.8 km N of Ngoma, Kafue National Park, fl. 14.xii.1962, *Mitchell* 15/99 (K, SRGH).

Widespread in West Africa and as far east as Uganda and SW Tanzania, and south to Angola. Scrambling over riverine forest or thicket; 1000–1600 m.

Several collections from SW Tanzania and Zambia are more slender than elsewhere in the species' range. The inflorescence appears less elongated, the corolla tube is slightly longer, and the tufts of hairs in the throat of the corolla are very well-developed.

4. **Marsdenia cynanchoides** Schltr. in Bot. Jahrb. Syst. **38**: 53 (1905). —Goyder in F.T.E.A., Apocynaceae (part 2): 197 (2012). Type: Mozambique, Dondo [25 Miles Station], Beira, 10.iv.1898, *Schlechter* 12243 (K lectotype, B†, MO), lectotype designated here.

> *Gongronema taylori* sensu Bullock in Kew Bull. **15**: 201 (1961), pro parte excl. type.
> *Marsdenia taylori* sensu Mapaura & Timberlake in Checkl. Zimbabwe Vasc. Pl.: 21 (2004), non Schltr. & Rendle.

Slender woody scrambler to c.3 m high, latex colour not recorded, arising from a cluster of fleshy fusiform roots; stems glabrous or pubescent, corky below. Leaves with petiole 0.5–1.5(2.5) cm long, pubescent on upper face; lamina 3–12 × 1.5–5.5 cm, oblong to elliptic, apex rounded to acute, usually shortly and abruptly acuminate, base cuneate or occasionally rounded, glabrous or sparsely pubescent above, glabrous except for a few short hairs along principal veins below. Inflorescences extra-axillary, subsessile, appearing umbelliform but flowers on two or more short branches, inflorescence generally shorter than the petioles of the adjacent leaf pair, glabrous or glabrescent below; bracts 0.5–1 mm long, ovate to lanceolate, glabrous but with ciliate margins; pedicels 3–5 mm long. Sepals 1–2 mm long, lanceolate to suborbicular with a rounded apex, glabrous, margins ciliate. Flower buds broadly cylindrical to subglobose, the corolla lobes overlapping to the right. Corolla cream or green, united into a tube for half to two-thirds of its length and completely enclosing the gynostegium; tube 1–2 mm long, glabrous outside, inner face pubescent, usually also with 5 patches of downward-pointing hairs near the base of the tube and a band of white hairs to 1 mm long in the throat; lobes spreading, 1–1.5 mm long, c.1 mm wide, oblong with a rounded apex, glabrous outside, glabrous or minutely papillose within beyond the tuft of long hairs in the throat. Corolline corona absent. Staminal corona lobes 1–2 mm long, fleshy, quadrate to triangular in section below and adnate to the staminal column for half their length, the upper half free and forming a slender tongue arching over the top of the stylar head. Anther wings c.0.7–0.8 mm long, much longer than broad; gynostegium ± sessile. Corpusculum c.0.2 mm long, subcylindrical; translator arms c.0.1 mm long; pollinia c.0.2 × 0.2 mm, broadly elliptic in outline, somewhat flattened. Stylar head not exserted beyond anther appendages. Follicles poorly preserved, c.7 cm long, 2 cm wide, smooth, glabrous. Mature seeds not seen.

Zimbabwe. E: Lower Sabi Valley, 12 km above Gudu's Pool, fl. 15.i.1960, *Pole-Evans* 5825 (K). **Mozambique**. N: c.25 km S of Palma on road to Olumbi, fl. 5.xii.2008, *Goyder et al.* 5090 (K, LMA, LMU, P). Z: Pebane, dunes close to the lighthouse, fl. 8.iii.1966, *Torre & Correia* 15089 (LISC). MS: 20 km S of Muanza, fl. 3.xii.1971, *Müller & Pope* 1856 (K, SRGH). GI: Panda Dist., Xivalo, Chihuane, fl. 29.i.2019, *Osborne et al.* 1559 (K, LMA, LMU).

Also recorded from SE Tanzania. Mixed deciduous woodland or thicket on sandy or stony ground; 20–700 m.

5. **Marsdenia chirindensis** Goyder, sp. nov. Resembles *M. cynanchoides*, but differs in its long-pedunculate and much branched inflorescence (rather than subsessile and sub-umbelliform). Type: Zimbabwe, Chirinda Forest, near Swynnerton's memorial, fl. 28.xii.2011, *Ballings & Würsten* 1690 (BR holotype).

Slender woody scrambler, latex colour not recorded; stems glabrous or sparsely pubescent. Leaves with petiole 2.5–3 cm long, pubescent; lamina 7–12 × 3.5–5 cm, oblong to elliptic, apex acute to attenuate, base cuneate, glabrous except for a few short hairs on the principal veins beneath. Inflorescences extra-axillary, simple or branched with flowers in umbelliform clusters, longer than the petioles of the adjacent leaf pair, pedunculate, the peduncle (to first branching point) 3–6 cm long, glabrous or pubescent along one line; bracts mostly 0.5–1 mm long, occasionally larger and more foliose; pedicels 10–15 mm long. Sepals 1.5–2.5 mm long, suborbicular with a rounded apex, glabrous, margins ciliate. Corolla united into a tube for around half of its length and completely enclosing the gynostegium; tube greenish, 2–2.5 mm long, glabrous outside, with a band of white hairs to 1 mm long inside the tube and throat; lobes yellowish with longitudinal reddish streaks within, 2–2.5 mm long, c.2 mm wide, suberect to spreading, oblong or broadly triangular with a subacute apex, glabrous outside, glabrous or minutely papillose within beyond the tuft of long hairs in the throat. Corolline corona absent. Staminal corona lobes c.2 mm long, fleshy, quadrate to triangular in section below and adnate to the staminal column for less than half their length, the upper free portion forming a dorsiventrally flattened tongue with a truncate or weakly emarginate apex arching over the top of the stylar head. Anther wings 0.8–0.9 mm long, much longer than broad; gynostegium ± sessile. Corpusculum c.0.2 mm long, subcylindrical; translator arms c.0.3 mm long; pollinia c.0.4 × 0.2 mm, broadly elliptic in outline, somewhat flattened. Stylar head not exserted beyond anther appendages. Fruit and seeds not known.

Zimbabwe. E: Chirinda Forest, near Swynnerton's memorial, fl. 28.xii.2011, *Ballings & Würsten* 1690 (BR holotype).

Known only from the type. Margins of evergreen forest; 1070 m.

6. **Marsdenia gazensis** S. Moore in J. Bot. **46**: 306 (1908). Type: Mozambique, Kurumadzi R., Jihu, 4.i.1906, *Swynnerton* 224 (BM holotype, K).
    Gongronema gazense (S. Moore) Bullock in Kew Bull. **15**: 199 (1961).

Woody scrambler to several metres; latex white; stems lenticellate or not, minutely pubescent. Leaves with petiole 2–4 cm long, densely pubescent when young, but becoming glabrescent; lamina 5–10 × 3–6 cm, broadly oblong to ovate, apex somewhat attenuate, base truncate to weakly cordate, both surfaces pubescent when young but rapidly becoming glabrescent with remaining indumentum restricted to principal veins. Inflorescences extra-axillary, pubescent, with several subumbelliform clusters of flowers on a simple or brached axis, peduncles up to 2.5 cm long; bracts c.1 mm long, triangular; pedicels 4–5 mm long. Sepals c.1.5 × 1 mm, ovate, sparsely or densely pubescent. Flower buds ovoid, the corolla lobes overlapping to the right. Corolla greenish, united basally for c.1 mm, leaving the gynostegium fully exposed; lobes spreading, ovate with a rounded or truncate apex, 2.5–3.5 mm long, 2 mm wide, glabrous except for a sparse tuft of hairs at the base of each lobe. Corolline corona absent. Staminal corona lobes c.1.5 mm long, adnate to the staminal column for around ¾ of their length, fleshy, and at least in herbarium material appearing laterally compressed in lower ¾, the apical portion expanded into an ovate tongue spreading from the column. Anther wings 0.9 mm long, gynostegium sessile. Corpusculum c.0.3 mm long, subcylindrical; translator arms c.0.2 mm long; pollinia c.0.4 × 0.1 mm, narrowly oblong and somewhat flattened. Stylar head domed, not or only partially exserted beyond anther appendages. Follicles not seen, but reported to have a thick mericarp with a smooth surface, tapering to a sharp point or rounded apically, bright green at first, turning black on dehiscence.

Zimbabwe. E: Chirinda Forest, near Swynnerton's old homestead, fl. xii.1961, *Goldsmith* 116/61 (K, SRGH). **Mozambique**. MS: Kurumadzi R., Jihu, fl. 4.i.1906, *Swynnerton* 224 (BM holotype, K).

Not known elsewhere. Evergreen forest or forest margins, among rocks; 1100–1800 m.

7. **Marsdenia macrantha** (Klotzsch) Schltr. in Bot. Jahrb. Syst. **51**: 143 (1913). — Goyder in F.T.E.A., Apocynaceae (part 2): 198 (2012). Type: Mozambique, Rios de Sena, *Peters* s.n. (B† syntype) & surroundings of Tete, *Peters* s.n. (B† syntype). FIGURES 7.3.**136D**, 7.3.**137**.

> *Dregea macrantha* Klotzsch in Peters, Naturw. Reise Mossambique 6(1): 272 (1861).
> *Periploca petersiana* Vatke in Oesterr. Bot. Z. **26**: 147 (1876). Types: Mozambique, near Tete, 26.xii.1844, *Peters* s.n. (B† syntype) & 11.i.1845 *Peters* s.n. (B† syntype); Rios de Sena, 1846, *Peters* s.n. (B† syntype).
> *Marsdenia zambesiaca* Schltr. in J. Bot. **33**: 338 (1895). —Brown in F.T.A. **4**(1): 420 (1903). Type: Malawi, Shire, Zambesi at Chiromo, i.1894, *Scott Elliot* 3791 (BM holotype, K).

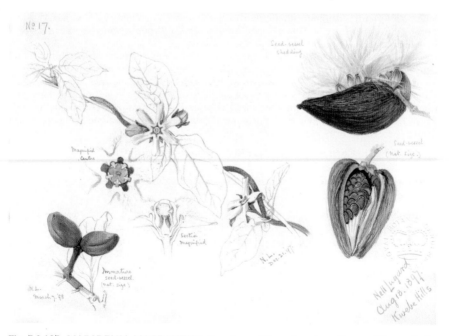

Fig. 7.3.**137**. MARSDENIA MACRANTHA. Drawings of flowering stem, dissected flower, fruits and seeds. Illustrated by Mrs. Nell Lugard.

Woody scrambler to c.10 m, but often much less; latex white; stems frequently corky and winged below, lenticellate above; very young shoots with short indumentum, rapidly becoming glabrescent. Leaves with petiole 1–3 cm long, densely rusty-puberulent; lamina 3–7 × 1–6 cm (but up to c.16 × 12 cm in older leaves), lanceolate to elliptic or suborbicular, apex usually acute, but sometimes rounded or obtuse and abruptly acuminate, base truncate to rounded or very shallowly cordate, younger leaves sometimes somewhat cuneate, both surfaces puberulent when young but rapidly becoming glabrescent. Inflorescences generally crowded on short lateral shoots, extra-axillary, ± umbelliform but with flowers opening successively, densely puberulent; peduncles up to 2 cm long; bracts c.3–6 mm long, foliaceous, lanceolate to elliptic; pedicels 5–20 mm long. Sepals foliaceous, 5–7 × 2–3 mm, lanceolate to elliptic, sparsely puberulent. Flowers unpleasantly scented; buds ovoid, the corolla lobes overlapping to the right. Corolla rotate, greenish cream, united into a glabrous tube for c.2–3 mm, partially enclosing the gynostegium; lobes spreading, slightly contorted, oblong with a rounded or truncate apex, 7–9 mm long, 2–4.5 mm wide, glabrous outside, with a sparse to dense pubescence of white hairs on the adaxial face. Corolline corona absent. Staminal corona lobes c.3 mm long, dorsiventrally flattened, adnate to the staminal column for just under half their length, the free portion narrowing gradually, spreading away from the column then arching over it as a dorsally flattened falcate tongue. Anther wings c.2 mm long, gynostegium ± sessile. Corpusculum c.0.6 mm long, subcylindrical; translator arms c.0.2 mm long; pollinia c.0.4 × 0.2 mm, ovate in outline, somewhat flattened. Stylar head ± flat, not exserted beyond anther appendages. Follicles occurring singly or in opposite pairs, 7–10 × 1–3 cm, narrowly ovoid to fusiform tapering gently to a rounded or acute apex, tough and horny in texture, surface yellowish and with fine longitudinal wrinkles when dry; seeds c.12 × 7–10 mm, flattened, obovate to suborbicular, smooth, with a shiny marginal rim; coma to c.3 cm long.

Botswana. N: Okavango Delta, Gcobega Lagoon, fl. 29.xii.2009, *A. & R. Heath* 1780 (GAB, K, PSUB). **Zambia**. C: Katondwe, fl. 4.xii.1969, *Mutimushi* 3875 (K, NDO). E: Jumbwe, Lwangwa valley, fl. 24.xi.1966, *Mutimushi* 1647 (K, NDO). S: Choma Dist., between Sinazongwe and mouth of R. Zongwe, fl. 29.xii.1958, *Robson & Angus* 1007 (K). **Zimbabwe**. N: Chipisa Hot Spring, Zambesi valley, fl. 24.xi.1953, *Wild* 4250 (K, SRGH). W: Sebungwe R. drift, 80 km NE of Kamativi Tin Mine, fr. 18.v.1955, *Plowes* 1839 (K, SRGH). E: Chimanimani (Melsetter) Dist., Hot Springs, fl. 16.xi.1952, *Chase* 4717 (K, SRGH). S: Gwanda Dist., Liebig's Ranch, fl. & fr. 26.x.1952, *Plowes* 1518 (K, SRGH). **Malawi**. N: Karonga Dist., Ngala, 26 kn N of Chilumba, fl. 28.xii.1976, *Pawek* 12063 (K, MAL). S: Foot of Thyolo Escarpment c.10 km ESE of Chikwawa, fl. 14.i.1992, *Goyder & Paton* 3518 (BR, K, MAL). **Mozambique**. N: Imala, 1 km from Namialo towards Netio, fl. 23.xi.1963, *Torre & Paiva* 9241 (LISC). T: N bank of Zambesi River opposite Msusa, fr. 25.vii.1950, *Chase* 2222 (K, SRGH). MS: Chemba, Chiou, C.I.C.A. Experimental Station, fr. 20.iv.1960, *Lemos & Macácua* 122 (K, LMA). GI: Chokwe (Guijá), old road towards Mabalane, fl. 20.xi.1957, *Barbosa & Lemos* 8210 (K, LMA).

Dry tropical regions of eastern and southern Africa from SE Kenya to northern Namibia and the Soutpansberg in the extreme north of South Africa. Dry acacia or mopane woodland; 100–1600 m.

8. **Marsdenia dregea** (Harv.) Schltr. in Bot. Jahrb. Syst. **51**: 143 (1913). Type: South Africa, Eastern Cape Province, "*Dregea floribunda* EM.a. Zuurbergen, in sylva prope Stroebels", *Drège* s.n. in herb. Benth. (K000305440 lectotype), designated here; "*Dregea floribunda* EM.a. Zuurbergen, in sylva prope Stroebels", *Drège* s.n., in herb. Hook. (HBG502852, K000305441 paralectotypes); "*Dregea floribunda* EM.b. Hoffmannskloof prope Enon", *Drège* s.n., in herb. Hook. (HEID701129, K000305442 paralectotypes), in herb. Benth. (K000305438 paralectotype); "*Dregea floribunda* EM.c. in collibus prope Katrivier", *Drège* s.n. (not traced).

Dregea floribunda E. Mey. in Comm. Pl. Afr. Austr.: 199 (Jan. 1838). Types as above.

Pterophora dregea Harv. in Gen. S. Afr. Pl.: 223 (Aug.–Dec. 1838). Types as for *Dregea floribunda*.

Marsdenia floribunda (E. Mey.) N.E. Br. in F.T.A. **4**(1): 422 (1903), illegitimate name, non (Brongn.) Schltr. (1899).

Woody twiner to 5 m with white latex; young shoots glabrescent, older stems grey-brown, lenticellate. Leaves with petiole 1–1.5 cm long; lamina 3–4 × 2–3 cm, broadly ovate, apex acute or shortly attenuate, base obtuse to sub-truncate, glabrous on both surfaces. Inflorescences extra-axillary, single, much shorter than the adjacent leaves, forming dense subumbelliform clusters of foetid-smelling flowers; peduncles 0.5–1 cm long, glabrescent; bracts 1–2 mm long, lanceolate; pedicels 2–3 mm long, glabrescent. Sepals 1.5 × 1 mm, ovate, glabrous. Corolla cream or white, glabrous except for a few long white hairs at the base of the lobes; tube c.1 mm long, ± completely enclosing the gynostegium; lobes suberect or spreading, 1.5–2 × 1–1.3 mm, oblong or ovate with a rounded or somewhat truncate or emarginate apex, margins sparsely ciliate. Corolline corona absent. Staminal corona of five fleshy ovoid vesicles c.0.4 mm long, adnate to the back of the anthers. Anther wings c.0.3 mm long. Pollinaria minute, corpusculum 0.2 mm long, ovoid; translator arms slender, c.0.15 mm long; pollinia c.0.1 mm long, circular in outline and somewhat compressed. Stylar head projecting for up to 0.5 mm beyond anther appendages, conical. Follicles single or in widely divergent pairs, 4–6 × 2–2.5 cm, tapering gently into a long attenuate tip, glabrous, with 4 longitudinally striate trangular wings running the length of the follicle. Seeds flattened, c.9–12 × 5–5 mm, ovate with a narrow margin; coma c.2 cm long.

Mozambique. M: Lebombo mountains c.1 km S of Goba Fronteira, fl. 23.xi.2001, *Goyder* 5002 (K, LMU).

Also occurs in eastern South Africa (KwaZulu-Natal, Eastern Cape) and most probably in neighbouring Swaziland. Forested gullies on rocky hillsides; 300–360 m.

9. **Marsdenia faulknerae** (Bullock) Omlor, Gen. Revis. Marsdenieae: 79 (1998). —Goyder in F.T.E.A., Apocynaceae (part 2): 202 (2012), as "*faulkneri*". Type: Tanzania, Korogwe, Magunga, 30.iv.1953, *Faulkner* 1189 (K holotype, EA, B). FIGURE 7.3.**136A**.

Dregea faulknerae Bullock in Kew. Bull. **11**: 520 (1958).

Woody scrambler to c.5 m, latex colour not recorded; young shoots densely pubescent, older wood with thick corky wings. Leaves with petiole 1.5–6 cm long, densely pubescent; lamina 6–13 × 4–10 cm, broadly ovate, apex acute to strongly acuminate, base weakly to strongly cordate, softly pubescent on both surfaces. Inflorescences lax, pedunculate, extra-axillary, irregularly-branched, densely pubescent; peduncles 1–3 cm long; bracts to c.5 mm long, lanceolate; pedicels 1–1.5 cm long. Sepals 4–5 × 1–1.5 mm, lanceolate, densely pubescent at least on mid-line. Flowers with foetid sweet odour; buds ovoid-conical, the corolla lobes overlapping to the right. Corolla greenish-cream, united into a tube for c.3 mm ± enclosing the gynostegium, the throat and tube sparsely pubescent within with long hairs; lobes 6–7 mm long, 1.5–2.5 mm wide, oblong with a rounded or truncate apex, weakly contorted, sparsely pubescent towards the mouth of the tube but otherwise glabrous on both faces. Corolline corona absent. Staminal corona lobes c.3 mm long, fleshy and quadrate in section below, adnate to the staminal column for c.1 mm, the upper ²/₃ free, dorsiventrally flattened, tapering gently to an acute apex, and reaching well beyond the anther appendages. Anther wings c.1.2 mm long, gynostegium sessile. Corpusculum c.0.4 mm long, ovoid-subcylindrical; translator arms c.0.1 mm long; pollinia c.0.3–0.4 × 0.2 mm, elliptic in outline, somewhat flattened. Stylar head rostrate, to c.3 mm long, exserted beyond the mouth of the corolla tube and somewhat contorted apically. Follicles occurring singly or in pairs held at an acute angle, c.6–8 × 1.5–2.5 cm, narrowly ovoid with an attenuate apex, woody, with 4 well-developed longitudinal wings running the entire length of the follicle, and fine longitudinal wrinkles over the whole surface, subglabrous; seeds not seen.

Mozambique. MS: Xiluvu Hills S of Beira [actually NW of Beira], fl. 2.iii.1962, *Pole-Evans* 6316 (K).

Known only from a handful of collections from central Mozambique, NE Tanzania and SE Kenya, but probably occurring sporadically through the intervening coastal regions. Forest margins; to c.400 m.

10. **Marsdenia abyssinica** (Hochst.) Schltr. in Bot. Jahrb. Syst. **51**: 143 (1913). —
Goyder in F.T.E.A., Apocynaceae (part 2): 204 (2012). Types: Ethiopia, Gondar,
near Mt Sabra, 19.vii.1838 (fl.), *Schimper* II: 1366 (K lectotype, B†, BR, M, WAG),
lectotype designated here, see note below; near Sabra, 9.iii.1840 (fr.), *Schimper* II:
1294 (B†, K, M paralectotypes). FIGURE 7.3.**136F**.

    Pterygocarpus abyssinicus Hochst. in Flora **26**: 78 (1843).

    Hoya africana Decne. in Candolle, Prodr. **8**: 639 (1844). Type as for *Pterygocarpus abyssinicus*.

    Dregea africana (Decne.) Martelli, Fl. Bogos.: 55 (1886).

    Dregea abyssinica (Hochst.) K. Schum. in Engler, Pflanzenw. Ost-Afrikas **C**: 326 (1895); in
Engler & Prantl, Nat. Pflanzenfam. **4**(2): 293 (1895).

    Marsdenia spissa S. Moore in J. Bot. **39**: 260 (1901). —Brown in F.T.A. **4**(1): 420 (1903).
Type: Kenya, near Lake Marsabit, 1898, *Delamere* s.n. (BM holotype, K).

    Marsdenia abyssinica f. *complicata* Bullock in Kew Bull. **7**: 423 (1952). Type: Tanzania, Ufipa
Dist.: Kate, 21.x.1949, *Silungwe* s.n. (K holotype).

Woody scrambler to several metres with white latex; older wood lenticellate, soft and
corky below, young shoots minutely puberulent. Leaves with petiole 1–2.5 cm long, minutely
puberulent; lamina 5–10 × 3–6 cm, ovate to broadly elliptic, apex acute to shortly attenuate
or rounded and abruptly acuminate, base broadly cuneate, rounded or truncate, glabrous on
both surfaces. Inflorescences extra-axillary, pedunculate, minutely puberulent, with congested,
irregularly branched, subumbelliform clusters of flowers; peduncles 1–1.5(2.5) cm long; bracts
to c.3 mm long, scarious, ovate; pedicels 0.7–1.4 cm long. Sepals c.1.5–3 × 1.5–3 mm, broadly
ovate to orbicular, densely puberulent at least towards the base. Flowers sweetly scented, buds
subglobose, corolla lobes overlapping to the right. Corolla white, cream or green, united for
1–2 mm and not obscuring the gynostegium; lobes 4–5 mm long, 2–3 mm wide, ovate or oblong
with a rounded or truncate apex, glabrous or sparsely pubescent abaxially, densely pubescent
or even bearded towards the apex, margins and throat adaxially, the central portion of the lobe
glabrous or minutely papillose. Corolline corona absent. Staminal corona lobes solid, fleshy,
2 mm long and c.1 mm diam., about as tall as the anthers, adnate to the staminal column for
c. half their length. Anther wings c.1 mm long, gynostegium sessile. Corpusculum c.0.3 mm
long, somewhat rhomboid; translator arms c.0.2 mm long; pollinia c.0.6 × 0.2 mm, elliptic in
outline, somewhat flattened. Stylar head obscured by the anther appendages. Follicles occurring
singly or in opposite pairs, 6–7(10) × 2–3 cm, ovoid and tapering to an acute apex, but the
outline generally obscured by the many well-developed and extremely convoluted longitudinal
wings which cover the surface, woody, subglabrous or minutely pubescent; seeds c.12 × 6 mm,
flattened, obovate, smooth, with a shiny marginal rim; coma c.4 cm long.

    Zambia. W: Ndola, fl. 29.xi.1953, *Fanshawe* 521 (K, NDO) & fr. 18.vi.1954, *Fanshawe*
1281 (K, NDO). C: Mt. Makulu Research Station near Chilanga, fr. 23.vii.1957, *Angus*
1642 (FHO, K). **Zimbabwe**. C: 65 km from Harare (Salisbury) on Shamva road, fl.
9.xii.1960, *Rutherford-Smith* 451 (K, SRGH). E: Mutare Dist., NW side of Murahwa's
Hill, fl. 3.xii.1964, *Chase* 8189 (K, SRGH). S: Lothian road, next to Canal, 20 km S
of Masvingo (Fort Victoria), fl. 8.xi.1972, *Cannell* 533 (K, SRGH). **Mozambique**. N:
Malema, Serra Murripa, fl. 15.xii.1967, *Torre & Correia* 16528 (LISC).

    Widespread in drier regions of tropical Africa, and the south-western parts of the
Arabian Peninsula. Acacia woodland and forest margins; c.600–1500 m.

    Bullock in Kew Bull. **11**: 517 (1957) designated *Schimper* 1573 as lectotype of
Pterygocarpus abyssinicus. This is inadmissible as the collection was not mentioned by
Hochstetter, and could not, therefore, have been a syntype.

11. **Marsdenia exellii** C. Norman in Exell, Cat. Vasc. Pl. S. Tomé: 244 (1944). —
Goyder in F.T.E.A., Apocynaceae (part 2): 205 (2012). Type: São Tomé, Vanhulst
(Macambrará), 29.x.1932, *Exell* 138 (BM holotype).

Slender wiry climber to several metres, latex colour not recorded; young shoots pubescent
with short hairs. Leaves with petiole c.0.5–1.5 cm long, sparsely to densely pubescent; lamina
5–13 × 1.5–5 cm, narrowly ovate-elliptic, apex acuminate, base rounded, cuneate or very weakly

cordate, both surfaces ± glabrous except for the prominent raised veins. Inflorescences extra-axillary, pedunculate, densely spreading-pubescent, umbelliform; peduncles 0.5–3 cm long; bracts to c.4 mm long, filiform; pedicels 1–2 cm long, ± equal in length within an inflorescence. Sepals 2–4 × 0.5–1 mm, linear or triangular, pubescent. Corolla colour highly variable, red, white or yellow, united basally into a shallowly campanulate bowl c.0.5–0.7 cm in diam., with a smaller pubescent cup or annulus around the base of the gynostegium, abaxial surface glabrous to sparsely pubescent, adaxial surface pubescent at least around the basal annulus, glabrous or minutely pubescent elsewhere; corolla lobes 3–5 mm long, 3–5 mm across at the base, broadly oblong, margins ciliate. Corolline corona absent. Staminal corona lobes solid, fleshy, ovoid drawn out into a weak point facing towards the column at least when dry, c.1.5 × 1 mm and about as tall as the anthers. Anther wings 0.5–0.7 mm long, gynostegium on stipe 1–2 mm long. Corpusculum c.0.3 mm long, rhomboid, translator arms c.0.1 mm long; pollinia c.0.4–0.5 × 0.1 mm, oblanceolate but slightly curved towards the base, flattened. Stylar head obscured by the anther appendages. Follicles not seen.

Malawi. S: Mt. Mulanje, Chisongeli Forest below Manene peak, fl. 29.xi.1957, *Jenkins* s.n. (K, MAL). **Mozambique**. N: Ribáuè, serra Mepáluè, fl. 9.xii.1967, *Torre & Correia* 16402 (LISC).

Known from very few but widely scattered collections in Guinea, São Tomé, eastern D.R. Congo, southern Tanzania, Malawi and Mozambique. Forest, possibly over shallow soil; 1300–1600 m.

I have treated this plant as a small-flowered variant of *Marsdenia exellii*, formerly thought to be restricted to São Tomé, where it occurs in montane forest. *Stolz* 1878 from the Southern Highlands of Tanzania and *Bytebier & Luke* 2899 from eastern D.R. Congo are also part of this complex. And it has also been collected recently in Guinea, West Africa. Vegetatively the specimens are almost identical, the ± glabrous leaves have prominent, arched, raised veins on the lower surface. The corollas of east African collections are much smaller than in São Tomé material, but have similarly distributed indumentum. Gynostegial characters differ only slightly, with a shorter stipe and a minor reduction in anther wing length in east African material. The differences therefore seem largely quantitative rather than qualitative. Flower colour, however, is reported to be yellow in São Tomé, reddish brown in the Congo collection, red in the Malawi material, and wine red in Mozambique.

As with other forest taxa, *Marsdenia exellii* is probably undercollected, and could perhaps be more widespread than the few localities suggest. In some localities it has been collected on inselbergs – it may be that shallow soil over rock is the preferred habitat.

43. TELOSMA Coville[15]

Telosma Coville in Contr. U.S. Natl. Herb. **9**: 384 (1905).

Slender herbaceous or somewhat woody twiners; latex white or clear. Leaves opposite, petiolate. Inflorescences extra-axillary; sessile or pedunculate, umbelliform. Corolla urceolate with an inflated base and contorted lobes. Corolline corona absent. Gynostegial corona of 5 flattened fleshy staminal lobes attached to anthers basally, and with an adaxial ligule on their inner faces. Pollinia erect, basally attached to translator arms. Apex of stylar head not exserted beyond the anther appendages. Follicles generally single, smooth. Seeds flattened, comose.

Five species, mostly in S.E. Asia, with one in tropical Africa.

Telosma africana (N.E. Br.) N.E. Br. in Fl. Cap. **4**(1): 776 (1908). —Goyder in F.T.E.A., Apocynaceae (part 2): 208 (2012). Type: Nigeria, Nupe, *Barter* s.n. (K lectotype), designated by Goyder in Kew Bull. **59**: 651 (2004). FIGURE 7.3.**138**.

---

[15] by D.J. Goyder

Fig. 7.3.**138**. TELOSMA AFRICANA. 1, habit (× 1); 2, flower (× 3); 3, flower bud (× 3); 4, gynostegium with staminal corona lobes (× 8); 5, gynostegium with corona lobes removed (× 8); 6, staminal corona lobe showing tooth on inner face (× 8); 7, stamen viewed from within (× 8); 8, gynoecium showing paired ovaried and stylar head with expanded stylar head appendage (× 8); 9, pollinarium (× 36). 1 from *Chandler* 551; 2–9 from *Maitland* 604. Drawn by D. Erasmus. Reproduced from Flora of Tropical East Africa (2012).

Pergularia africana N.E. Br. in Bull. Misc. Inform. Kew **1895**: 259 (1895); in F.T.A. **4**(1): 426 (1903).

Telosma unyorensis S. Moore in J. Bot. **46**: 307 (1908). Type: Uganda, Bunyoro [Unyoro], near Mruli, Victoria Nile, 28.iii.1907, *Bagshawe* 1558 (BM holotype).

Pergularia tacazzeana Chiov. in Ann. Bot. (Rome) **9**: 80 (1911). Type: Ethiopia, 'Scirè, boschi lungo le rive Tacazzè sotto Timchet', 28.vi.1909, *Chiovenda* 617 (FT holotype).

"*Telosma africanum* (N.E. Br.) Coville", comb. ined., in F.W.T.A. ed. 2, **2**: 97 (1963).

Slender woody twiner to c.5 m with clear or white latex; older stems somewhat lenticellate, glabrous. Leaves opposite, petiole 1.5–5 cm long, minutely pubescent; lamina 4–10 × 1.5–6 cm, broadly ovate to oblong, apex attenuate to acuminate, base mostly rounded or truncate, rarely cuneate, glabrous or subglabrous. Inflorescences extra-axillary, forming a dense sub-umbelliform cluster of flowers, subsessile or clearly pedunculate, minutely pubescent; peduncles to 10 mm long; pedicels 3–6 mm long. Sepals 2–4 × 1–1.5 mm, broadly ovate to triangular, glabrous or minutely pubescent. Corolla yellow or green, frequently flushed reddish purple outside, strongly twisted in bud; tube 4–6 mm long, urceolate, glabrous or sparsely pubescent outside, bearded with inward-pointing hairs in the throat and the base of the lobes; lobes spreading and somewhat contorted, 6–10 × 0.5–1 mm, narrowly oblong with a rounded or truncate apex, densely pubescent adaxially, glabrous or at most sparsely pubescent abaxially. Gynostegium concealed within corolla tube. Staminal corona lobes c.2–2.5 mm long, dorsiventrally compressed, broadly ovate, with a linear or narrowly triangular ligule on the inner face extending for a further 1 mm and overtopping the gynostegium. Anther wings c.1 mm long. Pollinaria with corpusculum c.0.2 mm long, ovoid; translator arms c.0.1 mm long; pollinia c.0.6 mm long, narrowly oblong. Stylar head conical, not projecting beyond anthers. Follicles generally single, c.6 cm long, subcylindrical, smooth. Seeds not seen.

Zimbabwe. E: Chimanimani (Melsetter) Dist., Haroni River, fl. 4.xii.1965, *Wild et al.* 6660 (K, SRGH). **Malawi**. S: Mingoli Farm estate, Likangali R., 11 km E of Zomba, fl. 29.xi.1977, *Brummitt et al.* 15223 (K, MAL).

Apparently common and widespread in tropical West Africa, but with remarkably few scattered records from the rest of the continent. Probably undercollected and more widespread than these records suggest. Riverine forest and wet forest margins; 400–800 m.

44. ORTHANTHERA Wight[16]

Orthanthera Wight in Contr. Bot. India: 48 (1834). —Brown in F.T.A. **4**(1): 433–435 (1903); in Fl. Cap. **4**(1): 784–786 (1908).
Barrowia Decne. in Candolle, Prodr. **8**: 629 (1844).

Erect shrubs, prostrate or scrambling lianas, or geoxylic suffrutices producing annual shoots from underground rhizomes; latex clear. Leaves opposite, well-developed or (not in our area) reduced to scales, persistent or deciduous. Corolla with an elongated tubular portion, often with a globose basal inflation; lobes erect or spreading. Corolline corona absent. Gynostegial corona of five arrow-head shaped lobes adnate to the staminal column except for their free wing-like margins. Pollinia held erect, with a pellucid germination zone at the apex; translator arms slender, attached to the base of the pollinia. Follicles usually single by abortion, if paired then not divergent, generally fusiform with a long attenuate apex, smooth; seeds flattened, with a coma of hairs.

Four species in south-western Africa, with a fifth in NW India.

Orthanthera jasminiflora (Decne.) Schinz in Verh. Bot. Vereins Prov. Brandenburg **30**: 265 (1889). Type: South Africa, Northern Cape Prov., Kuruman, 17.xi.1812, *Burchell* 2427 (G-DC holotype, K). FIGURE 7.3.**139**.

---

[16] by D.J. Goyder

Fig. 7.3.**139**. ORTHANTHERA JASMINIFLORA. Habit. 1, gynostegium showing characteristically arrow-shaped corona (a); 2, stamen; 3, pollinarium; 4, gynoecium with paired ovaries (a) and stylar head (b). Reproduced from Icones Selectae Plantarum (1846).

Barrowia jasminiflora Decne. in Candolle, Prodr. **8**: 630 (1844).

Orthanthera browniana Schinz in Verh. Bot. Vereins Prov. Brandenburg **30**: 264 (1889). Types: Namibia, Amboland, Omulonga, i.1886, *Schinz* 144 (K001208213 lectotype, B† GRA, K), lectotype designated here; Amboland, Oshiheke, 1886, *Schinz* s.n. (B† paralectotype, K).

Sprawling vine, stems to several m in length, mostly prostrate, pubescent, arising from a cylindrical grey taproot or tuber. Leaves with petiole 2–12 mm long, pubescent; lamina 1.5–5 × (0.1)0.3–2(4) cm, linear to broadly ovate, rarely suborbicular, apex rounded to acute, occasionally with a short apiculus, base truncate to weakly cordate, venation conspicuous at least in broader leaves, indumentum of scabrid hairs. Inflorescences umbelliform; peduncles (0.5)1–4 cm long, sparsely to densely pubescent; pedicels 4–10(14) mm long, densely pubescent. Sepals 3–6 × 1–1.5 mm, lanceolate, scabrid-pubescent. Corolla white, turning cream with age, greenish on the exterior; tube 7–10 mm long, narrowly cylindrical and with a subglobose basal inflation, pubescent; lobes spreading, but often appearing suberect in herbarium material, (6)8–12 mm long, 1.5–2 mm wide at the base, narrowly oblong to narrowly triangular, somewhat contorted apically, densely pubescent abaxially, adaxial face glabrous or sparsely pubescent. Gynostegium shortly stipitate; anther wings 1 mm long, triangular. Corona adnate to back of anthers, 1.5 mm long, lateral margins free, laminar and recurved. Follicles single or paired, not divergent, 10–14 × 1–2 cm, fusiform, tapering apically into a long beak and narrowed towards the base, smooth, glabrous; seeds not seen.

Botswana. N: Moremi Game Reserve, transit road North to South gate, fl. 29.xi.2007, *A. & R. Heath* 1468 (GAB, K, PSUB); SW: Nossop River, 10 km N of Twee Rivieren, fl. 17.iii.1969, *Rains & Yalala* 39 (K, SRGH). **Zambia**. B: Machili, fl. 21.ix.1969, *Mutimushi* 3795 (K, NDO); S: Simonga village c.20 km W of Livingstone, fl. 6.xi.1993, *Harder et al.* 1875 (K, MO). **Zimbabwe**. W: Victoria Falls, behind Falls Hotel, fl. 25.x.1966, *Simpathu* 56 (K, SRGH).

Also known from southern Angola, Namibia and northern provinces of South Africa. Mostly on Kalahari sand; 900–1000 m.

45. NEOSCHUMANNIA Schltr.[17]

Neoschumannia Schltr. in Bot. Jahrb. Syst. **38**: 38 (1905). —Meve in Pl. Syst. Evol. **197**: 233–242 (1995); in Bot. Jahrb. Syst. **119**: 427–435 (1997). —Harris & Goyder in Kew Bull. **52**: 733–735 (1997).

Neoschumannia Schltr. in Westafr. Kautschuk-Exped.: 310 (1900), nomen nudum.

Swynnertonia S. Moore in J. Bot. **46**: 308 (1908).

Glabrous woody climbers with slender stems; roots fibrous; latex clear, but turning into a white powder on drying. Leaves opposite, membranous. Inflorescences extra-axillary, pedunculate, with flowers clustered towards the tip of a long-lived racemose axis; bracts persistent. Calyx fused at the base. Corolla lobes fused only at their extreme base; corolline corona absent. Gynostegium stipitate; gynostegial corona of three series – downward pointing staminal lobes partially fused to form a skirt; a whorl of spreading or erect interstaminal lobes; and a further whorl of 5 erect dorsi-ventrally flattened staminal lobes with swollen bases. Pollinia erect in the anther sacs, with pellicid inner margins, attached to the minute corpusculum by very short, broad, translator-arms. Apex of stylar head flat or depressed, not exserted from the staminal column. Follicles paired, long, slender, fusiform. Seeds ovoid-oblong with a narrow wing and a long coma.

A genus of three species with relict distributions – the W African *Neoschumannia kamerunensis*, known from isolated forests in Ivory Coast, Cameroon and the Central African Republic, the East African *N. cardinea*, and the recently described *N. gishwatiensis* from Rwanda.

---

[17] by D.J. Goyder

Neoschumannia cardinea (S. Moore) Meve in Pl. Syst. Evol. **197**: 235 (1995). —Goyder
in F.T.E.A., Apocynaceae (part 2): 215 (2012). Type: Zimbabwe, Chirinda Forest,
29.x.1906, *Swynnerton* 1080 (BM holotype, K). FIGURE 7.3.**140**.

    Swynnertonia cardinea S. Moore in Journ. Bot. **46**: 309 (1908).

Large woody climber; glabrous in all parts except the flower. Stems slender, to c.4 mm diam.
in flowering portions. Leaves with petioles 1.5–2.5 cm long; lamina 5–15 × 3–8 cm, elliptic to
broadly ovate, apex acuminate, sometimes very shortly so, base rounded to truncate, glabrous.
Peduncles 1.5–6 cm long; bracts c.2 mm long, triangular, acute, sparsely pubescent; pedicels
filiform, 1–4 cm long. Calyx lobes 4–5 mm long, at most 1 mm broad at the base, very narrowly
triangular, glabrous or sparsely pubescent. Corolla fused only at extreme base; lobes 12–20 × 3–4
mm, narrowly lanceolate-elliptic, with an acute apex, abaxial face greenish, glabrous, adaxial
face purple with two white patches at the base of each lobe, glabrous except for the margins
which have a line of slender crisped hairs 2–3 mm long, the tips of the corolla lobes which
have a dense tuft of similar hairs, and the pale basal patches to the corolla lobes which have a
tuft of slightly more robust clavate hairs c.2 mm long. Gynostegial stipe c.2.5 mm long, mostly
concealed by the pendant white skirt of fused staminal corona lobes arising below the anthers;

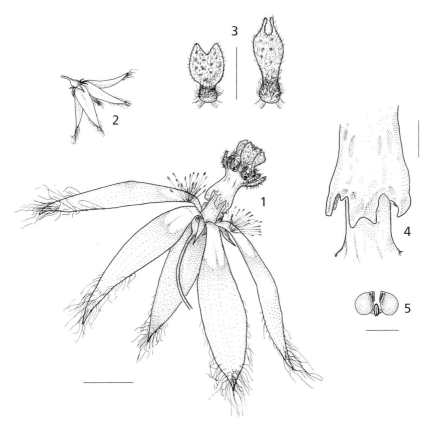

Fig. 7.3.**140**. NEOSCHUMANNIA CARDINEA. 1, flower in lateral view; 2, natural orientation of
flower with petals fallen forwards; 3, staminal corona lobes; 4, base of staminal column with coronal
skirt; 5, pollinarium. Scale bars: 1 = 5 mm; 3 = 2 mm; 4 = 1 mm; 5 = 0.5 mm. 1–4 from *Swynnerton*
1080; 5, 3 (corona lobe to the right) from *Rodgers & Hall* 1346. Drawn by U. Meve, mostly after
Swynnerton's sketches on the holotype. Adapted from Plant Systematics and Evolution (1995),

interstaminal lobes c.1.5 × 0.5 mm, flattened dorsiventrally, ± oblong with a somewhat bifid emarginate-truncate apex, spreading from the column initially, then turning erect, maroon, pubescent; innermost whorl of staminal corona lobes held erect, c.2–3 mm long, spathulate, with a truncate/emarginate apex from which two acute teeth may project a further 0.5 mm, maroon, pubescent. Anther wings c.0.4 mm long. Follicles and seeds not seen.

Zimbabwe. E: Chirinda Forest, 29.x.1906, *Swynnerton* 1080 (BM, K).

Also known from two localities in Tanzania. Moist submontane tropical forest; c.1200 m.

46. **RIOCREUXIA** Decne.[18]

Riocreuxia Decne. in Candolle, Prodr. **8**: 640 (1844). —Brown in F.T.A. **4**(1): 465–466 (1903); in Fl. Cap. **4**(1): 799–804 (1908). —Dyer in Codd *et al.*, Fl. S. Africa **27**(4): 83–88 (1980); in *Ceropegia, Brachystelma* & *Riocreuxia* in S. Africa: 227–238 (1983). —Masinde in Kew Bull. **60**: 401–434 (2005).

Ceropegia sect. *Riocreuxia* sensu Huber in Mem. Soc. Brot. **12**: 167–175 (1958), in part.

Perennial herbs with somewhat woody rootstock, producing a tuft of slightly fleshy or sub-fusiform roots; latex clear. Stems twining or sometimes suberect and tufted, sometimes becoming woody at base. Leaves opposite, petiolate, simple, ovate, apex acuminate or very rarely acute, base strongly cordate; interpetiolar stipules of tufts of whitish or brownish-translucent hairs. Flowers in laxly branched cymes or in fascicles. Sepals linear-subulate, acuminate. Corolla mostly tubular, straight, or campanulate, ± inflated at base; lobes erect, shorter or longer than tube, mostly connate at tips to form a subglobose cage-like structure; corolline corona absent. Gynostegial corona 1–2 seriate, variable, subsessile or distinctly stipitate. Anthers longer or shorter than staminal lobes; pollinia erect, oblong or elliptic with an acute pellucid margin at apex resembling a small beak, attached in pairs from near base by short caudicles to a small dark-brown corpusculum. Follicles paired, fusiform, acuminate, usually ± beaded from being constricted between the seeds, glabrous; seed dorsiventrally compressed, linear-elliptic or oblong, curling length-wise on drying to become convex-concave, with a narrow marginal wing, apex with a tuft of white hairs.

8 species worldwide restricted to Africa south of the Sahara.

1. Corolla campanulate, 7–10 mm long; lobes spreading . . . . **2.** *chrysochroma*
– Corolla urceolate or tubular, 10–28 mm long; lobes erect or united at tip to form a cage 2
2. Stems mostly with one line of pubescence; leaves sparsely pubescent or subglabrous; uplands. . . . .**1.** *polyantha*
– Stems uniformly spreading-pubescent; leaves densely pubescent; Lebombo Mountains . . . . **3.** *torulosa*

1. **Riocreuxia polyantha** Schltr. in J. Bot. **33**: 272 (1895). —Brown in Fl. Cap. **4**(1): 801 (1908). —Dyer in Fl. Pl. Africa **29**: t.1124 (1953). —Masinde in F.T.E.A., Apocynaceae (part 2): 216 (2012). Type: South Africa, Transkei, near Bashee River, i.1895, *Schlechter* 6291 (K lectotype, B†, BR, GRA, K, PRE), lectotype designated by Masinde in Kew Bull. **60**: 412 (2005), sheet stamped 17.i.1905. FIGURE 7.3.**141**.

Riocreuxia burchellii K. Schum. in Engler & Prantl, Nat. Pflanzenfam. **4**(2): 273 (1895). —Dyer in Codd *et al.*, Fl. S. Africa **27**(4): 85 (1980); in Bothalia **13**: 436 (1981); in *Ceropegia, Brachystelma & Riocreuxia* S. Africa: 233 (1983). Type: South Africa, Cape, Albany Division, between Blaauw Krantz and Kowie Poort, 12.ix.1813, *Burchell* 2668 (?B† holotype, K, M).

Riocreuxia profusa N.E. Br. in Bull Misc. Inform. Kew **1895**: 260 (1895); in F.T.A. **4**(1): 465 (1903). Type: South Africa, Shire Highlands, 750 m, xii.1881, *Buchanan* 205 (K holotype).

---

[18] by D.J. Goyder

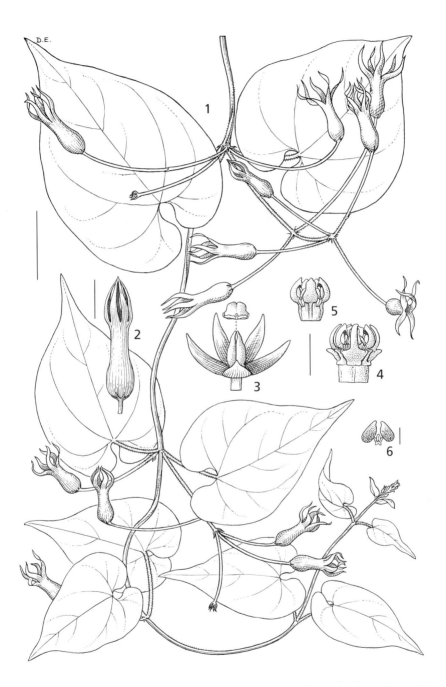

Fig. 7.3.**141**. RIOCREUXIA POLYANTHA. 1, flowering stem; 2, flower; 3, calyx and gynoecium showing glabrous carpels; 4, gynostegium; 5, gynostegium with outer corona removed; 6, pollinarium. Scale bars: 1 = 2 cm; 2 = 5 mm; 3–5 = 2 mm; 6 = 0.2 mm. All from *Milne-Redhead & Taylor* 9781. Drawn by D. Erasmus. Reproduced from Kew Bulletin (2005).

Ceropegia burchellii (K. Schum.) H. Huber in Mem. Soc. Brot. **12**: 171 (1957).
Ceropegia burchellii subsp. *profusa* (N.E. Br.) H. Huber in Mem. Soc. Brot. **12**: 172 (1957).
Riocreuxia flanaganii Schltr. subsp. *segregata* R.A. Dyer in Bothalia **12**: 632 (1979); in Codd
et al., Fl. S. Africa **27**(4): 87 (1980). Type: South Africa, Transvaal, Vryheid, Kastrol Nek near
Wakkerstroom, i.1925, *FitzSimons and van Dam* s.n. in *TRV* 25981 (PRE holotype).
Riocreuxia torulosa sensu Schltr. in Bot. Jahrb. Syst. **18**(Beibl. 45): 24 (1894), non Decne.
in Candolle, Prodr. **8**: 640 (1844).
Riocreuxia torulosa sensu K. Schum. in Engler, Pflanzenw. Ost-Afrikas **C**: 327 (1895), non
Decne. in Candolle, Prodr. **8**: 640 (1844).

A sparsely hairy, leafy climber, 1–3 m or more high. Stem annual, twining, freely branching, stout in older parts; nodes with a distinct ring of simple, minute whitish-translucent hairs; internodes mostly unifariously puberulous, sometimes bifarious. Leaves herbaceous, spreading; petiole (1)3–6(9) cm long, pubescent; lamina 3–9(13.5) × 2–8(18) cm, ovate or elliptic-ovate, apex acuminate, occasionally acute, base deeply cordate with a broad subtruncate sinus; sparsely and rather minutely pubescent on both sides or rarely glabrous above. Inflorescence of long, slender, branched, paniculate cymes up to 17 cm long; peduncles unifariously puberulous, branching at the node or at a distance into up to 4 branches, each branch in turn mostly dichotomously or trichotomously branching once or twice to produce the profusely flowering terminals; pseudoumbels 10–30+-flowered, many flowers opening simultaneously; pedicels (15)20–35 mm, glabrous. Sepals 2.5(4) mm long, often apically reflexed, glabrous. Corolla 10–28 mm long, glabrous; tube subcylindric or slightly inflated towards base, 7–14 mm long × 3–4(5) mm; exterior and interior white, yellow, cream or pale green occasionally with faint greenish striations; lobes erect, connate at apex to form an ellipsoid or subglobose cage-like structure 5(8) mm diam., occasionally breaking free, 4–8(13) mm long × 1–1.8(2.5) mm wide at the base, linear-lanceolate, apex acute to attenuate, sometimes orange-red/yellow or pale yellow. Corona biseriate, c.1.5 × 1.5 mm, subsessile or shortly stipitate; outer lobes emarginate or shortly bifid, with subhorizontally orientated teeth or lobules c.0.5 mm long; inner lobes 0.5–1.2 × c.0.3 mm, slightly longer or slightly shorter than anthers, closely applied to their backs and connivent over them, dorsiventrally compressed, linear-obtuse. Anther wings c.0.3 mm long. Follicles paired, acutely divergent, (9)15–17(21) × 0.2–0.4 cm, narrowly cylindrical, tapering apically; not constricted between the seeds; seeds 8–10 × c.2.5 mm, dorsiventrally compressed, oblong-elliptic, dark brown or blackish with a paler margin; coma c.20 mm long.

Zambia. N: Kalambo Stream Gorge, Siasi, fl. 24.iii.1955, *Richards* 5135 (K); W: Kitwe, fl. 2.ii.1961, *Fanshawe* 6167 (K, NDO); C: Chongwe Bridge, 50 km from Lusaka, fl. 29.i.1961, *Best* 269 (K); E: Nyika, fl. 31.xii.1962, *Fanshawe* 7368 (K, NDO); S: 65 km NW of Kalomo towards Kafue, fl. 11.iii.1997, *Schmidt et al.* 2494 (K, MO). **Zimbabwe**. C: Enterprise Dist., fl. iii, *Eyles* 6916 (K); E: Nyanga Distr., Iron Cliffs 16 km SSE of Juliasdale, fl. 12.ii.1997, *Brummitt & Pope* 19593 (EA, K, SRGH). **Malawi**. N: Chisenga, fl. 1.iii.1982, *Brummitt, Polhill & Banda* 16220 (K, MAL); C: Malingunde near Sinyala, fl. 6.iii.1968, *Salubeni* 999 (K, MAL); S: Chiradzulu Mountain, above Lisao Forest, fl. 13.iii.1977, *Brummitt, Seyani & Patel* 14859 (K, MAL).
Also recorded from Rwanda, Burundi, Tanzania, Angola, South Africa, Swaziland and Lesotho. Forest margins, including montane and riverine forests, wooded savannas and thickets; 1000–2100 m.

2. **Riocreuxia chrysochroma** (H. Huber) Radcl.-Sm. in Kew Bull. **21**: 298 (1967). —
Masinde in F.T.E.A., Apocynaceae (part 2): 219 (2012). Type: Tanzania, Njombe
Distr. Elton [Kitulo] Plateau, Nyarere River, 8.i.1957, *Richards* 7594 (K holotype).
Ceropegia chrysochroma H. Huber in Mem. Soc. Brot. **12**: 203 (1957).

A hairy leafy climber, 1–3 m high. Stem twining, branched, stout in older parts, dry stems hollow; nodes with indistinct ring of simple, minute brownish-translucent hairs; internodes pubescent all round. Leaves herbaceous, spreading; petiole (1)2–5 cm long, pubescent; lamina (1.8)4–7.5 × (1.3)3.5–5 cm, ovate or elliptic-ovate, apex acuminate or acute, base deeply or shallowly cordate with a broad subtruncate sinus; upper surface sparsely pubescent to more or less glabrous, lower surface minutely velvety pubescent especially along veins. Inflorescence of short, pedunculate cymes up to 8 cm long; peduncles pubescent all round, branched at the node into 2 or 3 peduncles then producing the umbellate cymes, or dichotomously branching once more before producing flowers; pseudoumbels up to 10-flowered, many flowers opening simultaneously; pedicels 5–10 mm, densely pubescent all round. Sepals c.1.5–2 mm long, subglabrous. Corolla 7–10 mm long, glabrous throughout; tube campanulate, 3–5.5 mm long × c.4 mm diam. in middle, mostly slightly longer than lobes; exterior and interior cream or whitish-yellow; lobes spreading, apically free, 2.5–4 × 1–1.5 mm wide at base, deltoid-linear, acute or obtuse, deep orange-red adaxially. Corona biseriate, c.1.8–2 × 1.8 mm, shortly stipitate; outer lobes deltoid-oblong, emarginate, teeth erect, c.0.2 mm long; inner lobes c.1 × 0.2 mm, slightly or much longer than anthers, closely applied to their backs and connivent over them, dorsiventrally compressed, linear-obtuse. Anther wings c.0.25 mm long long. Follicles and seeds not known.

Malawi. N: Nyika Plateau, Blue Heather Hill, fl. 8.xii.1975, *Phillips* 526 (K, MO).
Also known from southern Tanzania. Margins of montane forest and scrub; 2200–2400 m.

3. **Riocreuxia torulosa** Decne. in Candolle, Prodr. **8**: 640 (1844). Type: South Africa, Eastern Cape, King Williamstown Division, near Kachu, 6.i.1832, *Drège* 4944 (K lectotype), designated by Dyer in Codd *et al.*, Fl. S. Africa **27**(4): 84 (1980).

Ceropegia torulosa E. Mey. in Comm. Pl. Afr. Austr.: 194 (1838), illegitimate name, non Haw. in Revis. Pl. Succ.: 199 (1821).

Ceropegia effusa H. Huber in Mem. Soc. Brot. **12**: 173 (1957). Type as for *Riocreuxia torulosa*.

Riocreuxia torulosa var. *tomentosa* N.E. Br. in Fl .Cap. **4**(1): 803 (1908). Type: South Africa, Free State, Harrismith, xi.1904, *Sankey* 316 (K lectotype), designated by Masinde in Kew Bull. **60**: 417 (2005).

Riocreuxia torulosa var. *longidens* N.E. Br. in Fl. Cap. **4**(1): 803 (1908). Type: South Africa, Transvaal, slopes of Marovunye near Shilavane, *Junod* 724 (K holotype).

Riocreuxia torulosa var. *obsoleta* N.E. Br. in Fl. Cap. **4**(1): 803 (1908). Type: South Africa, Transvaal, by the Crocodile R., *Leendertz* 708 (K holotype).

A densely hairy, leafy climber, to 3 m or higher. Stem twining, branched, stout in older parts; nodes with a ring of simple, minute whitish-translucent hairs; internodes mostly pubescent all round. Leaves herbaceous, spreading; petiole 1.5–6(8) cm long, pubescent all round; lamina 3–9(18) × 1.7–6(14) cm, ovate to broadly ovate, apex acuminate, occasionally acute, base deeply cordate with a broad subtruncate sinus; pubescent on both sides. Inflorescence of slender, branched, paniculate cymes up to 15 cm long; peduncles puberulous, branching at the node or at a distance into up to 3 branches, each branch in turn mostly dichotomously or trichotomously branching once or twice to produce the profusely flowering terminals; pseudoumbels 6–15-flowered, many flowers opening simultaneously; pedicels 10–35 mm, mostly pubescent all round. Sepals 2–4 mm long, glabrous or pubescent. Corolla (10)12–20 mm long, glabrous; tube subcylindric or slightly inflated towards base, 6–10(12) mm long × 2–4 mm; pale yellow or cream; lobes erect, connate at apex to form an ellipsoid or subglobose cage-like structure 5 mm diam., occasionally breaking free, 5–8(10) mm long × (0.5)1 mm wide at the base, linear-attenuate, sometimes orange-yellow. Corona biseriate, 0.7–1 × 1–2.5 mm, stipe 0.5–1 mm long; outer lobes deltoid with long or short filiform or subulate processes, entire or bifid; inner lobes rudimentary, 0.1–0.2 × c.0.1 mm, reaching to the base of the anthers or half way up. Anther wings 0.2–0.4 mm long. Follicles paired, subparallel, 7.5–10(12) × 0.3 cm, narrowly cylindrical, tapering apically; constricted between the seeds; seeds 5–9 × c.2 mm, dorsiventrally compressed, oblong-elliptic, brown or blackish; coma c.20 mm long.

Mozambique. M: Namaacha, base of Monte Ponduine, fl. 2.ii.1982, *de Koning & Hiemstra* 9124 (LMU).
Distributed across eastern parts of South Africa and Swaziland. Rocky soil; 600 m.
Mozambican material is referable to var. *torulosa*.

47. SISYRANTHUS E. Mey.[19]

Sisyranthus E. Mey., Comm. Pl. Afr. Austr.: 197 (1838). —Brown in Fl. Cap. 4(1): 787–794 (1908).

Perennial herbs with a cluster of fleshy fusiform roots; latex clear. Stems slender, erect, mostly unbranched. Leaves opposite, linear to filiform. Flowers in extra-axillary and pedunculate or terminal and subsessile inflorescence. Corolla campanulate or urceolate enclosing the gynostegium, the mouth of the tube often obscured by a tuft of hairs; corolla lobes erect or suberect, shorter than or as long as the tube; corolline corona absent. Gynostegial corona of five free dorsiventrally flattened lobes. Anthers lacking a terminal appendage; pollinia erect, oblong or elliptic with pellucid germination zone on apical and inner margins, attached in pairs at the base by slender caudicles to the pale brown corpusculum. Follicles paired or occasionally single by abortion, fusiform, attenuate apically; seed dorsiventrally compressed, apex with a tuft of white hairs.

12 species in South Africa, Lesotho and Swaziland, with two species in the Flora Zambesiaca region. The genus is in need of critical revision.

Corolla tube 2–3 mm across, 1–1.5 mm deep, with 5 patches of inward or downward
  pointing hairs in throat . . . . **2**. *imberbis*
Corolla tube 3.5–4 mm across, 2 mm deep, entirely glabrous . . . . **1**. *rhodesicus*

1. **Sisyranthus rhodesicus** Weim. in Bot. Not. **1935**: 402 (1935). Type: Zimbabwe, foot of Mt Inyangani, fl. 6.xii.1930, *Friis, Norlindh & Weimarck* 3485 (LD holotype, BR, S).

Erect herb; annual stems 30 cm tall, simple, glabrous except for a few white hairs at the nodes. Leaves reduced at lower nodes, well-developed leaves 3–11 × 0.1 cm, glabrous except for sparse hairs on the midrib and margins. Inflorescence with 3–5 flowers, peduncle 1–1.5 cm long, glabrous or subglabrous; pedicels 1 cm long, glabrous or sparsely hairy. Calyx lobes 2–4 mm long, ovate to triangular with an attenuate apex, sparsely pubescent. Corolla white to yellowish purple, glabrous inside and out; tube 2 mm long, 3.5–4 mm across, broadly campanulate or slightly constricted at the mouth; lobes suberect, 2 × 1.5 mm, broadly ovate-triangular, acute, margins hyaline. Staminal corona lobes reaching to top of anther wings, rhomboid. Gynostegium with flared stipe topped by inward angled anther wings 0.3 mm long which form a conical structure beneath the fertile portion of the anthers. Follicles and seeds not seen.

Zimbabwe. E: Foot of Mt Inyangani, fl. 9.xii.1959, *Wild* 4886 (K, SRGH).
Apparently endemic to the eastern highlands of Zimbabwe and known only from Inyangani and the Nuza Plateau. Montane grassland at c.2100 m.

2. **Sisyranthus imberbis** Harv., Thes. Cap. **2**: 11 (1863); —Brown in Fl. Cap. 4(1): 793 (1908). Types: South Africa, Eastern Cape Province, Kreili's country, *Bowker* 299 (TCD lectotype, GRA, K), lectotype designated here; valleys SE of Grahamstown, xii.1859, *Hutton* s.n. (K paralectotype, S, TCD). FIGURE 7.3.**142**.

Erect herb; annual stems 30–90 cm tall, simple or little-branched, glabrous except for a few white hairs at the nodes. Leaves reduced or absent at lower nodes, well-developed leaves 3–8 × 0.1 cm, glabrous except for sparse hairs on the midrib and margins. Inflorescence with 3–5 flowers, peduncle 1 cm long, glabrous or subglabrous; pedicels 0.5–1 cm long, glabrous or sparsely hairy. Calyx lobes 1.3–2 mm long, ovate to triangular with an attenuate apex, glabrous. Corolla greenish or yellowish, glabrous inside and out except for 5 patches of inward or downward-pointing hairs in the throat; tube 1–1.5 mm long, 2–3 mm across, broadly campanulate or slightly constricted at the mouth; lobes suberect, 1–2 × 1 mm, broadly ovate-triangular, acute, margins hyaline. Staminal corona lobes reaching to top of anther wings, rhomboid. Gynostegium with flared stipe

---

[19] by D.J. Goyder

topped by inward angled anther wings 0.3 mm long which form a conical structure beneath the fertile portion of the anthers. Follicles and seeds not seen.

Zimbabwe. E: upper Bundu Plain, Chimanimani Mountains, fl. 29.xii.1959, *Goodier & Phipps* 325 (K, SRGH). **Mozambique**. MS: Mevumodzi Valley near Eastern Lakes, Eastern Chimanimani Mts, fl. 28.x.2014, *Wursten* 1059 (BR, K, LMA, SRGH).

Also known from Swaziland, Lesotho and eastern montane regions of South Africa as far south as the Eastern Cape. In the Flora region it is found in moist grassland at 900–1500 m.

A more delicate species in comparison with *Sisyranthus rhodesicus*, and with a bearded throat to the corolla.

Fig. 7.3.**142**. SISYRANTHUS IMBERBIS. 1, flowering shoot; 2, flower showing bearded throat. Photographed by B. Wursten.

48. CEROPEGIA L.[20]

Ceropegia L., Sp. Pl.: 211 (1753). —Brown in Dyer in F.T.A. **4**(1): 439–464 (1903); in Fl. Cap. **4**(1): 804–833 (1909). —Werdermann in Bot. Jahrb. Syst. **70**: 189–232 (1939). —Dyer in Codd *et al.*, Fl. S. Africa **27**(4): 43–82 (1980). —Meve, Ill. Handb. Succ. Pl. Asclepiadaceae: 88–107 (2002). —Masinde in F.T.E.A., Apocynaceae (part 2): 220–291 (2012). —Bruyns *et al.* in S. African J. Bot. **112**: 399–436 (2017).

    Brachystelma R. Br. in Bot. Mag. **49**: t.2343 (1822). —Brown in F.T.A. **4**(1): 466–472 (1903); in Fl. Cap. **4**(1): 834–865 (1909). —Dyer in Codd *et al.*, Fl. S. Africa **27**(4): 1–41 (1980). —Meve, Ill. Handb. Succ. Pl. Asclepiadaceae: 20–46 (2002). —Masinde in F.T.E.A., Apocynaceae (part 2): 292–310 (2012).

    Tenaris E. Mey., Comm. Pl. Afr. Austr.: 198 (1838). —Brown in Fl. Cap. **4**(1): 795–798 (1909). —Meve in Pl. Syst. Evol. **228**: 89–105 (2001).

    Macropetalum Decne. in Candolle, Prodr. **8**: 626 (1844). —Harvey, Gen. S. Afr. Pl. ed. 2: 241 (1868). —Brown in Fl. Cap. **4**(1): 798–799 (1909). —Schlechter in J. Bot. **1897**: 291 (1897). —Meve in Pl. Syst. Evol. **228**: 89–105 (2001).

    Decaceras Harv., Thes. Cap. **2**: 9 (1863).

    Dichaelia Harv., Gen. S. Afr. Pl., ed. 2: 241 (1868).

    Brachystelmaria Schltr. in Bot. Jahrb. Syst. **20**(Beibl. 51): 50 (1895).

    Siphonostelma Schltr. in Bot. Jahrb. Syst. **51**: 148, fig.3 (1913).

    Kinepetalum Schltr. in Bot. Jahrb. Syst. **51**: 149, fig.4 (1913).

Perennial herbs; rootstock a cluster of fusiform roots, a globose to disciform tuber or series of tubers or, rarely a rhizome or, in species with succulent stems, often with fibrous roots only; latex clear, rarely slightly cloudy in succulent stemmed species; indumentum of simple hairs. Stems erect, twining or trailing, herbaceous or succulent, sometimes slightly woody at base in more robust twining species. Leaves opposite, petiolate or sessile, sometimes reduced and scale like. Inflorescences extra-axillary, sometimes apparently terminal, pedunculate or sessile, mostly umbel-like cymes, less often raceme-like or paniculate, sometimes hysteranthous. Sepals free or almost so, filiform to lanceolate, acute. Corolla varying from divided almost to the base to forming a well-developed tube longer than the free parts of the corolla lobes, often with the base enlarged into a basal chamber; lobes varying from strongly reflexed to incurved and joined at their tips to form a cage, tips sometimes enlarged to form a parasol-like structure or extended into terminal beak or column, glabrous or hairy, hairs sometimes clavate and versatile. Gynostegial corona 2-seriate; outer series mostly two-lobed, sometimes joining laterally to form continuous cup, sometimes reduced to small teeth at bases of inner lobes or virtually absent; inner series ± linear, varying from about as long as anthers and horizontal to erect, often adhering to form central column, or reduced to small teeth on inside of cupular outer corona. Anthers reaching margin of gynostegial cap; pollinia ovoid to ± D-shaped, attached basally to the small translater arms and with well-developed germination crests on inner margins. Follicles usually paired, rarely solitary, parallel to widely divergent, narrowly to broadly fusiform, glabrous. Seeds many, flattened, ovate to lanceolate, distinctly concave on one side, pale to dark brown with distinct paler margin; coma white, longer than seed.

About 380 species, mainly in the drier parts of Africa south of the Sahara but extending from the Canary Islands east to Australia with a significant diversification in peninsular India and extensions into temperate India and China; 93 species plus six subspecies in the Flora area.

Molecular data have shown very clearly that *Brachystelma* is polyphyletic with at least 3 very distinct clades falling within widely separated parts of *Ceropegia* and with some morphologically classical species of *Ceropegia* being placed within one of these *Brachystelma* clades: Surveswaran *et al.* in Pl. Syst. Evol. **281**: 51–63 (2009); Bruyns *et al.* in Molec. Phylogen. Evol. **90**: 49–66 (2015); in S. African. J. Bot. **112**: 399–436 (2017). Moreover, there is no way of dividing the two genera morphologically as

---

[20] by M.G. Gilbert

there are exceptions to every possible combination of diagnostic characters. In these circumstances the decision had been taken to sink *Brachystelma* into *Ceropegia*.

These same analyses also show the stapeliads to form a very well defined monophyletic group nested within *Ceropegia*. However, the formal inclusion of this group within *Ceropegia* as proposed by Bruyns *et al.* (2017) is not accepted here. The stapeliads offer a very good example of a situation that makes a strong case for the acceptance of paraphyletic grouping as the alternatives of either following Bruyns *et al.* (2017) and lumping everything into the one genus, or of dividing up *Ceropegia* into what would have to be a very long series of totally new genera, are both impracticable as they would cause a major disruption in the connectivity of data within the group through the introduction of very many unfamiliar and thus uninformative names and thus serve very little useful purpose. The indications so far are that it is likely that such changes will not be accepted by the user community.

This account took as its starting point a set of unpublished preliminary accounts by F. Venter. The genus, as accepted here, is represented in the Flora area by the following sections:

Sect. **Phalaena** H. Huber in Mem. Soc. Brot. **12**: 30 (1957). Type: *C. aristolochioides* Decne.

> Subsect. *Aristolochioides* (H. Huber) Bruyns in S. African J. Bot. **112**: 411 (2017).
> Subsect. *Junceae* (H. Huber) Bruyns in S. African J. Bot. **112**: 411 (2017).
> Ser. *Arabicae* H. Huber in Mem. Soc. Brot. **12**: 30 (1957).

Slender climbers with fusiform or fibrous roots (sometimes arising from swollen tuber-like base of stem). Stems perennial, photosynthetic, usually succulent, usually twining (occasionally erect or sometimes creeping), glabrous. Leaves usually slightly fleshy, deciduous or caducous, ovate or cordate to lanceolate [or as small lanceolate to subulate rudiments], petiolate to sessile, glabrous except sometimes for fine marginal cilia. Inflorescences pedunculate to nearly sessile, usually glabrous, flowers 1–10 on sometimes fleshy pedicels. Corolla with slender tube inflated at base; lobes remaining joined at tips.

Distribution: 26 spp.; Africa, southern Arabian Peninsula (including Socotra), southern India. Species 1–5.

Sect. **Stenatae** Bruyns in S. African J. Bot. **112**: 412 (2017). Type: *C. tenuissifolia* Bruyns.

Slender, erect herbs from cluster of fusiform roots. Stem annual, herbaceous, erect and usually unbranched, glabrous. Leaves herbaceous, deciduous, linear to linear-lanceolate, ± sessile, glabrous. Inflorescences sessile, glabrous, flowers 1–3 on very slender often pendulous pedicels. Corolla with short tube; lobes free at tips and ± rotately spreading, very slender linear to linear-lanceolate.

Distribution: 4 spp.; tropical Africa south of the equator, from Angola to Tanzania. Species 6, 7.

Sect. **Convolvuloides** (H. Huber) Bruyns in S. African J. Bot. **112**: 430 (2017). Type: *C. convolvuloides* A. Rich.

> Ser. *Convolvuloides* H. Huber in Mem. Soc. Brot. **12**: 36 (1957).

Slender climber from cluster of fusiform roots. Stems annual, herbaceous, twining, pubescent. Leaves herbaceous, deciduous, usually cordate, petiolate, usually pubescent. Inflorescences pedunculate, pubescent, flowers 1–8. Corolla with slender tube inflated at base; lobes remaining joined at tips.

Distribution: 5 spp.; South tropical to North-east Africa. Species 8.

Sect. **Pseudoceropegiella** Bruyns in S. African J. Bot. **112**: 430 (2017). Type: *C. purpurascens* K. Schum.

Slender climber from one or more fleshy compressed and disc-like tubers often in vertical sequence. Stems partly deciduous, wiry, twining, glabrous. Leaves delicate and non-succulent, deciduous, ovate to ovate-lanceolate, petiolate, sparsely pubescent to glabrous. Inflorescences pedunculate, usually glabrous, flowers 1–5. Corolla with slender tube inflated at base; lobes remaining joined at tips, < 25 mm long.

Distribution: 1 sp.; tropical southern Africa to East Africa. Species 9.

Sect. **Speciosae** (H. Huber) Bruyns in S. African J. Bot. **112**: 430 (2017). Type: *C. speciosa* H. Huber.

> Ser. *Speciosae* H. Huber in Mem. Soc. Brot. **12**: 35 (1957).

Robust climber from rhizome with fibrous roots. Stems perennial, partly photosynthetic, succulent, twining, glabrous. Leaves herbaceous, deciduous, ovate, petiolate, glabrous. Inflorescences on slender ultimate growth, pedunculate, glabrous, flowers 1–4 on slightly fleshy pedicels. Corolla with slender tube inflated at base and lobes remaining joined at tips.

Distribution: 1 sp.; Malawi, Tanzania and Zambia. Species 10.

Sect. **Carnosae** Bruyns in S. African J. Bot. **112**: 430 (2017). Type: *C. carnosa* E. Mey.

Slender climber from one or more clusters of fusiform roots or with fibrous roots. Stems perennial, partly photosynthetic, wiry to slightly fleshy, twining, glabrous to sparsely pubescent. Leaves herbaceous and sometimes slightly leathery or fleshy, deciduous to persistent, ovate to elliptic, petiolate, glabrous to finely pubescent. Inflorescences pedunculate, glabrous to finely pubescent, flowers 1–6. Corolla with slender tube inflated at base and lobes remaining joined at tips.

Distribution: 6 spp.; South Africa to West and North-East Africa and on Madagascar. Species 11–15.

Sect. **Umbraticolae** (H. Huber) Bruyns in S. African J. Bot. **112**: 430 (2017). Type: *C. umbraticola* K. Schum.

> Ser. *Umbraticolae* H. Huber in Mem. Soc. Brot. **12**: 36 (1957).
> Ser. *Mirabiles* H. Huber in Mem. Soc. Brot. **12**: 36 (1957). Type: *C. mirabilis* H. Huber.

Slender erect herb with cluster of slender to stout fusiform roots. Stems annual, erect and often solitary, pubescent. Leaves herbaceous, deciduous, ovate to narrowly linear, petiolate, usually pubescent. Inflorescences ± sessile, usually pubescent, flowers 1–3. Corolla with tube inflated at base; lobes remaining joined at tips.

Distribution: 8 spp.; southern tropical to East Africa. Species 16–20.

Sect. **Laguncula** H. Huber in Mem. Soc. Brot. **12**: 35 (1957). Type: *C. abyssinica* Decne.

> Ser. *Nigrae* H. Huber in Mem. Soc. Brot. **12**: 36 (1957). Type: *C. nigra* N.E. Br.
> Ser. *Abyssinicae* H. Huber in Mem. Soc. Brot. **12**: 37 (1957). Type: *C. abyssinica.*
> Ser. *Ringentes* H. Huber in Mem. Soc. Brot. **12**: 37 (1957). Type: *C. ringens* A. Rich.
> Sect. *Astropegia* H. Huber in Mem. Soc. Brot. **12**: 38 (1957). Type: *C. furcata* Werderm.

Slender non-succulent geophytic climber or erect herb from small, firm starchy usually compressed and disclike tuber. Stem annual, wiry, twining to erect, sparsely to densely pubescent. Leaves delicate and non-succulent, deciduous, ovate to deltate to linear, petiolate, pubescent to nearly glabrous, entire or margins dentate or variously toothed (often variable in single population and in individuals). Inflorescences ± sessile, usually pubescent, flowers 1–many, sometimes opening ± simultaneously. Corolla with slender tube inflated at base; lobes mostly remaining joined at tips, sometimes free, often almost black adaxially on lobes and in mouth of tube, mostly < 25 mm long.

Distribution: 29 spp.; subtropical South Africa to West and North-east Africa (to Eritrea). Species 21–33.

Sect. **Coreosma** H. Huber in Mem. Soc. Brot. **12**: 31 (1957). Type: *C. stapeliiformis* Haw.

Creeping succulent with fibrous roots at nodes. Stems perennial, photosynthetic, highly succulent (to 15 mm thick), creeping and only rising up and twining to flower, tuberculate, glabrous. Leaves rudimentary and caducous, arising from tubercle, cordate-ovate to deltate, sessile, glabrous. Inflorescences on slender ultimate growth, pedunculate, glabrous, flowers 1–4 on fleshy pedicels. Corolla with slender tube inflated at base; lobes free at tips or joined at tips and alternating with extended sinus lobes.

Distribution: 2 spp.; Mozambique, South Africa and Swaziland. Species 34, 35.

Sect. **Radicantiores** Bruyns in S. African J. Bot. **112**: 431 (2017). Type: *C. radicans* Schltr.

Creeping succulent with clusters of fusiform roots at nodes. Stems perennial, photosynthetic, succulent, creeping and only rising up and twining to flower, glabrous. Leaves fleshy and lasting several years, circular to ovate-lanceolate (sometimes almost obsolete on some stems in *C. arenaria*), petiolate, glabrous. Inflorescences pedunculate, glabrous, flowers 1–4 on fleshy pedicels. Corolla with slender tube inflated at base; lobes remaining joined at tips.

Distribution: 4 spp.; southern Africa in Mozambique, South Africa and Swaziland. Species 36.

Sect. **Callopegia** H. Huber in Mem. Soc. Brot. **12**: 31 (1957). Type: *C. nilotica* Kotschy.
Ser. *Stenanthae* H. Huber in Mem. Soc. Brot. **12**: 32 (1957). Type: *C. stenantha* K. Schum.

Slender succulent climber from cluster of fusiform roots. Stems often deciduous above-ground in dry season and breaking up at nodes when drying out, photosynthetic, succulent and slightly angled in cross-section, usually twining (occasionally erect), glabrous. Leaves slightly to very fleshy, deciduous, ovate to lanceolate or linear, petiolate, glabrous except sometimes for fine marginal cilia, margins sometimes finely toothed. Inflorescences pedunculate (nearly sessile in *C. stenantha*), glabrous, flowers 1–7 on sometimes fleshy pedicels. Corolla with slender tube inflated at base, basal inflation sometimes divided into two by constriction or internal annulus; lobes remaining joined at tips.

Distribution: 8 spp.; South Africa to West and East Africa. Species 37–41.

Sect. **Ceropegiella** H. Huber in Mem. Soc. Brot. **12**: 32 (1957). Type: *C. africana* R. Br.
Ser. *Africanae* H. Huber in Mem. Soc. Brot. **12**: 32 (1957). Type: *C. africana*.
Ser. *Multiflorae* H. Huber in Mem. Soc. Brot. **12**: 32 (1957). Type: *C. multiflora* Bak.

Small succulent climbers or tiny erect to creeping or rhizomatous succulents from one or more fleshy usually compressed and disc-like tubers. Stems often many, annual to perennial, partly photosynthetic, fleshy, clambering to twining, glabrous to pubescent. Leaves succulent, deciduous to persistent, linear to ovate or cordate, petiolate, glabrous to pubescent. Inflorescences pedunculate, glabrous, flowers 1–many, sometimes opening ± simultaneously. Corolla with slender tube inflated at base; lobes remaining joined at tips, b 25 mm long.

Distribution: 19 spp.; Africa, from South Africa to Ethiopia. Species 42–51.

Sect. **Psilopegia** H. Huber in Mem. Soc. Brot. **12**: 34 (1957). Type: *C. zeyheri* Schltr.

Slender creeping, clambering or climbing succulent from cluster of fusiform roots. Stems perennial, photosynthetic, succulent, often only twining to flower, glabrous, sap cloudy. Leaves rudimentary and caducous, lanceolate to subulate, sessile, glabrous (or with minute marginal cilia). Inflorescences usually on more slender ultimate growth, sessile, glabrous, flowers 1–3 on fleshy pedicels. Corolla with slender tube inflated at base; lobes remaining joined at tips.

Distribution: 3 spp.; South Africa with *C. ampliata* extending to southern coastal Kenya and in southern Madagascar. Species 52.

Sect. **Chamaesiphon** H. Huber in Mem. Soc. Brot. **12**: 34 (1957). Type: *C. pygmaea* Schinz.

> *Brachystelma* R. Br. in Bot. Mag. **49**: t.2343 (1822), conserved name. Type: *Brachystelma tuberosum* (Meerb.) R. Br. as 'tuberosa' (*Ceropegia spathulata* (Lindl.) Bruyns).
> Ser. *Campanulatae* H. Huber in Mem. Soc. Brot. **12**: 33 (1957). Type: *C. campanulata* G. Don.

Small erect geophytic herbs arising from firm, starchy usually compressed and disc-like tuber (somewhat turnip-shaped in *C. blepharanthera*). Stem annual, herbaceous, not twining, pubescent to rarely glabrous. Leaves herbaceous, deciduous, very narrowly linear to oblong-lanceolate or ovate, petiolate to nearly sessile, pubescent to rarely glabrous. Inflorescences next to nodes among well-developed leaves, sessile, pubescent to glabrous, flowers 1–3 (rarely in dense simultaneously opening clusters of up to 20). Corolla usually with short (occasionally long) tube (but generally not divided nearly to base) or with slender tube inflated at base; lobes free or remaining joined at tips.

Distribution: 128 spp.; Africa, from South Africa north to West Africa and Eritrea but absent in the Mediterranean region and the Arabian Peninsula. Species 53–93.

The following species have been recorded from within the Flora region but it has not been possible to confirm these records. All of them occur in neighbouring countries and could well turn up so they have been included within the key to species.

Ceropegia arenaria R.A. Dyer in Bothalia **12**: 444 (1978).

Known from dune forests in KwaZulu-Natal. The species was recorded for Mozambique in Venter's unpublished preliminary Flora Zambesiaca account but the collection cited proved to be an undescribed species here described as *C. mucheevensis*. However, it does seem quite likely that *C. arenaria* could extend into southern Mozambique.

Ceropegia burchelliana Bruyns in S. African J. Bot. **112**: 433 (2017). Type: South Africa, Northern Cape Province, "Bechuanaland", near source of Kuruman River, *Burchell* 2498 (K000305395, K000305396, PRE0330560-0 syntypes), not *C. burchellii* (K. Schum.) H. Huber in Mem. Soc. Brot. **12**: 170 (1958).

> *Macropetalum burchellii* Decne. in Candolle, Prodr. **8**: 627 (1844).
> *Macropetalum burchellii* var. *grandiflora* N.E. Br. in Fl. Cap. **4**(1): 799 (1908).
> *Brachystelma burchellii* (Decne.) Peckover in Aloe **33**: 43 (1996), invalid name, without basionym data.
> *Brachystelma burchellii* var. *grandiflorum* (N.E. Br.) Meve in Pl. Syst. Evol. **228**: 103 (2001), invalid name.

This species is listed for Botswana by Setshogo, Prelim. Checkl. Pl. Botswana: 25 (2005) but no Botswana material has been located. The type collection *Burchell* 2498 was recorded from "Bechuanaland", the former name for Botswana but that name originally also included the northern part of Cape Province of South Africa. Burchell never collected within Botswana proper.

Ceropegia elongata (Schltr.) Bruyns in S. African J. Bot. **112**: 433 (2017). Type: South Africa, Cape Province, Somerset East Dist., on the Boschberg, *Schlechter* 2699, (B† holotype).

> *Dichaelia elongata* Schltr. in Bot. Jahrb. Syst. **18**(Beibl. 45): 35 (1894).
> *Brachystelma elongatum* (Schltr.) N.E. Br., Fl. Cap. **4**(1): 862 (1908).

This species is listed for Botswana by Setshogo, Prelim. Checkl. Pl. Botswana: 25 (2005), but no Botswana voucher material has yet been located.

Ceropegia minor (E.A. Bruce) Bruyns in S. Afric. J. Bot. 112: 434 (2017). Type: South Africa, Transvaal, Pietersburg Dist., on the downs, *Murray* in PRE 28365 (PRE holotype).

> *Brachystelma minus* E.A. Bruce in Fl. Pl. Africa **28**: t.1096A (1951).

There is an unsubstantiated internet record of this South African species from Zimbabwe (https://www.pinterest.co.uk/pin/860469072525285680/?lp=true). It is easily identified by the often acaulous rosette of prominently petiolate leaves, looking very like a *Saintpaulia*, and a corolla tube longer than the spreading lobes which are dull purple and prominently densely white setose.

Ceropegia rubella (E. Mey.) Bruyns in S. African J. Bot. **112**: 435 (2017). Types: South Africa, Eastern Cape, Uitenhage, Addo, *Drège a* 2227 (HAL0114334 syntype, K000305406, K000305409, S12-12354); Glenfilling, *Drège b* (HBG502767 syntype, REG000488, TCD0000996, TUB003599).

> *Tenaris rubella* E. Mey., Comm. Pl. Afr. Austr.: 198 (1838).
> *Tenaris rostrata* N.E. Br. in Gard. Chron., n.s. **24**: 39 (1885).
> *Tenaris volkensii* K. Schum. in Engler, Pflanzenw. Ost-Afrikas **C**: 327 (1895).
> *Tenaris simulans* N.E. Br., Fl. Cap. **4**(1): 796 (1908).
> *Brachystelma rubellum* (E. Mey.) Peckover in Aloe **33**: 43 (1996).

This very distinctive species is well represented both to the north and the south of the Flora area and it seems rather likely that it will turn up in the area between.

Ceropegia sandersonii Hook.f. in Bot. Mag. **95**: t.5792 (1869). Type: South Africa, [ex] Natal, iii.1869, *Sanderson* s.n. (K000305608 lectotype), designated here.

Records of this species from Mozambique were based on collections here treated as *C. monteiroae* Hook.f. which Huber (1958: 106) regarded as conspecific. It is found in northeast South Africa not far from the Mozambique border and it could be found within the Flora area.

This species was described from material sent by Mr. Sanderson and cultivated in Kew but there is no mention in the protologue of any preserved material. There is a herbarium sheet with a total of 8 fragmentary collections mounted on it, mostly detached flowers. Two of these are attributed to Sanderson and from Natal, one dated 1874, too late to be potential type material, the other dated '3/69' and thus possibly preserved at the time the illustration was prepared. It seems reasonable to designate this as the lectotype. In an interesting comment when publishing *C. monteiroae*, Joseph Hooker used the name 'C. sandersoniae' when mentioning that both species had been first collected by the wives of his correspondents and were named after them. However, this attempted correction has never been taken up.

Ceropegia vanderystii De Wild. in Bull. Jard. Bot. État. Bruxelles **4**: 393 (1914). Type: D.R. Congo, Kikwit, i.1914, *Vanderyst* 302 (BR holotype).

The unpublished account by F. Venter recorded this species for Zambia but it has not been possible substantiate this. The species is known from Angola and the D.R. Congo and could well occur within the Flora area.

1. Corolla tube shorter than wide, sometimes almost absent; stems never twining 2
– Corolla tube well-defined, longer than wide, often swollen at base; stems erect, spreading or often twining 47
2. Corolla lobes joined at tips to form cage 3
– Corolla lobes free. 11
3. Sepals 8–11 mm long; pedicel 10–30 mm long 4
– Sepals to 6 mm long; pedicel mostly 1–10(15) mm 5
4. Inflorescence apparently terminal, many-flowered; corolla adaxially uniformly reddish brown, tube 5–7 mm deep. . . . . **81.** *barberae*
– Inflorescence lateral, 1–5-flowered; corolla lobes adaxially green or yellow-green contrasting sharply with dark centre of corolla, tube (1.5)2–3 mm deep . . . .
 . . . . **62.** *megasepala*
5. Pedicel 12–15 mm; corolla lobes adaxially blackish-purple, often not joined at tips . . . . **89.** *floribundior*
– Pedicel 1–10 mm; corolla lobes adaxially reddish brown or paler in colour, rarely free at tips. 6
6. Corolla tube well-defined, 5–8 mm deep; sepals glabrous . . . .**79.** *stenifolia*
– Corolla tube very shallow, to 3 mm deep, usually less; sepals hairy 7
7. Leaf blade linear to filiform, 0.6–1 mm wide; corolla lobes adaxially densely white pilose . . . . **59.** *adamsiana*
– Leaf blade wider, 1.5–20 mm wide; corolla lobes adaxially glabrous or adpressed pubescent and ciliate with long, soft, white hairs 8
8. Stem normally unbranched, solitary, erect, 20–70 cm tall; leaf blade 40–90(120) mm long; pedicel 4–10 mm long; corolla lobes adaxially appressed pubescent and ciliate with long, soft, white hairs . . . . **58.** *gracilior*
– Stem branched, spreading to ascending, rarely strictly erect, 4–30 cm tall; leaf blade 5–35(50) mm long; pedicel to 5 mm long; corolla lobes adaxially glabrous 9
9. Sepals and corolla abaxially densely reddish-brown hairy; sepals 2–6 mm long . .
 . . . . **60.** *hirtella*
– Sepals and corolla abaxially subglabrous to white hairy; sepals 1.5–2 mm long. .10
10. Outer corona lobes shallowly bifid, almost entire; inner corona lobes much larger than outer lobes; corolla broadest at base . . . . **61.** *breviflora*
– Outer corona lobes deeply bifid; inner corona lobes shorter than outer lobes; corolla broadest at middle . . . .**57.** *circinata*
11. Inflorescence a lax terminal panicle; corolla uniformly white to pale pink (recorded from both north and south of the Flora area). . . . . *rubella*
– Inflorescence lateral or if terminal, a compact ± globose umbel; corolla various in colour, usually with darker markings, never as above 12
12. Inflorescence an apparently terminal, ± globose, umbel; flowers mostly opening simultaneously; outer corona lobe teeth often each with conspicuous dense tufts of stiff white hairs. 13
– Inflorescence lateral, 1–8(17)-flowered, not globose; flowers mostly opening in succession; outer corona lobe without teeth or teeth glabrous or with scattered inconspicuous hairs. 14

13. Pedicel 25–40 mm long; leaf blade 18–70 mm wide; corolla lobes adaxially glabrous . . . .**92.** *buchananii*
  – Pedicel 5–8(14) mm long; leaf blade 8–13(30) mm wide; corolla lobes adaxially pubescent . . . .**93.** *togoensis*
14. Corolla 4–10 mm diam., lobes < 5(6) mm long. 15
  – Corolla 10–40 mm diam., lobes more than 5 mm long 26
15. Stem often almost absent, leaves all in *Saintpaulia*-like rosette; corolla tube adaxially prominently uniformly white setose . . . . *minor*
  – Stem well-developed, leaves mostly distributed along stem; corolla adaxially glabrous or not as above 16
16. Stems prostrate, to 6 cm long; outer corona urceolate, margin clearly incurved and very shortly 10-toothed. . . . .**73.** *prostrata*
  – Stem erect, often little-branched, (5)6–45 cm high; outer corona divided into separate bifid lobes or cupular with margin erect or minutely inrolled 17
17. Rootstock a fascicle of fusiform roots; stem glabrous; inner corona lobes linear, meeting over gynostegium . . . . **7.** *tenuissifolia*
  – Rootstock a discoid, globose or napiform tuber; stem hispidulous, pubescent or laxly tomentose; inner corona lobes reduced to blunt teeth or cushionlike swellings shorter than anthers 18
18. Corolla adaxially with prominent ring of long pale purple vibratile hairs around corona. . . . .**76.** *incana*
  – Corolla adaxially glabrous, white pilose or with marginal vibratile hairs near tips of lobes 19
19. Pedicel (5)12–50 mm long; outer corona lobes deeply bifid 20
  – Pedicel to 10 mm long; outer corona lobes truncate and contiguous or emarginate, truncate or broadly rounded 21
20. Corolla lobes spreading-incurved, adaxially white pilose; leaf blade 6–10 cm long, abaxially glabrous . . . .**77.** *albipilosa*
  – Corolla lobes reflexed, adaxially glabrous; leaf blade 1–4 cm long, abaxially densely grey puberulous . . . .**70.** *arnotii*
21. Corolla black, usually with long vibratile marginal hairs near tips of lobes. . . . .**69.** *malawiensis*
  – Corolla yellowish or greenish with red mottling or dark maroon, glabrous or minutely puberulous 22
22. Corolla exterior minutely puberulous; outer corona lobes separated at least by narrow notch 23
  – Corolla exterior glabrous; outer corona lobes contiguous. 24
23. Corolla 4–6 mm diam.; outer corona lobes truncate, separated by narrow notches. . . . .**66.** *simplex*
  – Corolla 8–10 mm diam.; outer corona lobes rounded to shallowly emarginate, widely separated. . . . .**68.** *brevipedicellata*
24. Corolla tube slightly constricted at mouth, pure white, lobes narrowly spathulate with incurved tips, dull yellow tinged greenish at base . . . .**83.** *spatuliloba*
  – Corolla tube widest at mouth, interior greenish yellow or cream, rarely white or maroon, lobes triangular with acuminate to acute tips, same colour as tube . . 25
25. Pedicel 5–10 mm long; leaf blade to 30 mm long, apex acuminate, both surfaces usually glabrous . . . .**71.** *vahrmeijeri*
  – Pedicel 2–4 mm long; leaf blade 30–80 mm long, apex acute, both surfaces shortly pubescent . . . .**72.** *cupulata*
26. Rootstock a fascicle of fusiform roots; stem 60–100 cm tall, erect. .**6.** *neoarachnoidea*
  – Rootstock a discoid, globose or napiform tuber; stem to 50 cm tall, erect or spreading 27

27. Stems glabrous except sometimes for a few hairs at nodes 28
 – Stems puberulous or occasionally hispid to tomentose 32
28. Stem repeatedly dichomotously or trichotomously branched from base; corolla
    lobes narrowly linear-lanceolate to spathulate-ligulate, 2.5–5 mm wide, strongly
    incurved, ± bullate-rugose adaxially . . . . **78.** *plocamoides*
 – Stem simple or few-branched, erect; corolla lobes linear to filiform, 0.5–0.8 mm
    wide, straight, reflexed or spreading, smooth 29
29. Corolla lobes strongly reflexed and ± parallel with pedicel; outer corona absent,
    inner corona lobes erect, linear oblong, 1.5–2 mm long . . . . *burchelliana*
 – Corolla lobes spreading; outer corona lobes present, apices shortly 2-toothed or
    notched, inner corona lobes incumbent on anthers 30
30. Inflorescence pedunculate, racemelike with flowers spaced along distinct rachis,
    fascicles subtended by ovate bracts c.1.5 mm long . . . . **56.** *filifolia*
 – Inflorescence sessile, fasciculate, subtended by normal linear leaves 31
31. Leaf blade filiform, 0.5–1.5 mm wide; pedicel 7–15 mm long; corolla lobes 4.5–
    6.5 mm long . . . . **54.** *chlorantha*
 – Leaf blade linear lanceolate, 2–3 mm wide; pedicel 20–30 mm long; corolla lobes
    c.13 mm long . . . .**55.** *bikitaensis*
32. Corolla tube 4–10 mm deep 33
 – Corolla tube up to 3 mm deep 35
33. Leaf blade 15–25 mm wide; cyme 1-flowered; corolla lobes 4–5 mm long, as wide
    as long; corolla tube adaxially with continuous concentric green or brown band
    . . . . **84.** *chlorozona*
 – Leaf blade 3–10 mm wide; cyme 2- or more flowered; corolla lobes 10–35 mm
    long, much longer than wide; corolla tube adaxially brown and/or yellow,
    sometimes with discontinuous concentric markings 32
34. Pedicel 6–10 mm long; corolla lobes basally greenish-yellow spotted purple
    brown, tips dark purple brown . . . .**82.** *rehmannii*
 – Pedicel 12–15 mm long; corolla lobes blackish-purple throughout . . . .
    . . . . **89.** *floribundior*
35. Pedicel 4–30 mm long, flowers often pendent; corolla lobes with prominent
    coloured hairs, finely white pubescent or glabrous 36
 – Pedicel 1–4 mm long, straight, flowers not pendent; corolla lobes glabrous or
    with fine white hairs 41
36. Corolla lobes glabrous . . . .**67.** *discoidea*
 – Corolla lobes variously hairy 37
37. Corolla lobes with versatile purple hairs 38
 – Corolla lobes with fine white hairs 40
38. Pedicel 4–8 mm long, decurved with buds facing downwards; corolla lobes with
    6–10 mm long violet hairs covering half to whole lobe adaxially . . . . . . **74.** *tavalla*
 – Pedicel (6)8–30 mm long, flowers often pendent; corolla lobes with purple
    clavate hairs at base 39
39. Corolla lobes linear, 18–50 × (0.4)1–2 mm wide . . . . **85.** *schultzei*
 – Corolla lobes ovate to lanceolate, 4–6 × 2–4 mm wide . . . .**87.** *schinziata*
40. Pedicel bent just below apex so flower is nodding; corolla lobes incurved . . . . . .
    . . . . **80.** *nutans*
 – Pedicel uniformly curved, flower not nodding; corolla lobes spreading . . . .
    . . . . **58.** *gracilior*
41. Leaf blade 6–17 mm long 42
 – Leaf blade 23–100(120) mm long 43
42. Inflorescence 2- or 3-flowered; corolla exterior puberulent . . . . **75.** *lancasteri*
 – Inflorescence 1-flowered; corolla exterior glabrous . . . . **88.** *neofurcata*

43. Corolla interior uniformly sparsely hairy; inner corona lobes linear. . . . .
. . . . **65.** *daverichardsii*
– Corolla interior glabrous or with hairs at apices only. 44
44. Corolla lobes 3.5–6 mm long; inner corona lobes reduced to protuberances on inner wall of cupular outer corona. 45
– Corolla lobes 18–35 mm long; inner corona lobes equaling or longer than anthers 46
45. Apices of corolla lobes adaxially glabrous; outer corona deeply cupular, c.2 mm high, sometimes divided by slits c.0.5 mm deep . . . . **72.** *cupulata*
– Apices of corolla lobes adaxially hairy; outer corona lobes clearly separated, erect . . . . **86.** *punctifera*
46. Stem usually solitary, erect, branches when present ascending so plant higher than wide; leaves linear to obovate, 4–16 mm wide, finely puberulous to pubescent . .
. . . . **90.** *gracilidens*
– Stem branched at base, branches spreading so plant is wider than high; leaves filiform, c.1 mm wide, glabrous . . . . **91.** *cyperifolia*
47. Leaf blade succulent, margin minutely serrulate; corolla tube with basal chamber usually clearly constricted in middle (division internal in *C. mucheevensis*). . . . 48
– Leaf blade herbaceous or, if succulent, margin entire; corolla tube with basal chamber simple, sometimes poorly defined 51
48. Corolla lobes mostly dark with transverse white band above base, apex of cage broadly acute to rounded, sometimes truncate or emarginate . . . . **39.** *nilotica*
– Corolla lobes with white base and dark tips, apex of cage narrowly acute to attenuate to narrowly beaked 49
49. Leaf base cuneate; corolla 55–60 mm long, basal chamber uniformly cylindrical externally but divided internally by distinct annulus . . . . **41.** *mucheevensis*
– Leaf base rounded to subcordate; corolla 25–55 mm long, basal chamber very clearly externally constricted in middle with wider upper part 50
50. Corolla cage narrowly acute to attenuate, margins with short pale erect hairs. . .
. . . . **40.** *arenarioides*
– Corolla cage with clavate beak, margins with long purple vibratile hairs. . *arenaria*
51. Leaves scalelike, 2.5–3 mm long, shorter than stem is wide; stems very fleshy with leaves on low tubercles, initially trailing, becoming slender and twining when flowering. 52
– Leaves not scalelike, always much longer than stem is wide; stems erect or twining, less often trailing or pendent, mostly herbaceous, if succulent then narrower than leaf blades, never tuberculate 53
52. Sinuses between corolla lobes slightly spreading, much shorter than lobes proper, corolla lobes proper attenuate, as long as corolla tube . . . .
. . . . **34.** *stapeliiformis* subsp. *serpentina*
– Sinuses between corolla lobes produced into conspicuous rotately spreading lobes longer than lobes proper, corolla lobes triangular-acuminate, much shorter than corolla tube . . . . **35.** *cimiciodora*
53. Stems erect or spreading, never twining 54
– Stems twining, spreading, or trailing 68
54. Stems succulent, glabrous, spreading horizontally and often rooting adventitiously; roots fibrous, tuber absent; corolla lobes deeply revolute with keels ± meeting at their bases in the centre . . . . **3.** *aloicola*
– Stems herbaceous or if slightly succulent then hairy, erect, sometimes slightly lianescent; roots fusiform or rootstock a clearly defined, ± globose, tuber; corolla lobes shallowly revolute, keels poorly developed and not meeting below apex . . 55

55. Rootstock a fascicle of fleshy fusiform roots, tuber absent 56
  – Rootstock a globose to discoid tuber 61
56. Corolla 15–35 mm long. 57
  – Corolla 40–90(135) mm long 58
57. Stem glabrous; leaf blade subulate to filiform, 2–10(20) × 2–5 mm, uppermost reduced and almost scalelike . . . . **20.** *illegitima*
  – Stem hairy; leaf blade narrowly lanceolate to ovate, elliptic, elliptic-oblong to lanceolate, 13–80 × 6–46 mm, uppermost not reduced . . . . **12.** *racemosa*
58. Corolla cage with very prominent narrowly cylindrical to clavate beak 20–25(45) mm long, much longer than cage proper . . . . **19.** *mirabilis*
  – Corolla cage without appendage, apex at the most shortly acuminate 59
59. Leaf blade 25–38 mm wide; corolla with basal chamber 10–26 mm wide; exterior of corolla glabrous. . . . . **17.** *umbraticola*
  – Leaf blade 3–20 mm wide; corolla with basal chamber 4–10(15) mm wide; exterior of corolla minutely puberulent. 60
60. Corolla (4.6)8–16 mm wide at mouth; corolla lobes broadly oblong-ovate to obovate, forming cylindrical to obconical cage with terminal canopy (7)14–22 mm wide. . . . . **18.** *filipendula*
  – Corolla 2–3(4) mm wide at mouth; corolla lobes linear to filiform, forming narrowly cylindrical cage about as wide as corolla mouth . . . . **16.** *gilgiana*
61. Corolla lobes free. 62
  – Corolla lobe apices connate to form globose to fusiform cage 64
62. Corolla lobes glabrous for c.2 mm, then densely papillose-puberulent with patch of caducous purplish vibratile hairs near base; plants erect, unbranched, to 55 cm high. . . . . **53.** *robinsonii*
  – Corolla lobes glabrous throughout; plants usually branched, to 15 cm high . . 63
63. Leaf blade adaxially glabrous; corolla flask-shaped, obtusely 5-angled, 10–12 mm long, about as long as corolla lobes . . . . **63.** *mafekingensis*
  – Leaf blade adaxially pubescent; corolla strongly curved near base, subcylindrical, 12–50 mm long, longer than corolla lobes . . . . **64.** *pygmaea*
64. Stem glabrous; upper leaves filiform, c.0.3 mm wide; inflorescence 1-flowered. . . . . **51.** *cataphyllaris*
  – Stem hairy; leaves 0.7–40 mm wide; inflorescence (1)3–9-flowered 65
65. Corolla tube (17)22–35 mm long, lobes glabrous; inner corona lobes pilose at apex . . . . **24.** *bonafouxii*
  – Corolla tube 6–18 mm long, lobes ciliate at least near their bases; inner corona lobes glabrous at apex. 66
66. Pedicel (4)10–17(25) mm long; sepals subulate, hirsute to coarsely pilose; corolla exterior glabrous in Flora area . . . . **21.** *abyssinica*
  – Pedicel 3–10 mm long; sepals linear-lanceolate, pubescent; corolla exterior pubescent. 65
67. Stems 50–90 cm high; leaf base cuneate to rounded, rarely subcordate, margin entire; corolla lobes linear, uniformly green to brown . . . . **25a.** *achtenii* subsp. *achtenii*
  – Stems 2–40 cm long; leaf base cordate to sagittate, rarely narrowly cuneate, margin entire or irregularly toothed; corolla lobes oblong with broad, bright green tips and purplish bases . . . . **30.** *namuliensis*
68. Basal tuber absent, roots fibrous or more often fleshy and ± fusiform, rarely plants rhizomatous 69
  – Rootstock a napiform, globose or discoid tuber, sometimes more than one in a chain. 84

69. Corolla lobes broadly obovate and forming conspicuous, flat to slightly domed, canopy over mouth of corolla. 70
–   Corolla lobes linear to oblong, not forming canopy. 71
70. Margin of canopy down-curved, with conspicuous pendant, clavate, vibratile marginal hairs; canopy obscurely mottled, without cluster of projections on underside . . . . **36.** *monteiroae*
–   Margin of canopy spreading to slightly raised, with spreading white marginal hairs; canopy prominently spotted, with central cluster of 5 dark purple projections on underside . . . . *sandersonii*
71. Corolla cage narrower than mouth of tube, beaked or extended into slender column; stem ± succulent; roots fibrous or fusiform. 72
–   Corolla cage globose, ovoid or fusiform, not beaked; roots usually fleshy, fusiform (fibrous in *C. sankuruensis*) 77
72. Rootstock an elongated rhizome, 3–4 mm thick; corolla with basal chamber 10–13 mm wide, lobes (20)27–38 mm long, abruptly inflexed at base and forming narrow cylindrical beak or coherent into slender column for most of their length . . . . **10.** *speciosa*
–   Rootstock not rhizomatous; corolla with basal chamber 3–8 mm wide, lobes overall 3–24 mm long, if coherent into beak or column, this clavate or with knoblike tip 73
73. Leaf blade linear, 1.5–1.7 mm wide; corolla lobes inflexed at base to form cylindrical cage, narrower than mouth of corolla tube, apex obtuse. . . . .**15.** *chimanimaniensis*
–   Leaf blade elliptic to elliptic-oblong or oblong-lanceolate to ovate-oblong, 8–22(35) mm wide; corolla lobes not inflexed so cage ellipsoid, wider than corolla tube, apex acute to acuminate 74
74. Corolla straight, cage globose; inner corona lobes laterally compressed . . . . **4.** *zambesiaca*
–   Corolla bent through 90° above basal chamber, cage ovoid to distinctly beaked; inner corona lobes linear to narrowly spatulate, dorsiventrally compressed. . . 75
75. Corolla lobes with keels of plicate bases converging more gradually to form a cylindrical or clavate beak, margins not white-ciliate . . . . **1.** *lugardiae*
–   Corolla lobes with keels of plicate bases converging abruptly and meeting in centre of flower, then closely parallel to base of column 76
76. Stems twining; tip of corolla extended into a filiform column which is expanded at apex into globose 5-winged head with white-ciliate margins . . . . . .**2.** *haygarthii*
–   Stems trailing and rooting at nodes; tip of corolla extended into a very short dark beak . . . .**3.** *aloicola*
77. Leaves often scalelike, blade to 10 × 2 mm; inflorescence sessile; corolla tube cylindrical, only slightly widening at mouth . . . .**52.** *ampliata*
–   Leaves not scalelike, blade 10–150 × 3–52 mm; inflorescence pedunculate; corolla tube funnel-shaped, distinctly wider at mouth 78
78. Stem prostrate, mostly unbranched, to 10 cm long; leaf blade broadly ovate, base cordate; corolla lobes abaxially coarsely pubescent . . . . **13.** *cordifolia*
–   Stem twining, rarely trailing, up to 2 m or more long; leaf blade linear to ovate, base rounded to cuneate or attenuate; corolla lobes abaxially glabrous (margin sometimes ciliate) 79
79. Corolla bent at right angles above basal inflation, basal chamber 7–9 mm wide, abruptly narrowed into funnelform limb. 80
–   Corolla straight to gently curved, basal chamber 3–7 mm wide, more gradually narrowed into limb 81

80. Leaf blade herbaceous; corolla cage conical with subacute apex; roots slender, fibrous. . . . . **11.** *sankuruensis*
 – Leaf blade succulent; corolla cage cylindrical to ovoid with rounded to truncate apex; roots fusiform. . . . . **37.** *crassifolia*
81. Corolla tube narrowly cylindrical above basal chamber, abruptly widening into broadly cylindrical upper portion, lobes adaxially pale with maroon veins and lines of spots, margins conspicuously pilose/ciliate . . . .**14.** *carnosa*
 – Corolla tube funnelform above basal chamber, widening uniformly to mouth, lobes adaxially unmarked except for sometimes for darker tip, margins glabrous 82
82. Sepals with purplish tips or spots, glabrous; corolla lobes linear, revolute, forming fusiform to narrowly ellipsoid cage, glabrous; inflorescences often each with several open flowers. . . . .**38.** *stenantha*
 – Sepals uniformly green, pubescent; corolla lobes oblong, folded lengthwise, forming ovoid to cylindrical cage, with hairs on adaxial keel or margin; inflorescences rarely with more than one flower open. 83
83. Corolla lobes stiffly ciliate along adaxial keel, otherwise glabrous; corolla tube and lobes unmarked except for darker tip; base of leaf blade rounded, apex apex acute to obtusely apiculate . . . .**12.** *racemosa*
 – Corolla lobes prominently purplish pilose towards apex; corolla tube and lower part of lobes with darker reticulate markings; base of leaf blade cordate, apex acuminate, mucronate . . . . **5.** *volubilis*
84. Corolla lobes free at apices, not forming cage 85
 – Corolla lobes connate at apices to form cage 88
85. Buds straight; corolla radially symmetrical with rotately spreading lobes 86
 – Buds slightly S-shaped; corolla slightly zygomorphic with oblique mouth and inwardly curved lobes, exterior glabrous. 87
86. Corolla exterior pilose, throat blackish green, lobes adaxially glabrous, apices linear, straight . . . . **22.** *ringoetii*
 – Corolla exterior glabrous, throat pale green, lobes adaxially densely retrorsely white pubescent towards base, apices threadlike, often contorted . . .**44.** *multiflora*
87. Petiole 10–30 mm long; lamina ovate to ovate-oblong to cordate-ovate, 15–35 mm wide; corolla 12–15 mm long, lobes with margins long-ciliate towards base. . . . .
 . . . . *stenoloba*
 – Petiole 2–10 mm long, densely pubescent; leaf blade lanceolate to linear-lanceolate, 2–8 mm wide; corolla 17–30 mm long, lobes glabrous to puberulous, rarely margins sparsely short white setose near base. . . . .**33.** *schliebenii*
88. Leaf blade succulent 89
 – Leaf blade herbaceous 97
89. Corolla lobes with linear base and obovate apices which are joined laterally into an umbrella-like structure. . . . .**45.** *rendallii*
 – Corolla lobes uniformly narrow, joined only at very tip 90
90. Corolla lobes tapering to threadlike apex, basal part densely adpressed white hairy, forming globose to oblate cage with rounded to truncate apex . . . . . **44.** *multiflora*
 – Corolla lobes linear, glabrous or with erect purplish hairs. 91
91. Inflorescence sessile or nearly so; stems twining, occasionally flowering on short ascending stems before leaves expand; corolla cage with acute to acuminate apex 92
 – Inflorescence pedunculate; stems pendent or twining; corolla cage with rounded to subacute apex 93
92. Corolla 25–35 mm long; plants sometimes flowering on short ascending stems before leaves expand . . . .**49.** *conrathii*

–   Corolla 15–16 mm long; plants flowering on leafy twining stems............. .... **50.** *floribunda*

93.   Corolla lobes strongly revolute basally and almost flat apically 94

–   Corolla lobes revolute along entire length 95

94.   Stems and corolla exterior puberulent; corolla lobes adaxially pale with dark red band near base....................................... **42.** *pachystelma*

–   Stems and corolla exterior glabrous; corolla lobes adaxially uniformly pale yellow .... **43.** *inornata* subsp. *zambiana*

95.   Corolla limb almost cylindrical throughout, scarcely wider at mouth, sinuses acute, not raised in bud; cage wider than mouth .... **46.** *linearis*

–   Corolla limb funnel shaped, distinctly widest at mouth, sinuses rounded, slightly raised in bud; cage not wider than mouth........................... 96

96.   Corolla lobes not inflexed at base so cage is as wide as mouth of tube; leaf blade subterete, 1–2 mm wide....................................... **47.** *debilis*

–   Corolla lobes inflexed at base so cage is distinctly beaked and narrower than mouth of tube; leaf blade mostly cordate-ovate to reniform-cordate, sometimes narrower, 4–20 mm wide...................................... **48.** *woodii*

97   Corolla tube flask or bottle shaped, inflated for more than half length 98

–   Corolla tube inflated for less than 1/3 of length with globose to ovoid basal chamber ± abruptly contracted into longer cylindrical to funnelform limb . 102

98.   Corolla lobes linear-revolute for most of length, cage often oblate......... 99

–   Corolla lobes wider, plicate, cage globose to cylindrical................ 100

99.   Corolla 35–60 mm long, tube uniformly coloured, white sometimes suffused purple towards apex .... **23.** *meyeri*

–   Corolla 12–25 mm long, tube interior with well-defined transverse pale band, visible when dried .... **21.** *abyssinica*

100.   Pedicel 3–5 mm long; corolla 10–20 mm long, tube exterior glabrous........ .... **28.** *nephroloba*

–   Pedicel 10–20 mm long; corolla 27–35(40) mm long, tube exterior uniformly 101

101.   Petiole 5–10 mm long; mouth of corolla tube c.4 mm wide...... **24.** *bonafouxii*

–   Petiole 25–55 mm long; mouth of corolla tube 1–2 mm wide (unconfirmed record from W Zambia) .... *vanderystii*

102.   Corolla lobes ± oblong, folded lengthwise, apex abruptly differentiated in colour and texture from lower part.. 103

–   Corolla lobes linear, narrowly revolute, not clearly 2-coloured 107

103.   Corolla sinuses acute, flat; basal chamber 2–4(4.5) mm wide 104

–   Corolla sinuses rounded, auriculate; basal chamber (2.5)4–8 mm wide.... 105

104.   Robust twiner to 2 m or more long; petiole 30–40 mm long ....**29.** *claviloba*

–   Dwarf, often erect or trailing, plant to 40 cm long; petiole 2–10 mm long..... .... **30.** *namuliensis*

105.   Stem internodes with 2 lines of hairs; corolla exterior uniformly densely pubescent; inflorescence 1- or 2-flowered (Zambia)........ **31.** *mwinilungensis*

–   Stem internodes glabrous or uniformly hairy; corolla exterior glabrous or rarely minutely puberulent (Zimbabwe); inflorescence 2–20-flowered 106

106.   Calyx 3–8 mm long; corolla widening to 4–6 mm diam. at mouth; inflorescence up to 20-flowered...................................... **32.** *papillata*

–   Calyx 1.5–3 mm long; corolla widening to 2.5–3.5 mm at mouth; inflorescence 2–6-flowered...................................... **27.** *paricyma*

107.   Inflorescences clearly pedunculate, peduncle 10–35 mm long 108

–   Inflorescences mostly sessile, if pedunculate then peduncle less than 11 mm long.. 109

108. Leaf blade to 12–40 mm wide, base cordate to rounded . . . . **9.** *purpurascens*
  – Leaf blade to 3.5–8 mm wide, base cuneate to attenuate . . . .
  . . . . **43.** *inornata* subsp. *zambiana*
109. Corolla lobes 1.8–2 mm long, margins with very prominent purple cilia to 7 mm
  long; inflorescence 1-flowered; stems to 30 cm long with 1 or 2 lines of hairs
  along internodes . . . . **26.** *leptotes*
  – Corolla lobes 2.5–20 mm long; inflorescence 2–20-flowered; stems uniformly
  hairy or glabrous, rarely occasionally with 1 line of hairs and then stems much
  longer 110
110. Corolla often slightly zygomorphic with oblique mouth; corolla lobes long ciliate
  at base, glabrous distally . . . . *stenoloba*
  – Corolla mouth radially symmetrical; corolla lobes glabrous or uniformly shortly
  hairy 111
111. Petiole 12–40 mm long; base in lower leaf blades deeply cordate; sinuses between
  corolla lobes rounded, gaping . . . . **8.** *meyeri-johannis*
  – Petiole 1.7–8 mm long; base of leaf blade cuneate to rounded, rarely subcordate;
  sinuses between corolla lobes acute, often recurved-auriculate . . . .
  . . . . **25b.** *achtenii* subsp. *adolfi*

Sect. **Phalaena** H. Huber in Mem. Soc. Brot. **12**: 30 (1958).

1. **Ceropegia lugardiae** N.E. Br. in Gard. Chron., ser. 3 **30**: 302 (1901); in F.T.A. **4**(1):
455 (1903). —Werdermann in Bot. Jahrb. Syst. **70**: 220 (1939). —Dyer in Codd
et al., Fl. S. Africa **27**(4): 57 (1980). —Field in Kew Bull. **36**: 443–446 (1981). —
Archer, Kenya *Ceropegia* Scrapb.: 137, fig.27 (1992). —Meve, Ill. Handb. Succ.
Pl. Asclepiadaceae: 88 (2002). —Masinde in F.T.E.A., Apocynaceae (part 2): 269
(2012). Type: Botswana, Kwebe Hills, *Lugard* 262 (K000305466 holotype).

Roots clustered, fleshy, fusiform. Tuber absent. Stem twining, branched, fleshy, up to 5 m long,
1–2.5 mm thick, smooth or rugulose, glabrous. Leaves ± fleshy; petiole (5)7–16(20) mm long;
leaf blade oblong-ovate, less often oblong or ovate, (13)22–58(65) × (4)11–22(35) mm, base
usually cordate, sometimes truncate to rarely rounded, margin entire, minutely ciliolate when
young, apex shallowly retuse to rounded or occasionally acute, always prominently mucronate,
both surfaces glabrous. Inflorescence extra-axillary, peduncle (6)11–45 mm long, glabrous; cyme
umbel-like, 3–9-flowered; flowers opening in succession. Pedicel (4)6–15 mm. Sepals narrowly
lanceolate, (2)3–5.5 × (0.3)0.5–1.1(1.3) mm, glabrous to slightly pubescent. Bud with distinct
beak. Corolla (19)23–45 mm long, bent at right angles above basal inflation; tube 13–26(29)
mm, basal chamber ovoid, often indistinct, 3–8 mm wide, with slight depressions opposite apices
of sepals, limb cylindrical, 1.5–5.5(7) mm wide, widening to 7.5–15.5 mm at mouth, exterior
whitish or yellowish with purple spots, occasionally unmarked, glabrous or pubescent; interior
with basal inflation with raised longitudinal ridges and long hairs at top; sinuses rounded,
sometimes obscurely auriculate; lobes overall 6–24 mm long, triangular base plicate, 4–9(11)
mm long, converging gradually, apices connate to form beaked cage, adaxially pale at base
with purple veins, beak shortly cylindrical to clavate, 1.5–14 mm long, lower part of beak often
cohering into glabrous column, purplish red, glabrous or sparsely ciliate/pilose. Corona cup-
shaped, outer lobes triangular, joined to bases of inner lobes, ascending, narrowly notched/
divided at apex, 1.5–2 mm long, teeth convergent, purple with long white hairs adaxially; inner
lobes converging over gynostegium, 2–3 mm long, reddish, apices recurved. Follicles and seed
not recorded.

  This species has often been treated as part of *Ceropegia distincta* N.E. Br., along with
C. haygarthii Schltr., following Huber (1958). The view is taken here follows that of
Field (1981) in which *C. distincta* is restricted to material from Kenya and Tanzania,
most obviously distinguished by the larger sepals. Two subspecies are recognised for
the Flora area.

Corolla exterior usually glabrous; pedicel (7.5)10–15 mm; beak 9–14 mm long, often
clavate at least in bud, with slender paler base, often ± connate into column 3.5–
7.5 mm long with elliptic to clavate purplish tip 4–10 mm long . . . .
. . . . a) subsp. *lugardiae*
Corolla exterior usually puberulous; pedicel (4)6–10(15) mm; beak shortly cylindrical,
1.5–8 mm long, occasionally slightly clavate with differentiated tip on column to 2
mm long, usually not separating into open cage. . . . . b) subsp. *zimbabweensis*

a) Subsp. **lugardiae**. FIGURE 7.3.**143**.

> *Ceropegia apiculata* Schltr. in Bot. Jahrb. Syst. **51**: 152 (1913). Type: Namibia, Damaraland,
> Bei Aitsas und Otjituo (Omaheke), 19.xii.1908, *Dinter* 703 (SAM0070993-1 lectotype),
> designated here (see below); loc. cit., *Dinter* 703a (B† syntype); bei Epata (Omaheke), 1300
> m, *Seiner* 260 (B† syntype).
>
> *Ceropegia distincta* N.E. Br. subsp. *lugardiae* (N.E. Br.) H. Huber in Mem. Soc. Brot. **12**: 88
> (1958).
>
> *Ceropegia archeri* P.R.O. Bally in sched., invalid name, not *C. archeri* (L.C. Leach) Bruyns
> (2017).

Petiole (5)8–16(20) mm long; leaf blade oblong-ovate, less often oblong or ovate, (13)26–
42(60) × (6)12–21(35) mm, base cordate, rarely rounded. Peduncle (6)17–45 mm long. Pedicel
(7.5)10–15 mm. Sepals (2)3–5.5 × (0.3)0.7–1 mm. Corolla (25)36–45 mm long; tube (16)22–25
mm, basal chamber 4–8 mm wide, limb 2.2–5.5(7) mm wide, widening to 9–15.5 mm at mouth,
exterior usually glabrous (one collection pubescent); lobes overall 15–24 mm long; triangular
base (5)7–9 mm long, beak narrowly clavate, 9–14 mm long, lower part of beak ± cohering into
column 3.5–7.5 mm long, terminal 4–10 mm expanded into fusiform to clavate cage, glabrous
or sparsely ciliate/pilose.

Botswana. N: Ra, Mokgwebana area, near Isessabe Railway Station, 6.i.1974,
Ngoni 248 (K). **Zimbabwe**. N: Mazoe River, 7 km S of Umfatudzi-Mazoe junction, fl.
12.xii.1960, *Leach* 10584 (PRE). W: Hwange ['Wankie'] Game Reserve, 22.ii.1956, *Wild*
4789 (K, SRGH). E: Chimanimani ['Melsetter'] Dist., Hot Springs, 2000 ft, 23.xi.1952,
Chase 4728 (K). **Mozambique**. N: Cabo Delgado, between Pundanhara & Nangada,
156 m, 22.iii.2009, *J.E. & S.M. Burrows* 11292 (K).

Also known from Angola, Namibia, South Africa, Tanzania and Kenya. Forest and
thickets on Kalahari sand, granite outcrops; (156)600–1600 m.

Conservation status: LC, moderately frequently collected from over a wide area.

The only collection seen from Mozambique lacks the distinctive corolla beak that is
characteristic of this species but is otherwise a better match here than with any other
taxon.

Ceropegia verruculosa (R.A. Dyer) D.V. Fields is closely related but the views of Field
(1981) and Bruyns *et al.* (2017) are accepted and it is treated as a distinct species that
has not been recorded from the Flora area. It differs most obviously by the obviously
clavate, uniformly brownish purple, conspicuously white-ciliate corolla beak as well as
the verruculose stems after which it was named.

SAM0070993-1 is selected as the lectotype of *Ceropegia apiculata* on the basis that
it appears to be part of the only extant original material and has better preserved
flowers than the other surviving duplicate, SAM0070993-2.

b) Subsp. **zimbabweensis** M.G. Gilbert, stat. et nom. nov. Type: Zimbabwe, Victoria,
Monro 1452 (BM000528530 holotype). FIGURE 7.3.**143**.

> *Ceropegia distincta* f. *pubescens* H. Huber in Mem. Soc. Brot. **12**: 88 (1958).

Petiole 7–14(17) mm long; leaf blade oblong-lanceolate to oblong-ovate, rarely narrowly
oblong, (19)22–58(65) × (4)11–22(27) mm; base mostly cordate, sometimes truncate to rarely
rounded. Peduncle 11–28(40) mm, Pedicel (4)6–10(15) mm. Sepals 2.5–3.5(4.8) × 0.5–1.1(1.3)
mm. Corolla (19)23–36 mm long; tube 13–26(29) mm, basal chamber 3–6 mm wide, limb 1.5–

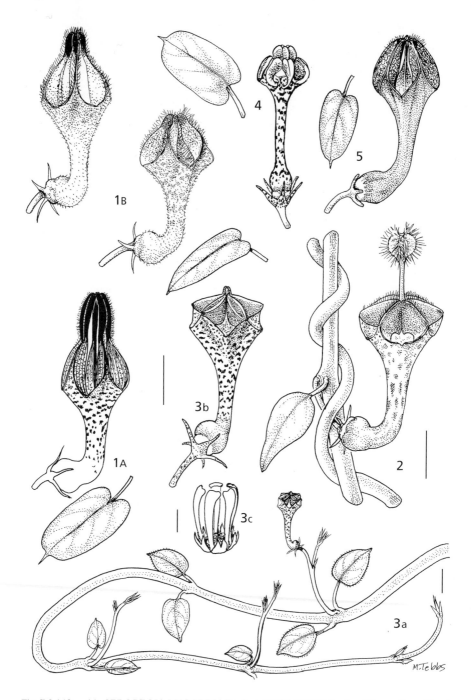

Fig. 7.3.**143**. —**1A**. CEROPEGIA LUGARDIAE subsp. LUGARDIAE, leaf and flower, from *Ngoni* 248. —**1B**. CEROPEGIA LUGARDIAE subsp. ZIMBABWEENSIS, two flower forms, from *Codd* P11 and from image taken by M. Bingham, Kafue, Zambia. —**2**. CEROPEGIA HAYGARTHII, section of stem with leaf and flower, from *Leach* 10729. —**3**. CEROPEGIA ALOICOLA, 3a flowering stem, 3b flower, 3c corona, from *Leach & Baylis* 11955. —**4**. CEROPEGIA ZAMBESIACA, leaf and flower, from *Bingham* s.n. —**5**. CEROPEGIA VOLUBILIS, leaf and flower, from *Leach* 10752. Scale bars = 10 mm, except for 3c = 1 mm. Drawn by Margaret Tebbs.

3.7(4) mm wide, widening to 7.5–16 mm wide at mouth, exterior pubescent, lobes 6–14(18) mm long, triangular base (2.5)4–7.5(11) mm long, beak cylindrical, occasionally narrowed basally and subclavate, 1.5–8 mm long, lower part of beak usually cylindrical, sometimes ± differentiated into short column to 2 mm long, tip usually not separating into open cage, pilose.

Zambia. E: Lusaka Dist., Kafue W, Kasusa R., 980 m, 30.xi.1997, *Bingham & Luwiika* 11501 (K). **Zimbabwe**. N: Mrewa, ix.1954, *Leach* 8197 (K). E: Umtali Dist., Zimunya Reserve, S of Chikuramadzwa River, path to Matekwatekwa Hill, 3500 ft, 12.ii.1961, *Chase* 7433 (BM, K, SRGH). S: Victoria, 3 miles E Zimbabwe, Glen Levit road, 22.x.1956, *Leach* 5909 (K).

Endemic. Most records from around granite outcrops, also along roadsides and in shady places in *Brachystegia* woodland on Kalahari sand; 900–1600 m.

Conservation status: LC/VU: at one stage collected frequently but mostly from a fairly small area and with an apparent drop off in the frequency of collecting.

This species was initially regarded as just a hairy form of *Ceropegia lugardiae* but it differs also in the length of the pedicel and form of the corolla lobes and in its distribution. It has been collected moderately frequently within Zimbabwe but only the one collection has been seen from Zambia, hence the choice of epithet.

2. **Ceropegia haygarthii** Schltr. in Bot. Jahrb. Syst. **38**: 46 (1905). —Brown in Fl. Cap. **4**(1): 813 (1908). —Phillips in Fl. Pl. South Africa **5**: t.191 (1925). —Field in Kew Bull. **36**: 446 (1981). —Meve, Ill. Handb. Succ. Pl. Asclepiadaceae: 81 (2002). Type: South Africa, Natal, sketch by Haygarth sent by J.M. Wood (B† iconotype); Stockenstrom Div., Maasdorp, *Scully* 196 (K neotype), designated by Huber in Mem. Soc. Brot. **12**: 90 (1958), but see note below. FIGURE 7.3.**143**.

    Ceropegia tristis Hutchinson in Fl. Pl. South Africa **2**: t.44 (1922). Type: t.44 of protologue, iconotype.
    Ceropegia distincta N.E. Br. subsp. *haygarthii* (Schltr.) H. Huber in Mem. Soc. Brot. **12**: 90 (1958). —Dyer in Codd *et al.*, Fl. S. Africa **27**(4): 56–57 (1980).

Roots fibrous. Tuber absent. Stem twining, branched, succulent, up to 10 meters long, 3–4 mm diam., glabrous. Leaves ± fleshy; petiole 3–8 mm long; leaf blade ovate-lanceolate, 8–60 × 3–25 mm, base cordate, apex acuminate, both surfaces glabrous. Inflorescence extra-axillary, pedunculate, glabrous; cyme 1- or 2-flowered. Pedicel 10–20 mm long, glabrous. Sepals narrowly lanceolate, 4–10 mm long, apex acuminate, glabrous, somewhat keeled. Corolla 30–50(70) mm long, bent at right angles above basal inflation; basal chamber slightly enlarged, c.4.5 mm wide, limb funnel-shaped, 4.5–6 mm wide, widening to 10–22 mm at mouth; exterior pink or yellowish with darker blotches except at base, glabrous, interior with long hairs in mouth and in basal chamber; lobes triangular 10–30 mm long, plicate, abruptly inflexed with almost square keels ± meeting in centre of tube, apices abruptly connivent into linear column with subglobose tip, abaxially glabrous, adaxially whitish or pale green with fine reticulate reddish lines, glabrous; column dark, glabrous, 5–15 mm long, sometimes absent, terminal head 5–6 mm wide, dark purple with prominently ciliate margins. Corona with outer lobes dentate, spreading, bifid, 1–2 mm long, with long hairs or glabrous, inner lobes converging over gynostegium, linear-spathulate, 2–3 mm long, apices sometimes 2-lobed, recurved, glabrous. Follicles strongly falcate, 12–14 cm long, recurved at apices. Seed not recorded.

Mozambique. M: Goba, fl. 11.iii.1967, *Carvalho* 886 (LMU).

Also in South Africa. Habitat not recorded; below 100 m.

Conservation status: LC in South Africa, only just extending into the Flora area. Very widely cultivated.

Schlechter published his new species based on a sketch made from a living plant by Mr. Haygarth and sent to him by J.M. Wood of Durban. This sketch is assumed to have been destroyed during the destruction of the Berlin herbarium in 1943. Huber (1958) designated "Scully 196" in K as a neotype but it has not been possible to locate this specimen. If this collection cannot be found then an alternative neotype with connections to the original illustration would be: South Africa, Natal, Sydenham, 3–400 ft, 15.iv.1913, *Haygarth* in *Wood* 12125 (K).

3. **Ceropegia aloicola** M.G. Gilbert, sp. nov. The succulent stem and leaves and the distinctive form of the lower parts of the corolla lobes indicate a clear relationship to *C. haygarthii* but this species is easily distinguished by the combination of its trailing habit (strongly twining in *C. haygarthii*) and the very short corolla extension which is reduced to a terminal dark cylindrical beak c.2.5 mm long, 1 mm wide (in contrast to the complex appendage seen in *C. haygarthii* most typically with a long slender column, rarely almost absent, ending in a globose, usually conspicuously ciliate, head). Type: Mozambique, Maputo, Namaacha Falls (cultivated in Nelspruit), *Leach & Bayliss* 11955 (K holotype, PRE, SRGH). FIGURE 7.3.**143**.

Roots fibrous. Tuber absent. Stem spreading, not twining, sometimes rooting adventitiously, branched, succulent, 3–4 mm diam., glabrous. Leaves ± fleshy; petiole 11 mm long; leaf blade ovate-lanceolate, c.19 × 12 mm, base cordate, apex acuminate, both surfaces glabrous. Inflorescence extra-axillary, peduncle c.26 mm, glabrous; cyme 1- or 2-flowered. Pedicel c.10 mm long, glabrous. Sepals narrowly lanceolate, c.4.5 mm long, 0.8 mm wide, apex acuminate, glabrous, somewhat keeled. Corolla c.35 mm long, bent at right angles above basal inflation; tube c.26 mm long; basal chamber slightly enlarged, c.7 × 6 mm wide, limb funnel-shaped, c.6.5 mm wide, widening to c.16 mm at mouth; exterior pale pink with darker spots, glabrous, interior with long hairs in mouth and in basal chamber; lobes triangular c.9 mm long, plicate, abruptly inflexed with somewhat rounded keels ± meeting in mouth of tube, apices abruptly connivent into small dark beak c.2.5 mm long, abaxially glabrous, adaxially pale pink, glabrous. Corona with outer lobes dentate, spreading, bifid, 1–2 mm long, with long hairs or glabrous, inner lobes converging over gynostegium, linear-obtriangular, c.3 mm long, apices truncate, c.0.9 mm wide, glabrous. Follicles and seeds not seen.

Mozambique. M: Naamacha Dist., Lebombos, fl. ii.2018, *Osborne* 1348 (LMA).

Endemic. Trailing over rocks, sometimes from under clumps of *Aloe*, c.400 m.

Conservation status: DD/VU. Known from just 2 collections from a small relatively poorly collected area.

The first collection of this species was treated as an abnormal form of *Ceropegia haygarthii* but the discovery of a second collection from the same area matching exactly both in habit and in corolla form justifies treating it as a distinct taxon.

4. **Ceropegia zambesiaca** Masinde & Meve in Kew Bull. **57**: 205 (2002). Type: Zambia, Putea Hill Farm near Lusaka, seed cultivated in greenhouses at Münster and Bayreuth, Germany, fl. vii.1999–i.2000, *Bingham* s.n. (K000305626 holotype, B 10 0153226, EA, MSUN, UBT). FIGURE 7.3.**143**.

Roots fibrous. Tuber absent. Stem twining, sparsely branched, succulent, to 2 m long, 3–5 mm diam., smooth, glabrous. Leaves ± fleshy; petiole 7–14(17) mm long; leaf blade ovate-oblong to broadly ovate, 28–65 × 12–30 mm, base cordate, margin entire, sometimes minutely ciliolate when young, apex acuminate to acute or rounded, mucronate, both surfaces glabrous. Inflorescence extra-axillary; peduncle 8–15 mm, glabrous; cyme umbel-like, 1–3(5)-flowered, flowers opening in succession. Pedicel 4–5 mm, glabrous. Sepals narrowly lanceolate, 4–5 × 1–1.5 mm glabrous. Corolla 25–32 mm long, straight or nearly so; tube 19–23 mm, basal chamber ovoid, 5–6 mm long, 5–6 mm wide, with slight depressions opposite apices of sepals, limb cylindrical, c.2.5 mm wide widening to 7–9 mm at mouth; exterior pale green to cream with progressively larger purple spots towards mouth, glabrous, interior white at base, upper part of basal chamber and lower part of tube dark purple; retrorsely ciliate on upper part of basal chamber and lower part of tube, then glabrous; sinuses acute, flat; lobes 7–9 mm long, triangular base plicate, 5–6 mm long, converging gradually, apices connate to form broadly beaked cage, adaxially pale yellow or cream with finely reticulate dark patch in middle, beak forming short cylindrical column, 2–3 mm long, margins pilose, tip purplish. Corona sessile, c.3.5 × 4 mm diam., outer corona cupular, lobes 2-toothed/ notched, 1.5–1.7 × 1.7–2 mm wide, deep maroon, margins densely ciliate; inner lobes converging over gynostegium, laterally compressed, 2.5–3 × c.0.4 mm wide near base, with purplish maroon speckling which is denser on thickened apices, apices recurved, bases dorsally with a few, short, whitish translucent hairs. Follicles linear-fusiform, 11–16 × 0.3–0.4 cm wide at middle, Seed dark brown, with narrow, paler margin, 11–12 × 3–4 mm; coma 30–40 mm long.

Zambia. C: cultivated ex Putea Hill Farm near Lusaka. *Bingham* s.n. (K, B, EA, MSUN, UBT).

Endemic around Lusaka. Growing in hedgerow.

Conservation status: VU. Very rarely collected from a relatively well known region.

Apparently still only known in herbaria from the original collection but see www.zambiaflora.com for further images.

5. **Ceropegia volubilis** N.E. Br. in Bull. Misc. Inform. Kew **1895**: 261 (1895). —Huber in Mem. Soc. Brot. **12**: 99 (1958). —Meve, Ill. Handb. Succ. Pl. Asclepiadaceae: 106 (2002). Type: Angola, *Welwitsch* 4272 (BM000930078 lectotype, COI00071476, K000305463, P00109700), lectotypified by Huber (1958), see note below. FIGURE 7.3.**143**.

Ceropegia scandens N.E. Br. in Bull. Misc. Inform. Kew **1895**: 262 (1895). Type: Angola, Cuanzo Norte, near Sange, *Welwitsch* 4273 (BM000930079, BM000930080, K000305462, LISU220341 syntypes).

Ceropegia dewevrei De Wild. in Ann. Mus. Congo Belge, Bot. sér. 5 **1**: 192 (1904). Type: D.R. Congo, Bas Congo, between Tshoa and Tshie, *Dewevre* 199 (BR0000008863249 holotype).

Roots clustered, fleshy, fusiform. Tuber absent. Stem twining, fleshy, glabrous. Leaves ± fleshy; petiole 7–18 mm long; leaf blade cordate-ovate, 20–80 × 12–34 mm, base cordate, margin ciliolate, apex acuminate, mucronate, both surfaces glabrous. Inflorescence extra-axillary; peduncle 12–30 mm long, glabrous; cymes 2–4-flowered, flowers opening in succession. Pedicel 6–10 mm long. Sepals subulate, c.3 mm long. Corolla 17–45 mm long; bent at right angles above basal inflation; tube 12–30 mm long, basal inflation globose, 5–7 mm wide, limb gradually expanding to mouth, exterior whitish speckled purple; glabrous, interior with long hairs in basal inflation then glabrous then with band of long hairs just below mouth; lobes deltoid-oblong, 5–10 mm long, plicate, converging gradually, apices connate to form beaked cage, adaxially densely purplish pilose towards apex. Corona sessile; outer corona lobes deeply bifid, ciliate; inner lobes converging over gynostegium, linear, glabrous. Follicles and seed not recorded.

Botswana. N: near Namibia border fence, 20°11'S 21°00'E, 25.iii.1980, *P.A. Smith* 3334 (K, PSUB). **Zimbabwe**. W: Hwange (Wankie), fl. 13.i.1953, *Levy* 1000 (PRE). C: Headlands, fl. xii, *Eyles* 3486 (K, SRGH). **Malawi**. N: Rumphi Gorge, x.1967, *Williamson & Simon* 963 (K).

Also in Angola, D.R. Congo. On *Euphorbia matabelensis, Acacia, Boscia albitrunca*, etc. in mixed deciduous scrub woodland on sand; 700–1500 m.

Conservation status: probably LC: widely but sparsely distributed.

Huber (1958) stated that the sheet of *Welwitsch* 4272 in BM was the holotype of *Ceropegia volubilis* but there is nothing on this sheet to indicate that this, and not the duplicate in K, was the designated holotype and it seems safer to treat Huber's choice as an effective lectotypification.

Sect. **Stenatae** Bruyns in S. African J. Bot. **112**: 412 (2017).

6. **Ceropegia neoarachnoidea** Bruyns in S. African J. Bot. **112**: 412 (2017), new name, non *Ceropegia arachnoidea* (P.R.O. Bally) Bruyns (2017). Type: Tanzania, Sumbawanga, Tatanda Mission, *Bidgood, Mbago & Vollesen* 2474 (K000197087 holotype, DSM, NHT). FIGURE 7.3.**150**.

Brachystelma arachnoideum Masinde in Kew Bull. **62**: 76, 78, fig.15 (2007); in F.T.E.A., Apocynaceae (part 2): 269 (2012).

Roots clustered, fleshy, fusiform, 3–4 mm thick. Tuber absent. Stem solitary, erect, unbranched, herbaceous, 60–100 cm high, internodes (1.7)12–30 cm long, 1–1.5(2) mm thick, mostly leafless or with 1–2 pairs of leaves when in flower, glabrous. Leaves far apart, herbaceous, spreading, upper leaves progressively smaller, sessile or subsessile; leaf blade linear-lanceolate, 40–120 × 1–5 mm, base rounded, margin entire, very narrowly revolute, apex acute, both surfaces glabrous. Inflorescences from uppermost nodes, extra-axillary, sessile, cyme umbel-like, 1–4-flowered, flowers often opening simultaneously. Pedicel 10–22 mm long, c.0.2 mm thick, wiry, curved downwards at anthesis, purplish, glabrous. Sepals linear-lanceolate or lanceolate-attenuate, 2–3 × 0.3–0.4 mm at the base, abaxially pubescent or ± glabrous. Corolla c.40 mm in diam., divided almost to base, tube 0.7–1 mm deep, 1–1.5 mm wide, exterior similar in colour to interior but paler, interior dark purple, with a dense cover of purple/maroon hairs; lobes rotate, gradually tapering into long filiform spreading revolute lobes, 20–27 × 1.5–2 mm wide at base, bases triangular, revolute, apices free, slightly incurved, abaxially glabrous, adaxially green at base, then green or dark-purple, covered with purple hairs (0.5–1 mm long). Corona subsessile, globose, 1–1.5 × 1–2 mm in diam., purple, brown or pale yellow, glabrous throughout; outer lobes cupular, only slightly raised from bases of inner lobes thus quite low, very shallowly emarginate to form a wave or tiny V-shaped sinus, teeth absent or deltoid and minute to c.0.12 mm long, glabrous; inner lobes incumbent on anthers, oblong, 0.2–0.4 mm long, c.0.2 mm broad, apices obtuse, glabrous. Follicles and seed not recorded.

Zambia. S: Mapanza, Choma fl. 27.xii.1958, *Robinson* 2936 (K). **Malawi**. C: Kasungu, Chipala Hill, 1000 m, fl. 14.i.1959, *Robson* 1176 (K). **Mozambique**. N: Niassa Marupa, estrada para Lichinga, fl. 17.i.1981, *Nuvunga* 549 (K).

Also in Tanzania. Near rocks in *Brachystegia* woodland; 900–1200 m.

Conservation status: LC? Sparsely collected over quite a large area but probably often overlooked.

Molecular data (Bruyns *et al.* in S. African J. Bot. **112**: 412, 2017) places this and the next species as members of one of the earlier diverging groups within *Ceropegia* and far removed from other groups of former species of *Brachystelma* emphasizing the unsustainable polyphyletic nature of that genus.

7. **Ceropegia tenuissifolia** Bruyns, S. African. J. Bot. **112**: 412 (2017), new name, non *Ceropegia tenuissima* S. Moore (1905). Type: Tanzania, Mpwapwa, Kiboriana Hills, *Bruyns* 9641 (BOL150387 holotype).

Brachystelma tenuissimum Bruyns in Novon **19**: 19, fig.2 (2009).

Roots clustered, fleshy, fusiform. Tuber absent. Stem solitary, erect, unbranched, herbaceous, 15–40 cm high, 1–2 mm thick, glabrous. Leaves herbaceous, upper leaves progressively smaller, sessile; leaf blade filiform, 20–65 × 2–4 mm, slightly folded upwards, margin entire, apex acute, both surfaces glabrous or subglabrous. Inflorescences 1–5 from upper nodes, extra-axillary, peduncle absent or to 3 mm long; cyme fasciculate, 3–6-flowered, flowers opening in succession. Pedicel 12–17 mm long, decurved, sparsely puberulous. Sepals narrowly lanceolate, c.1.5 × 0.5 mm, adpressed to corolla, sparsely puberulous. Bud narrowly ovoid, with sinuses flat, acute. Corolla 5–6 mm long, 8–10 mm diam., divided almost to base; tube vestigial; lobes ascending to slightly spreading, linear-triangular, 5–6 × 0.8–1 mm wide at base, tapering from base, revolute, apices free, slightly incurved, abaxially green, glabrous, margins finely purple pubescent, hairs to 0.5 mm long, adaxially purple, glabrous. Corona very shortly stipitate, c.1.8 mm deep, 2.25 mm in diam., outer lobes almost free, deeply bifid, horns widely spreading, linear, c.1 mm long, glabrous; inner lobes incumbent on anthers, linear, c.1 mm long, meeting in centre, apices obtuse, glabrous. Follicles and seed not recorded.

Zambia. N: Near Kalombo Falls, 1200 m, fl. 6.xii.2003, *Bruyns* 9603 (BOL).

Also in Tanzania. *Brachystegia* woodland, on sloping stony ground under trees; c.1200 m.

Conservation status: DD, very easily overlooked and likely more common than the very few collections indicate.

Sect. **Convolvuloides** (H. Huber) Bruyns in S. African J. Bot. **112**: 430 (2017).

8. **Ceropegia meyeri-johannis** Engl. in Abh. Könjgl. Akad. Wiss. Berlin **1891**: 343 (1892). —Schumann in Engler & Prantl, Nat. Pflanzenfam. **4**(2): 272, fig.C (1895). —Brown in F.T.A. **4**(1): 449 (1903). —Werdermann in Bot. Jahrb. Syst. **70**: 222 (1939). —Bullock in Kew Bull. **9**: 590 (1955). —Huber in Mem. Soc. Brot. **12**: 154 (1958). —Bally in Fl. Pl. Africa **35**: t.1371 (1962). —Archer in Upland Kenya Wild. Fl.: 391 (1974); in Kenya *Ceropegia* Scrapb.: 53, fig.11 (1992). —Masinde in F.T.E.A., Apocynaceae (part 2): 231 (2012). —Meve, Ill. Handb. Succ. Pl. Asclepiadaceae: 88 (2002). Type: Tanzania, steppes between Samburi, Moshi and Marangu, *Meyer* 196 (B† holotype); Kilimanjaro, 6000 ft, *Johnson* s.n. (BM000528529 neotype, K000305483), neotype designated by Huber in Mem. Soc. Brot. **12**: 155 (1958). FIGURE 7.3.**144**.

Ceropegia calcarata N.E. Br. in F.T.A **4**(1): 453 (1903). Type: Malawi ['British Central Africa'], Zomba, cult. K, 26.ix.1899, *Mahon* s.n. (K000305635 holotype).

Ceropegia verdickii De Wild. in Ann. Mus. Congo Belge, Bot. sér. 4: 109 (1903); in Contrib. Fl. Katanga, Suppl. **1**: 73 (1927). Type: D.R. Congo, Lukafu, *Verdick* 389 (BR0000008861337 holotype).

Ceropegia angiensis De Wild., Pl. Bequaert. **4**: 358 (1928). Type: D.R. Congo, Angi, *Bequaert* 5781 (BR0000008861320, BR0000008861351 'holotype').

Ceropegia criniticaulis Werderm. in Bull. Jard. Bot. État. Bruxelles **15**: 232 (1938). Type: D.R. Congo, Virunga, Nyamlagira, *Lebrun* 1873 (BR0000008861344 holotype).

Ceropegia meyeri-johannis var. *angiensis* (De Wild.) H. Huber in Mem. Soc. Brot. **12**: 155 (1958).

Ceropegia meyeri-johannis var. *verdickii* (De Wild.) H. Huber in Mem. Soc. Brot. **12**: 155 (1958).

Ceropegia dubia R.A. Dyer in Fl. S. Africa **27**(4): 64 (1980). Type: South Africa, Cape, between Coega and Uitenhage, *Bayliss* 7280 (PRE0659383-0 holotype; NBG0117470-1, GRA0002438-1–3, MO-166308, MO-166307, MO-166311, NBG0114020-0, NBG0117470-2).

Roots clustered, fleshy, fusiform. Tuber absent. Stem twining, herbaceous, up to 2 m long, 1 mm diam., pubescent. Leaves herbaceous; petiole 12–40 mm long, pilose; leaf blade ovate, 30–60 × 15–35 mm, base in lower leaf blades deeply cordate, shallower in upper leaves, margin entire, apex acute to acuminate, both surfaces thinly pubescent to scabrid. Inflorescence extra-axillary, sessile to pedunculate; cyme umbel-like, many-flowered, flowers opening in succession. Pedicel 4–12 mm long, glabrous to pilose. Sepals linear-lanceolate to subulate, 3–5 × 0.5–1 mm, apex acute, glabrous to sparsely pilose. Corolla 18–25 mm long; slightly curved immediately above inflated base; tube 8–18 mm long, basal chamber pear-shaped, 4–5 × 5–7 mm wide, abruptly narrowed apically, limb funnel-shaped, 1–1.5 mm wide at base, widening to 4–6 mm diam. at mouth, exterior whitish to dull purple, glabrous, interior dark purple to dark green in lower part, above and at mouth white veined dark green, with many longitudinal tuberculate ribs, mouth thinly pubescent; sinuses rounded, spreading and gaping, margins flat; lobes linear from short triangular base, 3–6 mm long, straight, replicate, apices connate to form cylindrical to ovoid cage narrower than mouth, apex ± rounded, abaxially glabrous, margins and keel densely deflexed pilose, adaxially dark green or greenish-yellow with brown apices. Corona stipitate; 2–3.5 × 2–3.5 mm diam.; outer corona lobes oblong, erect, deeply bifid, sometimes with central tooth, to 2 mm long, white or cream, glabrous; inner lobes converging over gynostegium, linear, overtopping outer lobes, white with recurved red apices, glabrous. Follicles narrowly divergent, linear-fusiform, (8.5)10–13 × 0.3–0.4 cm wide at middle. Seed black throughout, c.12 × 2 mm; coma c.24 mm long.

Zambia. W: Mwinilunga just S of Matonchi Farm, 13.ii.1938, *Milne-Redhead* 3943A (K). N: Kasama Dist., Chibutubutu woodland, close to Lukulu River, 1320 m, 25.ii.1960, *Richards* 12589 (K). S: 13 km NE from road junction, 26.1 km from Livingstone on

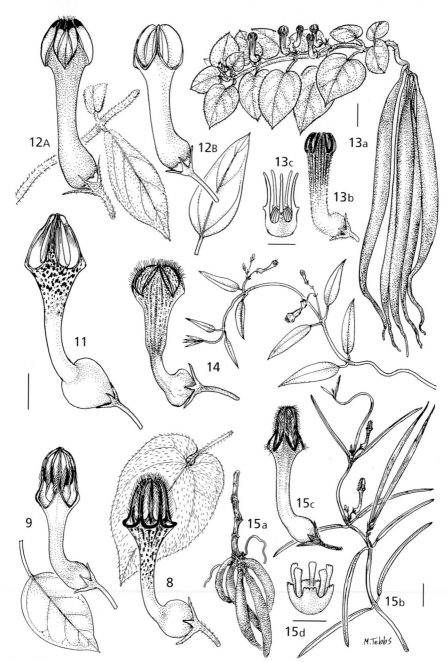

Fig. 7.3.**144**. —**8**. CEROPEGIA MEYERI-JOHANNIS, leaf and flower, from *Mutimushi* 588. —**9**. CEROPEGIA PURPURASCENS, leaf and flower, from *Brummitt et al.* 14237. —**11**. CEROPEGIA SANKURUENSIS, flower, drawn from image taken by Ulrich Meve. —**12A**. CEROPEGIA RACEMOSA subsp. SECAMONOIDES, stem with leaves and flower, from *Richards* 12609. —**12B**. CEROPEGIA RACEMOSA subsp. GLABRA, leaf and flower, drawn from image by George Schatz. —**13**. CEROPEGIA CORDIFOLIA, 13a habit, from *Wild* 3912, 13b flower and 13c corona, from *Goyder & Paton* 4102. —**14**. CEROPEGIA CARNOSA, flowering stem and flower, from *Goyder* 5027. —**15**. CEROPEGIA CHIMANIMANIENSIS, 15a rootstock from *Ballings & Wursten* 2289, 15b flowering stem, 15c flower and 15d corona, from *Osborn* 1190. Scale bars: habits = 10 mm; flowers and leaves = 5 mm; 13c, 15d =1 mm. Drawn by Margaret Tebbs.

road to Lusaka, 1140 m, 23.ii.1997, *Zimba et al.* 966 (K). **Zimbabwe**. N: Viladale Farm, Mazowe. 20.vii.1983, *Best* 1959 (SRGH). C: Enterprise, Harare, 9.iii.1946, *Greatrex* 923 (SRGH). E: Inyanga Dist., Inyanga, banks of Inyangombi River below Rhodes Hotel, *Chase* 600 (BM). **Malawi**. C: Lilongwe Dist., Mkhoma, 2.iv.1958, *Reynecke* P6129 (K). E: Umwumwumwu River Gorge, 19.ii.1964, *Chase* 8129 (K).

Also in Kenya, Tanzania, Uganda, D.R. Congo. Mixed evergreen forest margins, riverine woodland; 1100–1350 m.

Conservation status: LC, widely distributed and quite frequently collected.

Sect. **Pseudoceropegiella** Bruyns in S. African J. Bot. **112**: 430 (2017).

9. **Ceropegia purpurascens** K. Schum. in Bot. Jahrb. Syst. **17**: 152 (1893). —Brown in F.T.A. **4**(1): 450 (1903). —Huber in Mem. Soc. Brot. **12**: 110 (1958). — Malaisse in Bull. Jard. Bot. Natl. Belg. **54**: 222 (1984). —Malaisse & Schaijes in Asklepios **58**: 28 fig.4,15 (1993). —Meve, Ill. Handb. Succ. Pl. Asclepiadaceae: 96 (2002). —Masinde in F.T.E.A., Apocynaceae (part 2): 248 (2012). Type: Angola, Cuanza Norte by Pungo Andongo, 1879, *von Mechow* 122 (Z000001600 holotype). FIGURE 7.3.**144**.

> *Ceropegia kwebensis* N.E. Br. in F.T.A. **4**(1): 456 (1903). Type: Botswana, Kwebe Hills, *Lugard* 116 (K000305460 holotype).
> *Ceropegia kaessneri* S. Moore in J. Bot. **48**: 256 (1910). Type: D.R. Congo, Kitimbo, *Kassner* 2349 (K000305533 holotype, BM000930073, BR0000008863041, E00193246, P00109667).
> *Ceropegia thysanotos* Werderm. in Bot. Jahrb. Syst. **70**: 225 (1939). Type: Tanzania, Morogoro, Uluguru Mts., *Schlieben* 3818 (B† holotype, M0110219 lectotype, BM000930072, BR0000008863089, G, P00109668, Z), lectotype designated by Huber in Mem. Soc. Brot. **12**: 110 (1958).
> *Ceropegia purpurascens* subsp. *thysanotos* (Werderm.) H. Huber in Mem. Soc. Brot. **12**: 112 (1958).

Roots fibrous. Tuber depressed globose, to 6 cm diam., sometimes in chains along rhizome. Stem twining, sparsely branched, herbaceous, up to 3 meters long, internodes to 13 cm long, 1–2 mm diam., glabrous to puberulous. Leaves herbaceous; petiole 5–30 mm long, puberulous; leaf blade oblong to ovate-oblong, (17)25–70 × 12–40 mm, base cordate to rounded, margin entire, ciliolate, apex acute to acuminate, apiculate, both surfaces glabrous to puberulous. Inflorescence extra-axillary, peduncle descending, 10–35 mm long, glabrous to thinly pubescent; cyme umbel-like, 3–10(16)-flowered, flowers opening in succession. Pedicel 4–15 mm long. Sepals subulate-lanceolate, purplish, 1–2 × c.0.5 mm, apex acuminate, glabrous. Bud with raised sinuses and narrowly clavate beak. Corolla 18–22(25) mm long; straight or bent at right angle above basal chamber; tube 8–18 mm long; basal chamber 4–5 × 2–5 mm wide, limb cylindrical above, 1.5–2 mm diam., widening to 6–7 mm at mouth, exterior greenish to purplish white, glabrous or pubescent, interior purplish below mouth, sparsely purplish pilose in throat; sinuses rounded-acuminate, spreading and gaping, raised in bud, not or very slightly auriculate; lobes linear from short triangular base, 6–20 × c.0.5 mm wide, revolute, apices connate to form slightly beaked cage, slightly narrower than mouth, apex ± rounded to subacute, abaxially similar to tube, adaxially throat and lobes pale yellow except for dull purple mark above triangular base, long, purple or white ciliate. Corona stipitate; outer corona lobes shallowly cupular, broadly bifid, c.1 mm high, teeth c.0.5 mm long, diverging, sparsely ciliolate; inner lobes linear-spathulate to falcate, c.2 mm long, much exceeding staminal column, laterally compressed, apices obtuse, recurved, glabrous. Pollinia ellipsoid, bright orange. Follicles obtusely diverging to slightly reflexed, linear-fusiform, 10–15 × 0.2–0.3 cm wide at middle, glabrous. Seed not recorded.

Caprivi. c.20 mi from Singalamwe on WNLA road to Katima Mulilo, 3300 ft, fl. 3.i.1959, *Killick & Leistner* 3267 (K). **Botswana**. N: Kwebe Hills, fl. 19.i.1898, *Lugard* s.n. (K). **Zambia**. W: Mwinilunga, R. Luao - Matonchi Junction fl. 27.xii.1937, *Milne-Redhead* 3834 (K). B: Barotse Prov. Sesheke distr, Masese Forest Station, 1050 m, fl. 2.ii.1975, *Brummitt et al.* 14237 (K). **Zimbabwe**. N: Mensa Pan, 11 mi ESE of Chirundu

Bridge, fl. 3.ii.1958, *Drummond* 5449 (K). **Malawi**. C: Lilongwe Nature Sanctuary Zone 4, fl. 29.i.1985, *Patel & Banda* 1994 (K). N: Chitipa Misulu Hills, Mughese Rain forest, 1750 m, fl. 8.iv.1969, *Pawek* 2012 (K). **Mozambique**. MS: Guro, just SW of town fl. 7.i.1999, *Bruyns* 7761 (K).

Also known from Namibia and Tanzania. Rock outcrops, *Brachystegia* woodland, forest and thicket fringes, often in relatively damp shady situations; 750–1450(1750) m.

Conservation status: LC, quite frequently collected over a very wide area.

Sect. **Speciosae** (H. Huber) Bruyns in S. African J. Bot. **112**: 430 (2017).

10. **Ceropegia speciosa** H. Huber in Mem. Soc. Brot. **12**: 144 (1958). —Meve, Ill. Handb. Succ. Pl. Asclepiadaceae: 101 (2002). —Masinde in F.T.E.A., Apocynaceae (part 2): 273 (2012). Type: Tanzania, Mpwapwa, Kiboriana Hills, *Burtt* 4632 (K000305486 holotype, EA000001881, K000305487). FIGURE 7.3.**145**.

Roots fibrous. Tuber absent, rootstock a horizontal rhizome, 3–4 mm thick. Stem twining or sometimes trailing, branching, herbaceous, to 4 m long, 1.5–5 mm thick, sometimes reaching 15 mm thick at base, glaucous green, glabrous. Leaves herbaceous; petiole 10–35 mm long, glabrous; leaf blade ovate to elliptical, 60–110 × (12)16–52 mm, base cuneate, apex acuminate, both surfaces glabrous. Inflorescence extra-axillary, peduncle descending, 30–40 mm long, slender; cyme 1- or 2-flowered. Pedicel 17–38 mm long, slender, descending, glabrous. Sepals linear-triangular, 4.5–6(11) × c.1 mm wide, apex straight or recurved, glabrous. Corolla (44)66–82 mm long, strongly curved above basal chamber so upper part is vertical; tube to 40 mm long, basal chamber obovoid to ellipsoid, 19–23 × 10–13 mm wide, narrowing to neck, limb obconical, 1.7–3 mm wide at base, widening to 13–16 mm at mouth, exterior greenish white, glabrous, interior cream with purple lines on upper part of basal chamber and lower part of limb, glabrous; sinuses rounded, spreading and gaping, not revolute, lobes linear, (20)27–38 mm long, abruptly converging at base, revolute, apices connate to form narrow cage, 1.5–2.5 mm wide, much narrower than mouth, often coherent into column, both surfaces glabrous, sometimes very minutely papillose at apex, margins with very prominent purple cilia to 7 mm long, adaxially purplish to dull red. Corona stipitate, outer corona lobes narrowly triangular, erect, bifid, teeth parallel to slightly convergent, slightly shorter than inner lobes, adaxially densely ciliate; inner lobes erect, linear, laterally compressed, cream, apices densely minutely hairy. Immature follicles widely divergent, linear, to 19 × 0.4 cm. Seed not recorded.

Zambia. C: Serenje Dist., Kundalila Falls, 53 km ENE Serenje, fl. 4.ii.1973, *Strid* 2820 (K). **Malawi**. S: Mulanje Dist., Mt. Mchese Forest Reserve, W slopes above Ulolo, 1000 m, fl. 14.ii.1992, *Goyder & Paton* 3649 (K), Mt. Mulanje, middle western slopes above Likhubala, 1000 m, fl. 18.ii.1982, *Hepper* 7365 (K).

Also known from Tanzania. Riverine forest and along gulley margins, less often *Brachystegia* woodland on sandy soil. 1000–1750 m.

Conservation status: LC, rare but fairly widespread.

Sect. **Carnosae** Bruyns in S. African J. Bot. **112**: 430 (2017).

11. **Ceropegia sankuruensis** Schltr. in Bot. Jahrb. Syst. **51**: 155 (1913) (as "sankurnensis"). —De Wilde. in Bull. Jard. Bot. État. Bruxelles **7**: 29 (1920). —Werdermann in Bot. Jahrb. Syst. **70**: 233 (1939). —Huber in Mem. Soc. Brot. **12**: 79 (1958). —Bullock in F.W.T.A. ed. 2 **2**: 102 (1963). —Archer, Kenya *Ceropegia* Scrapb.: 155, fig.30 (1992); in Upland Kenya Wild. Fl. ed. 2: 183 (1994). —Masinde in F.T.E.A., Apocynaceae (part 2): 271 (2012). —Gilbert in Hedberg & Edwards, Fl. Ethiopia Eritrea **4**(1): 165 fig.140.37.16 & 17 (2003). Type: D.R. Congo, Sankuru, *Ledermann* 59 (B† holotype); same locality, *Luja* s.n. (BR0000009826281 neotype) designated by Huber in Mem. Soc. Brot. **12**: 80 (1958). FIGURE 7.3.**144**.

Roots fibrous. Tuber absent. Stem twining, sparsely branched, filiform, to 2(4) m long, flexuose, laxly leaved, glabrous. Leaves herbaceous; petiole 8–40 mm long, slender; leaf blade obovate-elliptic, 40–100 × 12–45 mm, base cuneate to rounded, margin entire, apex acuminate, both surfaces glabrous (or very sparsely pilose). Inflorescence extra-axillary, peduncle as long as petiole, glabrous; cyme umbel-like, 3–6-flowered, flowers opening in succession. Pedicel 3–10(25) mm long, filiform, minutely puberulous. Sepals lanceolate, 2–3(4) × 0.5–1 mm, apex acuminate, puberulous. Corolla 25–30 mm long, bent at right angles above basal inflation; tube 19–20 mm long; basal chamber globose to obovoid, 7–8 mm wide, abruptly narrowed into funnelform limb, 1.8–2.5 mm wide near base, widening to 8–15 mm at mouth, exterior white, often with darker longitudinal lines on basal chamber and purplish spots or blotches towards mouth, glabrous; sinuses acute, slightly raised in bud; lobes lanceolate, elongate-ovate to linear, 6–10 mm long, plicate, apices coherent to form conical, slightly beaked, cage, apex acute, abaxially shortly ciliate, margins undulate, sometimes purplish ciliate, adaxially dark purple at base, then paler and green, yellow or brown at apex. Corona sessile, c.2.5 mm high, outer corona lobes joined laterally, erect deeply bifid, c.2 mm long, higher than inner lobes, teeth triangular, with a few long, stiff, hairs; inner lobes incumbent over anthers, then erect and cohering into column, linear, c.2 mm long, much exceeding staminal column, glabrous. Pollinia ovoid. Follicles and seed not recorded.

Zimbabwe. E: 25 km north of Mutare, fl. 1.iii.1970, *Plowes* 3425 (SRGH); Mutare, Himalayas, 6000 ft, fl. 11.xii.1954, *Wild* 4640 (K, SRGH).

Ethiopia, Kenya, Uganda, Tanzania, Sierra Leone, Liberia, Nigeria, Cameroon, D.R. Congo. Forest margins; 1800 m.

Conservation status: LC. A widely distributed West African species only just extending into Zimbabwe.

12. **Ceropegia racemosa** N.E. Br. in Bull. Misc. Inform. Kew **1895**: 262 (1895); in F.T.A. **4**(1): 456 (1903). —Werdermann in Bot. Jahrb. Syst. **70**: 207 (1939). —Bullock in Kew Bull. **9**: 591 (1955); in F.W.T.A. ed. 2 **2**: 102 (1963). —Huber in Mem. Soc. Brot. **12**: 93–97 (1958). —Archer, Kenya *Ceropegia* Scrapb.: 43, IX (1992); in Upland Kenya Wild. Fl. ed. 2: 184 (1974). —Dyer in Codd *et al.*, Fl. S. Africa **27**(4): 63–64 (1980). —Gilbert in Hedberg & Edwards, Fl. Ethiopia Eritrea **4**(1): 166 fig.140.36.1–3 (2003); in Fl. Somalia **3**: 173 (2006). —Masinde in F.T.E.A., Apocynaceae (part 2): 261 (2012). Type: South Sudan, Bahr el Ghazal, Seriba Jur Ghattas, *Schweinfurth* 2105 (K000305524 holotype).

Roots clustered, fleshy, fusiform. Tuber absent. Stem usually twining, occasionally erect, slightly fleshy, up to 3 meters long, glabrous to uniformly sparsely puberulent. Leaves ± fleshy; petiole 3–20 mm long, hairy; leaf blade green to brownish-green, narrowly oblong-oblanceolate, elliptic, elliptic-oblong or lanceolate, 13–80 × 6–46 mm, widest below, at, or above middle, base cuneate to rounded or minutely cordate, margin entire, scaberulous, apex acute to obtusely apiculate, both surfaces glabrous to pubescent. Inflorescence extra-axillary, peduncle 2–35(100) mm long; cyme umbellate to racemelike, rachis absent or to 65 mm long with internodes longer than pedicels, flowers opening in succession. Pedicel 3–8.5(12) mm long, pubescent. Sepals linear-lanceolate to ovate-lanceolate, 2–4.3 mm long, apex acuminate, abaxially pubescent. Corolla 11–35 mm long, straight to slightly curved; tube (7.5)12–30 mm long, basal chamber sometimes indistinct, 3–6.8 mm wide, gradually narrowed into funnelform limb, 1.8–4.5 mm diam. at base, widening to 3.2–7.5 mm wide at mouth, exterior yellow sometimes deeply suffused with dark red, glabrous or pubescent, interior glabrous except for long-ciliate mouth; sinuses acute, flat or very slightly raised; lobes oblong-lanceolate, 3–14 mm long, plicate, apices connate to form conical to ovoid cage as wide or wider than mouth, apex rounded to subacute, abaxially glabrous, margins often white ciliate near apex, adaxially yellowish with short dark red tip, sometimes uniformly flushed dark reddish, glabrous except for long-ciliate keels. Corona sessile, c.4 × 3.5–3.7 mm diam.; outer corona shallowly cupular, lobes triangular-ovate, laterally confluent with bases of inner lobes, erect, bifid, lower than inner lobes, teeth triangular, 1–1.75 mm long, incurved, white with dark purple marks, white ciliate within; inner lobes incumbent over anthers, erect, converging over gynostegium, linear-filiform, much exceeding staminal column, recurved, glabrous. Follicles divergeant at up to 180°, linear-fusiform, 5–13 × 0.25–0.5 cm, glabrous. Seed not recorded.

The species as a whole is distributed from South Africa north to Guinea in the west and Eritrea in the east.

Huber's division of this species into a series of subspecies has not been taken up with any consistency. However, there is a definite geographical basis to the variation which does deserve a more detailed analysis. The species as a whole was named after the well-developed, somewhat racemelike, rachis producing flowers from 2 or more clearly separated nodes but such a rachis is lacking in most of the material from Zambia where the peduncle is up to 15 mm long with a rachis up to 3 mm long and often lacking. These collections match the Angolan type of subsp. *secamonoides*. Some of the collections from Malawi differ by their glabrous stems and have distinctively longer peduncles, 24–35 mm long, approaching the Madagascan subsp. *glabra*, a similarity taken further by *Grosvener* 384 (K) from Mozambique which has a well-developed rachis as seen in the Madagascan material. Material from south of the Flora area mostly has distinctively larger leaves, a well-developed rachis and corolla with reticulate markings, and was treated by Dyer as subsp. *setifera*. These plants have not been matched from within the Flora area. Collections from north of the Flora area corresponding to subsp. *racemosa* have narrow leaves, a well-developed rachis and a slightly differently shaped corolla cage, widest at the base unlike the southern forms where the cage is widest above the base. Plants with an erect habit are quite widely distributed including within the Flora area. Such a plant was described by Masinde as var. *tanganyikensis* Masinde. This plant otherwise matches subsp. *racemosa* whereas such plants from the Flora area match subsp. *secamonoides* suggesting that the erect habit is better regarded as a sporadic growth form of no taxonomic significance.

Meve (Ill. Handb. Succ. Pl. Asclepiadaceae: 64, 2002) included this species within *Ceropegia affinis* Vatke but that species differs in the form of the corolla lobes which form a globose, not ovoid, cage and by the densely dark spotted and veined corolla. The two species are also separated by molecular data (Bruyns *et al.*, 2015).

Bruyns 7420, from the Vumba Mountains, eastern Zimbabwe, resembles this complex in flower morphology but differs markedly from the rest by its sessile, 1-flowered inflorescences and rhizomatous habit.

Stem puberulent; leaf blade widest at or above middle; peduncle 2–18 mm long; inflorescence umbel-like with rachis absent or very short, to 3 mm long . . . .
. . . . .a) subsp. *secamonoides*
Stem glabrous; leaf blade widest below middle; peduncle 19–35 mm long; inflorescence often raceme-like with distinct rachis and flowers at 2 or more separate nodes. .
. . . . .b) subsp. *glabra*

a) Subsp. **secamonoides** (S. Moore) H. Huber in Mem. Soc. Brot. **12**: 97 (1958); in Merxmüller, Prodr. Fl. Sudwestafr. **114**: 25 (1967). Type: Angola, Laussingua, *Gossweiler* 2552 (BM000528528 holotype, K000305464, COI00070586). FIGURE 7.3.**144**.

> *Ceropegia secamonoides* S. Moore in J. Bot. **50**: 364 (1912).
> *Ceropegia cynanchoides* Schltr. in Bot. Jahrb. Syst. **51**: 153 (1913). Type: Namibia Damaraland, Gaub, *Dinter* 2410 (SAM0070996-0 lectotype), designated by Huber in Mem. Soc. Brot. **12**: 97 (1958).

Stem uniformly sparsely puberulent. Petiole 3–7 mm long; leaf blade narrowly oblong-oblanceolate, elliptic, or elliptic-oblong 36–75 × 7–22(36) mm, widest at or above middle. Peduncle 2–18 mm long; cyme subumbellate, rachis very short to absent. Pedicel 3–7 mm long. Sepals 3–4.3 mm long, apex acuminate. Corolla 16–25 mm long; tube 13–18 mm long; basal chamber 3.7–6.8 mm wide, narrowest part 1.8–4.5 mm diam., widening to (3.2)4.2–7.5 mm wide at mouth, exterior usually glabrous, rarely pubescent, lobes 3–7.5 mm long, cage wider than mouth, widest above its base.

Zambia. W: 2 miles SW of Ndola, 4000 ft, fl. 5.i.1953, *Draper* 20 (K). N: Kasama Dist., 30 km S of Kasama on road to Mpika, 1250 m, fl. 13.i.1975, *Brummitt & Polhill* 13759 (K). C: Mkushi Fiwila, 4500 ft, fl. 4.i.1958, *Robinson* 2609 (K); Lusaka Dist., Lazy J Ranch, 20 km SE Lusaka, 1300 m, fl. 4.ii.1995, *Bingham* 10365 (K). **Malawi**. S: Blantyre Dist. Michiru Mts, below Blantyre, fl. i.1990, *Jenkins* 2 (K). C: Ft. Manning Dist., near Tamanda Mission, 1400 m, fl. 8.i.1959, *Robson* 1098 (K).

Also in Angola, Namibia, Tanzania. *Brachystegia* woodland, rock outcrops; 1100–1400 m.

Conservation status: LC, quite widely distributed in Zambia and neighbouring countries.

b) Subsp. **glabra** H. Huber in Mem. Soc. Brot. **12**: 96 (1958). Type: Madagascar, Ambilobe Dist., Massif de Marivarohona SW of Manambato, Haute Mahavavy du Nord, *Humbert & Capuron* 25664 (P00114437 holotype). FIGURE 7.3.**144**.

Stem glabrous. Petiole 6.5–11 mm long; leaf blade lanceolate, 24–50 × 9–16 mm, widest below middle. Peduncle 19–35 mm long; cyme umbellate to raceme-like, rachis 0–32(65) mm long. Pedicel 2.5–8.5 mm long. Sepals 2–3.3 mm long, apex acuminate. Corolla 11–26 mm long; tube 7.5–19 mm long; basal chamber 3.7–4.6 mm wide, narrowest part 2.1–2.4 mm diam. widening to 3.2–6 mm wide at mouth, exterior glabrous, lobes 3.2–7 mm long.

Malawi. S: Zomba Plateau, 1 km below Ku Chawe, 1470 m, fl. 7.iv.1984, *Brummitt et al.* 17133 (K); Ku Chawe, above Zomba town, fl. 13.ii.1982, *Hepper* 7336 (K); Chimanimani Mts., near St. Georges Cave between saddle & Poacher's cave, 5000 ft, fl. 12.iv.1967, *Grosvener* 384 (K). **Mozambique**. MS: Chimanimani Mts., 1640 m, fl. 5.v.2016, *Osborne* 1206 (LMA).

Also in Madagascar. *Colophospermum mopane* woodland, evergreen thickets, amongst rocks; 1450–1650 m.

Conservation status: LC: Rare in the Flora area but frequently collected in Madagascar.

Subsp. *racemosa* is widely distributed from D.R. Congo and Tanzania north to Guinea and Eritrea; subsp. *setifera* is found in Namibia and South Africa.

13. **Ceropegia cordifolia** M.G. Gilbert, sp. nov. Most closely related to *C. emdenpienaarii* Bruyns, but differing by the uniformly minutely puberulent stems and leaves (not glabrous) and the distinctly cordate leaf base (not truncate to subcordate); these two species are related to *C. racemosa* but differ most obviously by the short prostrate stems, broadly ovate to cordate leaves and almost flat corolla lobes which are abaxially coarsely adpressed hairy. Type: Zimbabwe, N: Mazoe, Umvukwes, Ruorka Ranch, fl. 16.xii.1952, *Wild* 3912 (K001400202 holotype). FIGURE 7.3.**144**.

Roots clustered, fleshy, fusiform, up to 11 cm long, c.5 mm thick. Tuber absent. Stem prostrate, mostly unbranched, to 10 cm long, uniformly sparsely minutely puberulent. Leaves ± fleshy; petiole to 10 mm long, puberulous; leaf blade broadly ovate, to 27 × 17 mm, base cordate, margin entire, apex acuminate, abaxial surface paler, both surfaces uniformly minutely puberulent. Inflorescence mostly concealed beneath leaves, extra-axillary; peduncle to 5 mm long, sparsely puberulent, cyme umbel-like, rachis very short to absent; 2- or 3-flowered, flowers opening in succession. Pedicel 4–8 mm long, occasionally with a few hairs, often apparently glabrous. Sepals linear-lanceolate to lanceolate, c.2.5 × 0.7 mm wide, slightly recurved, pilose. Bud cylindrical, with flat sinuses. Corolla c.14 mm long, abruptly curved at right angle above basal inflation; tube c.11.5 mm long, basal chamber indistinct, c.4 mm wide, limb cylindrical, c.2.3 mm wide, widening very slightly at mouth; exterior mostly whitish with faint purplish stripes, denser towards apex, darkening with age, pubescent towards mouth, interior glabrous; sinuses acute, flat; lobes oblong, c.2.5 mm long, mostly flat with shallowly revolute margins, apices connate to form subglobose cage, slightly wider than mouth, abaxially dull purple, coarsely adpressed pubescent, adaxially glabrous. Corona subsessile, taller than wide; outer corona lobes laterally confluent with bases of inner lobes, deeply bifid, almost to base, teeth triangular, erect, slightly incurved, densely ciliate; inner lobes incumbent-erect, linear, much exceeding staminal column, apices recurved, glabrous. Follicles and seed not recorded.

Zimbabwe. N: Umvukwe Range, Vanad Pass, fl. 5.ii.1997, *Goyder & Paton* 4102 (K).
Endemic (but see note below). Shallow soil among rocks on top of chromium rich dyke; 1600–1650 m.

Conservation status: VU: Apparently of very restricted distribution and thus potentially vulnerable particularly to any development of mining.

Originally included within *C. racemosa*, this species is very easily distinguished by the short, prostrate stems bearing cordate, almost orbicular, leaves that are densely puberulous abaxially, and flowers with the corolla tube exterior white progressively streaked dull purplish with age and the corolla lobes only slightly plicate, abaxially dull purplish and coarsely hairy.

Dyer in Codd *et al.*, Fl. S. Africa **27**(4): 64 (1980) refers to an anomalous collection from northern Transvaal (*Percy-Lancaster* 200) that he regarded as a good match with a collection by Wild from Zimbabwe (under *Leach* 5907) that was probably the same collection as the type of this species. The recently described *Ceropegia emdenpienaarii* Bruyns included both the South African and Zimbabwe material but the description given seems to have been based only on the South African material which differs from the Zimbabwe material by the glabrous stems and leaves and the subtruncate leaf base. The two groups are here treated as separate species.

14. **Ceropegia carnosa** E. Mey., Comm. Pl. Afr. Austr.: 193 (1838). —Brown in Fl. Cap. **4**(1): 822 (1908). —Huber in Mem. Soc. Brot. **12**: 98 (1958). —Bruyns in Bradleya **3**: 16–18 (1985). —Meve, Ill. Handb. Succ. Pl. Asclepiadaceae: 71 (2002). Type: South Africa, Cape: Bathurst, "inter Kovi et Kaprivier", *Drège* 4946 (P00109634 holotype). FIGURE 7.3.**144**.

Roots clustered, fleshy, up to 15 cm long. Tuber absent. Stem twining, up to 2 m or more long, glabrous. Leaves ± fleshy; petiole 5–15 mm long, sparsely puberulous; leaf blade ovate to lanceolate, 15–30 × 4.5–20 mm, base rounded, margin entire, ciliolate, apex (obtuse to) acuminate, both surfaces glabrous. Inflorescence extra-axillary; peduncle 10–25 mm long, slender; cyme 2–5-flowered, flowers opening in succession. Pedicel 5–12 mm long, glabrous. Sepals subulate-deltoid, 2–4 mm long, straight, glabrous. Corolla 15–25 mm long; erect to ascending, tube 10–15 mm long, basal chamber indistinct, limb narrowly cylindrical above inflation then widening to broadly cylindrical upper part, lower part c.1.8 mm wide, upper part c.3 mm wide, widening to 4.5(9) mm at mouth, exterior creamy-white, glabrous, interior dark (?purplish) above basal chamber, with maroon stripes along broader upper part, pubescent from above inflation to mouth; sinuses acute, raised in bud; lobes oblong-ovate, 3–8 mm long, plicate, apices connate to form an oblate cage wider than mouth, abaxially glabrous, margins conspicuously densely pilose/ ciliate, adaxially pale with maroon veins and lines of spots, keel ciliate with long white hairs. Corona sessile; outer corona very reduced, lobes laterally confluent with bases of inner lobes, deeply divided, almost to base, by broad, rounded, sinus teeth 1–2 mm long, incurved, sometimes with long white hairs within; inner lobes incumbent-erect, linear, 2–3 mm long, apices recurved, slightly thickened and finely papillate, base with hairs. Follicles and seed not recorded.

Mozambique. M: Maputo, Licuati Forest Reserve, 60 m, fl. 4.xii.2001, *Goyder* 5027 (K); Naamacha Dist., Lebombos, fl. ii.2018, *Osborne* 1388 (LMA).

Also in South Africa and Swaziland. Margin of low forest on old sand dunes; 60–200 m.

Conservation status: LC in South Africa, only just extending into southern Mozambique.

Bruyns in S. African J. Bot. **112**: 430 (2017) included *C. racemosa* within this species but the two seem adequately distinguished by the distinctively shaped corolla tube of *C. carnosa* and the difference in corolla coloration and indumentum. Images from N

Mozambique (Zambesia: Serra de Gurue below Namuli Peak, 1340 m, Bart Wursten) have been seen of what looks to be a striking colour form with a corolla with a more strongly curved, almost white tube and pure yellow lobes with fewer, longer hairs. The status of these plants needs further investigation.

15. **Ceropegia chimanimaniensis** M.G. Gilbert, sp. nov. Easily recognised by the combination of the rootstock with a cluster of fleshy roots, the glabrous stem, linear leaves, 1.5–1.7 mm wide, abaxially apressed pubescent and adaxially glabrous, the sessile 2- or 3-flowered infloresence and the straight corolla with a cylindrical cage distinctly narrower than the mouth of the corolla and with densely purplish pilose margins. Type: Mozambique, Manica, Chimanimani Mts., slopes towards Mt. Namadima ['Nhamudimu'], 1723 m, fl. 21.iv.2014, *Ballings & Wursten* PB2289 (BR0000015256386V holotype). FIGURE 7.3.**144**.

Rootstock a fascicle of fleshy roots. Stem twining, solitary, slightly succulent, c.30 cm long, c.1 mm wide, smooth, glabrous except for a few hairs at nodes. Leaves slightly succulent; petiole 2.5–4 mm long, sparsely minutely puberulent; leaf blade linear, to 38 × 1.5–1.7 mm, base cuneate, margin revolute, apex acute, abaxial surface adpressed puberulent, adaxial surface glabrous. Inflorescence extra-axillary, sessile; cyme 1- or 2-flowered. Pedicel c.4 mm long, red, glabrous. Sepals narrowly triangular-lanceolate, c.1.9 × 0.6 mm, acute, sparsely puberulent abaxially. Bud with slightly raised sinuses and short beak. Corolla 16–17 mm long, straight or nearly so; tube 12–13 mm, basal chamber ± globose, c.3 mm wide, limb cylindrical, c.2.4 mm wide, widening to 4.5 mm at mouth, exterior spotted/blotched with purple, uniformly purple when dried, glabrous; sinuses acute, very narrowly auriculate; lobes upper part narrowly oblong, c.4.5 mm long, triangular base c.1.5 mm long, plicate, apices connate to form cylindrical cage narrower than mouth; margins densely purplish pilose towards apex, adaxially uniformly dark purplish, keel pilose. Corona sessile, c.2 × 2.2 mm diam.; outer corona cupular, lobes prominent, joined across base of inner lobes, erect, subentire, c.1 mm, higher than staminal column, glabrous; inner lobes erect, spathulate, c.1.5 mm long, 0.3 mm wide at tips, dorsiventrally compressed, glabrous. Follicles and seeds not seen.

Zimbabwe. E: between Muhohwa Falls and Raphia Pool, The Corner, 1200 m, fl. 1.ii.2006, *Ballings & Wursten* PB143 (BR); Chimanimani Mts., 1 mile NE of Mt Hut on slopes of Turret Towers and above Bundi R., 5.iv.1969, *Kelly* 74 (SRGH); near St. George's Cave between the saddle and Poacher's Cave, *Grosvenor* 395 (SRGH). **Mozambique**. MS: lower southern slopes of Mt Nhamdimo, N Chimanimani, 1707 m, 4.v.2016, *Osborne* 1150 (LMA).

Endemic. Twining through low heath vegetation and grasses on quartzite outcrops; 1200–1700 m.

Conservation status: V; all records are of solitary plants from a rather small area.

Images of this plant were named as *Ceropegia linophylla* H. Huber by De Kock, The Genus *Ceropegia*: 31, figs.190,191 (2017), but that West African species differs most clearly by the pedunculate inflorescence. The nature of the rootstock is uncertain. Plants grow in narrow rock crevices making its recovery extremely difficult. It was first assumed on overall morphology, particularly corolla form which is similar to that of *C. woodii*, that this species was a member of sect. *Ceropegiella* but the only stem base seen indicates that the rootstock is a fascicle of thickened roots which would suggest rather that it more likely a member of sect. *Carnosae*.

Sect. **Umbraticolae** (H. Huber) Bruyns in S. African J. Bot. **112**: 430 (2017).

16. **Ceropegia gilgiana** Werderm. in Bot. Jahrb. Syst. **70**: 205 (1939). —Huber in Mem. Soc. Brot. **12**: 150 (1958). —Meve, Ill. Handb. Succ. Pl. Asclepiadaceae: 80 (2002). —Masinde in F.T.E.A., Apocynaceae (part 2): 254 (2012). Type: Tanzania, Tabora: Unyamwezi, km. 991.6, E of Kombe, *Peter* 35763 (B† holotype); Ufipa,

Sumbawanga, 30.i.1950, *Bullock* 2363 (K000305485 neotype), designated by Huber in Mem. Soc. Brot. **12**: 150 (1958). FIGURE 7.3.**145**.

Roots clustered, fleshy, fusiform, to 11 cm long, c.5 mm thick. Tuber absent. Stem erect, solitary, herbaceous, to 30 cm tall, uniformly minutely pubescent or with 1 or 2 lines of hairs. Leaves herbaceous, sometimes subsessile; petiole to 2 mm long, densely pubescent; leaf blade linear (to narrowly lanceolate), 10–60 × 3–9 mm, base cuneate, apex acute, both surfaces puberulous. Inflorescence from middle and upper nodes, extra-axillary; sessile, cyme (1)2-flowered, flowers opening in succession. Pedicel 3–5(10) mm long, pubescent. Sepals linear-lanceolate, 4–5 × c.1 mm, ± straight, shortly puberulous. Corolla (25)55–65(75) mm long, straight, very slender, tube (20)35–40 mm long, basal chamber ovoid to subglobose, 5(8) × 4–5 mm wide, limb gradually narrowed to neck, 1–1.5 mm wide, then widening very gradually to 2–3(4) mm wide at mouth; exterior minutely puberulent, interior glabrous; sinuses acute, flat; lobes linear to filiform, 20–40 mm long, revolute, apices connate to form narrowly cylindrical cage about as wide as mouth, abaxially pubescent, adaxially sparsely pubescent. Corona shortly stipitate; outer lobes each forming a shallow pocket at base, divided for ½ length, teeth rounded, margins vaguely papillate; inner lobes erect, converging over gynostegium, ± spathulate, c.2.5 mm long, dorsiventrally compressed. Pollinia bright yellow. Immature follicles acutely divergent, relatively thick.

Zambia. C: Lusaka Dist., 6 mi E of Lusaka, 4200 ft, fl. 24.i.1958, *King* 408 (K). Also known from Tanzania. "Red clay at edge of swamp"; 1400 m.

Conservation status: Known only from the one collection in the Flora area but fairly widely distributed in southern Tanzania.

Unusual for its preference for wet habitats.

17. **Ceropegia umbraticola** K. Schum. in Bot. Jahrb. Syst. **17**: 153 (1893); in Engler & Prantl, Nat. Pflanzenfam. **4**(2): 272, fig.F (1895). —Brown in F.T.A. **4**(1): 461 (1903). —Norman in J. Bot. **67**(suppl. 1 & 2): 99 (1929). —Huber in Mem. Soc. Brot. **12**: 148 (1958). —Stopp in Bot. Jahrb. Syst. **83**: 117 (1964). —Lisowski & Malaisse in Bull. Jard. Bot. Natl. Belg. **44**: 407–409, fig.3 (1974). —Malaisse in Bull. Jard. Bot. Natl. Belg. **54**: 215 (1984). —Malaisse & Schaijes in Asklepios **58**: 27, figs.3,8 (1993). —Meve, Ill. Handb. Succ. Pl. Asclepiadaceae: 105 (2002). —Masinde in F.T.E.A., Apocynaceae (part 2): 258 (2012). Type: Angola, Malange, *von Meechow* 370 (B† holotype, K000305642 lectotype) designated here. FIGURE 7.3.**145**.

    Ceropegia wellmanii N.E. Br. in Bull. Misc. Inform. Kew **1908**: 408 (1908). Type: Angola, Benguela, Ulondo Mts., *Wellman* 1781 (K000305643 holotype).

    Ceropegia rostrata E.A. Bruce in Kew Bull. **3**: 464 (1949). Type: Zambia, Mwinilunga, just SE of Dobeka Bridge, *Milne-Redhead* 3702 (K000305645 holotype).

    Ceropegia chipiaensis Stopp in Bot. Jahrb. Syst. **83**: 119 (1964). Type: Angola, Chipia, *Stopp* BO105 (K000305641 holotype).

    Ceropegia schaijesiorum Malaisse in Bull. Jard. Bot. Natl. Belg. **56**: 491 (1986). Type: D.R. Congo, Shaba, Nzilo-Kyamasumba, km 11, vallée de la Luilu, 10.30° S 25.24° E, zone Kolwezi, *Malaisse* 13692 (BR0000008863102 holotype).

Roots clustered, fleshy, fusiform, to 18 cm long. Tuber absent. Stem erect, solitary, herbaceous, to 30 cm high, uniformly densely pubescent, glabrescent towards base. Leaves herbaceous, petiole 5(18) mm long; leaf blade linear-lanceolate to broadly ovate or rounded, 26–48 × 25–38 mm, base broadly cuneate to subcordate, apex acute to subacute, both surfaces densely minutely pubescent. Inflorescence subterminal, sessile; cyme 1(2)-flowered. Pedicel 8–20 mm long. Sepals lanceolate, 5–6 × c.1 mm wide, puberulous to pubescent. Bud with minutely auriculate sinuses and prominent linear beak. Corolla (50)64–90(110) mm long, erect, straight or slightly bent at top of basal chamber; tube 30–70 mm long, basal chamber urceolate, 20–30 × 10–26 mm wide, limb cylindrical, 9–17 mm wide, slightly narrowed at mouth; exterior white or pale green with green or purple striations suffused brown towards apex, sometimes with distinct narrow pale band around apex of tube and base of of lobes; both surfaces glabrous; interior whitish with deep green to purple lines on inside, irregular dark band at top of basal chamber, sometimes purple-reticulate towards

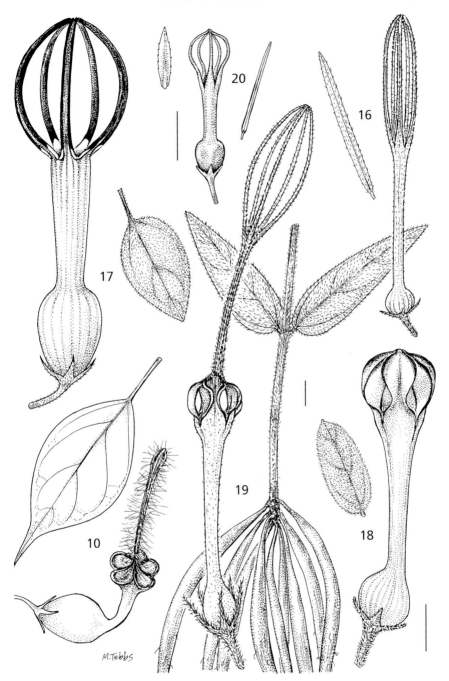

Fig. 7.3.**145**. —**10**. CEROPEGIA SPECIOSA, leaf and flower, from *Goyder & Paton* 3649. —**16**. CEROPEGIA GILGIANA, leaf and flower, from *King* 408. —**17**. CEROPEGIA UMBRATICOLA, leaf and flower, from *Merrett* 1082. —**18**. CEROPEGIA FILIPENDULA, leaf and flower, from *Werner* s.n. —**19**. CEROPEGIA MIRABILIS, basal parts and flower, from *Goyder et al.* 3598. —**20**. CEROPEGIA ILLEGITIMA, leaves and flower, from *Robson* 1162. Scale bars = 10 mm. Drawn by Margaret Tebbs.

mouth; sinuses subacute, narrowly auriculate; lobes abruptly narrowed from base then linear, 20–40 × 1–2(4) mm wide, revolute, apices connate to form ellipsoidal to conical cage, wider than tube or not, abaxially often whitish or pale yellow for up to 6 mm at base, then purplish brown to bright yellow-green, often tinged with brown, very sparsely minutely setulose to glabrous, margins usually sparsely very minutely setulose, adaxially minutely pubescent and often white setulose. Corona cupular, 5–8.5 × 6–7 mm diam.; outer corona with 10 equal teeth, teeth linear, 2–3 mm long, inner lobes incumbent on anthers, vestigial, glabrous. Follicles very narrowly divergent, 2–4 × 0.4–0.6 cm wide at middle, glabrous. Seed brownish, c.8 × 4 mm; coma c.30 mm long.

Zambia. W: Kitwe, fl. 9.i.1958, *Fanshaw* 4194 (K). N: Kawambwa, fl. 24.x.1962, *Banda* 750 (K); Abercorne Tunduma road to Mambwa, 1500 m, fl. 8.xii.1964, *Richards* 19305 (K). E: Chadiza, 850 m, fl. 30.xi.1958, *Robson* 788 (K). **Malawi**. N: Chitipa, 5 mi E of Chendo, 4500 ft, fl. 2.i.1977, *Pawek* 12199 (K).

Also known from Angola, D.R. Congo and Tanzania. Open areas within miombo woodland; 800–1500 m.

Conservation status: LC: quite frequently collected from a wide area.

When designating a neotype of *Ceropegia umbraticola* (D.R. Congo, Station de Keyberg, 8 km SE of Lumumbashi [Elisabethville], *Schmitz* 1055 (BR)), Huber (1958: 148) and Stopp (1964: 123) overlooked the fact that original material: a flower detached (with permission) from the Berlin holotype, is preserved in Kew, along with a pencil sketch to show the habit. This material invalidates any neotypification. In view of the fragmentary nature of the lectotype, the neotype could be treated as an epitype.

18. **Ceropegia filipendula** K. Schum. in Bot. Jahrb. Syst. **17**: 150 (1893); in Engler & Prantl, Nat. Pflanzenfam. **4**(2): 272, fig.F (1895). —Brown in F.T.A. **4**(1): 462 (1903). —Huber in Mem. Soc. Brot. **12**: 150 (1958). —Stopp in Bot. Jahrb. Syst. **83**: 123 (1964). —Malaisse in Bull. Jard. Bot. Natl. Belg. **54**: 218 (1984). —Malaisse & Schaijes in Asklepios **58**: 27 (1993). —Meve, Ill. Handb. Succ. Pl. Asclepiadaceae: 78 (2002), excl. syn. *C. mirabilis*. —Masinde in F.T.E.A., Apocynaceae (part 2): 257 (2012), excl. syn. *C. mirabilis*. Type: Angola, Cuanze Norte, Cissacola, River Coanga, *von Meechow* 553B (B† holotype); Ganda, *Damann* 1228 (K000449183 neotype) designated by Stopp in Bot. Jahrb. Syst. **83**: 123 (1964). FIGURE 7.3.**145**.

    Ceropegia medoensis N.E. Br. in Bull. Misc. Inform. Kew **1895**: 263 (1895). —Huber in Mem. Soc. Brot. **12**: 149 (1958). Type: Mozambique, Medo country, between Lugenda River & Ibo, *Last* s.n. (K000305637 holotype).

    Ceropegia dichroantha K. Schum. in Bot. Jahrb. Syst. **30**: 385 (1901). Type: Tanzania, Njombe, Ukangu Mt. near Lubila [Langenburg] *Goetze* 839 (B† holotype, K000305484 lectotype), designated by Masinde in F.T.E.A., Apocynaceae (part 2): 258 (2012) from a fragment of the holotype.

    Ceropegia peteri Werderm. in Bot. Jahrb. Syst. **70**: 204 (1939). —Huber in Mem. Soc. Brot. **12**: 150 (1958). —Stopp in Bot. Jahrb. Syst. **83**: 123 (1964). Type: Tanzania, Tabora: near Kombe, *Peter* 35383 (B† holotype); Angola, Sandu-Tjigaka, *Damann* 1225 (K neotype), designated by Stopp in Bot. Jahrb. Syst. **83**: 123 (1964).

    Ceropegia renzii Stopp in Bot. Jahrb. Syst. **83**: 124 (1964). Type: Angola, Huila and Moçâmedes, 6 km SE of Nova Lisboa, *Stopp* BO106 (K000305639 holotype).

Roots clustered, fleshy, fusiform, 7–12 cm long, to 12 mm thick. Tuber absent. Stem erect, solitary, herbaceous, 13–35(48) cm tall, to 2.5 mm diam., puberulous to sub-scabrid. Leaves 3–5(10) pairs, herbaceous, sometimes subsessile; petiole to 3 mm long, hirsute; leaf blade ovate-oblong, 25–40 × 6–20 mm, base rounded to subcordate, apex subacute to obtuse, mucronulate, both surfaces scabrid-puberulous. Inflorescence extra-axillary, mostly subterminal, sessile, cyme 1-flowered. Pedicel 4–10 mm long, strongly puberulous. Sepals oblong-lanceolate, subulate, 4–10 × 0.9–1.7 mm wide, erect or ± spreading, shortly puberulous. Corolla (33)40–80(95) mm long, straight, tube (18)25–65 mm long, basal chamber ovoid to obovoid, (5)9–17 × 4–10(15) mm wide, limb gradually narrowed to neck, (<2)2.3–5(6.7) mm wide, then widening gradually to (4.6)8–16 mm wide at mouth, exterior greenish yellow or yellow flushed purple, to maroon, minutely sparsely puberulent, interior glabrous; sinuses acute, flat; lobes broadly oblong-ovate to obovate, 13–29 × 3–8 mm wide folded, plicate, apices connate to form cylindrical to obconical cage, often widest near apex, with umbrella-like canopy at the top which is distinctly raised in centre, (7)14–22 mm wide, abaxially very sparsely pubescent, adaxially pale yellowish green flushed pink at tip to maroon, usually glabrous, sometimes densely pubescent. Outer corona lobes with triangular bases, bifid, 2–3 mm long, higher than gynostegium; inner lobes converging over gynostegium, subulate, 2–4 mm long, apices touching above gynostegium. Follicles erect, spreading to the sides, fusiform, 6–7 × 0.3–0.4 cm wide, flesh-coloured with purple streaks, apices slightly obtuse and curved inwards. Seed not recorded.

Zambia. N: Abercorne, road to Locust HQ, 1500 m, fl. 27.i.1955, *Richards* 4258A (K). C: Lusaka, 15 km E of Bell-Cross Farm, 23.i.1993, *Bingham* 8793 (K, MO). S: Chirundu Escarp. 70 km WNW of Chirundu, 850 m, 24.i.1965, *Robinson* 6366 (K). **Malawi**. C: Chitala escarpment, 950 m, fl. 2.xii.1959, *Robson* 1562 (K); Zomba Dist. Domasi Mission, 900 m, fl. 1.xii.1992, *Goyder & Paton* 3508 (K). **Mozambique**. N: Medo Country between Lugenda River & Ibo, 1888, *Last* s.n. (K).

Also in Angola and D.R. Congo. *Brachystegia* woodland, often on slopes, often with *B. boehmii*, on red to brown sandy soils; 900–1800 m.

Conservation status: LC. Moderately frequently collected from a wide area.

19. **Ceropegia mirabilis** H. Huber in Mem. Soc. Brot. **12**: 149 (1958). —Malaisse in Bull. Jard. Bot. Natl. Belg. **54**: 218 (1984). Type: Malawi, 40 km W of Karonga, *Williamson* 200 (BM000559156 holotype). FIGURE 7.3.**145**.

Roots clustered, fleshy, fusiform, to 7 mm thick. Tuber absent. Stem erect, solitary, herbaceous, 15–30(45) cm tall, c.3 mm thick, uniformly densely pubescent to hirsute. Leaves 6–10 pairs, herbaceous; petiole 2–4 mm long; hirsute; leaf blade narrowly ovate to oblong, 30–60 × 12–26 mm, base rounded, apex subacute to obtuse, mucronulate, both surfaces pubescent. Inflorescence at upper nodes, extra-axillary, sessile; cyme 1-flowered. Pedicel 5–12 mm long, slender, erect, densely pubescent to pilose. Sepals linear-triangular, 6–10(14) × 0.7–1 mm wide, ± straight, abaxially densely pilose, adaxially glabrous. Corolla 90–135 mm long, straight, tube 35–66 mm long, basal chamber ovoid, 10–14 × to 8.2 mm wide, limb gradually narrowed to neck, 2–2.9 mm wide, widening gradually to 7–12 mm wide at mouth, exterior green or initially greenish-yellow then yellow and later pinkish-brown, minutely puberulent, interior 'velvety purple', glabrous; sinuses acute, flat to very minutely revolute; lobes oblong-obovate at base then linear, 60–70 mm long overall, base plicate, narrowly divergent, apices connate to form cylindrical cage topped by canopy to 12 mm wide, abaxially yellow, pubescent, adaxially yellow, minutely pubescent and often white setulose, apex of cage extended into linear column or narrow cage 20–25(45) mm long, flushed reddish, apex sometimes expanded into fusiform second cage. Corona c.3.5 × 5 mm diam.; outer lobes erect, narrowly triangular, deeply 2-lobed, c.3 mm long, gynostegium higher than base of U-shaped cleft, glabrous with inside of cleft papillose; inner lobes converging over gynostegium, linear from a broadly deltoid base, c.5 mm long, overtopping the staminal column. Follicles fusiform, 10–11 × 0.4–0.6 cm wide, glabrous, apices obtuse. Seed not recorded.

Malawi. Karonga Dist., Thulwe Hills, 35 km W of Karonga, 800 m, fl. 1.ii.1992, *Goyder et al.* 3598 (K). S: Machinga Dist., Liwonde National Park, fl. 11.i.1984, *Patel & Balaka* 1405 (MAL); Mzimba, 5000 ft., 1948, *Benson* 1463 (K).

Also in Tanzania and D.R. Congo. *Brachystegia* woodland on sandy soils; 800–1500 m.

Conservation status: DD: Apparently of rather sporadic occurrence over a fairly wide area.

When including this species within *C. filipendula*, Meve (Ill. Handb. Succ. Pl. Asclepiadaceae: 78, 2002) argued that the double cage so very characteristic of this species was simply variation within *C. filipendula* analogous to that seen in *C. somalensis* Chiov. but this is very clearly not the case: in *C. somalensis*, the overall length of the corolla lobes is the same in flowers with a simple single cage and those where the lobes remain twisted together in the middle but separate above to form a second cage, the two forms sometimes occurring on the one plant. In *C. mirabilis* the lower cage is very similar in size to the simple cage seen in *C. filipendula* whilst the spectacular upper cage is formed from massive extensions of the corolla lobes, quite absent in *C. filipendula*. No material has been seen that could be interpreted as intermediate.

20. **Ceropegia illegitima** H. Huber in Mitt. Bot. Staatssaml. München Heft **12**: 72 (1955); in Mem. Soc. Brot. **12**: 125 (1958). —Lisowski & Malaisse in Bull. Jard. Bot. Natl. Belg. **44**: 409–411, fig.4 (1974). Type: D.R. Congo, Station de Keyberg, 7 km SSE of Elizabethville, *Schmitz* 2116 (BR0000008861429 holotype, YBI126770232). FIGURE 7.3.**145**.

Roots clustered, fleshy, fusiform, to 5 mm diam. Tuber absent. Stems erect, 1–3, somewhat succulent, 20–60 cm tall, middle internodes 6–16 cm long, 1–2.5 mm diam. at base, glabrous. Leaves herbaceous, sessile, upper and lower leaves reduced and almost scalelike, middle leaves with leaf blade subulate to filiform, 2–10(20) × 2–5 mm, both surfaces glabrous. Inflorescences at upper nodes, extra-axillary, peduncle 2.5–9 cm long, slender, glabrous; cyme 1–3(6 or more)-flowered; flowers opening in succession. Pedicel 2–4(6) mm long, filiform. glabrous. Sepals (subulate)lanceolate, c.(1.5)1.8 × 0.9 mm, glabrous. Bud with slender beak. Corolla 15–17(20) mm long, straight, tube 8–12(14) mm long, basal chamber ovoid, 4–5 × 2–3 mm wide, limb shortly cylindrical, (1)2 mm diam., slightly dilated to (1.5)2.5–3 mm wide at mouth, exterior (bright pink to) pinkish purple, both surfaces glabrous; sinuses subacute, flat; lobes linear(-spathulate), 4(9) × 0.2–0.4 mm wide, base triangular, c.1 mm wide, slightly revolute, apices connate to form slightly beaked to conical cage, abaxially same colour as tube, glabrous, margins glabrous (or with short white hairs), adaxially dull orange, glabrous. Outer corona lobes cupular, emarginate to minutely 2-toothed, hyaline, glabrous or with short hairs inside; inner lobes converging over gynostegium, linear. Follicles narrowly divergent, linear-fusiform, 5–7 × c.0.2 cm, slightly incurved glabrous. Seed not recorded.

Malawi. C: nr. Kasungu Hill, 1100 m, fl. 14.i.1959, *Robson & Jackson* 1162 (BM, K). Also in D.R. Congo and Central African Republic. Dambo; c.1100 m.
Conservation status: DD: Rarely collected.

Though the Malawi collection is geographically far removed from the original collections of *Ceropegia illegitima* it is a good match morphologically with only slightly shorter and broader flowers with broader sepals and follicles in pairs.

Sect. **Laguncula** H. Huber in Mem. Soc. Brot. **12**: 35 (1958).

21. **Ceropegia abyssinica** Decne. in Candolle, Prodr. **8**: 644 (1844). —Richard, Tent. Fl. Abyss. **2**: 46 (1851). —Brown in F.T.A. **4**(1): 462 (1903). —Werdermann in Bot. Jahrb. Syst. **70**: 206 (1939). —Bullock in Kew Bull. **3**: 424 (1952) excl. syn. *C. achtenii*. —Huber in Mem. Soc. Brot. **12**: 161–164 (1958). —Lisowski & Malaisse in Bull. Jard. Bot. Natl. Belg. **44**: 406–407, fig.2 (1974). —Archer in Upland Kenya Wild. Fl.: 391 (1974); in Kenya *Ceropegia* Scrapb.: 155 (1992); in Upland Kenya Wild. Fl., ed. 2: 183 (1994). —Meve, Ill. Handb. Succ. Pl. Asclepiadaceae: 63–64 (2002). —Gilbert in Hedberg & Edwards, Fl. Ethiopia Eritrea **4**(1): 160, figs.140.36.14–16 (2003). —Masinde in F.T.E.A., Apocynaceae (part 2): 250 (2012). Type: Ethiopia, Tigray, Gafta, *Schimper* II:1416 (G00022149 lectotype, FI000529, K000305508, S-G-1292, TUB003571, TUB003572, W) designated as holotype by Huber in Mem. Soc. Brot. **12**: 163 (1958). FIGURE 7.3.**146**.

Ceropegia steudneri Vatke in Linnaea **40**: 217 (1876). Type: Eritrea, Keren, *Steudner* 765 (B† holotype).
Ceropegia giletii De Wild. & T. Durand in Bull. Soc. Roy. Bot. Belgique **38**: 95 (1899). Type: D.R. Congo, lower Congo, Ndembo [Dembo], 1898, *Gillet* s.n. (BR0000008861504 holotype).
Ceropegia hispidipes S. Moore in J. Bot. **46**: 309 (1908). Type: Zimbabwe, Gazaland, near Chirinda, *Swynnerton* 1137 (K000305629 holotype, BM000930137).

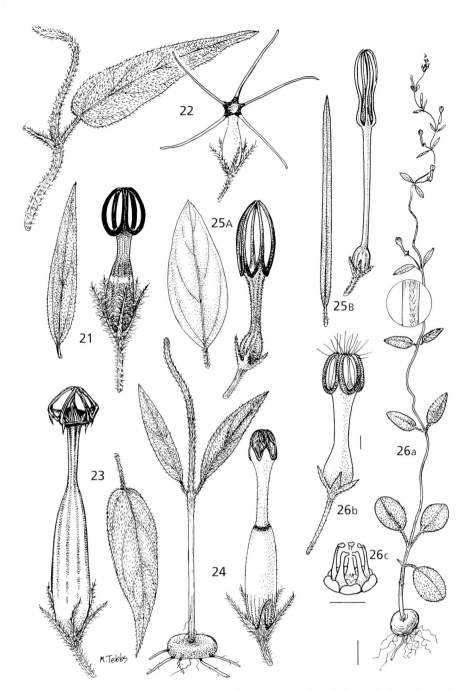

Fig. 7.3.**146**. —**21**. CEROPEGIA ABYSSINICA, leaf and flower, from *Milne-Redhead* 4298A. —**22**. CEROPEGIA RINGOETII, stem with leaf and flower, from *Richards* 864. —**23**. CEROPEGIA MEYERI, leaf and flower, from *Levy* 1172. —**24**. CEROPEGIA BONAFOUXII, base of plant and flower, from *Brummitt et al.* 14114. — **25A**. CEROPEGIA ACHTENII ssp. ACHTENII, leaf and flower, from *Richards* 4222. —**25B**. CEROPEGIA ACHTENII ssp. ADOLFI, leaf and flower, from *Richards* 4137. —**26**. CEROPEGIA LEPTOTES, 26a habit with inset of stem indumentum, 26b flower, 26c corona, from *Richards* 10700. Scale bars = 10 mm except for 26b and 26c = 1 mm. Drawn by Margaret Tebbs.

Ceropegia bequaertii De Wild. in Rev. Zool. Bot. Africaines **13**(Suppl. Bot.): 1 (1920). Type: D.R. Congo, Bogoro-Mboga, *Bequaert* 4975 (BR0000008861535 lectotype), designated by Huber in Mem. Soc. Brot. **12**: 163 (1958).

Ceropegia filicalyx Bullock in Bull. Misc. Inform. Kew **1933**: 145 (1933). Type: Tanzania, Kondoa: Kikori, *Burtt* 2755 (K000305477 holotype, K000305478).

Ceropegia bonafouxii var. *linearifolia* Stopp in Bot. Jahrb. Syst. **90**: 474 (1971). Type: Angola, Huila: Sá de Bandeira, am Ufer des Bergsees Ivantala, *Torre* 8625B (LISC holotype).

Roots not fusiform. Tuber globose to slightly oblate, to 5 cm diam., usually smaller. Stem erect, rarely twining, usually solitary, little branched, herbaceous, to 60 cm high (or to 1 m long), 1.5–2.5 mm thick, hirsute to pilose. Leaves herbaceous; petiole 3–5(10) mm long, pilose; leaf blade (linear), lanceolate, or oblong-oblanceolate to narrowly ovate, (22)46–72(130) × 7–22(27) mm, base tapering to cuneate (or sometimes subcordate), apex acute, both surfaces hirsute Inflorescence extra-axillary, mostly subterminal, sessile, cyme 3–9-flowered, flowers opening in succession. Pedicel (4)10–17(25) mm long, slender, erect, hirsute to villous. Sepals linear-subulate to subulate, 4.5–7.5(10) × < 0.5 mm wide, ± straight, abaxially hirsute to coarsely pilose, adaxially glabrous. Buds with slightly raised sinuses and narrowly cylindrical beak. Corolla 12–25 mm long, straight; tube flask-shaped, 6.5–15 mm long, basal c.²⁄₃ inflated, 4–10 mm long, 2.3–4 mm wide, gradually narrowed at apex, limb ± cylindrical, 1–2 mm wide, limb expanded to 1.6–5 mm diam. at mouth, exterior greyish white to greyish purple with longitudinal purplish lines, sometimes drying brownish-purple to purple, glabrous (or uniformly puberulent north of Flora area); interior pale at base, then purple, often with distinct narrow to sometimes broad pale band just below middle of tube, neck paler, mouth dark with whitish longitudinal lines, smooth, glabrous; sinuses obtuse, narrowly auriculate, lobes linear, subspathulate or narrowly oblong, 4–10(15) mm long, revolute to subplicate, apices connate to form broadly ovoid to oblate cage, abaxially pale grey-green, glabrous, adaxially margin and apex dark green to almost black, minutely velvety papillose. Corona 1.2–2 mm long; outer lobes each forming a pocket, Follicles erect, subparallel, 6.5–11 × 0.2–0.3 cm wide, glabrous. Seed 3–4(6) mm long, c.2 mm wide; coma to 9 mm long.

Zambia. W: Mwinilunga, Luakera Forest, 13 km N of Mwinilunga, fl. 20.i.1975, *Brummitt et al.* 13903 (K); Kitwe, fl. 7.ii.1964, *Mutimushi* 606 (NDO). **Zimbabwe**. C: Makoni Dist., near Duniden, 1800 m, fl. 9.ii.1931, *Norlindh & Weimark* 4951 (K). E: Umtali Dist., Banti Forest, 1830 m, fl. 4.ii.1955, *Excell, Mendonca & Wild* 181 (BM, SRGH). **Malawi**. C: Dedza, Chongoni Forest Reserve, fl. 22.ii.1968, *Salubeni* 975 (MAL). N: Mzimba Dist., 22 mi W of Mzuzu, Kasitu R., fl. 2.iii.1974, *Pawek* 8165 (K). S: Soche, fl. i.1990, *Jenkins* 1 (K).

Also in D.R. Congo, Central African Republic, Tanzania, Kenya, Ethiopia, Eritrea. Typically in open deciduous woodland and mopane scrub on sandy soils; 1300–1850 m.

Conservation status: LC. Very widely distributed and frequently collected.

The earliest name applied to this taxon was '*C. hirsuta*' by Hochstetter in his printed exsiccata label for *Schimper* (*sectio secundum*) 1416 but that was a later homonym of an Indian species. Decaisne did not designate a holotype for *C. abyssinica* so Huber's listing of the sheet of *Schimper* II:1416 as the holotype is here treated as an effective lectotypification. Authors have followed Huber in designating *Bequaert* 4975 as the holotype of *C. bequaertii* but the protologue also listed *Bequaert* 5531 so Huber's designation is again taken as a lectotypification.

There is significant variation within this species. Material from Ethiopia and Kenya, including the type of the accepted name, can be immediately distinguished from all material from the Flora Zambesiaca area by the hairy exterior to the corolla, which also lacks the transverse pale band which is obvious in dried flowers of the southern material. There are more subtle differences in the habit with the northern material flowering from lower down the stem. However, these differences break down in Tanzania where it does not seem possible to make a division between the two forms. Plants are mostly strictly erect but a few collections from the Flora area, including the type of *C. hispidipes*, show a lianescent habit as can cultivated plants in more shaded conditions.

Most material from Angola differs very sharply and uniformly from material from the rest of the range. Typical *C. abyssinica* has narrow leaves, cuneate at the base into short to almost non-existent petioles whereas the Angolan plants are all robust twiners with much broader leaves with well-defined petioles: petiole 17–39 mm long; lamina ovate to oblong ovate, 5–7 × 2–4.2 cm, base truncate to mostly shallowly cordate, apex acuminate. It is proposed that this material be excluded and treated as the distinct species *C. leucotaenia* K. Schum.

22. **Ceropegia ringoetii** De Wild. in Bull. Jard. Bot. État. Bruxelles **4**: 394 (1914). —Werdermann in Bot. Jahrb. Syst. **70**: 206 (1939). —Huber in Mem. Soc. Brot. **12**: 166 (1958). —Lisowski & Malaisse in Bull. Jard. Bot. Natl. Belg. **44**: 411, fig.2 (1974). —Meve, Ill. Handb. Succ. Pl. Asclepiadaceae: 97 (2002). —Masinde in F.T.E.A., Apocynaceae (part 2): 239 (2012). Type: D.R. Congo, Shinsenda, *Ringoet* in *Homblé* 553 (BR886309 lectotype, BR886315), lectotype designated here. FIGURE 7.3.**146**.

   Ceropegia schlechteriana Werderm. in Bot. Jahrb. Syst. **70**: 196 (1939). Type: Tanzania, Morogoro, Uluguru Mts., *Schlieben* 3762 (B† holotype, BM000528532, BR0000008863133, P00109685, Z-000001604).
   Ceropegia abyssinica var. *songeensis* H. Huber in Mem. Soc. Brot. **12**: 202 (1958). —Masinde in F.T.E.A., Apocynaceae (part 2): 251 (2012). Type: Tanzania, Songea, Chandamara Hill, *Milne-Redhead & Taylor* 8807 (K000305479 holotype, BR0000014569623, EA).
   Ceropegia wilmsiana Schltr., in sched., invalid name.

Roots not fusiform. Tuber depressed globose to discoid, to 3 cm diam. Stem twining, herbaceous, up to 1.5 m long, internodes 5–12 cm long, 1–3 mm diam., uniformly pilose to hirsute. Leaves herbaceous; petiole 2–20 mm long, hirsute; leaf blade linear-ovate to lanceolate, 25–90(110) × 3–35(40) mm, upper leaves sometimes much smaller, base rounded to subcordate, margin ciliate, apex acuminate, both surfaces glabrous to hirsute-pilose. Inflorescence extra-axillary, subsessile; cyme 5–12-flowered, flowers opening in succession. Pedicel 4–13 mm long, slender, hirsute. Sepals linear-subulate, c.5 mm long, acute, densely pilose. Flower bud with very slightly raised sinuses and long linear beak. Corolla 20–30(45) mm long, straight; tube 5–9 mm long, narrowly ovoid without distinct basal chamber, 2–3 mm diam., limb narrowing to 1–2 mm wide, widening to 3 mm diam. at mouth, exterior creamy-white, pilose, interior with mouth and base of lobes blackish green, glabrous; lobes filiform with triangular base, 10–20 mm long, apices free, rotately spreading, abaxially cream-coloured, pilose, adaxially velutinous, blackish at base with filiform part cream, glabrous. Corona stipitate, cupular, c.2 × 1.3 mm diam.; outer lobes forming small pouches, hyaline, ciliate; inner lobes converging above gynoecium, S-shaped, meeting at tips, narrowly ligulate, 1.2–1.5 mm long, base pilose, apices shortly hirsute-papillose. Pollinia yellow. Follicles erect, ± parallel, 8–8.5 cm long, glabrous, tapered. Seed not recorded.

Zambia. W: Solwezi Dist., Meheba R, Solwezi-Mwinilunga road, 1400 m, fl. 15.ii.1975, *Hooper & Townsend* 50 (K). N: Tasker Deviation, 4900 ft., 2.iii.1952, *Richards* 864 (K); Mbala [Abercorne], Katuka, 5000 ft, fl. 12.iii.1950, *Bullock* 2622 (K). **Malawi**. N: Mzimba Dist., Mzuzu, Marymount, 4500 ft, fl. 5.vi.1969, *Pawek* 2455 (K) & fl. 10.iv.1974, *Pawek* 8317 (K, MAL).

Also in D.R. Congo and Tanzania. *Brachystegia* woodland sometimes near rocks or in dense bush or tall grass; 1230–1500 m.

Conservation status: LC, moderately frequently collected over quite a wide area.

A second stage lectotypification is required as there are two sheets of *Ringoet* in *Homblé* 553 in BR, both labelled as 'TYPUS'; the better preserved sheet has been selected.

23. **Ceropegia meyeri** Decne. in Candolle, Prodr. **8**: 645 (1844). —Brown in F.T.A. **4**(1): 462 (1903); in Fl. Cap. **4**(1): 828 (1908). —Phillips in Fl. Pl. South Africa **1**: t.30 (1921). —Huber in Mem. Soc. Brot. **12**: 160 (1958). —Dyer in Codd *et al.*, Fl. S. Africa **27**(4): 79, fig.17 (1980). —Bruyns in Bradleya **3**: 31–32 (1985); in

Strelitzia **34**: 87 (2014). —Meve, Ill. Handb. Succ. Pl. Asclepiadaceae: 90 (2002). Type: South Africa, Transkei, *Drège* 4945 (P holotype). Based on *C. pubescens* E. Mey. FIGURE 7.3.**146**.

Ceropegia pubescens E. Mey., Comm. Pl. Afr. Austr.: 193 (1838), illegitimate name, not Wallich (1831).

Roots not fusiform. Tuber discoid, 4–7 cm diam. Stem twining, herbaceous, to 2 m long, 3 mm thick, puberulous. Leaves herbaceous; petiole 6–30 mm long, pubescent; leaf blade variable in outline, mostly ovate cordate to ovate-lanceolate, 20–84 × 8–45 mm, base cuneate to cordate, margin usually entire, sometimes toothed or lobed to laciniate, ciliate, apex acuminate, both surfaces pubescent, venation abaxially prominent. Inflorescence extra-axillary, sessile, cyme umbel-like, 2–8-flowered, flowers opening in succession. Pedicel 5–20 mm long, ascending, villous. Sepals subulate, 7–10 mm long, straight, coarsely pubescent to pilose. Bud with raised sinuses. Corolla 35–60 mm long, straight, tube flask-shaped, 25–40 mm long, basal c.²/₃ inflated, 7–10 mm wide, gradually narrowed at apex, limb ± cylindrical, 2–3 mm wide, abruptly dilated to 4–7 mm wide at mouth. exterior white or greenish-white with or without purplish lines and dots on upper part, both surfaces glabrous, except for a few hairs in mouth, interior with basal ¼ dark, sinuses rounded, auriculate, recurved in bud; lobes linear, to 10 mm long, revolute, apices connate to form oblate-globose cage to 5 mm high, 12 mm wide, abaxially pale, glabrous, adaxially dark green to black, minutely velvety papillose and pilose with prominent dark crinkled hairs to 3 mm long. Corona subsessile; outer lobes separate, narrowly deltoid to deltoid, spreading, entire, c.1 mm long, white, glabrous; inner lobes incumbent at base then erect, linear, c.2 mm long, apices inwardly curved minutely papillate, glabrous. Follicles erect, subparallel, 8–10 × 0.2–0.3 cm wide, glabrous. Seed not recorded.

Caprivi. Mpalile Island, *Bruyns* 2303 (NBG). **Zambia**. S: Livingstone, fl. iii, *Fairweather* 9752 (K). **Zimbabwe**. N: Mount Darwin, 14.xii.1987, *Percy-Lancaster* 2082 (SRGH). W: Shabi Camp, Hwange National Park, fl. 23.ii.1967, *Rushworth* 133 (K, PRE, SRGH). C: Empress Mine, Zhombe Communal Land, 7.vi.1981, *Pope* 1943 (SRGH). E: 15km N. of Mutare, 19.xii.1985, *Percy-Lancaster* 1545 (SRGH). **Mozambique**. T: Boroma, *Menyhardt* 706 (K).

Also in Namibia and South Africa. Most records from grassland, also open *Colophospermum mopane* or *Combretum* bushland, on sandy soils; 1000–1100 m.

Conservation status: LC, moderately frequently collected from quite a wide area.

24. **Ceropegia bonafouxii** K. Schum. in Bot. Jahrb. Syst. **33**: 327 (1903). —Huber in Mem. Soc. Brot. **12**: 161 (1958). —Dyer in Codd *et al.*, Fl. S. Africa **27**(4): 81 (1980). —Meve, Ill. Handb. Succ. Pl. Asclepiadaceae: 69 (2002). Type: Angola, Huila, *Antunes & Dekindt* 42 (B† holotype); Kubango, *Gossweiler* 2439 (BM000930062 neotype), designated by Huber in Mem. Soc. Brot. **12**: 161 (1958). FIGURE 7.3.**146**.

Ceropegia saxatilis S. Moore in J. Bot. 46: 310 (1908), illegitimate name, non Jum. & H. Perrier (1908). Type: Angola, Kubango, east of R. Kutchi, *Gossweiler* 2439 (BM000930062 holotype).

Roots not fusiform. Tuber 1–3 cm diam. Stem initially stiffly erect, then twining, few-branched, herbaceous, up to 2 m long, pubescent. Leaves herbaceous; petiole 5–10 mm long, densely pubescent; leaf blade oblong- to ovate-elliptic, 30–70 × 14–40 mm, base cuneate to sub-cordate, apex bluntly acute, both surfaces pubescent, denser abaxially, adaxial hairs sometimes with swollen bases. Inflorescence extra-axillary, sessile, cyme 1–3-flowered, flowers opening in succession. Pedicel 10–20 mm long. Sepals subulate, 5–7.5 mm long, straight, coarsely pubescent to pilose. Corolla 27–35(40) mm long; straight; tube flask-shaped, (17)22–35 mm long, basal chamber cylindrical-ovoid, more than half length of tube, 15–24 mm long, 5–8 mm wide, ± abruptly narrowed at apex, limb c.2 mm wide, limb cylindrical, expanded to c.4 mm wide at mouth, exterior white, often longitudinally purple-striate, minutely puberulent, interior white with dark green veins at base and purple band round base of neck; sinuses acute, narrowly auriculate; lobes narrowly oblong, 5–8 mm long, plicate, apices connate form a cylindrical cage c.7 mm high, c.5 mm wide, abaxially pale with dark margins and tip, more coarsely pubescent than tube, adaxially prominently dark veined, margins and apex dark green, minutely velvety papillose. Outer corona lobes each forming a pocket c.1 mm deep, c.0.7 × 0.6 mm; inner lobes incumbent-erect, c.1.2 mm long, apices pilose. Pollinia oblong-pyriform. Follicles 6–10 × c.3.5 mm wide. Seed not recorded.

Caprivi. Okavango Native Territory, 15.8 miles S of Kapupahedi Camp on track to Tamso, *De Winter & Marais* 4633 (BOL). **Zambia**. B: Mongu Dist., Lealui, fl. 19.ii.1952, White 2099 (K); Mongu Dist., Looma, Nanganda, 1060 m, fl. 21.xii.1993, *Bingham* 9905 (K). W: Mwinilunga 60 km S of Mwinilunga on road to Kabompo, fl. 25.i.1975 *Brummitt et al.* 14114 (K).

Also in Angola. Mixed woodland on Kalahari sand; 1100–1200 m.

Conservation status: LC? Apparently thinly distributed over a fairly wide area.

Listed for Botswana by Setshogo, Prelim. Checkl. Pl. Botswana (2005), but no voucher listed. Records of this species from Namibia were referred to *C. meyeri* by Bruyns in Strelitzia **34**: 87 (2014). The two species have been confused but the differences in the shape of the corolla cage, cylindrical with relatively wide corolla lobes in this species, oblate-sphaeroid with more slender lobes in *C. meyeri*, are clear cut.

25. **Ceropegia achtenii** De Wild., Pl. Bequaert. **4**: 356 (1928). —Werdermann in Bot. Jahrb. Syst. **70**: 231 (1939). —Huber in Mem. Soc. Brot. **12**: 156–158 (1958). —Meve, Ill. Handb. Succ. Pl. Asclepiadaceae: 64 (2002). —Masinde in F.T.E.A., Apocynaceae (part 2): 251 (2012). Type: D.R. Congo, Katende, *Achten* 589 (BR000000855287 lectotype, BR000000855333, BR000000855285), lectotype designated here.

Roots fibrous. Tuber depressed globose, to 2 cm diam. Stem erect, twining or trailing, little branched, herbaceous, usually uniformly hairy, occasionally glabrous. Leaves herbaceous; often ± sessile; petiole 1.5–8 mm long, hairy; leaf blade obovate to linear-lanceolate, (21)28–110 × 1.7–21(26) mm, 2–more than 10 × as long as broad, base attenuate to rounded, rarely subcordate, margin smooth, rarely sinuate, ciliate, apex acute, abaxially paler, hairy at least on midrib, adaxially glabrous or hairy. Inflorescence extra-axillary, peduncle very short to absent; cyme umbel-like, 2–8-flowered, flowers opening in succession. Pedicel 3–9 mm long, hairy. Sepals triangular-lanceolate, 2–4.5(5.5) × 0.3–0.7 mm, abaxially puberulous. Bud with flat or slightly raised sinuses and narrowly cylindrical beak. Corolla 11–36 mm long, straight or with asymmetric basal chamber; tube 6–23 mm long, basal chamber ovoid, sometimes ill-defined, 2.5–7 mm long, (1.6)2.2–5.5 mm wide, limb 0.8–2 mm diam. above swelling, slightly expanded at mouth to 1.9–4.4(7.5) mm diam., exterior greenish to deep purple or purplish-brown, glabrous to puberulous, sinuses acute, flat to narrowly auriculate, lobes linear to oblong-linear from triangular base, (4)6–19 mm long, replicate, apices connate to form slightly beaked to clavate cage, 1.5–5(8.5) mm wide, abaxially glabrous or setulose, margins ciliate, adaxially uniformly brown or pale to vivid green merging into paler base, glabrous or setulose, particularly near sinuses. Corona stipitate, cupular, 2–3 mm long; outer lobes forming shallow pockets, broadly rectangular, 3-toothed, hyaline, glabrous; inner lobes converging over gynostegium, narrowly linguiform or linear clavate to spathulate, 1.2–1.5 mm long, hyaline or purplish, glabrous. Follicles erect or narrowly divergeant, linear-fusiform, (6)7.5–15 × 0.2–0.3 cm wide. Seed not recorded.

The species as a whole is recorded from Zambia, Zimbabwe, Malawi and Mozambique in the Flora area plus Angola, D.R. Congo, Tanzania and Togo outside this area.

The type collection in BR includes 3 sheets, each with 2 or more plants and marked as 'TYPUS' and so a second step lectotypification is proposed here. The selected individual has the best preserved flowers.

Most collections seen fall clearly and easily into two subspecies. The following problem collections are each anomalous in several characters and could prove to represent distinct taxa.

Williamson 2292 is a diminutive twiner with relatively large, ± ovate, basal leaves contrasting with the much smaller upper leaves, stems with lines of hairs, solitary flowers with glabrous exteriors, truncate outer corona lobes and narrowly oblong, not spatulate, inner corona lobes.

Hooper & Townsend 738 resembles subsp. *achtenii* in habit and leaf form but has distinctively small corollas (c.11 mm long) with glabrous exteriors and long slender calyx lobes half as long as the corolla, rather reminiscent of those of *C. abyssinica* and may be the product of introgression from that species. All other collections of *C. achtenii* s. lat. have the calyx less than 1/5 as long as the corolla.

Goyder & Adams 3232 is possibly related but it is only known from one collection with only the one detached flower. This has a distinctive slender corolla with probably rather conspicuous longitudinal lines and long, coloured, hairs along the lower 2/3 of the lobe margins.

Leaf blade 2.1–9.6 × as long as broad; corolla 11–25 mm long, exterior of tube puberulous, rarely glabrous; stems mostly erect, sometimes twining . . . . a) subsp. *achtenii*
Leaf blade 11–40 × as long as broad; corolla (21)24–36 mm, exterior of tube usually glabrous; stems mostly twining, rarely erect . . . . b) subsp. *adolfi*

a) Subsp. **achtenii**. —Huber in Mem. Soc. Brot. **12**: 156–158 (1958). —Lisowski & Malaisse in Bull. Jard. Bot. Natl. Belg. **44**: 403–405, fig.1 (1974). FIGURE 7.3.**146**.

Stem erect (or twining), simple, sometimes branched above, 50–90 cm tall, usually pubescent, occasionally glabrous. Petiole 1.5–5 mm long, pilose; leaf blade obovate to lanceolate, (21)28–77 × (3.5)9–21(26) mm, 2.1–9.6 × as long as broad, lower leaves shorter and broader, diminishing in size towards stem apex, base cuneate to rounded, rarely subcordate, abaxially pubescent, adaxially glabrous. Pedicel 3–7 mm long, ciliate. Bud with slightly raised sinuses. Corolla 11–25 mm long, tube 6–11 mm long, basal chamber 3–7 mm long, (1.6)2.2–4 mm wide, exterior greenish to deep purple, puberulous, rarely glabrous, lobes (4)6–12 mm long, apices connate to form fusiform to clavate cage, 1.5–4.4 mm wide. Inner corona lobes linear clavate to spatulate, 1.2–1.4 mm long, partly purplish to carmine.

Zambia. N: Abercorne, road to Kaka village, 1740 m, fl. 20.ii.1960, *Richards* 12518 (K). W: Mwinilunga Dist., near Luao River, fl. 27.xii.1937, *Milne-Redhead* 3835 (BM, K). E: Lusaka Dist., 6 mi E of Lusaka, 4200 ft, fl. 24.i.1958, *A.E. King* 409 (K). **Malawi**. S: Mt. Mlanje, Chambe cableway path, fl. 25.i.1967, *Hilliard & Burtt* 4623 (K, NU). N: Chitipa Dist., nr Mtikafundi on Stevenson road, S of Misuku Hills, 1220 m, fl. 4.iii.1982 *Brummitt et al.* 16315 (K). **Mozambique**. N: Ribáuè, Hamateze Mountain, fl. 16.ii.1962, *Carvalho* 510 (K, LMU).

Also in Angola, D.R. Congo, Tanzania, Togo. Mostly among grasses in woodland, mostly *Brachystegia*, also *Cryptosepalum*, *Isoberlinia* or *Acacia*, usually on sandy soils, rarely shallow soil over rocks; 1050–1750 m.

Conservation status: LC, moderately frequently collected from quite a wide area.

The only collection seen from Mozambique *Carvalho* 510 (K) matches quite well apart from the glabrous exterior and prominent pale transverse band on the corolla tube reminiscent of some southern forms of *Ceropegia abyssinica*.

b) Subsp. **adolfi** (Werderm.) H. Huber in Mem. Soc. Brot. **12**: 158 (1958). Type: Tanzania, Rungwe: Kyimbila, *Stolz* 1215 (B† holotype, K000305482 lectotype, M0110206, S-G-1293, P00109625, WAG0000247, Z000001589), lectotype designated here. FIGURE 7.3.**146**.

Ceropegia adolfi Werderm. in Bot. Jahrb. Syst. **70**: 214 (1939). Type: Tanzania, Rungwe, Kyimbila, *Stolz* 1215 (B† K000305482 M0110206, S-G-1293, P00109625, WAG0000247, Z000001589).

Ceropegia adolfi var. *gracillima* Werderm. in Bull. Jard. Bot. État. Bruxelles **15**: 232 (1938). Type: D.R. Congo, upper Katanga, Kisamba, Ferme Selemo, *Quarré* 2404 (BR holotype), invalid name (Art. 35.1).

Stem twining or trailing, little branched, internodes 4–15 cm long, 0.5–1.5 mm diam., uniformly subhirsute/puberulent. Petiole 2–8 mm long, laxly hirsute; leaf blade lanceolate to linear-lanceolate, 50–110 × 1.7–4.4(6.7) mm, more than 10 × as long as broad, base attenuate to narrowly cuneate, abaxially grey-green, puberulent on midrib only, adaxially scabrid-puberulous. Peduncle very short. Pedicel 4–9 mm long, pubescent. Bud with almost flat sinuses. Corolla (21)24–34(36) mm long; tube (9)13–23 mm long, basal chamber 2.5–6 mm long, 2.5–5.5 mm wide, exterior purplish-brown, glabrous, rarely puberulous, lobes 10–19 mm long, apices connate to form slightly beaked to fusiform cage, 2–5(8.5) mm wide. Inner corona lobes narrowly linguiform to subspathulate, c.1.5 mm long, hyaline.

Zambia. N: Abercorne, Kawimbe Rocks, old Sumbawanga road, 1500 m, 31.xii.1962, *Richards* 17081 (K). **Malawi**. N: 45 km west of Karonga, fl. 12.iii.1953, *Williamson* 201 (BM). S: Mt. Mlanje, Chambe - Tuchila path, fl. 23.i.1967, *Hilliard & Burtt* 4601 (K, NU); Zomba Plateau, path to Namitembo View Point, from Chingwe's Hole, 6100 ft, fl. 1.iv.1980 *Blackmore* 1179 (BM, K).

Also in Tanzania. With grasses on rocky outcrops, less often in *Brachystegia* woodland on sandy soil; 1250–2100 m.

Conservation status: LC, moderately frequently collected from quite a wide area.

26. **Ceropegia leptotes** M.G. Gilbert, sp. nov. A member of sect. *Laguncula* on the basis of its basal tuber, non-succulent stems and leaves and sessile inflorescence of small flowers with adaxially minutely velvety papillose upper part of the corolla lobes, distinguished from all other members of the group by its very delicate habit and solitary small flowers with short corolla lobes ornamented by purplish marginal hairs almost twice as long as the lobes themselves. Type: Zambia, Chinsali Dist., N Machipira Hill, Shiwa-Ngandu, 1500 m, 16.i.1959, *Richards* 10700 (K001400201 holotype). FIGURE 7.3.**146**.

Roots fibrous. Tuber globose, to 1 cm diam. Stem twining, unbranched, herbaceous, to 30 cm long, < 0.5 mm thick, with 1 or 2 minutely puberulous lines along internodes. Leaves herbaceous; basal leaves: petiole 2–6 mm long, margins puberulent; leaf blade ovate to suborbicular, 12–18 × 6–14 mm, upper leaves progressively smaller, base tapering to cuneate, margin entire, apex acute to obtuse, abaxial surface very sparsely puberulent on major veins, otherwise glabrous, adaxially and margins shortly pubescent. Inflorescence extra-axillary, sessile, cyme 1-flowered. Pedicel 7–8 mm long, slender, minutely puberulent. Sepals lanceolate, c.1.7 × 0.5 mm wide, adpressed to corolla, midrib puberulent. Corolla c.10 mm, straight or slightly bent above basal inflation; tube 7–8 mm long, basal chamber subglobose, c.2 mm long, 2.5 mm wide, ± abruptly narrowed at apex, limb cylindrical, c.1 mm wide, limb widening gradually to 2 mm wide at mouth; exterior unmarked, both surfaces glabrous, interior with scattered darker spots, sinuses obtuse, narrowly auriculate; lobes 1.8–2 mm long, revolute towards apex, apices connate to form broadly ovoid to cylindrical cage, abaxially pale adaxially dark brown(?), minutely velvety papillose, with very prominent purple cilia to 7 mm long. Corona apparently sessile; outer lobes forming shallow pockets, spreading, rounded, entire, lower than anthers, pale, glabrous; inner lobes erect, narrowly linguiform to spathulate, c.1.1 mm long, dark, glabrous. Follicles and seed not recorded.

Endemic, known only from the type collection. Twining up grasses growing in crevices of large rocks on mountain side; 1500 m.

Conservation status: DD, known from a single collection from a relatively undercollected area.

The only collection seen included 5 plants, all very delicate twiners hence the choice of epithet. The very short corolla lobes with marginal cilia up to more than three times as long as the lobes are immediately diagnostic.

27. **Ceropegia paricyma** N.E. Br. in Bull. Misc. Inform. Kew **1898**: 309 (1898); in F.T.A. **4**(1): 457 (1903). —Werdermann in Bot. Jahrb. Syst. **70**: 218 (1939). — Huber in Mem. Soc. Brot. **12**: 152–153 (1958). —Meve, Ill. Handb. Succ. Pl., Asclepiadaceae: 94 (2002). —Masinde in F.T.E.A., Apocynaceae (part 2): 238 (2012). Type: Malawi, Lake Malawi ["Nyassa"], *Simons* s.n. (BM holotype, not found, K000305633). FIGURE 7.3.**147**.

    Ceropegia dentata N.E. Br. in Bull. Misc. Inform. Kew **1909**: 327 (1909). Type: Mozambique, Macome, Madanda, *Johnson* 100 (K000305630 holotype).

    Ceropegia mutabilis Werderm. in Bot. Jahrb. Syst. **70**: 218 (1939). Type: Tanzania, Lindi, c.65 km W of Lindi, *Schlieben* 6152 (B† holotype).

Roots not fusiform. Tuber globose, 1–2 cm diam. Stem twining, up to 2 meters long, pilose to glabrous. Leaves membranous or slightly succulent; petiole 8–50 mm long; leaf blade oblong-ovate, occasionally hastate to ± 3-lobed, 25–120 × 12–60 mm, base tapering to cordate, margin mostly entire, sometimes with 1 or 2 large teeth at base, apex obtusely acuminate to acuminate, both surfaces pubescent. Inflorescence extra-axillary, subsessile; cyme umbel-like, 2–6-flowered, flowers opening in succession. Pedicel 4–10 mm long, glabrous to pubescent. Sepals subulate-lanceolate, 1.5–3 mm long, glabrous to pubescent. Bud with recurved sinuses. Corolla (10)16–22 mm long; straight or nearly so; tube 7–15 mm long, basal chamber ovoid, (2.5)4.7–8 mm wide, narrowed at apex, limb cylindrical, 1.2–1.5 mm wide, widening to 2.5–3.5 mm at mouth; exterior whitish, glabrous, interior with lower part of basal chamber dark, otherwise pale; sinuses rounded, obviously auriculate; lobes linear-spathulate from short triangular base, 2.5–8 mm long, erect, replicate, apices connate to form cylindrical cage, adaxially with lower part black or purple black, upper part green, sparsely pilose. Corona with outer lobes connate at base, spreading, bifid, teeth subulate to deltoid, glabrous; inner lobes erect, linear, apices sometimes slightly hooked. Pollinia bright yellowish. Follicles narrowly divergent, 7–8 × 0.2–0.3 cm wide at middle, glabrous. Seed not recorded.

Caprivi. Impalele Island, fl. 29.iv.1981, *Bruyns* 2302 (NBG). **Zambia**. N: Mfuwe, Mpika, 2000 ft, fl. 29.i.1969, *Astle* 5439 (K). C: Lusaka, Leopard's Hill Road, fl. 22.ii.1967, *Nath Nair* 20436 (NDO). S: Choma Dist., 10 km from Sinazongwe to mouth of Zongwe River, fl. 29.xii.1958, *Robson & Angus* 1012 (BM, K, PRE). **Malawi**. S: Mt. Mlanje, Chambe basin, fl. 21.i.1967, *Hilliard & Burtt* 4570 (NU). C: Chintembwe Mission, nr. Nchisi, 1370 m, fl. 22.ii.1959, *Robson & Steele* 1712 (BM). **Mozambique**. MS: Macone, Madanda, 70 m, fl. i, *Johnson* 100 (K).

Also in Namibia and Tanzania. Growing in deep shade in thiclets with *Acacia* or *Diospyros senensis* (Namibia: among dense bushes and trees; F.T.E.A. area, open *Millettia–Cussonia* forest and thickets on hills and in gulleys); 60–1400 m.

Conservation status: LC, moderately frequently collected from quite a wide area.

The holotype of *Ceropegia paricyma* could not be found in BM where it would be expected to be. Masinde (2012) stated that the sheet in K was the holotype but this consists of only a detached flower which makes this extremely unlikely as the author clearly had more complete material available to him and the main set of Simons collections is in BM. If the BM sheet remains lost, then the fragmentary K sheet would have to be the lectotype and it would be advisable to designate an epitype.

28. **Ceropegia nephroloba** (H. Huber) Bruyns in S. African J. Bot. **112**: 431 (2017). Type: Tanzania, Matagoro Hills, S of Songea, *Milne-Redhead & Taylor* 8863 (K000305488 holotype, K spirit 3408.000), EA000001839, G). FIGURE 7.3.**147**.

    Ceropegia sobolifera var. *nephroloba* H. Huber in Mem. Soc. Brot. **12**: 201 (1958). —Masinde in F.T.E.A., Apocynaceae (part 2): 238 (2012).

Roots somewhat fleshy. Tuber globose (to oblate), to 1(2.5) cm diam. Stem twining or sometimes trailing, branching, herbaceous, to 60 cm or more long, delicate, slender with 1 or 2 puberulous lines along internodes. Leaves herbaceous; petiole 5–10(22) mm long, minutely

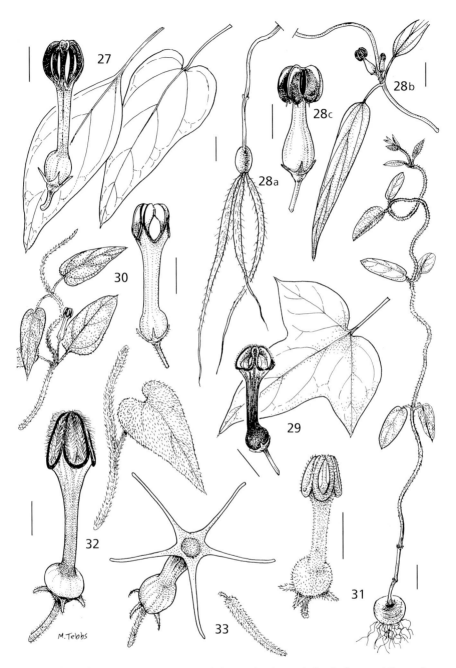

Fig. 7.3.**147**. —**27**. CEROPEGIA PARICYMA, leaves showing variation in form and flower, from *Astle* 5439. —**28**. CEROPEGIA NEPHROLOBA, 28a rootstock, 28b part of flowering stem, 28c flower, from *Robson & Steel* 1588. —**29**. CEROPEGIA CLAVILOBA, lobed leaf, from *Robson & Steel* 1712, and flower, from *Zimba et al.* 581. —**30**. CEROPEGIA NAMULIENSIS, part of flowering stem and flower, from *Brummitt et al.* 16393. —**31**. CEROPEGIA MWINILUNGENSIS, habit and flower, from *Milne-Redhead* 4377. —**32**. CEROPEGIA PAPILLATA, part of flowering stem, from *Jackson* 1655, and flower, from *Brummitt* 14352. —**33**. CEROPEGIA SCHLIEBENII, leaf and flower, from *Fanshaw* 12171. Scale bars: habits and stems = 10 mm; flowers = 5 mm. Drawn by Margaret Tebbs.

ciliate; leaf blade mostly narrowly lanceolate, sometimes ovate-hastate, almost as broad as long, 10–140 × 5–20 mm, base narrowly cuneate to truncate or shallowly cordate, margin usually entire, broad-leafed forms with margin irregularly deeply lobed, sparsely, minutely ciliolate, apex acute to subacute, both surfaces glabrous except sometimes for main veins abaxially. Inflorescence extra-axillary, sessile, cyme 1–3(5)-flowered. Pedicel 3–5 mm long. Sepals linear-lanceolate, c.2.5 × 0.7 mm wide, abaxially minutely puberulent. Corolla 10–20 mm long, straight; tube urceolate, 6–12 mm long, basal c. ⅔ inflated, 2.5–4 mm wide, tapering to 1.5–2 mm wide near mouth; exterior yellowish-green to greenish white, both surfaces glabrous, interior pinkish-purple; sinuses auriculate; lobes crescent-shaped in side view, 3–7 × c.2 mm wide, strongly plicate, apices connate to form broadly ovoid cage, abaxially yellowish green, mostly hidden, glabrous, margins sometimes shortly ciliate, adaxially purple maroon. Corona c.3.5 × c.3.5 mm diam.; outer corona lobes rectangular, rotately spreading; inner lobes converging over gynostegium, linear, c.3 mm long, lower part maroon, glabrous, Follicles and seed not recorded.

Zambia. E: Lutembwe River just below Great East Road bridge, fl. 9.i.1959, *Robson* 1111 (BM, K, PRE). **Zimbabwe**. N: Gokwe Dist., Sengwa Research Station, fl. 20.i.1969, *Jacobsen* 4201 (PRE). C: 20 km on Sanyati Rd, north of Kadoma, fl. 2.i.1989, *Percy-Lancaster* 2480 (SRGH). **Malawi**. C: Junction of Salima and Lower Chitala Roads, fl. 13.ii.1959, *Robson & Steele* 1588 (BM, K, MAL). **Mozambique**. Z: 12 km W of Malenga, fl. 25.xii.1998, *Bruyns* 7718 (K).

Also in Tanzania. *Brachystegia* woodland, among rocks; 750–900 m.

Conservation status: LC, moderately frequently collected from quite a wide area.

The elevation of Huber's variety to full species rank was first proposed by Peter Bally in notes on herbarium specimens but he never validated the new name. The differences in rootstock, habit and corolla form (both tube and lobes) are very consistent and there is rather little to connect this species with *Ceropegia sobolifera*, endemic to the Semien Mountains of N Ethiopia, which is strictly rhizomatous, lacking tubers, with robust trailing to pendent stems, a corolla tube with an obovoid basal chamber sharply delineated from the rest of the tube and corolla lobes not auriculate at their bases and remaining joined for the apical third to form an apical cap.

29. **Ceropegia claviloba** Werderm. in Bot. Jahrb. Syst. **70**: 221 (1939) – Huber in Mem. Soc. Brot. **12**: 153–154 (1958). —Meve, Ill. Handb. Succ. Pl. Asclepiadaceae: 72 (2002). —Masinde in F.T.E.A., Apocynaceae (part 2): 236 (2012). Type: Tanzania, Singida/Dodoma Dist., Turu, *Peter* 33836 (B† holotype); Zimbabwe, Umtali, Engwa, 1830 m, 3.ii.1955, *Exell, Mendonça & Wild* 147 (BM001217354 neotype), designated by Huber in Mem. Soc. Brot. **12**: 154 (1958). FIGURES 7.3.**147**, 7.3.**148**.

Roots fibrous. Tuber depressed globose. 1.2–3.2 cm diam. Stem twining, sparsely branched, herbaceous or rarely slightly succulent, up to at least 2 meters long, internodes c.8–15 cm long, c.0.5–1.5 mm thick, mostly glabrous, sometimes minutely puberulent on one side at base of internode. Leaves herbaceous or rarely slightly succulent; petiole 30–40 mm long, pubescent; leaf blade cordate or subsagitate (to 3-lobed), 30–70 × 20–30(40) mm, base cordate, margin in the upper part subsinuate to entire or basal lobes sometimes with acute, lateral teeth, ciliolate, apex subacuminate to long acuminate, abaxial surface quite glabrous except short hairs on veins, adaxial surface sparely laxly pilose. Inflorescence extra-axillary, sessile to subsessile, peduncle to 3.5 mm long, laxly pilose, cyme umbel-like, 4–8-flowered, flowers opening in succession. Pedicel 4–8 mm long, almost glabrous. Sepals subulate deltoid, 1–2 × c.0.5 mm, apex recurved, glabrous or with a few minute hairs. Corolla 10–16(17) mm long, straight or slightly curved above basal inflation; tube 7–11.5 mm, basal chamber inconspicuous, 3–4 mm wide, gradually narrowed at apex, limb cylindrical, 1.1–2.2 mm wide, widening to 2–4 mm at mouth; exterior pale to bottle green except for narrow whitish base, glabrous, interior pale green with dark markings in throat; sinuses rounded, ± flat; lobes bases ± deltoid, (2.5)4–7 mm long, flattened, then narrowed and plicate, upper part clavate-dilatate, apices connate into obovoid cage, wider than mouth, adaxially with base yellow-green with bright green margins, upper half and keel black, margins and to lesser extent keels adaxially with spreading dark purple hairs. Corona shallowly

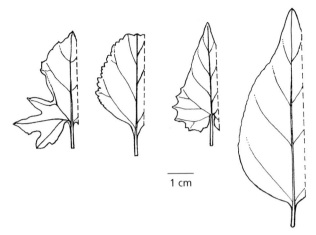

Fig. 7.3.**148**. Variation in leaf form of **29**. CEROPEGIA CLAVILOBA, left to right from *Chase* 1993, *Robson & Steel* 1650, *Stolz* s.n. and *Chase* 3556. Drawn by Margaret Tebbs.

cupshaped, outer lobes connate, ± deltoid, hyaline, Follicles narrowly divergent, 5.5–12 × 0.2–0.3 cm wide at middle. Seed blackish with brown margin, 6–7(9) × c.3 mm; coma c.12 mm long.

Zambia. W: Mwinilunga Dist., Matonchi Farm, fl. 24.xii.1937, *Milne-Redhead* 3848 (K). **Zimbabwe**. E: Umtali Dist., Engwa 1830 m, fl. 3.ii.1955, *Excell, Mendonca & Wild* 147 (BM); Umtali Dist., Commonage, S slope of Inyamatshera Mt. range, 3900 ft, fl. 23.ii.1950, *Chase* 1993 (BM); Inyanga Dist., Nyamziwa Falls, 5400 ft, 12.i.1951, *Chase* 3556 (BM). **Malawi**. N: Mugesse, Misuka, *Jackson* 1164 (K). C: Chintembwe Mission near Nchisi, fl. 22.ii.1959, *Robson & Steele* 1712 (K, PRE). S: Mt. Mlanje, Chambe basin, 6000 ft, fl. 21.i.1967, *Hilliard & Burtt* 4570 (K).

Also in D.R. Congo and Tanzania. Mostly in evergreen forest, also *Brachystegia* woodland; 1250–2000 m.

Conservation status: LC, moderately frequently collected from quite a wide area.

Leaf shape is very variable in this species with some forms developing coarse marginal teeth on the basal lobes. Two collections deserve species mention: *Chase* 1993 has very striking deeply divided leaves but can easily be interpreted as an extreme development of the more frequent shallowly lobed forms; *Robson & Steele* 1650 is more problematic with the lowermost leaves obovate with cuneate bases, margins mostly dentate except at the base and rounded apices. However the very young upper leaves look as if they will be quite similar to those of the preceding collection, the collector states 'leaves entire or toothed' and it seems likely that this collection is just an unusual variant of *C. claviloba*. A not dissimilar range of variation has been observed within a population of the Ethiopian endemic *C. ringens* A. Rich.

30. **Ceropegia namuliensis** Bruyns in Aloe **41**: 76 (2004). —Masinde in F.T.E.A., Apocynaceae (part 2): 237 (2012). Type: Mozambique, Zambesia, N flank of Namuli, *Bruyns* 9725 (BOL140270 holotype, K001236077). FIGURE 7.3.**147**.

Roots not fusiform. Tuber depressed globose, 1–3 cm diam., 1–1.5 cm deep. Stem erect or twining, 2–40 cm long sometimes with very short internodes, wiry, uniformly sparsely pubescent. Leaves herbaceous; petiole 2–10 mm long; leaf blade narrowly to broadly ovate, sometimes hastate, rarely oblanceolate, (10)14–65 × (5)10–35 mm, base cordate to sagittate, rarely narrowly cuneate, margin entire or irregularly toothed, apex acuminate to rounded or shallowly emarginate, sometimes mucronate, both surfaces glabrous to densely finely pubescent.

Inflorescence extra-axillary, mostly sessile, occasionally pedunculate, peduncle to 11 mm, pubescent, cyme umbel-like, 2–4(10)-flowered. Pedicel 5–10 mm long, glabrous or densely hairy. Sepals linear-lanceolate, 1.5–3 × 0.5–1 mm, acute, abaxially puberulous. Corolla 11–22 mm long, slightly curved; tube 7–18 mm, basal chamber globose to ovoid, 2–3.5(4.5) mm wide, gradually narrowed at apex, limb cylindrical, 1.5–2 mm wide, widening to c.4.5 mm wide at mouth, exterior translucent cream at base of inflation, then uniformly purplish or spotted purplish, glabrous or occasionally sparsely pilose, interior pale purple at throat, glabrous; sinuses acute, flat; lobes linear from triangular base, 3–5 × c.1 mm wide, plicate almost to base, apices connate to form broadly ovoid cage, marginal hairs purplish, adaxially base whitish to dark purple, plicate tip bright green, apical part uniformly pale pilose. Corona stipitate, c.3 × 2.8 mm diam., outer corona deeply cupular, lobes bifid, adaxially purplish, ciliate; inner lobes converging over gynostegium, c.2 mm long, dorsiventrally compressed, purplish at base, paler above, Pollinia pale orange. Follicles narrowly divergent, (4)5–7 × 0.2–0.3 cm wide at middle. Seed shiny brown with paler margin, 4–4.5 × c.2 mm; coma c.12 mm long.

Zambia. W: Bancroft, fl. 30.i.1964, *Fanshaw* 8242 (K). C: Lusaka Dist. Kanyanja, Kafue Powerline road 5 km S, then 2 km W 1350 m, fl. 25.ii.1996, *Harder & Bingham* 3514 (K). E: Lutembwe R. just below Gt. East Road bridge, 900 m, 9.i.1959, *Robson* 1115 (K). **Malawi**. C: Kongwe Mt., nr Dowa, 1525 m, fl. 18.ii.1959, *Robson* 1650 (K); Mulanje Dist., Mt. Mulanje, Litchenya path 1700 m, fl. 18.ii.1982, *Brummitt et al.* 15946 (K). **Mozambique**. Z: Zambesia, northern slopes of Namuli, 1200 m, fl. 4.i.2004, *Bruyns* 9725 (BOL, K).

Also in Tanzania. *Brachystegia* woodland, evergreen forests; (300)900–1700 m.

Conservation status: LC, moderately frequently collected from quite a wide area.

31. **Ceropegia mwinilungensis** M.G. Gilbert, sp. nov. A member of sect. *Laguncula* on the basis of its basal tuber, non-succulent stems and leaves and sessile inflorescence of small flowers with adaxially minutely velvety papillose upper part of the corolla lobes, distinguished from other members of the section by the combination of the 2 lines of hairs along each stem internode, the very densely puberulous exterior of the corolla and the rounded, strongly auriculate sinuses between the oblong, shallowly plicate, corolla lobes. Type: Zambia, Mwinilunga, near Matonchi Farm, 27.i.1938, *Milne-Redhead* 4377 (K spirit 6714 holotype). FIGURE 7.3.**147**.

Tuber subglobose, c.1 cm diam. Stem twining, unbranched, herbaceous, c.30 cm long, c.1 mm thick, with 2 lines of hairs along internodes. Leaves herbaceous; petiole to 6 mm long; leaf blade oblong-elliptic, to 23 × 11 mm, base rounded, margin entire, apex slightly acuminate, abaxial surface sparsely minutely stiffly hairy, denser on main veins. adaxial surface dull green, glabrous. Inflorescence extra-axillary, subsessile; cyme 1- or 2-flowered. Pedicel short, densely pubescent. Sepals lanceolate, c.3 mm long, densely pubescent. Corolla c.14 mm long, tube c.10 mm long, slightly curved above basal chamber; basal chamber subglobose, c.4 × 4 mm wide, abruptly narrowed at apex, limb cylindrical, c.1.5 mm wide, widening to c.3.5 mm at mouth, exterior greenish at base, tube mainly dull purplish, densely minutely pubescent, interior glabrous, basal chamber with lines of minute tubercles along veins; sinuses rounded, strongly auriculate; lobes oblong, c.4 mm long, shallowly plicate, apices connate to form ovoid-conical cage widest at base, abaxially pubescent, adaxially dull purplish black, glabrous. Corona sessile; outer lobes oblong, spreading, truncate to shallowly emarginate or obscurely 2-toothed, almost as high as inner lobes, glabrous; inner lobes erect, linear-oblong, glabrous. Follicles not seen.

Known only from the type specimen. *Brachystegia* woodland, among grass.

Conservation status: DD, known only from the one collection from an undercollected region.

This species is based on a single small plant preserved in spirit. It was initially identified as '*C. achtenii* var. *achtenii*' but differs in so many features that it is here described as a distinct species. It is worth noting that two other collections from the same area by the same collector in 1937-8: *Ceropegia cataphyllaris* Bullock and *C. prostrata* (E.A. Bruce) Bruyns, are still only known from the original collections.

32. **Ceropegia papillata** N.E. Br. in Bull. Misc. Inform. Kew **1898**: 308 (1898); in F.T.A. **4**(1): 452 (1903). —Werdermann in Bot. Jahrb. Syst. **70**: 219 (1939). —Huber in Mem. Soc. Brot. **12**: 152 (1958), excl. *C. cordiloba.* —Cribb & Leedal, Mountain Fl. S. Tanzania: 101, t.23a (1982). —Meve, Ill. Handb. Succ. Pl. Asclepiadaceae: 94 (2002), excl. *C. cordiloba.* —Masinde in F.T.E.A., Apocynaceae (part 2): 234–235 (2012). Type: Malawi, Zomba Mts., 5500 ft., *Whyte* s.n. (K000305636 holotype). FIGURE 7.3.**147**.

    Ceropegia stolzii Schltr., invalid name, in sched.

Roots not fusiform. Tuber subglobose, to 4 cm diam. Stem twining, herbaceous, up to 3 m tall, uniformly pubescent. Leaves herbaceous; petiole 30–50 mm long, pubescent; leaf blade narrowly to broadly ovate, rarely obscurely 3-lobed to hastate, 30–80(120) × 12–65(75) mm, base deeply cordate with rounded sinus, margin usually entire, rarely shallowly toothed, apex acuminate, abaxial surface pale, densely pubescent to subtomentose, adaxial surface darker green, pubescent. Inflorescences usually many, sometimes paired at a node, extra-axillary, sessile or subsessile, peduncle to 3 mm long, cyme umbel-like, up to 20-flowered, several flowers open at once. Pedicel 8–10(12) mm long, pubescent. Sepals linear-lanceolate to lanceolate, 3–8 × to 1.5 mm wide, acute, abaxially sparsely to densely pubescent, adaxially glabrous. Bud with sinuses raised and reflexed. Corolla 19–24 mm long, straight or slightly curved; tube 12–18 mm long, basal chamber ovoid, 4–7 × 4–5 mm, abruptly narrowed at apex, limb cylindrical, c.1 mm wide, widening to 4–6 mm at mouth, exterior with basal inflation white, and limb white or pale to lime green, rarely flushed red, glabrous or rarely minutely puberulent (Zimbabwe); interior purple at top of basal inflation and base of limb, otherwise pale, glabrous (papillate at base), sinuses rounded, spreading and gaping, strongly narrowly auriculate; lobes linear to oblong-linear from triangular base, 4–8 mm long, plicate, apices loosely connate to form ovoid-conical cage distinctly narrower to slightly broader than mouth, abaxially glabrous, margins and sometimes inner keel white villose, adaxially white with minutely papillose blackish green or purple tip. Corona stipitate, c.2.5 × 2.5 mm diam., mostly white; outer lobes each forming a pocket, lobes oblong ovate, not connate laterally, erect, shortly bifid, glabrous. Follicles narrowly divergent, linear-fusiform, 7.5–12 × 0.2–0.3 cm wide, glabrous. Seed dark brown or blackish with pale margin, 5–6 × 2–2.5 mm; coma 10–15 mm long.

Zambia. N: Abercorne, Ndundu, 1740 m, fl. 7.iii.1962. *Richards* 16225 (K). W: 2 mi SW of Ndola in forest reserve 4000 ft, 24.i.1953, *Draper* 24 (K). C: Mt. Makulu Res. Station, fl. 1.ii.1960, *Angus* 2131 (K). E: Lusaka Dist., 7 mi E Lusaka, 1280 m, fl. 9.ii.1958 *King* 420 (K). **Zimbabwe**. E: Inyanga, Inyangombi River, fl. 19.i.1948, *Chase* 600 (PRE); near fence at ram tank Clogheen, farm W of Old Mutare, 950 m, fl. 26.iii.1962, *Evans* 6335 (K). **Malawi**. N: Chombe Plateau, Mlanje Mts., 22.iii.1958, *Jackson* 2164 (K). C: Namitete River below bridge on Lilongwe–Chipata (Fort Jameson) road, 1150 m, fl. 5.ii.1959, *Robson* 1454 (BM, K, PRE). S: Mount Mulanje, Chambe basin, ± 3000 m, fl. 19.i.1967, *Hilliard & Burtt* 4509 (NU).

Also in D.R. Congo and Tanzania. Forest and woodland margins, often twining up tall grasses, sometimes on rock outcrops or termite mounds; 950–1850 m.

Conservation status: LC, moderately frequently collected from quite a wide area.

33. **Ceropegia schliebenii** Markgr. in Notizbl. Bot. Gart. Mus. Berlin-Dahlem **11**: 404 (1932). —Werdermann in Bot. Jahrb. Syst. **70**: 195 (1939). —Huber in Mem. Soc. Brot. **12**: 165–166 (1958). Type: Tanzania, Iringa, Lupembe, near Ruhudje, *Schlieben* 143 (B† holotype, M0110218 lectotype, BM000528531, BR0000008863218, BR0000008863676, G00022143, K000305474, P00109686, P00109687, S-G, Z) lectotype designated by Huber in Mem. Soc. Brot. **12**: 166 (1958). FIGURE 7.3.**147**.

    Ceropegia stenoloba var. *schliebenii* (Markgr.) Masinde in F.T.E.A., Apocynaceae (part 2): 242 (2012).

Roots not fusiform. Tuber present, Stem trailing or twining, many branched, herbaceous, c.1 mm thick, angular, with 1 or 2 puberulous lines along internodes. Leaves herbaceous; petiole 2–10 mm long, densely pubescent; leaf blade lanceolate to linear-lanceolate, 10–30 × 2–8 mm, leaves on lateral branches often distinctly shorter, base tapering to cuneate or sometimes subcordate, margin entire, setulose-ciliate, apex bluntly acute, both surfaces densely pubescent. Inflorescence extra-axillary, sessile; cyme 1–3(5)-flowered. Pedicel 6–20 mm long, slender, pubescent. Sepals linear, 1–3 × to 0.6 mm wide, mostly somewhat recurved near apex, pubescent. Corolla 17–30 mm long, usually slightly S-shaped with oblique mouth; tube 7–9 mm long, basal chamber often not enlarged, c.½ as long as tube, 2.5–4 mm wide, limb cylindrical, widening to 1.8–3 mm wide at oblique mouth, exterior greenish white, glabrous, interior conspicuously dark striate in lower half, then suffused dull red or purplish, glabrous; sinuses obtuse, flat to very minutely revolute; lobes linear, (4)10–18 × c.0.5 mm wide, revolute, apices free, rotately spreading (or occasionally remaining coherent at apex) abaxially uniformly pale green to reddish with darker veins, both surfaces glabrous (to puberulous), rarely margins sparsely short white setose near base. Outer corona lobes each forming a pocket, bluntly 2-toothed; inner lobes erect, loriform, c.2 mm long, Pollinia subglobose. Follicles and seeds not recorded.

Zambia. N: Nyika Plateau, i.1974, *Laing* s.n. (image in www.zambiaflora.com). **Malawi**. N: Rumphi Dist., Chelinda bridge rocks, 7500 ft, fl. 8.i.1974, *Pawek* 7896 (K); same locality, fl. 10.i.1967, *Hilliard & Burtt* 4390 (K); same locality, 2225 m, fl. 27.ii.1982. *Brummitt et al.* 16148 (K).

Also in Tanzania, Rwanda and Burundi. Shallow soils overlying granitic rock outcrops within montane grassland; 2200–2300 m.

Conservation status: LC?, locally frequent, known from quite a wide area.

Records of *Ceropegia stenoloba* Chiov. from the Flora area seem to be based on material better placed within this species. The two taxa are closely related but *C. stenoloba* can be distinguished by several characters such as the uniformly hairy to glabrous stems, and broader leaves, 15–35 mm wide.

Sect. **Coreosma** H. Huber in Mem. Soc. Brot. **12**: 31 (1958).

34. **Ceropegia stapeliiformis** Haw. in Philos. Mag. Ann. Chem. **1**: 121 (1827), as "stapeliaeformis". —Huber in Mem. Soc. Brot. **12**: 107–108 (1958). —Dyer in Codd *et al.*, Fl. S. Africa **27**(4): 50–51, fig.10 (1980). —Meve, Ill. Handb. Succ. Pl. Asclepiadaceae: 72 (2002). Type: ?, not designated.

Roots fibrous. Tuber absent. Stem initially decumbent or trailing and rooting adventitously, branches constricted at bases, very fleshy, 8–15 mm thick, smooth with raised tubercles, dull green mottled and often tinged with purple, glabrous, flowering stems more slender and twining. Leaves inserted on basal tubercles, scalelike, sessile; leaf blade cordate-ovate or triangular, to 3 mm long, base with with 2 pairs of golden yellow glands, apex acute, glabrous except for a few short marginal hairs near apex. Inflorescences extra-axillary, peduncle 2–20 mm long, glabrous; rachis producing successive cymes, Pedicel up to 10 mm long, glabrous. Sepals ovate-lanceolate, to 4 mm long. Buds with raised sinuses and slender beak. Corolla to 70 mm long, slightly curved; divided to about halfway down; basal chamber ovoid, c.7 mm wide, gradually narrowed at apex, limb cylindrical, c.3.5 mm wide, widening fairly abruptly to mouth, exterior mottled grey with larger purplish blotches towards mouth, glabrous, interior hairy in basal chamber and around mouth, otherwise glabrous; sinuses rounded, spreading; lobes triangular to triangular-attenuate, to 35 mm long, plicate, apices recurved or erect, free or twisted together, abaxially glabrous, adaxially white with margins and tips purplish, purplish brown or greenish yellow, finely pilose. Corona shortly stipitate or sessile, cupular, c.5 × 3.5 mm white or brownish, outer lobes confluent with bases of inner lobes, bifid, 2–2.5 mm high, pubescent and often with some longer hairs adaxially and at tips; inner lobes incumbent-erect. Follicles widely divergeant, thickly fusiform, c.10 cm long, grey-green, glabrous, verruculose. Seed not recorded.

Subsp. **serpentina** (E.A. Bruce) R.A. Dyer in Codd *et al.*, Fl. S. Africa **27**(4): 51,
   fig.10.2 (1980). —Meve, Ill. Handb. Succ. Pl. Asclepiadaceae: 72 (2002). Type:
   South Africa, Transvaal, Hammanskraal, north of Pretoria, *Erens & Phillips*
   2176 (PRE0632469-0 holotype, K000305618, K000305619, PRE0330435-0,
   SRGH0106582-0). FIGURE 7.3.**149**.

   Ceropegia serpentina E.A. Bruce in Fl. Pl. Africa **27**: t.1072 (1949).
   Ceropegia stapeliiformis var. *serpentina* (E.A. Bruce) H. Huber in Mem. Soc. Brot. **12**: 108
   (1958).

Plants twining readily. Corolla lobes triangular-attenuate, to 35 mm long, plicate, apices erect,
free or twisted together.

Mozambique. GI: Planicic Davetawe, south of Mambone, near the mouth of the
Save River, 20 m, 15.xii.1998, *Bruyns* 7682 (BOL, E).
   Also in South Africa, Swaziland, north of 28°S. In loose leaf litter under bushes;
c.20 m.
   Conservation status: LC in South Africa, known from just the one collection within
the Flora area.
   The typical subspecies, subsp. *stapeliiformis*, is found in the Eastern Cape region of
South Africa south of 31° S. It has much less prominent stipular glands, less extensively
twining flowering stems and shorter, usually somewhat recurved corolla lobes.

35. **Ceropegia cimiciodora** Oberm. in Fl. Pl. South Africa **13**: t.488 (1933). —Huber
   in Mem. Soc. Brot. **12**: 108 (1958). —Dyer in Codd *et al.*, Fl. S. Africa **27**(4):
   56, figs.9,4a-c (1980). —Meve, Ill. Handb. Succ. Pl. Asclepiadaceae: 72 (2002).
   Type: South Africa, Transvaal, Soutpansberg, *Obermeyer et al.* 322 (PRE0330434-0
   holotype). FIGURE 7.3.**149**.

Roots fibrous. Tuber absent. Stem initially very succulent, sprawling, then more slender and
twining, sparsely branched, to several meters long, grey mottled purplish, glossy, glabrous.
Leaves scalelike, sessile; leaf blade triangular to subcordate, c.2.5 × 3 mm, both surfaces glabrous.
Inflorescence extra-axillary, shortly pedunculate, cyme 1- or 2-flowered, flowers opening in
succession. Pedicel to 8 mm long, glabrous. Sepals broadly lanceolate, c.2 mm long, glabrous.
Bud with sinuses prominently raised and reflexed. Corolla c.50 mm long, overall c.20 mm
diam., curved; tube with basal chamber ovoid c.8 × 6 mm; limb cylindrical, c.3.5 mm wide,
widening fairly abruptly to mouth, exterior with white basal chamber, then grey mottled with
purple, more densely so towards mouth (some cultivated forms predominantly yellow-green),
glabrous, interior hairy; sinuses produced into prominent rotately spreading oblong-lanceolate
lobes, longer than corolla lobes proper, adaxially beige finely mottled with purple, apex acute
or rounded, stiffly dark pilose, lobes triangular-acuminate, only slightly exerted above mouth,
plicate, apices usually free, erect, adaxially dark purple, finely pilose. Corona shortly stipitate,
cupular, c.6 × 4 mm white or brownish, outer lobes cupular, erect, deeply bifid, teeth subulate,
c.2 mm long, purple-brown, white ciliate, adaxially pubescent; inner lobes arising from within
outer corona, incumbent-erect, c.3 mm long, apices papillose. Pollinia ovoid. Follicles erect,
thickly fusiform, c.10 cm long, grey-green, glabrous, finely tuberculate. Seed not recorded.

Mozambique. LM: Moamba, 15 km from Moamba on road to Sabie, flowered in
cultivation, *Carvalho* 743 (K).
   Also in South Africa and Swaziland. Growing in sandy soil; 80 m.
   Conservation status: VU in South Africa, only just extending into the Flora area.

Sect. **Radicantiores** Bruyns in S. African J. Bot. **112**: 431 (2017).

36. **Ceropegia monteiroae** Hook.f. in Bot. Mag. **113**: t.6927 (1887). —Brown in Fl.
   Cap. **4**(1): 815 (1908). Type: Mozambique, Delagoa Bay, *Monteiro* s.n. (K000305461
   holotype). FIGURE 7.3.**149**.

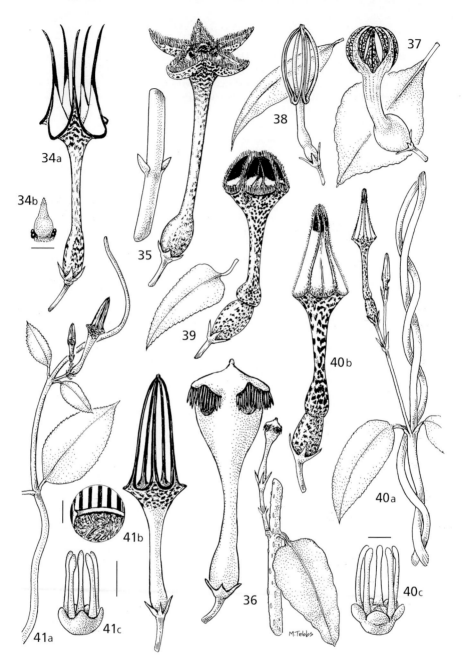

Fig. 7.3.**149.** —**34.** CEROPEGIA STAPELIIFORMIS subsp. SERPENTINA, scale leaf with stipular glands and flower, from *Codd* 46. —**35.** CEROPEGIA CIMICIODORA, section of stem and flower, from *Carvalho* 743. —**36.** CEROPEGIA MONTEIROAE, section of flowering stem and flower, from *Mrs Monteiro* s.n. —**37.** CEROPEGIA CRASSIFOLIA, leaf and flower, from *Carvalho* 763. —**38.** CEROPEGIA STENANTHA, leaf and flower, from *Harder & Bingham* 2592. —**39.** CEROPEGIA NILOTICA, leaf and flower, from *de Winter* 9162. —**40.** CEROPEGIA ARENARIOIDES, 40a section of flowering stem, 40b flower, 40c corona, from *Carvalho* 735. —**41.** CEROPEGIA MUCHEEVENSIS, 41a section of flowering stem, 41b flower with inset of internal annulus of basal chamber, 41c corona, from *Carvalho* 602. Scale bars = 10 mm except for 41b = 2 mm; 34b, 40c, 41c = 1 mm. Drawn by Margaret Tebbs.

Roots clustered, fleshy, white. Tuber absent. Stem twining, robust, succulent, pale green mottled with brown verrucose patches, glabrous. Leaves succulent; petiole short and stout; leaf blade oblong-ovate, 50–70 mm long, base rounded to cordate, margin undulate, sometimes reddish, apex obtuse to subacute, both surfaces glabrous. Inflorescence extra-axillary, peduncle 12–16 mm, cyme c.3-flowered, flowers opening in succession. Pedicel to 12 mm, glabrous. Sepals linear-oblong, c.6 mm long, spreading, glabrous. Bud with acute sinuses, apex deeply lobed with acuminate 5-toothed tip. Corolla 50–75 mm long, erect to slightly bent above basal chamber; tube 40–60 mm long; basal chamber cylindrical-ovoid, 12–17 × 9–11 mm wide, limb narrowly cylindrical at base, 3–4 mm wide, chalice-shaped above, 17–23 mm wide with slightly constricted mouth, exterior uniformly pale green, sometimes obscurely darker striped above, glabrous; interior hairy; sinuses truncate; lobes ovate with narrowed bases, apices connate to form domed 10-lobed canopy slightly wider than mouth, canopy margins down-curved, with pendent, purple, vibratile hairs covering openings, apex apiculate with minutely 5-pointed appendage, abaxially pale green with faint reticulate markings, mostly glabrous, adaxially uniformly mottled green, glabrous. Corona shortly stipitate; outer lobes each forming a minute pocket, laterally confluent with base of inner lobes; obscurely 2-toothed, inner lobes converging over gynostegium, linear, recurved at tips, hairy at base. Follicles and seeds not recorded.

Mozambique. M: Delagoa Bay, 16.vii.1886, *Monteiro* s.n. (K); Maputo Dist., Inhaca Island, fl. 20.x, *Mogg* 31912 (K); Lourenço Marques, Umbeluzi, near OP formans house at km. 18, fl. 12.iii.1964, *de Carvalho* 740 (K); Muntanhane, 28.ii.1971, *Balsinhas* 1787 (LMA).

Also in South Africa. Coastal dunes.

Conservation status: VU. A fairly rare local species in southern Mozambique, just extending into NE South Africa.

Ceropegia monteiroae was included within *C. sandersonii* Hook.f. by Huber but it is easily distinguished from that species by the fact that it has a more slender flower with a narrower more dome-shaped canopy with downcurved margins, the openings between the lobes filled with pendent purple clavate hairs, in contrast to the flatter and wider canopy with spreading to upcurved margins and spreading white marginal hairs of *C. sandersonii*. No collection of *C. sandersonii* s. str. has been seen from Mozambique.

Sect. **Callopegia** H. Huber in Mem. Soc. Brot. **12**: 31 (1958).

37. **Ceropegia crassifolia** Schltr. in J. Bot. **33**: 273 (1895). Brown in Fl. Cap. **4**(1): 818 (1908). —Dyer in Fl. Pl. South Africa **24**: t.924 (1944); in Codd *et al.*, Fl. S. Africa **27**(4): 62–63 (1980). —Huber in Mem. Soc. Brot. **12**: 100 (1958); in Merxmüller, Prodr. Fl. Sudwestafr. **114**: 22 (1967). —Bruyns in Bradleya **3**: 18–19 (1985); in Strelitzia **34**: 85 (2014). —Meve, Ill. Handb. Succ. Pl. Asclepiadaceae: 73 (2002). —Masinde in F.T.E.A., Apocynaceae (part 2): 280–281 (2012). Type: South Africa, Cape, King William's Town, *Sim* 312 (B† holotype, BOL137968, NU0015530-0). FIGURE 7.3.**149**.

Ceropegia brachyceras Schltr. in Bot. Jahrb. Syst. **38**: 45 (1905). Type: South Africa, "Bechuanaland", Kalahari: Maritzani, *Duparquet* 432 (?B† holotype).

Ceropegia crispata N.E. Br. in Fl. Cap. **4**(1): 819 (1908). Type: South Africa, Kalahari Region, Orange River Colony or Griqualand West, near the Vaal River, *Barber* 675 (K000305598 holotype).

Ceropegia thorncroftii N.E. Br. in Bot. Mag. **138**: t.8458 (1912). Type: Cambridge Botanic Garden, raised from seed sent from Barberton, Transvaal by M.G.Thorncroft (K000305594–K000305596 syntypes).

Ceropegia tuberculata Dinter, Repert. Spec. Nov. Regni Veg. **19**: 178 (1923). Type: Namibia, Auasberge, *Dinter* 4423 (location not known).

Roots clustered, fleshy, fusiform. Tuber absent. Stem twining, succulent, up to 3 m tall, glabrous. Leaves succulent; petiole 1–5 mm long; leaf blade linear-lanceolate to broadly ovate, up to 50–80 × 12–45 mm, margin ciliate, sometimes crispate, midrib abaxially prominent, both surfaces glabrous. Inflorescence extra-axillary, peduncle 5–30 mm long; cyme 2–7-flowered, flowers opening in succession. Pedicel 5–7 mm long. Sepals linear-lanceolate to lanceolate, 3–5 mm long, glabrous. Corolla 25–50 mm long, bent above basal inflation; basal chamber globose, to 9 mm wide, limb funnel-shaped, expanded to c.9 mm diam. at mouth, exterior pale green usually with distinct purplish blotches, glabrous, interior with long deflexed hairs; sinuses acute, flat; lobes oblong ovate, to 15 mm long, tardily plicate, apices connate to form cylindrical to ovoid cage with rounded to truncate apex, cage 8–15 mm diam., adaxially closely purplish-brown reticulate, with long purple hairs on basal part of keel. Corona subsessile, outer lobes erect, emarginate to shallowly bifid, 0.5–1.0 mm long, equalling or overtopping the staminal column; minutely pubescent; inner lobes incumbent on anthers linear, 1.5–2 mm long, apices straight, meeting in centre, glabrous. Pollinia subglobose. Follicles nearly horizontally spreading to decurved, 10–12 cm long, glabrous. Seed not recorded.

Botswana. N: 20.1°S 24.6°E, 14.i.1991, *Barnard* s.n. (PRE, not seen). **Zimbabwe**. N: Gokwe, fl. 19.xii.1963, *Bingham* 984 (K). **Mozambique**. M: Lourenço Marques, Umbeluzi, near the Institute by the lake, fl. 6.iii.1965, *Carvalho* 763 (K).

Also in Namibia, South Africa and Swaziland. Ecology in the Flora area not recorded but in Namibia recorded from among trees on rocky slopes and sandy flats; 1200–2000 m.

Conservation status: LC. Widespread in South Africa with isolated extenions into the Flora area.

Leaf shape is very variable and some of the more narrowly leafed plants from the southern part of the distribution, including *Bruynes* 12380 (BOL, NBG) have been identified as var. *copleyae* (E.A. Bruce & P.R.O. Bally) H. Huber, based on material from Kenya with leaves only 3–5 mm wide and smaller flowers.

38. **Ceropegia stenantha** K. Schum. in Bot. Jahrb. Syst. **17**: 152 (1893). —Brown in F.T.A. **4**(1): 459 (1903). —Werdermann in Bot. Jahrb. Syst. **70**: 223 (1939). —Bullock in Kew Bull. **7**: 425 (1953). —Huber in Mem. Soc. Brot. **12**: 123 (1958); in Merxmüller, Prodr. Fl. Sudwestafr. **114**: 25 (1967). —Lisowski & Malaisse in Bull. Jard. Bot. Natl. Belg. **44**: 413–415, fig.6 (1974). —Malaisse, in Bull. Jard. Bot. Natl. Belg. **54**: 231, fig.9 (1984). —Archer, Kenya *Ceropegia* Scrapb.: 61, XIII (1992); in Upland Kenya Wild. Fl., ed. 2: 183 (1994). —Meve, Ill. Handb. Succ. Pl. Asclepiadaceae: 101–102 (2002). —Masinde in F.T.E.A., Apocynaceae (part 2): 282–283 (2012). Type: South Sudan, Bahr el Ghazal, Djur Ghattas, *Schweinfurth* 2104 (B† holotype, K lectotype), designated by Huber in Mem. Soc. Brot. **12**: 125 (1958); (K000305521 lectotype, K000305522, MEL710747) second step lectotype designated here. FIGURE 7.3.**149**.

Riocreuxia longiflora K. Schum. in Bot. Jahrb. Syst. **28**: 459 (1900), not *Ceropegia longiflora* Poir. (1786). Type: Tanzania, Iringa, between Ukutu [Khutii] & Uhehe near River Mloa, *Goetze* 495 (B† holotype).

Ceropegia infausta N.E. Br. in F.T.A **4**(1): 459 (1903). Type: as for *Riocreuxia longiflora*.

Ceropegia stenantha var. *parvifolia* N.E. Br. in F.T.A. **4**(1): 459 (1903). Types: German East Africa, Coast Dist., *Hannington* s.n. (K syntype, not found); British Central Africa, Rhodesia, Central Leshumo Valley, *Holub* s.n. (K000305647 syntype).

Ceropegia tenuissima S. Moore in J. Linn. Soc., Bot. **37**: 185 (1904). Type: Uganda, Ankobe: 'Wazinga' near Mulema, *Bagshawe* 254 (BM000930066 holotype).

Ceropegia mazoensis S. Moore in J. Bot. **46**: 309 (1908). Type: Zimbabwe, Mazoe, *Eyles* 518 (BM000930065 holotype).

Ceropegia quarrei De Wild., Contr. Fl. Katanga, Suppl. **1**: 71 (1927). Type: D.R. Congo, Katanga: Katuba, *Quarré* 80 (BR0000008863225 holotype).

Roots clustered, fleshy, fusiform, up to 15 cm long, c.5 mm thick. Tuber absent. Stem twining or sometimes trailing, branched, succulent, up to 8 meters long, 2–5 mm thick, glabrous. Leaves succulent; petiole (0)2–8 mm long, glabrous; leaf blade lanceolate to linear, 25–150 × 2–18 mm, base and apex tapering, margin sometimes ciliolate, midrib abaxially prominent, both surfaces glabrous. Inflorescences often many, extra-axillary, peduncle (0)3–12 mm long; cyme umbel-like, 3–8(10)-flowered, up to 4 flowers open at once. Pedicel 1–5(10) mm long, glabrous. Sepals subulate, mottled purple or with purplish tip, 2–3 × 0.5–1 mm wide, glabrous. Bud with minutely recurved sinuses and linear-oblong beak. Corolla 20–30(35) mm long, straight to slightly curved; tube 10–20 mm long, basal chamber cylindrical-inflated, 4–8 mm long, 2–3 mm wide, narrowed above, not dilated at mouth, exterior pale yellow, cream or white, often greenish, rarely pinkish, glabrous, interior blotched purple at base, thinly pilose with fine hairs inside at base and middle of tube; sinuses acute, narrowly but distinctly auriculate; lobes linear, 10–20 mm long, ± as long as tube, somewhat revolute, apices usually connate to form narrowly ellipsoid to fusiform cage with acute apex, some lobes occasionally becoming detached at their tip, abaxially pale, both surfaces glabrous, adaxially pale, rarely brighter, yellow, sometimes darker at tip. Corona stipitate, outer lobes forming very shallow pouches, laterally confluent with base of inner lobes; truncate, inner lobes incumbent-erect, linear-lanceolate or linear-oblong, c.1.5 mm long, apices obtuse, sometimes recurved. Follicles erect, narrowly divergent, linear-fusiform, (4.5)9–11 × 0.2–0.3 cm wide at middle, glabrous. Seed dark brown, 6–7 × 1.5–2 mm wide; coma 20–30 mm long.

Botswana. N/W: Zambesi, central Leshume (Lesomo) Valley, 1883, *Holub* s.n. (K). **Zambia**. B: Barotse Prov., Sesheka Dist., Meses Forest station, 1050 m, fl. 2.ii.1975, *Brummitt et al.* 14244 (K). N: Abercorne Dist., Mbulu River, 1500 m, fl. 5.ii.1957, *Richards* 8067 (K). E: Lusaka Dist., Chipongwe Cave, 28 km SSW Lusaka, fl. 26.ii.1995, *Bingham* 10428 (K). **Zimbabwe**. C: Domboshawa, 16.ii.1958, *Leach* 5937 (K, SRGH). E: Maranke Reserve, Mutare, 3500 ft, 11.ii.1953, *Chase* 4800 (K, SRGH). N: Chibakwe River, Mrewa, 6.ii.1958, *Leach* 5902 (K, SRGH). S: Turn off to Zimbabwe Ruins, Masvingo, 12.xii.1987, *Percy-Lancaster* 2022 (SRGH). W: Khami Rangemore, Bulawayo, 1450 m, 16.iii.1969, *Cannell* 72 (K, SRGH). **Malawi**. N: Mzimba Dist., 20 mi W of Mzuzu, Lunyangwa R bridge, 3500 ft, fl. 2.iii.1974, *Pawek* 8145 (K). C: Dedza Dist., Nchinji Hill, Kachere N.A., 1500 m, fl. 22.i.1959, *Robson & Jackson* 1294 (BM, K). S: Mangochi Dist., Chowe, 14 km NE Mangochi, 920 m, fl. 21.ii.1982, *Brummitt & Polhill* 16015 (K). **Mozambique**. N: Niassa 36 km NNW of Naulixa, 441 m, fl. 15.iii.2009, *Lotter & Turpin* 1691 (K).

Also in Namibia, South Africa, D.R. Congo, Rwanda, Tanzania, Kenya, Uganda and South Sudan. on rock outcrops, in *Colophospermum mopane*, *Brachystegia* or riverine woodland, on sandy, loamy or clay soils, often associated with termite mounds, often in shade; 40–1650 m.

Conservation status: LC. The most frequently collected member of the genus in the Flora area.

Timberlake & Gold 5899 (Mozambique. M: Salamanga, 2.5 km S of Santaca village above W bank of Rio Matupo) seems to be a small flowered form with very delicate corollas c.21 mm long and the outer corona lobes reduced to V-shaped notches between the anthers.

Ceropegia stenantha 'var. *parviflora* N.E. Br.' in Masinde, F.T.E.A, Apocynaceae (part 2): 283 (2012) is an orthographic error for 'var. *parvifolia*'.

39. **Ceropegia nilotica** Kotschy, Sitzungsber. Kaiserl. Akad. Wiss., Math.-Naturwiss. Cl., Abt. 1, **51**: 356 (1865). —Brown in F.T.A. 4(1): 447 (1903). —Werdermann in Bot. Jahrb. Syst. **70**: 202 (1939). —Bullock in Kew Bull. **8**: 58 (1954). —Huber in Mem. Soc. Brot. **12**: 103 (1958), pro parte. —Lisowski & Malaisse in Bull. Jard. Bot. Natl. Belg. **44**: 417, figs.7,8f (1974). —Dyer in Codd *et al.*, Fl. S. Africa **27** (4): 59, fig.12.3 (1980). —Malaisse, in Bull. Jard. Bot. Natl. Belg. **54**: 232 (1984). —Archer, Kenya *Ceropegia* Scrapb.: 61, XIII (1992); in Upland Kenya Wild. Fl., ed. 2: 183 (1994). —Masinde in F.T.E.A., Apocynaceae (part 2): 283–285 (2012). —Meve, Ill. Handb. Succ. Pl. Asclepiadaceae: 92 (2002). —Gilbert in Hedberg & Edwards, Fl. Ethiopia Eritrea **4**(1): 165, fig.140.37.4 (2003). Type: South Sudan, Gondokoro, *Knoblecher* 5 (W holotype). FIGURE 7.3.**149**.

Ceropegia constricta N.E. Br. in Bull. Misc. Inform. Kew **1895**: 260 (1895). Type: D.R. Congo, Kawal Islands in Lake Tanganyika, *Carson* 35 pro parte (K 000305543 holotype).

Ceropegia mozambicensis Schltr. in J. Bot. **33**: 273 (1895). Types: Mozambique, mouth of Pungwe R., S of Zambezi R., fl. iv.1895, *Schlechter* 7106 (B† syntype); Kenya, Mombassa Island, *Taylor* s.n. (?B† syntype); Mozambique, 23 miles from Beira, 2.iii.1909, *Johnson* 303 (K neotype), designated here.

Ceropegia gemmifera K. Schum., Bot. Jahrb. Syst. 33: 328 (1903). Type: Togo, near Lome, *Warnecke* 242 (B† holotype, BM000930076 lectotype), designated here.

Ceropegia boussingaultiifolia Dinter, Neue Pfl. Deutsch-Sudwest-Afr.: 21 (1914). Type: Namibia, Otjiwarongo, cult. in Okahandja, *Dinter* 2780 (SAM lectotype), designated by Bruyns in Strelitzia **34**: 89 (2014).

Ceropegia gossweileri S. Moore in J. Bot. **67**(Suppl. 2): 99 (1929). Type: Angola, banks of Caringa R. near to where it enters river Mumbeje–N'Dalatando–Cazengo, *Gossweiler* s.n. (BM000930077 holotype).

Ceropegia plicata E.A. Bruce in Fl. Pl. South Africa **17**: t.675 (1937). Type: South Africa, Muden Valley, near Greytown, *Cronwright* 16 (PRE0331516-0 holotype).

Ceropegia mozambicensis var. *ulugurensis* Werderm. in Bot. Jahrb. Syst. **70**: 202 (1939). Type: Tanzania, Morogoro: Uluguru Mts., *Schlieben* 3822 (B† holotype, BM000930074, BR0000008863027, BR0000008863010, BR0000008863058, BR0000008863034, G00022494, M0110221, P00109658, PRE0332894-0).

Ceropegia decumbens P.R.O. Bally in Succulentes (France) **53**: 133 (1957). Type: Kenya, Machakos, E slope of Chyulu Hills, *Bally* 7933 (EA000001838 holotype, K000814151).

Ceropegia nilotica var. *plicata* (E.A. Bruce) H. Huber in Mem. Soc. Brot. **12**: 105 (1958).

Roots fibrous to somewhat fleshy, to 4.5 mm thick. Tuber absent. Stem twining, somewhat branched above, succulent, sometimes longitudinaly ridged to ± 4-sided, up to 8 meters long, glabrous. Leaves semi-succulent to succulent; petiole to 10 mm long, sometimes attenuate to base and leaf effectively sessile, glabrous; leaf blade lanceolate to broadly ovate, 30–80 × 10–43(60) mm, base cuneate to attenuate, margin minutely serrulate, less often subentire, apex acute to ± acuminate, both surfaces glabrous. Inflorescence extra-axillary, peduncle spreading, 10–50(70) mm long, sometimes with pair of scale-like bracts part way along, cymes sometimes produced in succession along short rachis, few-flowered. Pedicel 5–8(10) mm long, glabrous. Sepals subulate to linear-lanceolate, 3–4.5 × 0.7–1 mm, spreading, glabrous. Bud with prominent, projecting, ± recurved sinuses. Corolla 27–47(57) mm long; straight or nearly so, tube (15)23–37(46) mm long, basal chamber double, lower ellipsoid to obovoid, 6–11 mm long 3.8–6.6(7.3) mm wide, then abruptly expanded into a second subglobose inflation 3–7 mm diam., gradually narrowed at apex, limb cylindrical, 1.7–3.8 mm wide, widening abruptly to 8–15(19.5) mm at mouth, exterior pale cream with purple spots and/or streaks, glabrous; interior with collar-like annulus at constriction between basal chambers, upper part of lower chamber and lower part of tube pilose; sinuses rounded, spreading and gaping, flat; lobes triangular, 4–14(20) mm long, plicate, apices very shortly connate to form a broadly conical to rounded or sometimes almost flat cage, apex acute to rounded to sunken, abaxially usually uniformly stiffly pilose especially on margins, rarely glabrous. Corona sessile, basin-shaped; outer lobes each forming a shallow pouch, glabrous. Follicles widely divergeant, linear, to 18 cm long, slender, glabrous. Seed not recorded.

Botswana. N: Xanatshaa, 19.ii.1975, *P.A. Smith* 1251 (PSUB). **Zambia**. B: Barotse Prov. Sesheke distr, Masese Forest reserve, 10 km NE Masese, 1050 m, fl. 3.ii.1975, *Brummitt et al.* 14249 (K). E: Lusaka Dist., Luano Valley, Shikabeta, 560 m, fl. 25.i.2004, *Bingham* 12732 (K). S: Gwembe, Siavonga area, Mutulanganga PFA, 420 m, fl. 14.iii.1997, *Zimba et al.* 1144 (K). **Zimbabwe**. W: Wankie, fl. 16.iii.1954, *Levy* 1001 (K). N: Urungwe, Mensa Pan, 11 mi ESE of Chirundu Bridge, 460 m, fl. 3.ii.1958, *Drummond* 5448 (K). E: Landsdowne Farm, 8 miles E of Chipinge, fl. 6.iv.1974, *Percy-Lancaster* 106 (SRGH). S: Fort Victoria, 3 miles E Zimbabwe, Glen Levit road, fl. 24.ii.1961, *Leach* 10728 (K). **Malawi**. N: Rumphi, Njakwe gorge, S Rukuru river, 2 mi E of Rumphi, fl. 20.v.1973, *Pawek* 6751 (K). **Mozambique**. M: Lourenço Marques Umbeluzi, Quinta Olsa 15.iv.1965 *Carvalho* 765 (K). MS: Beira, fl. 2.iii.1909, *Johnson* 303 (K).

Also in Namibia, Angola, D.R. Congo, South Africa, Tanzania, Kenya, Ethiopia, South Sudan, Ghana, Togo. *Colophospermum mopane* and *Brachystegia* woodland, among rocks, on Kalahari sand or black clay soils; 20–1350 m.

Conservation status: LC. A common and widely distributed species.

Ceropegia nilotica is very variable, hence the number of synonyms, but whilst the extremes are certainly very different from each other, there seems to be continuous variation between them. Peckover in Aloe **30**: 21 (1993) presents evidence that *C. grandis* E.A. Bruce in Fl. Pl. Africa **28**: t.1113 (1951), is a hybrid between *C. nilotica* and *C. sandersonii*.

It appears that there is no surving type material of one of the most distinctive synonyms, *Ceropegia mozambicensis*, so a neotype showing the characteristic very short corolla lobes, *W.H. Johnson* 303, is designated here.

Huber (1958) recognised three varieties within *Ceropegia nilotica* but this author follows Masinde in treating var. *simplex* Huber as the distinct species *C. denticulata* K. Schum., and Dyer (1980) in not separating var. *plicata*.

40. **Ceropegia arenarioides** M.G. Gilbert, sp. nov. Closely related to *C. nilotica* but easily separated by the shape and colouration of the corolla lobes which are longer (c.21 mm vs. 4–14(20) mm in *C. nilotica*), coming to a narrowly acute to attenuate point (vs. bluntly acute to rounded or even impressed) and with the lower half pure white below a uniformly purplish brown tip (vs. transversely banded: dark at the base, then with a relatively narrow band of near white, followed by another dark band and a clear bottle green or brown tip); also similar in the overall shape and colouration to *C. arenaria* from which it differs by the narrowly acute to attenuate corolla apex, inconspicuously shortly white pilose (vs. a distinctly clavate corolla apex with conspicuous purplish clavate hairs in *C. arenaria*). Type: Zimbabwe, W, Insiza, 5 mi N of Filabusi, 1200 m, 18.ii.1971, *Wild* 7821 (K001400222 holotype). FIGURE 7.3.**149**.

Basal parts not recorded. Stems twining, semisucculent, glabrous. Leaves somewhat succulent; petiole c.8 mm long, glabrous; leaf blade ovate, to 42 × 18 mm, margin serrulate/denticulate, both surfaces glabrous. Inflorescence extra-axillary, peduncle to 30 mm long, cyme racemelike, with rachis to 23 mm long, few-flowered, flowers opening in succession. Pedicel c.6 mm long, glabrous. Sepals subulate to linear-lanceolate, c.4.5 × 1.1 mm, glabrous. Bud with flat sinuses and long beak. Corolla 49–55 mm long, slightly curved; tube 28–34 mm long, basal chamber double, lower ovoid, 8–8.5 × 3.8–5.5 mm wide, then abruptly expanded into a second subglobose inflation 4.2–6.2 mm diam., gradually narrowed at apex, limb cylindrical, 2.4–3.1 mm wide, widening abruptly to 10–13 mm at mouth, exterior pale cream uniformly covered with purple spots, glabrous; sinuses rounded, spreading and gaping; lobes triangular-attenuate, c.21 mm long, plicate, apices converging to form a narrowly conical cage, apex bluntly acute to acuminate, abaxially glabrous, margin shortly pilose, adaxially white for more than half length, tip dark purplish brown, uniformly sparsely stiffly hairy. Corona subsessile, 3.5–4 mm high, 2.5–3 mm diam.; outer lobes each forming a shallow pouch, entire, glabrous; inner lobes erect, converging over gynostegium, linear, c.3 mm long, apices obtuse, glabrous. Pollinia yellow. Follicles and seed not recorded.

Caprivi. Sabinda area, 40 miles W of Katima Mulilo, 1000–1500 ft 14.ii.1969, *De Winter* 9162 (K). **Botswana**. SE: 52 km N of Lephepe, 1000 m, 18.xii.1995, *Bruyns* 6425 (BOL); Francistown, 7 km along road to Sebina, 1050 m, 26.xii.2012, *Bruyns* 12343 (BOL). **Zambia**. E: Lusaka Province, Luano Valley, Shikabeta, i.2004, *Bingham* (image, www.zambiaflora.com). **Zimbabwe**. W: Insiza, 5 mi N of Filabusi, 1200 m, 18.ii.1971, *Wild* 7821 (K). N: Pumba Safari Lodge, Gokwe, *Wursten* (image, www.zambiaflora.com). **Mozambique**. T: Tete, Mucangádzi, 5.ii.1974, [illegible] 5545

(LMA). N: Montepuez, 15 km S of Alua, 310 m, 12.xi.2000, *Bruyns* 8549 (BOL), form with fusiform roots and filiform leaves.

Also in Namibia [Kaokoveld, 17 miles W of Otjiwero, 12.iv.1957, *de Winter & Leistner* 5518 (K); Okavango, 16.7 miles E of Rundu [Runtu] on road to Sambiu, 10.ii.1956, *de Winter & Marais* 4560 (K)]. Among bushes, including *Acacia tortilis* and *Colophospermum mopane*, and piles of twigs on reddish ground and hard brown loam; 300–1200 m.

Conservation status: Probably LC. Fairly widely distributed and probably under recorded because of confusion with *C. nilotica.*

A number of collections that have hitherto been included within *Ceropegia nilotica* differ very clearly in the form and colouring of their corolla lobes which have more in common with the South African species *C. arenaria* R.A. Dyer. They resemble *C. nilotica* quite closely in vegetative morphology but differ by the longer corolla lobes which extend into a slender, acuminately tapered beak. In *C. nilotica* the shorter lobes converge abruptly to form a broadly acute to rounded or even impressed apex. The other difference is in the colour pattern of the corolla lobes: in *C. nilotica* there are three or four transverse bands of colour, a basal blackish green or brown zone, a middle white band, sometimes quite narrow but always present, and a coloured apex, often similar to the basal area but then with a clearly delineated green or brown apex whereas in this species there are only two zones, a large, white basal zone taking up c. $2/3$ of the lobe and the purplish brown tip. *Ceropegia arenaria* has a similarly coloured corolla but differs most obviously by the largely creeping stems with often reduced leaves and by the corolla lobes with linear to slightly spathulate tips which converge slightly at the base to form a narrowly clavate beak with vibratile purplish clavate marginal hairs contrasting strongly with the much shorter erect hairs below the beak. In *C. nilotica* and *C. arenarioides* the lobe margins are uniformly shortly pilose. *Bruyns* 8549 (BOL) has the same distinctive corolla but has filiform leaves and fusiform roots. The status of this form deserves further investigation.

41. **Ceropegia mucheveensis** M.G. Gilbert, sp. nov. Related to *Ceropegia arenaria* R.A. Dyer, *C. arenarioides* M.G. Gilbert and *C. nilotica* Kotschy, differing most obviously by the apparently simple cylindrical basal chamber of the corolla in contrast to the double chamber of the other species, which has a well-developed internal annulus and is pilose internally below this annulus (as opposed to glabrous and often without an internal annulus); corolla lobes forming ± conical beak similar in colouring and shape to those of *C. arenaria* and *C. arenarioides* but their apices without purple vibratile hairs characteristic of *C. arenaria* and differing from both by their vibratile-pilose bases. Type: Mozambique. MS: Buzi, Mucheve Forest Reserve, 20°34'S 33°50'E, fl. 26.iv.1962, *Carvalho* 602 (K001400200 holotype, LMU). FIGURE 7.3.**149**.

Basal parts not recorded. Stem twining. c.1.5 mm diam., smooth, glabrous. Leaves somewhat succulent; petiole to 10 mm long, glabrous; leaf blade elliptic-ovate, to 77 × 35 mm, base cuneate-attenuate, margin minutely denticulate-serrulate, apex acuminate, both surfaces glabrous. Inflorescence extra-axillary, peduncle to 30 mm long, glabrous; cyme very shortly racemelike, rachis to 3 mm long, to 5-flowered, flowers opening in succession. Pedicel to 8.5 mm long, glabrous. Sepals narrowly lanceolate, c.5.5 × 0.9 mm subtending a number of elongated scales. Bud with spreading sinuses and long slender beak. Corolla 55–60 mm long, straight or nearly so; tube 23–28 mm long, basal chamber cylindrical, 8–9 × 3.8–4 mm wide, with internal annulus part way along but not constricted into 2 chambers, gradually narrowed at apex, limb cylindrical, 2.2–2.4 mm wide, widening to 14–18 mm at mouth, exterior uniformly pale with copious mottling on expanded upper part, glabrous; interior of basal chamber uniformly dark (purplish?) with narrow pale internal annulus, upper part with longitudinal dark veins, pilose below annulus, glabrous above; sinuses rounded, spreading and gaping; lobes c.26 mm long, base triangular, c.6 mm long, then linear, apices connate to form broadly beaked cage, abaxially glabrous, adaxially with whitish triangular base, margin and upper parts dark purplish, base vibratile-pilose, hairs with narrowly attached rounded bases, mostly pale, some purplish, dark upper parts subglabrous with a few scattered hairs. Corona subsessile, cup-shaped, outer corona cupular, margin erect, with 5 blunt teeth opposite anthers, shallowly U-shaped between, c.1 mm high, glabrous; inner lobes incumbent-erect, linear-lanceolate, c.2 mm long, apices bluntly acute, Follicles diverging by 180°, gently curved, linear, to 18 × 0.5 cm wide. Seed dark brown with paler margins, c.6 × 2 mm; coma to 22 mm long.

Known only from the type collection. Habitat not recorded but general area mostly with open woodland; 130–150 m.

Conservation status: DD, Known only from the one collection from a relatively undercollected region.

Superficially this species resembles *Ceropegia arenaria* R.A. Dyer, particularly in the size and form of the corolla, but it differs in many details such as the twining stems, well developed leaves with denticulate margins, corolla with an apparently simple cylindrical basal chamber which has a well-developed annulus on the inside and is pilose internally below the annulus, corolla lobes forming ± conical beak, their apices without purple vibratile hairs and a continuous cupular outer corona with the margin erect, bluntly toothed opposite the anthers and ± U-shaped between (*C. arenaria* has thicker, prostrate stems, smaller, often scalelike fleshy leaves with entire margins, two distinct basal chambers not separated by an internal annulus and glabrous internally, corolla lobes forming a narrower, clavate beak with purple vibratile hairs along margins and separate outer corona lobes, each with a low blunt central tooth and two taller, acute, lateral teeth).

Sect. **Ceropegiella** H. Huber in Mem. Soc. Brot. **12**: 32 (1958).

42. **Ceropegia pachystelma** Schltr. in Bot. Jahrb. Syst. **20**(Beibl. 51): 47 (1895). — Brown in F.T.A. **4**(1): 457 (1903); in Fl. Cap. **4**(1): 827 (1908). —Huber in Mem. Soc. Brot. **12**: 119 (1958); in Merxmüller, Prodr. Fl. Sudwestafr. **114**: 24 (1967). —Dyer in Codd *et al.*, Fl. S. Africa **27**(4): 79 (1980). —Meve, Ill. Handb. Succ. Pl. Asclepiadaceae: 93–94 (2002). —Bruyns in Strelitzia **34**: 91 (2014). Type: South Africa, Transvaal, Mailas, Kop R., *Schlechter* 4511 (B† holotype, BOL137972 lectotype), designated by Dyer (1980).

Ceropegia undulata N.E. Br. in Fl. Cap. **4**(1): 826 (1908). Type: South Africa, Natal, Tugela, *Gerrard* 1799 (K000305557 holotype, BM000645908).

Ceropegia obscura N.E. Br. in Fl. Cap. **4**(1): 827 (1908). Type: Mozambique, Delagoa Bay, *Mrs. Monteiro* s.n. (K000305455 holotype).

Ceropegia acacietorum Dinter, Neue Pfl. Sudw.-Afr.: 20 (1914); in Repert. Spec. Nov. Regni Veg. **15**: 426 (1919). Type: Namibia, Okasuma, 25.i.1913, *Dinter* 2723 (SAM0071001-0 holotype).

Ceropegia boerhaaviifolia Schinz in Vierteljahrsschr. Naturf. Ges. Zürich **71**: 139 (1926), illegitimate name, non Defl. (1896). Type: Namibia, Okasewa, *Dinter* 2723 (SAM lectotype), designated by Bruyns in Strelitzia **34**: 91 (2014).

Ceropegia schinziana Bullock in Kew Bull. **9**: 626 (1956), new name. Type: as for *C. boerhaaviifolia* Schinz.

Ceropegia pachystelma subsp. *undulata* (N.E. Br.) H. Huber in Mem. Soc. Brot. **12**: 120 (1958).

Tuber depressed globose, 4–12 cm diam., upper surface warty. Stem twining, up to 4 meters long, 2–3 mm thick, pubescent. Leaves semi-succulent to succulent; petiole 2–10 mm long, hirsute; leaf blade lanceolate to ovate-oblong, 20–50 × 7–25 mm, margin sometimes undulate, apex acuminate, surfaces glabrous to pubescent. Inflorescence extra-axillary, peduncle 10–12 mm long; cyme umbel-like, up to 10-flowered, usually fewer, flowers opening in succession, up to 4 open at once. Pedicel 3–9 mm long, puberulous. Sepals deltoid-lanceolate to subulate, puberulous. Corolla 20–30 mm long, straight to curved, tube 15–25 mm long, basal chamber broadly ovoid to subglobose, 4–5 mm wide, narrowed at apex, limb cylindrical, 1.5–1.8 mm wide, widening to c.3 mm at mouth, exterior pale purplish-brown to green, often mottled, densely pubescent, interior pale, pilose, basal chamber papillate; sinuses acute, flat; lobes linear-oblong, 5–12 mm long, shortly revolute near base, otherwise flat, apices connate to form broadly ovoid cage wider than mouth, abaxially pilose, adaxially pale with dark red band near base, glabrous or tomentose-villous. Corona sub-campanulate; outer lobes each forming a pouch c.0.75 mm deep; spreading, inner lobes converging over gynostegium, lanceolate-acuminate in upper half, 1.5–2 mm long, c.0.5 mm broad, apices slender, attenuate, recurved. Follicles narrowly divergent, 6–14 cm long, slender, glabrous. Seed not recorded.

Mozambique. M: Maputo Bay ['Delagoa Bay'], *Monteiro* s.n. (K).

Also in Namibia, South Africa and Swaziland. Habitat in the Flora area not recorded but in Namibia given as 'among stones and bushes or in sand between trees; 1000–1600 m.

Conservation status: Possibly extinct in the Flora area as it has not been recollected since the 19th Century. A common South African species.

This species is also recorded for Botswana and Zimbabwe by Bruyns (2014) but no voucher material has been seen.

43. **Ceropegia inornata** Masinde in Kew Bull. **53**: 949 (1998). Type: Kenya, Taita Dist., Mt. Kasigau, *Masinde et al.* 838 (EA holotype, K000305470, MSUN).

Tuber oblate, 18–25(80) mm diam., sometimes formed at successive nodes. Stem twining, unbranched, herbaceous, to 200 cm long, usually less, c.0.7 mm thick, glabrous. Leaves slightly succulent; petiole to 6 mm long, sometimes ill-defined; leaf blade narrowly lanceolate to cordate, 20–60 × 5–23(40) mm, upper leaves much narrower, base attenuate into winged petiole or cuneate to rounded or shallowly cordate, margin entire, sometimes densely ciliolate, apex obtuse, abaxial surface sparsely puberulous, adaxial surface glabrous or sparsely pubescent. Inflorescence extra-axillary, peduncle 1.8–30(35) mm long, glabrous; cyme (1)2–6-flowered, flowers opening in succession. Pedicel 2.5–8(15) mm long, glabrous. Sepals narrowly triangular, 2–2.5 × 0.25–0.6 mm, acute, glabrous, ciliate or puberulous, abaxially very minutely pale papillose. Bud with flat sinuses and slender clavate beak. Corolla 15–27(32) mm long, straight or curved; tube 15–20 mm long, basal chamber cylindrical-ovoid, 4–9 mm long, 3–5 mm wide, limb 1–2.5 mm wide, widening to 3–5 mm at mouth, both surfaces apparently unmarked, glabrous; sinuses acute, flat; lobes linear-spatulate, 6–13 mm long, c.0.2 mm wide for most of length, shallowly revolute except for flat, 1–1.5 mm wide, apices, apices connate to form cylindrical cage, narrower than mouth, abaxially glabrous, "tips pale yellow". Corona sessile to shortly stipitate, outer corona cupular, lobes prominent, triangular, erect or spreading, obtuse, c.1 mm long, with long white hairs within; inner lobes incumbent-erect, linear, about as long as to distinctly longer than outer corona, glabrous. Follicles paired, divergeant at 20–90°, 7–16 × 0.2–0.3 cm, pale green with reddish streaks and spots; seeds brown with paler margin, c.6 × 1.5 mm; coma 25(40) mm long.

Subsp. **zambiana** M.G. Gilbert, subsp. nov. Distinguished from subsp. *inornata* by the narrower leaf blades, to 8 mm wide (not up to 40 mm wide), with the base attenuate into an obscure winged petiole to 2 mm long (not broadly cuneate to rounded or shallowly cordate with a well defined petiole 2–6 mm long, shorter corolla lobes, c.6 mm long (not 10–13 mm long) and erect outer corona lobes about as high as the inner corona (not spreading and clearly shorter). Type: Zambia, W: Mwinilunga, 6 km N of Kalene Hill, fl. 12.xii.1963, *Robinson* 5966 (K001400199 holotype). FIGURE 7.3.**150**.

Base not recorded. Stem unbranched, to 60 cm long, glabrous. Leaves slightly succulent; petiole ill-defined, to 2 mm long; leaf blade narrowly lanceolate, to 40 × 8 mm, upper leaves much narrower, base attenuate into winged petiole, margin entire, densely ciliolate, apex obtuse, abaxial surface sparsely puberulous, adaxial surface glabrous. Inflorescence extra-axillary, peduncle to 30 mm long, glabrous; cyme (1)2-flowered. Pedicel to 3.5 mm long, glabrous. Sepals

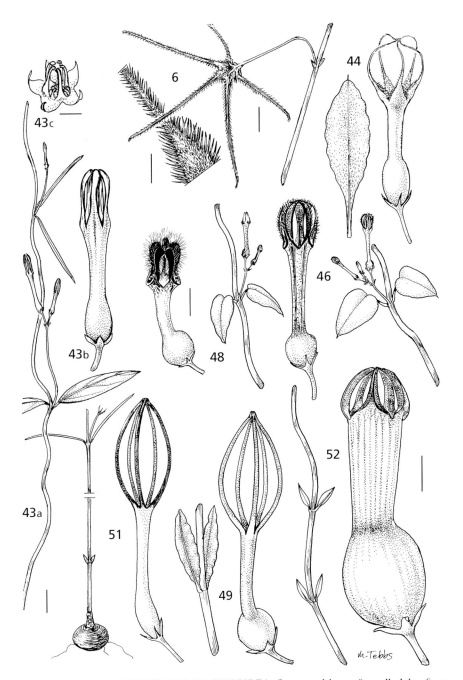

Fig. 7.3.**150**. —**6**. CEROPEGIA NEOARACHNOIDEA, flower and base of corolla lobe, from *Richards* 10361. —**43**. CEROPEGIA INORNATA subsp. ZAMBIANA, 43a flowering stem, 43b flower, 43c corona, from *Richards* 16188. —**44**. CEROPEGIA MULTIFLORA, leaf and flower, from *Carvalho* 767. —**46**. CEROPEGIA LINEARIS, section of flowering stem and flower, from *Marais* 965. —**48**. CEROPEGIA WOODII, section of flowering stem and flower, from *Codd* 8106. —**49**. CEROPEGIA CONRATHII, leaves and flower, from *Conrath* s.n. —**51**. CEROPEGIA CATAPHYLLARIS, rootstock and stem, flower, from *Milne-Redhead* 3421. —**52**. CEROPEGIA AMPLIATA, section of stem and flower, from *Macuacua* 92. Scale bars: stems = 10 mm; flowers and leaves = 5 mm; 43c = 1 mm. Drawn by Margaret Tebbs.

narrowly triangular, c.2.2 × 0.6 mm, acute, glabrous, abaxially very minutely pale papillose. Corolla c.26 mm long; straight; tube c.16 mm long, basal chamber cylindrical-ovoid, c.6 mm long, 3.5 mm wide, limb c.2.4 mm wide, widening to c.3 mm at mouth, both surfaces apparently unmarked, glabrous; lobes c.6 mm long, 0.2 mm wide for most of length except for flat, 1 mm wide, apices, abaxially glabrous, "tips pale yellow". Corona sessile, outer corona lobes prominent, triangular, erect, obtuse, c.1 mm long; inner lobes incumbent-erect, linear, about as long as outer corona, glabrous. Follicles and seeds not seen.

Zambia. W: Mwinilunga, 6 km N of Kalene Hill, fl. 12.xii.1963, *Robinson* 5966 (K). N: Kasama Dist., Mungwi-Kasama road, 1320 m, fl. 26.ii.1962, *Richards* 16188 (K).

Endemic. Habitat not recorded; 1200–1350 m.

Conservation status: DD, Known only from 2 collections c.800 km apart.

This taxon was first identified as *C. bulbosa* Roxb., first described from India and subsequently extended to include northeast African material placed within *C. vignaldiana* A. Rich. However, that species differs markedly in corolla form from the Zambian collections, most obviously by the widely gaping rounded sinuses between the densely hairy, uniformly slender corolla lobes. The slender completely glabrous corolla with flat, acute sinuses between the distinctly spatulate corolla lobes are a very good match for those of *C. inornata* known from Ethiopia, Kenya and Tanzania. They differ from that material by the much narrower leaves with attenuate bases and distinctively densely ciliolate margins and a sparsely hairy abaxial surface.

44. **Ceropegia multiflora** Baker in Saunders, Refug. Bot.: t.10. (1869). —Brown in Fl. Cap. **4**(1): 829 (1908). —Huber in Mem. Soc. Brot. **12**: 121 (1958). —Dyer in Codd *et al.*, Fl. S. Africa **27**(4): 69, fig.14 (1980). —Bruyns in Strelitzia **34**: 89 (2014). Type: t.10 of protologue, designated as lectotype by Huber in Mem. Soc. Brot. **12**: 122 (1958). FIGURE 7.3.**150**.

    Ceropegia tentaculata N.E. Br. in Bull. Misc. Inform. Kew **1895**: 260 (1895). —Brown in F.T.A. **4**(1): 443 (1903). Type: Angola, Loanda, by Alto das Cruzes, *Welwitsch* 4277 (BM000930067 lectotype, BM000930068–70, C10000298, K000305453, K000305454, LISU220329, LISU220330, LISU220331, P00109659), lectotype designated by Huber in Mem. Soc. Brot. **12**: 12 (1958); second step lectotype designated here (see note below).

    Ceropegia tentaculata var. *puberula* Hiern, Cat. Afr. Pl. **1**: 695 (1896). Type: Angola, Moçâmedes, *Welwitsch* 4268 (BM000930071 lectotype, LISU220333), designated by Huber in Mem. Soc. Brot. **12**: 122 (1958).

    Ceropegia multiflora var. *latifolia* N.E. Br. in Fl. Cap. **4**(1): 829 (1908). Type: cultivated plant of unrecorded origin (K000305583 holotype).

    Ceropegia multiflora f. *pubescens* H. Huber in Mem. Soc. Brot. **12**: 122 (1958). Type: South Africa, Transvaal: near Pretoria, 21.i.1914, *Pott* 4740 (K000865588 holotype).

    Ceropegia multiflora subsp. *tentaculata* (N.E. Br.) H. Huber in Mem. Soc. Brot. **12**: 122 (1958). —Dyer in Codd *et al.*, Fl. S. Africa **27**(4): 69 (1980).

    Ceropegia multiflora subsp. *tentaculata* f. *puberula* (Hiern) H. Huber in Mem. Soc. Brot. **12**: 123 (1958).

Tuber discoid, to 12 cm diam., upper surface often concave, sometimes warty. Stem twining, branched above, herbaceous, up to 3 meters long, 1–3 mm diam., glabrous to uniformly hirsute. Leaves succulent, those at upper flowering nodes reduced; petiole to 8 mm long, sometimes lamina attenuate to base and leaf effectively sessile; leaf blade linear-oblong, ovate, ovate-lanceolate to spathulate-obovate, 10–88 × 3–30 mm, base cuneate to attenuate, margin entire to sometimes undulate, ciliate, apex acuminate to obtuse, both surfaces glabrous to puberulous. Inflorescences usually many, sometimes paired at a node, extra-axillary, sessile or nearly so; cyme umbel-like, many-flowered, flowers opening in succession. Pedicel 1.5–10 mm long, glabrous. Sepals lanceolate, 2–4 mm long, acute, glabrous. Corolla to 35 mm long, straight; tube 7–22 mm long, basal chamber obovoid to subglobose, 2–7 mm wide, limb 1–2 mm diam., slightly expanded at mouth, exterior greenish, glabrous, interior thinly tomentose, basal chamber with longitudinal lines of purple papillae; sinuses acute, flat; lobes 5–12 mm long, base triangular to lanceolate, abruptly tapering into thread-like tips, apices often free, spreading, subrotate, or apices connate to form globose or oblate cage wider than mouth, apex obtuse, adaxially pale green sometimes suffused pinkish towards apices, lower part with dense short retrorse white hairs. Corona cupular, c.1 mm high; outer lobes each forming a deep pouch, outer margin truncate or minutely toothed; inner lobes incumbent-erect, linear to spathulate. 1–1.5 mm long, Follicles fusiform, 12–16 cm long. Seed not recorded.

Botswana. N: Tsessebe, fl. 10.i.1992, *Venter, Archer & Hahn* 170 (UNIN). **Zimbabwe**. W: Hwange Falls Road, 40 miles S of Hwange, fl. 20.ii.1956, *Wild* 4752 (SRGH). N: 112km N of Kwekwe, Gokwe, fl. 2.i.1986, *Percy-Lancaster* 1697 (SRGH). C: Gweru, fl. 22.ii.1972, *Biegel* 3825 (SRGH). S: Masvingo (Fort Victoria), *Munro* 889 (BM). **Mozambique**. M: Lourenço Marques, Mahotas (near the district of C de Ferro), fl. 15.iv.1965, *Carvalho* 767 (LMU, K).

Also in Angola, Namibia and South Africa. Woodland with *Brachystegia* or *Acacia tortilis* on sandy or stoney soil, grassland; 50–1400(1600) m.

Conservation status: LC: Moderately frequent over a wide area.

The protologue of *Ceropegia multiflora* was based on living material sent by Dr. Arnot to Kew. Dyer in Codd *et al.*, Fl. S. Africa **27**(4): 69 (1980) indicated the type as '*Arnot* (K, holo.!)' but no such specimen can be found and it seems rather likely that no herbarium material was preserved when the illustration was prepared. In his protologue, Baker also indicated that he had seen herbarium material in Dr. Burchell's herbarium that had been given the manuscript name 'Systrephus multiflora'. Such a specimen is in the Kew herbarium and could have been chosen as lectotype.

Two subspecies have been traditionally recognized differing on whether the corolla lobes remain attached at their tips to form a globose to oblate cage (subsp. *multiflora*) or whether they are free and spread ± rotately (subsp. *tentaculata*). However, as Peckover in Aloe **30**: 25 (1993) has pointed out, the two subspecies overlap considerably in their distributions and sometimes grow together and their separation does not seem justifiable. This is reinforced by the fact that both forms can have either glabrous or puberulous exteriors to the corolla.

Huber (1958) indicated that the holotype of *C. multiflora* subsp. *tentaculata* was the collection *Welwitsch* 4277 in BM. This must be treated as a lectotypification as two collections were cited in the protologue, *Welwitsch* 4277 and *Schinz* s.n. from Ondongu. There are 4 sheets of *Welwitsch* 4277 in BM, so an additional, second stage, lectotypifocation is needed.

The name 'Desmostemma anisophyllum' is a manuscript name proposed by Welwitsch but never taken up.

45. **Ceropegia rendallii** N.E. Br. in Bull. Misc. Inform. Kew **1894**: 106 (1894). —Brown in Fl. Cap. **4**(1): 814 (1908). —Phillips in Fl. Pl. South Africa **1**: t.39 (1921). — Huber in Mem. Soc. Brot. **12**: 114 (1958). —Dyer in Codd *et al.*, Fl. S. Africa **27**(4): 71, fig.15 (1980). —Meve, Ill. Handb. Succ. Pl. Asclepiadaceae: 96–97 (2002). Type: South Africa, Transvaal, prob. Barberton, *Rendall* s.n. (K000305578 holotype).

Ceropegia galpinii Schltr. in Bot. Jahrb. Syst. **18**(Beibl. 45): 23 (1894). Type: South Africa, Transvaal, Barberton, *Galpin* 1251 (B† holotype, PRE).

Tuber slightly depressed, 1–3 cm diam., sometimes forming series of adventitious tubers. Stem twining, up to 1 m long, glabrous. Leaves succulent; petiole 2–5 mm long; leaf blade linear-oblong or elliptic to ovate, 18–38 × 3–20 mm, apex acute, surfaces glabrous to scabrid. Inflorescence extra-axillary, peduncle 6–18 mm long; cyme 1–3-flowered. Pedicel 3–5 mm long. Sepals subulate, 2–3 mm long. Corolla 15–25 mm long, curved to straight; tube 12–20 mm long, basal chamber subglobose, c.3 mm wide, narrowed into funnel shaped limb, c.1.5 mm wide at base, widening to 5–6 mm diam. at mouth, abruptly narrowed at apex, exterior white with slender green or purple venation, glabrous, interior sparsely pilose; sinuses rounded; lobes deeply cordate with claw-like base, apices connate to form spreading 10-lobed canopy, 8–10 mm diam., canopy margins flat or slightly up-curved, ciliate, apex apiculate, abaxially dull green or pale brown, mostly glabrous. Corona forming a 5-fluted tube, c.2 × 2.5–3 mm diam., outer lobes each forming a deep pocket, entire, shorter than inner lobes, finely papillate; inner lobes falcately spreading above, laterally compressed, c.1 mm tall. Follicles and seeds not recorded.

Botswana. N: Tsessebe, *Venter, Archer & Hahn* 399 (UNIN); 8 km N of Francistown, *Bruyns* 12381 (BOL).

Also in South Africa, Swaziland. "in loose soil or leaf mould"; c.1000 m.

Conservation status: Overall status LC, rarely collected within the Flora area.

Dyer (1980) mentions that it has also been recorded from Mozambique but no voucher has been recorded.

Masinde described *Ceropegia rendallii* subsp. *mutongaensis* from central Kenya. It differs by the smaller terminal canopy of the corolla, only 4–5 mm wide, the corolla lobe keels and bases adaxially with needlelike, purple, vibratile hairs, and the dorsiventrally compressed inner corona lobes. The relationship to *C. rendallii*, particularly when coupled with the considerable geographical isolation, needed reconsideration and Bruyns has transferred it to *C. collaricorona* as *C. collaricorona* subsp. *mutongaensis* (Masinde) Bruyns in S. African J. Bot. **112**: 432 (2017).

46. **Ceropegia linearis** E. Mey., Comm. Pl. Afr. Austr.: 194 (1838). —Decaisne in Candolle, Prodr. **8**: 644 (1844). —Brown in Fl. Cap. **4**(1): 832 (1908). —Huber in Mem. Soc. Brot. **12**: 117 (1958). —Dyer in Codd *et al.*, Fl. S. Africa **27**(4): 78 (1980). Type: South Africa, Cape, *Drège* 4947 (P00109650–P00109652 syntypes). FIGURE 7.3.**150**.

    Ceropegia caffrorum Schltr. in J. Bot. **32**: 358 (1894). Type: South Africa, Natal: near Durban, *Wood* 5376 (B† holotype, NH0006627-0 lectotype), designated here.
    Ceropegia caffrorum var. *dubia* N.E. Br. in Fl. Cap. **4**(1): 824 (1908). —Bruyns in Bradleya **3**: 37–40, fig.19 (1985). —Meve, Ill. Handb. Succ. Pl. Asclepiadaceae: 96 (2002). Type: Mozambique, Delagoa Bay, specimens cultivated at Kew, sent by Mrs. Monteiro (K000305562 holotype).

Tuber subglobose, 2–3 cm diam., adventitious tubers often forming at nodes along branches. Stems several from base, usually pendant, less often twining, occasionally branched, up to 2 m long, c.1 mm diam., glabrous. Leaves succulent; petiole 2–7 mm long; leaf blade linear-lanceolate to ovate-lanceolate, distally narrower, to linear, 22–50 × 4–11 mm, base rounded to subcordate, apex acute, often with paler veins, both surfaces glabrous. Inflorescence extra-axillary, peduncle 6–12 mm long; glabrous; cyme 2- or 3-flowered, flowers opening in succession. Pedicel 3–6 mm long, glabrous. Sepals narrowly-lanceolate to lanceolate-subulate, 2–3 mm long, acute, glabrous. Bud cylindrical, with flat sinuses. Corolla (13)15–23 mm long, straight or nearly so; tube 9–17.5 mm long, basal chamber ovoid, 2–4 mm wide, narrowed into cylindrical limb, 0.7–1(1.5) mm wide, scarcely widening to 0.7–1.5(2.5) mm at mouth, exterior pale purplish to green with longitudinal purple stripes, glabrous, sometimes slightly ridged, interior covered with very fine long hairs; sinuses acute, flat; lobes linear, 6–10 mm long, replicate, apices connate to form subglobose to ellipsoidal cage wider than mouth, adaxially purplish, ciliate with short dark purple hairs on margins, otherwise glabrous. Corona subsessile, white; outer lobes pouch-like, joined across base of inner lobes, spreading, entire, to 1 mm long, lower than staminal column, glabrous; inner lobes converging over gynostegium, spathulate-lanceolate, 1.5–2 mm long, slightly dorsiventrally compressed, apices acute, recurved, glabrous to minutely papillate. Follicles and seeds not recorded.

Mozambique. LM: Maputo, Marracuene, fl. 20.v.1981, *De Koning & Boane* 8774 (LMU).

Also in South Africa and Swaziland. No information for the Flora area but the species is common in NE South Africa where it mostly occurs on rock outcrops; 20–650 m.

Conservation status: Overall status LC, only just extending into the Flora area.

Huber (1958), followed by Bruyns (1985) and Meve (2002), took a wide view of this species and included the little known *C. debilis* and the very well known *C. woodii* within it as subspecies, defining these three subspecies on vegetative characters, primarily leaf shape. Huber's treatment was rejected by Dyer (1980) who, correctly in the view of this author, emphasised corolla form and maintained these taxa as separate species.

The three sheets of the type collection of *Ceropegia linearis* in P need careful examination before designating a lectotype.

47. **Ceropegia debilis** N.E. Br. in Gard. Chron. **2**: 358 (1895). —Brown in F.T.A. **4**(1): 458 (1903). —Werdermann in Bot. Jahrb. Syst. **70**: 222 (1939). Type: Malawi, Zomba (cult. London), *Buchanan* s.n. (K000305456 holotype).

> *Ceropegia linearis* subsp. *debilis* (N.E. Br.) H. Huber in Mem. Soc. Brot. **12**: 118 (1958). —Meve, Ill. Handb. Succ. Pl. Asclepiadaceae: 96 (2002).

Tuber subglobose, depressed with short woody neck, adventitious tubers often at nodes along branches. Stems several from base, pendent, occasionally twining, up to 1 m long, c.1 mm diam., glabrous. Leaves succulent; petiole c.2 mm long; leaf blade subterete, 12–25 × 1–2 mm, base cuneate, apex acute, light green, both surfaces glabrous. Inflorescence extra-axillary, peduncle 5–12 mm long; glabrous; cyme 1–3-flowered, flowers opening in succession. Pedicel c.4 mm long, glabrous. Sepals lanceolate-acute, c.2 mm long, acute, glabrous. Flower bud with raised sinuses and linear beak. Corolla 20–26 mm long, straight or nearly so; tube 15–19 mm long, basal chamber ovoid, 3–6 mm wide, abruptly narrowed at apex, limb cylindrical, 1.7–1.9 mm wide, widening to 3.5–5 mm at mouth; exterior pale purplish to green with purple longitudinal stripes, glabrous, sometimes slightly ridged, interior sparsely pilose; sinuses rounded, spreading; lobes linear-spathulate, 5.5–7 mm long, plicate, apices connate to form ovoid-conical cage as wide as mouth, adaxially greenish with blackish-purple keel, with long, curly purple hairs. Corona subsessile, white; outer lobes joined laterally across bases of inner lobes, spreading, entire, broadly U-shaped, c.1 mm long, lower than staminal column, glabrous; inner lobes dorsally connected at base to outer lobes, erect, converging over gynostegium, narrowly oblanceolate, 2–2.5 mm long, dorsiventrally compressed, apices acute, recurved, sometimes lobed, glabrous to minutely papillate. Follicles and seeds not recorded.

Zimbabwe. N: Rusape, fl. 5.ii.1949, *Munch* 164 (K). **Malawi**. S: ex Zomba, *Buchanan* s.n. (K).

Endemic.

Conservation status: DD, known from very few collections, only one since the nineteenth century.

48. **Ceropegia woodii** Schltr. in Bot. Jahrb. Syst. **18**(Beibl. 45): 45 (1894). —Brown in Bot. Mag. **126**: t.7704 (1900); in Fl. Cap. **4**(1): 832 (1908). —Wood & Evans, Natal Pl. **4**: t.357 (1905). —Phillips in Fl. Pl. South Africa **10**: t.382 (1930). —Dyer in Codd *et al.*, Fl. S. Africa **27**(4): 76–77, fig.16, 3a-c (1980). Type: South Africa, Natal, Groenberg, *Wood* 1317 (B† holotype, NH0004120-0 lectotype), designated here. FIGURE 7.3.**150**.

> *Ceropegia euryacme* Schltr. in Bot. Jahrb. Syst. **38**: 46 (1905). Type: South Africa, Transvaal, Houtboschberg R., *Schlechter* 4402 (B† holotype).
>
> *Ceropegia leptocarpa* Schltr. in Bot. Jahrb. Syst. **38**: 47 (1905). Type: South Africa, Natal, bushland near Maramkene ['zwischen Gebusch bei Maramkene'], Delagoa Bay Dist., *Schlechter* 12077 (B† holotype).
>
> *Ceropegia barbertonensis* N.E. Br. in Fl. Cap. **4**(1): 1132 (1909). Types: South Africa, from near Barberton, cultivated Kew, *Gumbleton* s.n. (K000305567 lectotype), designated here; Transvaal, Woodbush, *Swierstra* 3990 (PRE syntype).
>
> *Ceropegia hastata* N.E. Br. in Bull. Misc. Inform. Kew **1909**: 327 (1909). Type: South Africa, Cape, Bethelsdorp, *Patterson* s.n. (K000305569 holotype, GRA0002451-0).

Ceropegia schoenlandii N.E. Br. in Bull. Misc. Inform. Kew **1913**: 303 (1913). Type: South Africa, Redhouse, *Patterson* 1049 (K000305564 holotype, GRA0002456-0).

Ceropegia linearis subsp. *woodii* (Schltr.) H. Huber in Mem. Soc. Brot. **12**: 118 (1958). — Bruyns in Bradleya **3**: 40, fig.20 (1985). —Meve, Ill. Handb. Succ. Pl. Asclepiadaceae: 97 (2002).

Roots fibrous. Tuber subglobose, to 2.5 cm diam., adventitious tubers often at nodes of branches. Stems several from base, usually pendant, less often twining, somewhat branched, to 1 m long, c.1 mm diam., glabrous. Leaves succulent; petiole 1–12 mm long, glabrous; leaf blade mostly cordate-ovate to reniform-cordate, sometimes narrower, 6–18(35?) × 4–20 mm, base subcordate, margin entire, apex shortly acuminate, often with greyish variegated veins, less often uniformly green, adaxially often with distinct, coloured veins, both surfaces glabrous to sparsely hirsute. Inflorescence extra-axillary, peduncle 4–15 mm long; glabrous; sometimes producing successive cymes along short fleshy rachis, 1–3-flowered, flowers opening in succession. Pedicel 2.5–10 mm long, suberect, glabrous. Sepals linear-lanceolate to lanceolate, 1.2–3 mm long, acute, glabrous. Flower bud with raised sinuses and clavate beak. Corolla (10)16.5–25(27.5) mm long, straight or nearly so; tube 10–17(20) mm long, basal chamber globose, 2.6–5 mm wide, abruptly narrowed at apex, limb cylindrical, 1.3–1.7(2.5) mm wide, widening to (2.6)3.4–5 mm at mouth, exterior pale purplish or green with purple longitudinal stripes, glabrous, sometimes slightly ridged, interior hairy; sinuses rounded, spreading and gaping; lobes linear-spathulate, 3.8–7.5 mm long, inflexed at base, S-shaped, plicate, apices connate to form cylindrical-clavate cage narrower than mouth, adaxially uniformly dark purplish, densely purple pilose. Corona white; outer lobes pouch-like, oval, sometimes joined across base of inner lobes, spreading, entire, c.0.5 mm long, lower than staminal column, glabrous; inner lobes converging over gynostegium, linear-lanceolate to spathulate-lanceolate, 1.5–2.5 mm long, far longer than outer lobes, Follicles diverging, narrowly fusiform, 5–7 cm long, glabrous. Seed not recorded.

Zimbabwe. W: Bulawayo, fl. vii.1942, *Walsh* s.n. sub PRE59245 (PRE). C: Rusape, Between Rusape and Rusape Motel, near river, 25.ii.1993, *Archer et al.* 281 (SRGH). E: East bridge, Umvumvumwu R Gorge, fl. 19.ii.1964, *Chase* 8129 (K). **Mozambique**. M: ± 50 km east of Ribue, fl. 17.v.1961, *Leach & Rutherford-Smith* 10907 (PRE). Z: Zambezia Prov., Errego, Mt Muli, near tea factory, fl. 6.i.2004, *Bruyns* 9744 (BOL).

Also South Africa, Swaziland. *Brachystegia–Pterocarpus rotundifolius* woodland. Sandy soil over granite, locally abundant on termitaria; 550–1100 m.

Conservation status: Overall status LC, infrequently collected within the Flora area but common in South Africa and Swaziland.

This is by far and away the most widely cultivated member of the genus and it is possible that some of the records from the Flora area are escapes from cultivation.

The holotype of *Ceropegia woodii* was destroyed in 1943 so the surviving isotype in Durban is here designated as the lectotype. The protologue of *C. barbertonensis* listed 2 syntypes. The specimen in K is clearly marked as 'type specimen' by N.E. Brown and is here designated as the lectotype.

49. **Ceropegia conrathii** Schltr. in Bot. Jahrb. Syst. **38**: 45 (1905). —Brown in Fl. Cap. **4**(1): 831 (1908). —Dyer in Fl. Pl. Africa **32**: t.1246 (1957); in Codd *et al.*, Fl. S. Africa **27**(4): 67–68, fig.13 (1980). —Huber in Mem. Soc. Brot. **12**: 118 (1958). —Meve, Ill. Handb. Succ. Pl. Asclepiadaceae: 72–73 (2002). Type: South Africa, Transvaal, Modderfontein, *Conrath* 1008 (GZU000260960 holotype, K000305584). FIGURE 7.3.**150**.

Tuber discoid 3–10 cm diam., upper surface concave, Stems 1 or 2, branched above, 4–12 cm long, Leaves succulent; petiole 1–3 mm long; leaf blade oblong-lanceolate to ovate-oblong, 20–35 × 10–13 mm, margin undulate to crisped, apex acuminate, both surfaces sparsely ciliate to glabrous. Inflorescence usually produced before leaves develop, extra-axillary, sessile to obscurely pedunculate; 1–6-flowered, opening ± together. Pedicel 6–15 mm long, glabrous. Sepals lanceolate, 2–3 mm long, glabrous. Corolla 25–35 mm long; slightly curved; tube 12–23 mm long, basal inflation 3–5 mm wide, limb subcylindrical, c.2 mm diam. at mouth; exterior pale buff to greenish yellow with lines of faint spots, glabrous, interior of limb with scattered long white hairs above, basal chamber with scattered dark tipped papillae; sinuses acute, flat; lobes linear, 5–12 mm long, apices connate to form ellipsoidal cage, wider than mouth, apex acute, both surfaces same colour as tube, glabrous. Corona with outer lobes forming pockets c.0.5 mm deep, outer margin broadly U-shaped, c.0.5 mm long; inner lobes linear-spathulate to spathulate, c.2 mm long, glabrous to minutely papillate; Follicles erect, slender, 8–10 × c.0.4 cm, glabrous, sharply tapering to apices. Seed not recorded.

Botswana. N: Between Tsessebe and Tsamaya, *Venter et al.* s.n. (UNIN).

Also from South Africa (Transvaal and Natal). In Kalahari *Acacia–Combretum* woodland.

Conservation status: Overall status LC, fairly common in South Africa but only just extending into the Flora area where it is known from only the one collection listed by Venter.

50. **Ceropegia floribunda** N.E. Br. in F.T.A. **4**: 460 (1903). —Huber in Mem. Soc. Brot. **12**: 118 (1958), excl. syn. *C. conrathii*. —Dyer in Codd *et al.*, Fl. S. Africa **27**(4): 81–82 (1980). —Meve, Ill. Handb. Succ. Pl., Asclepiadaceae: 79 (2002). Type: Botswana, Kwebe Hills, *Lugard* s.n. [161] (K000357605 holotype).

Tuber 1–1.5 cm diam. Stem twining, 1–2 mm diam., glabrous to slightly hirsute. Leaves succulent, those at upper flowering nodes reduced; petiole 2–6 mm long; leaf blade elliptic to elliptic-oblong, 12–25 × 8–18 mm, base rounded to cuneate, margin ciliolate, apex apiculate, abaxial surface minutely and sparsely pubescent, adaxial surface glabrous. Inflorescences usually many, extra-axillary, sessile to obscurely pedunculate; cyme umbel-like, (sometimes on very reduced lateral branches) to 6- or more flowered, often several flowers open together. Pedicel 6–15 mm long, glabrous. Sepals lanceolate, c.2.5 × 0.6 mm, acute, glabrous. Bud slender with slightly raised sinuses and cylindrical beak. Corolla 15–16 mm long, straight or nearly so, tube 11(13) mm long, basal chamber globose to ovoid, c.3.5 mm wide, abruptly narrowed at apex, limb cylindrical, c.1.5 mm wide, widening to c.2 mm at mouth, exterior green or yellowish tinged purplish at base, glabrous, interior sparsely hairy; sinuses acute, flat; lobes linear-triangular, narrowing gradually or sometimes oblong, c.5 mm long, flat, apices connate to form ellipsoidal cage, wider than mouth, apex acute, less often cage subcylindrical with obtuse apex, both surfaces same colour as tube, glabrous. Corona cup-shaped, obtusely pentagonal, outer lobes truncate, c.0.6 mm long, white, glabrous; inner lobes converging over gynostegium, linear-spathulate, c.1 mm long, overtopping the staminal column, white, apices obtuse, slightly ciliate along the margin within, otherwise glabrous. Follicles spreading, linear, 10–15 × c.0.6 cm wide at middle, green spotted purple, tapering to an acute apex. Seed not recorded.

Botswana. N: Kwebe Hills, 900 m, fl. 7.ii.1898, *Lugard* 161 (K).

Also in Namibia. In the Flora area recorded by P.A. Smith from disturbed woodland, elsewhere among bushes and trees on sand; 1000–1100 m.

Conservation status: Overall status VU. Known only from the type collection plus a small number of collections from the immediately adjacent part of Namibia.

No collection number was given in the protologue but the only collection by Mrs. Lugard has been given the number '161'.

51. **Ceropegia cataphyllaris** Bullock in Kew Bull. **10**: 625 (1956). —Huber in Mem. Soc. Brot. **12**: 134 (1958). —Meve, Ill. Handb. Succ. Pl. Asclepiadaceae: 72 (2002). Type: Zambia, Mwinilunga, SW of Matochi Farm, *Milne-Redhead* 3241 (K000305646 holotype). FIGURE 7.3.**150**.

Ceropegia filiformis E.A. Bruce in Kew Bull. **3**: 463 (1949), illegitimate name, not (Burch.) Schltr. (1896).

Tuber slightly oblate, 1.5–2.5 cm diam. Stem erect, solitary, herbaceous, 18–28 cm high, glabrous. Leaves herbaceous, lowermost scale-like, c.2 × 0.7 mm; uppermost sessile, blade linear to filiform, to 30 × c.0.3 mm, margin revolute, glabrous except for sparsely, minutely, ciliolate margins. Inflorescence subterminal, sessile, 1-flowered. Pedicel 15–20 mm long, slender, reddish.

Sepals linear-triangular, reddish, 1.8–2.2 × c.0.5 mm, glabrous. Corolla to 35 mm long, ± straight; tube 15–17 mm long, basal chamber obovoid, 7–8 × to 5 mm wide, gradually narrowed at apex, limb cylindrical, c.1.5 mm wide, widening to c.2.5 mm at mouth, exterior white, tinged pinkish, both surfaces glabrous; sinuses flat; lobes linear, c.20 × 1 mm wide, revolute, apices connate to form fusiform cage, c.8 mm wide, adaxially with basal c.1 mm whitish, then olive-green or brown, glabrous. Outer corona lobes bifid, teeth linear, to 1 mm long, apices each with 1 or 2 white setae c.1 mm long; inner lobes erect, linear, 1.5–2 mm long, apices shortly recurved, glabrous. Follicles and seeds not recorded.

Known only from the type. Endemic. Shallow soil overlying laterite; 1300 m.

Conservation status: DD: known only from the type collection from an area with other taxa known only from type collections.

Sect. **Psilopegia** H. Huber in Mem. Soc. Brot. **12**: 34 (1958).

52. **Ceropegia ampliata** E. Mey., Comm. Pl. Afr. Austr.: 194 (1838). —Decaisne in Candolle, Prodr. **8**: 645 (1844). —Brown in Fl. Cap. **4**(1): 817 (1908). —Phillips in Fl. Pl. South Africa **4**: t.140 (1924). —Huber in Mem. Soc. Brot. **12**: 143 (1958). —Dyer in Codd *et al.*, Fl. S. Africa **27** (4): 78 (1980). —Bruyns in Bradleya **3**: 21–23 (1985). —Archer, Kenya *Ceropegia* Scrapb.: 89, XIX (1992). —Masinde in F.T.E.A., Apocynaceae (part 2): 287 (2012). —Meve, Ill. Handb. Succ. Pl. Asclepiadaceae: 67 (2002). Type: South Africa, Cape, Fish River valley, hills near Trumpeter's Drift, *Drège* 4949 (P00109631 holotype, P00109630, HAL0114327, K000305616, K000305631, MEL710750, W). FIGURE 7.3.**150**.

　　Ceropegia ampliata var. *oxyloba* H. Huber in Mem. Soc. Brot. **12**: 143 (1958). Type: Tanzania, Uzaramo, Msasani, N of Dar es Salam, *Vaughn* 2806 (BM holotype, EA).

　　Ceropegia ampliata var. *madagascariensis* Lavranos in Adansonia, n.s. **13**(1): 71 (1973). Type: Madagascar, Fianarantsoa, Près de Zazafotsy (Ihosy), *Lavranos* 9590 (P00724205 holotype, K, P00442720, PRE0332896-0, TAN000091–92, W).

Roots clustered, fleshy, fusiform, white. Tuber absent. Stem twining, succulent, c.1.5 mm diam., glabrous. Leaves succulent, often scalelike, petiole 2 mm long, glabrous; leaf blade linear-lanceolate, to 10 × 2 mm, base cuneate, apex acute, both surfaces glabrous. Inflorescence extra-axillary, sessile, cyme 2- or 3-flowered, flowers opening in succession. Pedicel 18–20 mm long, stiffly straight, glabrous. Sepals linear-triangular, c.3 mm long, mostly somewhat recurved near apex, glabrous. Bud with pointed sinuses. Corolla 30–70 mm long, straight; tube 23–50 mm long, basal chamber obovoid, 10–30 × 6.5–25 mm wide, slightly but distinctly narrowed at apex, limb cylindrical, 7–15 mm wide, slightly narrowed at mouth, exterior greenish white with darker longitudinal lines, glabrous, interior similar in colour to exterior except for narrow purple band at junction of basal chamber and tube, basal chamber very sparsely hairy, tube densely retrorsely pilose; sinuses obtuse, narrowly auriculate; lobes oblong-lanceolate, 6–20 mm long, plicate, apices loosely connate to form ovoid-conical cage, widest at base, glabrous, adaxially pale at base then abruptly green, glabrous. Corona sessile, outer basin or saucer shaped, lobes bifid to base, spreading, teeth triangular so corona subequally 10-toothed, c.1 mm long, adaxially pilose; inner lobes slightly incumbent on anthers and then erect, filiform, to 5 mm long, glabrous. Follicles not recorded.

Botswana. SE: Gaborone Campus, fl. 25.viii.1986, *Turton* s.n. (K). **Mozambique**. M: Lourenço Marques, Marracuene, Costa do Sol, fl. 18.iv.1961, *Macuacua* 92 (BM, K).

Also in Kenya, Tanzania, Madagascar and South Africa. Little information for Flora area, only description gave 'degraded forest c.15 m high, mainly *Afzelia quanzensis*, also *Syzigium*, *Mimusops caffra*, *Dialium schlechteri*.

Conservation status: LC: A widely distributed species, common south of the Flora area, and well established in cultivation. The records from Botswana are probably based on escapes from cultivation.

Huber (1958: 144), followed by Masinde (2012: 287), indicated that the holotype of this species was in P, as has been assumed for most of E. Meyer's taxa. There are two sheets of *Drège* 4949 in P: one with no locality data and annotated by Huber as an isotype; the other with good locality data and what might be Meyer's own dermination label but not annotated by Huber. This is here treated as the holotype.

Sect. **Chamaesiphon** H. Huber in Mem. Soc. Brot. **12**: 34 (1958).

53. **Ceropegia robinsonii** M.G. Gilbert, sp. nov. Resembles *Ceropegia dinteri* Schltr. in overall habit and corolla form and indumentum but differs by its more slender habit with smaller leaves, to 63 × < 1 mm, shorter than the internodes, (120–160 × c.4 mm and longer than the internodes in *C. dinteri*), and much smaller flowers, 20–24 mm long (40–100 mm long in *C. dinteri*) with relatively longer outer corona lobes more than half as long as inner corona lobes (c. one third as long in *C. dinteri*). Type: Zambia, N: Kasama Dist., Chambeshi Flats, 50 km SE of Kasama, fl. 23.i.1961, *Robinson* 4306 (K001400197 holotype). FIGURE 7.3.**151**.

Tuber subglobose, to 1.5 × 0.9 cm deep. Stem solitary, erect, unbranched, herbaceous or possibly slightly succulent, to 55 cm high, < 2 mm thick, internodes to 11 cm long, glabrous. Leaves herbaceous, sessile; leaf blade linear, to 6.3 cm × < 1 mm, base attenuate, apex acute, both surfaces glabrous. Inflorescence extra-axillary, terminal or subterminal, sessile; cyme 1- or 2-flowered. Pedicel 8–15 mm long, slender, straight, glabrous. Sepals linear-triangular, 2.5–3 × c.0.4 mm, glabrous. Bud cylindrical, with flat sinuses. Corolla 20–24 mm long, straight or nearly so, basal chamber indistinct, fusiform, c.3 mm wide, tube c.15 mm long, narrowing gradually to c.2 mm wide, 3 mm wide at mouth; exterior colour not recorded, apparently uniformly dark ?purplish, glabrous, interior glabrous; sinuses acute, flat; lobes c.8 × 0.8 mm wide, suberect, slightly recurved, linear, apices free, abaxially glabrous, adaxially glabrous for c.2 mm, then densely papillose-puberulent with patch of purplish, caduceus, vibratile hairs to 2 mm long near base. Corona stipitate, stipe pale; outer lobes erect, deeply bifid almost to base, teeth linear, c.½ as long as inner lobes, divided dark (purple?), sparsely ciliate; inner lobes incumbent on anthers and produced above them, linear, c.1.5 mm long, dark (purple?), glabrous.

Known only from the type. Endemic. Growing on dry causeway through flood-plain; 1200–1300 m.

Conservation status: DD, known only from the type collection.

This collection was initially identified as *Ceropegia dinteri* Schtr., otherwise only known from Namibia (the nearest record is separated by nearly 10° of latitude and 14° of longitude). Its habitat also appears to be quite different: *C. dinteri* grows in association with dolomitic limestone outcrops.

54. **Ceropegia chlorantha** (Schltr.) Bruyns in S. African J. Bot. **112**: 433 (2017). Type: South Africa, Transvaal, prope Kl. Olifant-Rivier, *Schlechter* 3812 (B† holotype); prope Pretoria, *Schlechter* 4152 (BOL140269 lectotype, Z000001778), designated by Bruyns (2017). FIGURE 7.3.**151**.

Tenaris chlorantha Schltr. in Bot. Jahrb. Syst. **20**(Beibl. 51): 45 (1895).

Brachystelma chloranthum (Schltr.) Peckover in Aloe **33**: 43 (1996). —Meve, Ill. Handb. Succ. Pl. Asclepiadaceae: 25 (2002).

Tuber depressed globose, 15–25 mm diam. Stem solitary, erect, unbranched to branched above, lower part without leaves 20–55 cm long, glabrous. Leaves herbaceous, erect-spreading, subsessile; leaf blade lanceolate to linear-filiform, 50–110 × 0.5–1.5 mm, upper smaller, both surfaces glabrous. Inflorescence extra-axillary, cyme 3–7-flowered, flowers opening in succession, facing ± horizontally or slightly below. Pedicel 7–15 mm long, filiform, straight, glabrous. Sepals linear-lanceolate, 1–1.5 mm long, glabrous. Bud ovoid at base with long linear beak. Corolla 6–8.5 mm long, divided nearly to base; tube campanulate, 1–1.5 mm deep; exterior glabrous; interior and base of lobes with scattered minute papillae, nearly glabrous; lobes shortly lanceolate-ligulate to linear-filiform, 5–7 × c.1 mm wide at base, 0.5 mm wide above, geniculate near base, strongly replicate, apices free, straight with slightly recurved tips, margins recurved, apex somewhat obtuse, abaxially glabrous, adaxially reddish- or olive-brown or yellow with obscure longitudinal lines at base, rest yellow to reddish-brown, minutely purple-papillate. Corona with outer lobes shortly ovate, apices obtuse to subquadrate, shortly emarginate-sinuate, c.0.3 mm long, inner lobes incumbent on anthers, lanceolate, c.0.3 mm long, shorter or as long as the anthers, apices acute or somewhat obtuse. Follicles sub-fusiform, 4.5–6 cm long.

Zimbabwe. C: Harare ['Salisbury'], Cranborne, 4900 ft, fl. 1.xii.1945, *Wild* 445 (K); Ruwa River, 4900 ft, fl. 10.i.1948, *Wild* 2278 (K). W: Matobo Dist., Besna Kabila Farm, 4900 ft, fl. xii.1961, *Miller* 8066 (K).

Also in South Africa. *Brachystegia spiciformis* veldt, in shallow soil over granite; 1450 m.

Conservation status: LC, of sporadic occurrence over a wide area but inconspicuous and likely to be overlooked.

55. **Ceropegia bikitaensis** (Peckover) Bruyns in S. African J. Bot. **112**: 433 (2017). Type: Zimbabwe, Glencova, Bikita Mine, *Peckover* 242 (PRE0632631-0 holotype).

　　Brachystelma bikitaense Peckover in Aloe **32**: 78 (1995). —Meve, Ill. Handb. Succ. Pl. Asclepiadaceae: 22 (2002).

　　Tenaris bikitaensis (Peckover) J.E. Victor & Nicholas in S. African J. Bot. **64**: 207 (1998).

Tuber 3–5 cm diam., 2 cm deep. Stem solitary, erect, to 50 cm tall, internodes to 10 cm long, 2 mm thick at base, glabrous. Leaves herbaceous, subsessile; leaf blade linear lanceolate, 70–90 × 2–3 mm, margin entire, both surfaces glabrous. Inflorescence extra-axillary, sessile, cyme fasciculate, up to 10-flowered, flowers opening successively, facing ± downwards. Pedicel 20–30 mm long, glabrous. Sepals linear lanceolate, c.3 × 0.8 mm, glabrous. Bud globose at base with long linear beak. Corolla c.30 mm diam., divided nearly to base; tube almost absent, interior and base of lobes yellowish, minutely puberulent; lobes linear above slightly enlarged base, c.13 × 0.8 mm wide, strongly replicate, somewhat incurved, adaxially pale brown, minutely puberulous. Corona sessile, bowl-shaped, c.1 mm high, 3 mm diam.; greenish, outer lobes reduced to small pockets, apices shortly 2-toothed, c.1 mm wide; inner lobes incumbent on anthers, slender. Follicles diverging at 45°, 4–7.5 × c.0.2 cm. Seeds 15–20, almost black, 6–7 × c.1.5 mm, coma 15–20 mm.

Known only from the type collection. Endemic. In north facing miombo woodland in dappled shade on greyish sandy loam; 1200 m.

Conservation status: V?, known only from the type collection from a relatively well collected area, but inconspicuous and very likely to be overlooked.

56. **Ceropegia filifolia** (Schltr.) Bruyns in S. African J. Bot. **112**: 433 (2017). Type: South Africa, Transvaal, Komati-Poort, *Schlechter* 11733 (B† holotype, BR0000008864024 lectotype, E00279502, E00279505, HBG502768, K000305403), lectotype designated by Bruyns (2017). FIGURE 7.3.**151**.

　　Macropetalum filifolium Schltr. in Bot. Jahrb. Syst. **38**: 36 (1905).

　　Tenaris filifolia (Schltr.) N.E. Br. in Fl. Cap. **4**(1): 797 (1908).

　　Brachystelma filifolium (Schltr.) Peckover in Aloe **33**: 43 (1996). —Meve, Ill. Handb. Succ. Pl. Asclepiadaceae: 29 (2002).

Tuber ± depressed globose, 3–4 cm diam. Stem solitary, erect, simple to sparingly branched above, herbaceous, 45–75 cm long, slender, glabrous. Leaves herbaceous, erect to erect-spreading; leaf blade filiform, 30–85 mm long, the lower ones the longest, both surfaces glabrous. Inflorescence extra-axillary, slender, racemelike, 75–140 mm long, with ovate bracts c.1 mm long; cymes fasciculate, 1–3-flowered, flowers opening in succession. Pedicel 4–10 mm long, glabrous. Sepals lanceolate to ovate, 1–1.5 × c.0.8 mm, glabrous. Bud ovoid at base with long linear beak. Corolla 9–10.5 mm long, divided nearly to base; tube campanulate, c.1.5 × 1.5 mm, both surfaces pale, glabrous; lobes filiform from deltoid-ovate base, 8–9 mm long, erect-spreading, abaxially greenish, both surfaces glabrous, adaxially purplish-brown. Corona stipitate, c.0.8 mm high, outer lobes minute and pouch-like, inner lobes incumbent on anthers, linear, c.0.5 mm long, apices connivent-erect, glabrous. Follicles diverging, linear-terete, 5–6.5 × 0.2–0.3 cm, glabrous.

Fig. 7.3.**151**. —**53**. CEROPEGIA ROBINSONII, 53a habit, 53b flower, 53c corona, from *Robinson* 4306. —**54**. CEROPEGIA CHLORANTHA, leaf and flower, drawn from image taken by Bart Wursten. —**56**. CEROPEGIA FILIFOLIA, leaf and flower, from *Pedro & Pedrogao* 423. —**59**. CEROPEGIA ADAMSIANA, 59a flowering branch, 59b flower, 59c corona, from *Richards* 7289. —**61**. CEROPEGIA BREVIFLORA, flowering branch and flower, from *Chase* 7895. —**62**. CEROPEGIA MEGASEPALA, leafy stem and flower, from *Richards* 10425. —**64**. CEROPEGIA PYGMAEA, flowering stem and flower, from *Whellan* in SRGH 34864. —**68**. CEROPEGIA BREVIPEDICELLATA, leaf and flower, from image taken by Martin Heigan. Scale bars: stems = 10 mm; flowers (except 59b) and leaves = 5 mm; 59b and coronas = 1 mm. Drawn by Margaret Tebbs.

Mozambique. M: Lourenço Marques, Goba Fronteira, 8.i.1947, *Barbosa* 6921 (LMA, K fragment & image); Maputo, Goba Fronteira, 8.i.1947, *Pedro & Pedrogao* 423 (LMA, K fragment & image).

Also in South Africa and Swaziland. No information on habitat.

Conservation status: LC, of sporadic occurrence over a wide area but inconspicuous and very likely to be overlooked.

57. **Ceropegia circinata** (E. Mey.) Bruyns in S. African J. Bot. **112**: 433 (2017). Type: South Africa, Cape, nr Aliwal North, at foot of Witteberg ["in collibus graminosis prope Rietvalei ad radices montium Withebergen"], *Drège* s.n. (B† holotype, K000305820 in part lectotype, ?MEL2406192, ?S12-11943) lectotype designated by Dyer in Codd *et al.*, Fl. S. Africa **27**(4): 33 (1980), mounted with type of *Brachystelma commixtum* N.E. Br.; probable isotypes (locality details missing): MEL2406192, S12-11943). FIGURES 7.3.**152**, 7.3.**153**.

   Brachystelma circinatum E. Mey., Comm. Pl. Afr. Austr. **2**: 196 (1838). —Brown in Fl. Cap. **4**(1): 833–865 (1908). —Dyer in Bothalia **10**: 374 (1971); in Codd *et al.*, Fl. S. Africa **27**(4): 33–34, fig.6 (1980). —Meve, Ill. Handb. Succ. Pl. Asclepiadaceae: 26 (2002).

   Dichaelia circinata (E. Mey.) Schltr. in Bot. Jahrb. Syst. **21**(Beibl. 54): 13 (1896).

Tuber 3–10 cm diam., upper surface sometimes depressed. Stems 1 to few, much branched, herbaceous, 7–30 cm high, densely puberulous with recurved hairs. Leaves herbaceous, shortly petiolate; leaf blade oblong to ovate, 5–15(20) × 5–12 mm, margins sometimes undulate, abaxial surface puberulous, adaxially puberulous or sometimes glabrous. Inflorescence extra-axillary, sessile; cyme 1–8-flowered, flowers opening in succession or several together. Pedicel to 5 mm

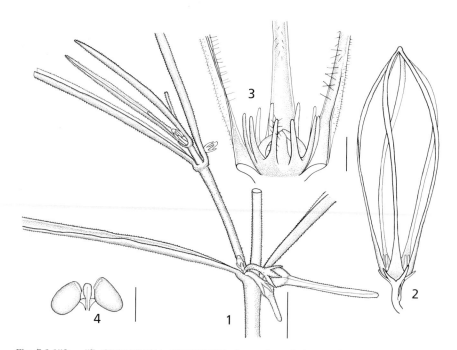

Fig. 7.3.**152**. —**57**. CEROPEGIA CIRCINATA. 1, portion of plant with inflorescence; 2, side view of flower; 3, side view of centre of dissected flower; 4, pollinarium. Scale bars: 1, 2 = 4 mm; 3 = 1 mm; 4 = 0.25 mm. All from *Bruyns* 8604. Drawn by Peter Bruyns. Reproduced from Phytotaxa (2018).

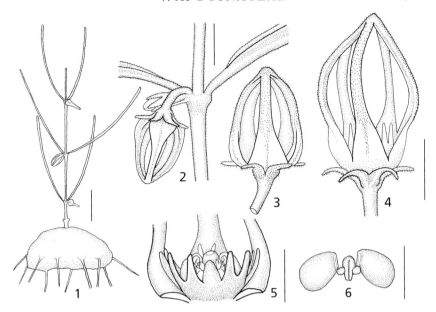

Fig. 7.3.**153**. —**57**. CEROPEGIA CIRCINATA (with smaller flowers). 1, plant; 2, portion of plant with inflorescence; 3, 4, side view of flower; 5, side view of centre of dissected flower; 6, pollinarium. Scale bars: 1 = 10 mm; 2–4 = 2 mm; 5 = 1 mm; 6 = 0.25 mm. 1–3, 5 from *Bruyns* 9743; 4, 6 from *Bruyns* 7698. Drawn by Peter Bruyns. Reproduced from Phytotaxa (2018).

long. Sepals lanceolate, joined for ¼ of length, c.1.5 mm long, puberulous. Bud with slender, cylindrical beak. Corolla 5–25 mm long, divided nearly to base; tube flattish or saucer-shaped, rarely to 3 mm deep; lobes linear to filiform, 5–25 mm long, revolute, apices connate to form ellipsoidal cage, abaxially pale green, puberulous, adaxially white or pale yellow at base, dull purple to reddish brown towards tips, glabrous. Corona subsessile, predominantly yellow; outer lobes erect, variably bifid, sometimes nearly to base, teeth linear to subulate, shorter than to distinctly longer than gynoecium, glabrous or with a few hairs, inner lobes incumbent on anthers, linear, sometimes meeting over centre, shorter than outer lobes, glabrous. Follicles and seeds not seen.

Botswana. SE: 3 km west of Pontdrift, 547 m, 5.x.1993, *Venter* 13629 (UNIN). **Mozambique**. N: Niassa, Nampula, 8 km east of Malema, 700 m, 19.xi.2000, *Bruyns* 8604 (BOL, K, MO). Z: Zambézia, 25 km north of Errego, 550 m, 6.i.2004, *Bruyns* 9746 (BOL); Mt Muli, 550 m, 6.i.2004, *Bruyns* 9743 (BOL).

Also in Namibia, South Africa, Lesotho, Swaziland. In patches of shallow soil overlying granite, among trees and bushes in Kalahari sand; 550–1250(1800) m.

Conservation status: LC, widely distributed and very frequently collected, especially in South Africa.

Dyer (1980: 33) gives the type as "Drège 3440, partly, (K!)" but no collection number was given in the protolog. The material in K has been annotated by N.E. Brown as having been compared with the type in E. Meyer's herbarium (mostly destroyed in Berlin, 1943) and so Dyer is here taken to have effectively lectotypified this species. The number 3440 has been added on a separate label. Possible isotypes in MEL and S lack locality details.

This species is most wide spread and variable within South Africa where some dozen further taxa, now regarded as synonyms have been described. These are fully listed in Dyer (1980).

58. **Ceropegia gracilior** Bruyns in S. African J. Bot. **112**: 434 (2017), new name, not *Ceropegia gracilis* Bedd. (1864). Type: Zimbabwe, Plumtree, *Porter* s.n. in PRE27227 (PRE0468611-0 holotype, K000305673).

> *Brachystelma gracile* E.A. Bruce in Fl. Pl. Africa **27**: t.1077 (1949), as "gracilis". —Dyer in Codd *et al.*, Fl. S. Africa **27**(4): 30 (1980). —Percy-Lancaster in Excelsa **13**: 66–67 (1988). —Boele in Excelsa **14**: 47 (1989). —Adams in Seyani & Chikuni, Proc. XIIIth Plenary Meeting AETFAT, Malawi **1**: 473–480 (1994). —Meve, Ill. Handb. Succ. Pl. Asclepiadaceae: 31 (2002).

Tuber oblate-spherical, 4–10 cm diam. Stem solitary, erect, usually unbranched, herbaceous, 20–70 cm tall, internodes 2–6 cm long, 1.5–2 mm thick, terminal region sometimes pendulous, minutely reflexed-pubescent. Leaves herbaceous, subsessile; leaf blade linear to linear-lanceolate, 40–90(120) × 1.5–8 mm, margins strongly incurved, apex acute, both surfaces minutely ascending-appressed-pubescent with stiff white hairs, sparser adaxially. Inflorescence extra-axillary, peduncle short to almost absent; cyme 1–4-flowered, flowers usually facing downwards. Pedicel 4–10 mm long, slender, pubescent. Sepals linear-lanceolate, 1–2 mm long, dorsally pubescent. Bud with broad, 5-lobed, reddish-purple base and a long, straight, green, terete beak. Corolla 10–17(30) mm long, divided nearly to base; tube flat, lobes linear, 10–17(30) × to 1 mm wide at base, bases rather broader, revolute, apices connate to form conical-ellipsoid cage, abaxially sparsely minutely pubescent, adaxially green, tinged reddish-purple towards base, sparsely appressed pubescent and ciliate with long, soft, white hairs. Corona subsessile, 3–5 × 3–4 mm diam., outer lobes spreading, bifid, c.2 mm long, basal undivided portion reddish, 0.75 mm long, teeth slender, terete, c.1.25 mm long, dorsally white-pubescent, inner lobes incumbent on anthers and produced above them, spathulate, 2.5–3 mm long, blackish-red, apices obtuse to emarginate, dorsally pubescent. Pollinia yellow. Follicles spreading, subfusiform, gradually narrowed to the apex, 8–13 × 0.3–0.4 cm, light greenish-brown, glabrous.

Botswana. N: Tsessebe, *Venter et al.* 403 (UNIN). **Zimbabwe**. W: Bulawayo, *Brain* 5038 (SRGH). C: Harare, on Harare–Muhrewa road, *Venter et al.* 233 (UNIN). S: Masvingo, *Munro* 2186 (BM). **Mozambique**. N: 8 km E of Malema, 700 m, 19.xi.2000, *Bruyns* 8604 (K).

According to GBIF recorded from Zambia: Lusaka North, 29.xi.1987, *Bingham* 4159 (MO). Also in South Africa, Namibia?, D.R. Congo?, Tanzania? Grassland, shallow soil over granitic slabs; 50–1600 m.

Conservation status: LC, fairly frequently collected by more expert collectors.

The considerable variation in morphology seen within this species has been described in some detail by Adams (1994). Typical material has leaves 1–2(4) mm wide, nodding flowers with the corolla lobes 10–30 mm long and remaining coherent at their tips and erect inner corona lobes much exceeding the gynostegium. Adams recognised 3 further variants from within Zimbabwe: 1) Leaves 5–6 mm wide, flowers erect, less than 10 mm long, otherwise as in the type (Zimbabwe. N: Sable Park, 8 km NE Kwekwe, 12.ii.1988, *Percy-Lancaster* 2207 (SRGH)); 2) Leaves as in type, flowers erect with rotately spreading lobes and inner corona lobes ± incumbent on the anthers (Zimbabwe. N: Turnoff to Murehwa Township, 12.xii.1987, *Percy-Lancaster* 2044 (SRGH)); 3) Leaves to 8 mm wide, corolla erect, with rotately spreading, wider lobes, the corona distinctly stipitate with well developed outer lobes much exceeding the inner lobes which are incumbent on the anthers (Zimbabwe. E: 4 km NE of Bonda Mission on Chambowa road, 20.i.1990, *M.J. & R.C. Kimberley* 112 (SRGH)). Peckover in Cact. Succ. J. (Los Angeles) **69**: 155 (1997) reported a hybrid between this and *C. albipilosa*.

The dividing line from *C. circinata* is not well defined and the two could prove to be better merged. The record of this species from Tanzania (Masinde in Kew Bull. **62**: 80, fig.17 (2007) and in F.T.E.A., Apocynaceae (part 2): 310 (2012)) needs confirmation, based as it was on photographs.

59. **Ceropegia adamsiana** M.G. Gilbert, sp. nov. Most closely resembling *Ceropegia gracilior* Bruyns but differing by the shorter, much more densely branched stems (simple or very few branched in *C. gracilior*) and the smaller corollas, to 8 mm long with densely white pilose adaxial surface (10–17(30) mm long with sparsely appressed pubescent adaxial surface and ciliate with long, soft, white hairs in *C. gracilior*). Type: Zambia, N: Kawimbe, 1680 m, 15.xii.1956, *Richards* 7289 (K001400196 holotype). FIGURE 7.3.**151**.

Tuber 'large and flat.' Stem solitary, much branched from near base, herbaceous, to 30 cm tall, longest internodes near base, to 6 cm, to 3 mm wide, uniformly minutely pubescent. Leaves herbaceous, spreading to ± recurved, sessile; leaf blade linear to filiform, 40–60 × 0.6–1 mm, revolute, base attenuate, margin entire, apex acute, both surfaces minutely puberulent. Inflorescences at most nodes, extra-axillary, sessile; cyme to 4-flowered, flowers opening in succession. Pedicel to 1.5 mm long, minutely puberulent. Sepals triangular-lanceolate, 1.5–2 × c.0.3 mm, acute, abaxially minutely hairy. Corolla to 8 mm long, divided nearly to base; tube shorter than corona, almost flat, interior glabrous; lobes linear, to 8 mm long, revolute, apices connate to form conical cage, apex subacute, abaxially minutely puberulent, adaxially brown and green, densely white pilose. Corona sessile; outer lobes separate, erect, deeply bifid, teeth linear, erect, overtopping anthers, sparsely hairy; inner lobes converging over gynostegium, linear, glabrous. Follicles and seeds not seen.

Zambia. N: Mbala (Abercorn), Old Isanya Road, 5000 ft, fl. 22.xii.1954, *Richards* 3718 (K); Kawimbe, 1680 m, 15.xii.1956, *Richards* 7289 (K); Kawimbe, nr. Leper settlement, 1800 m, 22.xii.1958, *Richards* 10358 (K).

Endemic. Open bush on sandy soil, among flat rocks; 1500–1800 m.

Conservation status: DD, possibly V: only 3 collections from a small but relatively undercollected area.

Material of this species was initially identified as the previous taxon, but differs by the shorter, densely branched stems, short pedicels and smaller corollas with a distinctive adaxial indumentum. The epithet commemorates Bryan Adams who first suggested that this material might be an undescribed species in Seyani & Chikuni, Proc. XIIIth Plenary Meeting AETFAT, Malawi **1**: 473–480 (1994).

60. **Ceropegia hirtella** (Weim.) Bruyns in S. African J. Bot. **112**: 434 (2017). Type: Zimbabwe, Inyanga, Cheshire, fl. 15.i.1931, *Norlindh & Weimarck* 4399 (LD1214831 holotype).

    Brachystelma hirtellum Weim. in Bot. Not. **1935**: 406 (1935). —Boele in Excelsa **14**: 47–48 (1989). —Meve, Ill. Handb. Succ. Pl. Asclepiadaceae: 31 (2002).

    Dichaelia hirtella (Weim.) Bullock in Kew Bull. **8**: 359 (1953).

    Brachystelma pilosum R.A. Dyer in Bothalia **6**: 541 (1956). —Dyer in Codd *et al.*, Fl. S. Africa **27**(4): 32 (1980). Type: South Africa, Transvaal, near Naboomspruit, *Galpin* 11573 (PRE holotype).

Tuber present. Stem stiffly erect, few branched from near base, herbaceous, 25–30 cm tall, internodes 2–4 cm long, 3–5 mm thick at base, densely hirsute to pilose, hairs 1–1.5 mm long. Leaves herbaceous, sessile to shortly petiolate; leaf blade linear-oblong to ovate-oblong, 10–35 × 3–20 mm, longitudinally folded, base cuneate, margin slightly undulate, apex obtuse to apiculate, both surfaces sparsely appressed pilose to nearly glabrous or with long reddish-brown hairs. Inflorescences from upper nodes, extra-axillary, sessile; cyme (1)2–3-flowered. Pedicel c.1 mm long, pilose. Sepals linear, 2–6 mm long, revolute, densely reddish-brown hairy, ciliate. Corolla 16–22 mm long; divided nearly to base; tube almost flat, lobes linear, 16–25 × c.1 mm wide, revolute, apices connate to form cage, abaxially with reddish-brown hairs, margin ciliolate, adaxially glabrous. Corona stipitate, stipe c.0.75 mm high, outer lobes deeply bifid, teeth linear, 1.5–2.5 mm long, inner lobes incumbent on anthers, as long as or just longer than them, slightly thickened on backs near base. Follicles and seeds not seen.

Zimbabwe. W: Nyamandhlovu, farm Glencurragh, Chesa Vlei, fl. xii.1955, *Plowes* 1909 (PRE, SRGH). C: Gweru, 6.x.1967, *Biegel* 2285 (SRGH). E: Inyanga, Cheshire, fl. 15.i.1931, *Norlindh & Weimarck* 4399 (LD).

Also in South Africa. Wooded grassland.

Conservation status: DD: rarely collected from widely dispersed localities, treated as NT in South Africa.

61. **Ceropegia breviflora** (Schltr.) Bruyns in S. African J. Bot. **112**: 433. (2017). Type: South Africa, Transvaal, Mooifontein nr. Heidelberg R., 22.x.1893, *Schlechter* 3568 (B† holotype); Natal, Vernon Crooke's Reserve, 10.i.1991, *Bruyns* 4426 (BOL neotype, PRE), neotype designated by Bruyns in S. African J. Bot. **112**: 433 (2017). FIGURE 7.3.**151**.

Dichaelia breviflora Schltr. in Bot. Jahrb. Syst. **20**(Beibl. 51): 49 (1895).

Dichaelia pygmaea Schltr. in J. Bot. **32**: 262 (1894), not *Ceropegia pygmaea* Schinz, (1888). Type: South Africa, ['*in regionibus orientalibus Coloniae Capensis, versimiliter Kaffrariae*'], ?Transkei, *Barber* s.n. (?B† holotype, MEL710754 lectotype), designated by Foster (1995/05/06) on sheet.

Brachystelma pygmaeum (Schltr.) N.E. Br. in Fl. Cap. **4**(1): 857 (1908). —Dyer in Codd *et al.*, Fl. S. Africa **27**(4): 28, fig.5, 1a–d (1980), excl. subsp. *flavidum* (Schltr.) R. A. Dyer. —Percy-Lancaster in Excelsa **13**: 67–68 (1988). —Boele in Excelsa **14**: 48 (1989). —Meve, Ill. Handb. Succ. Pl. Asclepiadaceae: 28 (2002), excluding subsp. *flavidum* (Schltr.) R.A. Dyer.

Brachystelma pygmaeum var. *breviflorum* (Schltr.) N.E. Br. in Fl. Cap. **4**(1): 857 (1908). —Dyer & Bruce in Fl. Pl. Africa **28**: t.1088 (1950).

Tuber subglobose to discoid, 4–10 cm diam., slightly depressed. Stems 1 to several, erect to spreading, mostly branched, herbaceous, 4–8 cm long, puberulous. Leaves sometimes undeveloped at anthesis, herbaceous, shortly petiolate; leaf blade linear-spathulate to oblanceolate, 5–20 × 5–10 mm, base cuneate, margin entire, apex subacute, both surfaces glabrous or with a few hairs on veins and margins, sometimes shortly ciliate. Inflorescence extra-axillary, sessile; cyme 1–3-flowered, flowers opening in succession. Pedicel 3–8 mm long, puberulous. Sepals linear-lanceolate, 1.5–2 mm long, puberulous. Corolla 6–10 mm long, divided nearly to base; tube very shallow, c.1.5 mm deep, exterior glabrous, interior and base of lobes white or yellow, sometimes with medifixed hairs; lobes linear from deltoid base, 4–8 mm long, revolute for most of length, apices connate to form ellipsoidal to slightly conical cage, abaxially glabrous or puberulous towards apices, adaxially uniformly yellow or white at base and yellowish to purplish-green towards tips, glabrous. Corona stipitate, outer lobes subcylindrical, subquadrate, very shortly bifid, inner lobes incumbent on anthers, linear-oblong, much larger than outer lobes, apices obtuse or notched, glabrous. Follicles and seeds not seen.

Zimbabwe. E: Inyanga, near summit of Inyangani Mountain, 1700 m, fl. 9.xi.1968, *Grosvenor* 454 (K, PRE, SRGH). **Mozambique**. MS: Gorongo Mountain, Nyamakwarara Valley, fl. 2.xi.1967, *Mavi* 441 (SRGH).

Also in South Africa. In rock crevices, sides of grass tussocks and amongst mosses on granitic outcrops; 1850–2500 m.

Conservation status: LC: well collected from a large area.

The presumed holotype of *Dichaelia pygmaea* is presumed to have been destroyed in 1943 but there is a duplicate of that collection in MEL, which has been designated as the lectotype.

Plants from South Africa with broader, radially spreading, uniformly bright chrome yellow, corolla lobes and larger corolla tube, first described as *Brachystelma flavidum* Schltr. have been treated as *C. breviflora* subsp. *flavida* (Schltr.) Bruyns. There seems to be a good case for this material to be treated as a good species, not present in the Flora area:

Ceropegia flavida (Schltr.) M.G. Gilbert, comb. nov. Type: South Africa, "Natal: an trockenen, sehr steinigen Abhangen bei Fairfield, Alexandra-County, c.750 m ü M. (H. Rudatis n. 68. — Bluhend im August 1905)" (B† holotype); " Abhange b.

Fairfield", 750 m, 8.viii.1909, *Rudatis* 675 (K000305828 neotype), designated here.

Brachystelma flavidum Schltr. in Bot. Jahrb. Syst. **40**(1): 94 (1907).

Brachystelma pygmaeum (Schltr.) N.E. Br. subsp. *flavidum* (Schltr.) R.A. Dyer in Bothalia **12**(4): 629 (1979).

It is possible, as was assumed by Dyer, that the Kew sheet here designated as the neotype was part of the original type collection, especially as it has been named by Schlechter, but the differences in collection number and date of collection (apparently later than the protologue) from those given in the protologue can not be ignored and it seems more correct to treat it as a neotype.

62. **Ceropegia megasepala** (Peckover) Bruyns in S. African J. Bot. **112**: 434 (2017). Type: Tanzania, Ruvuma, S of Mpepo near Mozambique border, cultivated in South Africa, *M. & E. Specks* 385 (PRE holotype). FIGURE 7.3.**151**.

Brachystelma megasepalum Peckover in Kakteen And. Sukk. **47**: 250 (1996). —Masinde in Kew Bull. **62**: 50 (2007); in F.T.E.A., Apocynaceae (part 2): 295 (2012). —Meve, Ill. Handb. Succ. Pl. Asclepiadaceae: 31 (2002).

Tuber discoid, 5–6 cm in diam. Stems 1–several, spreading to procumbent, few branched from near base of stem, to 20 cm high, internodes 6–20 mm long, 1–1.5 mm in diam., purplish-green, uniformly minutely hairy, hairs simple, colourless, spreading or slightly reflexed as in rest of the plant. Leaves herbaceous, spreading; petiole 2–3(15) mm long; leaf blade obovate or oblong, (10)20–30 × 10–18 mm, base rounded, margin entire, undulate or not, ciliolate, apex rounded to subacute, apiculate; both surfaces minutely pubescent especially abaxially on veins and midrib. Inflorescence subterminal, extra-axillary, sessile, cyme umbel-like, up to 4-flowered, flowers almost all opening simultaneously, emitting a putrid smell reminiscent of dung. Pedicel 14–30 mm long, densely pubescent. Sepals lanceolate, 8–10 × 2–3 mm at the base, spreading, abaxially densely pubescent at least on midvein and towards apex, adaxially more sparsely hairy. Corolla 25–38 mm long, to c.70 mm diam., divided nearly to base; tube campanulate, to shallowly dish-shaped, (1.5)2–3 × 6–10(12) mm diam. at mouth, partially enclosing corona; exterior similar in colour to interior but paler, pubescent; interior and base of lobes yellow-green with brown-violet spots, glabrous; lobes narrowing abruptly into conspicuously long linear-oblong lobes, 25–38 × 2.5–3 mm at the base, bases triangular-ovate, revolute, margins minutely ciliolate, apices often connate to form broadly ovoid cage, apex subacute, less often lobes free and spreading, abaxially minutely hairy, adaxially green or yellow-green, glabrous. Corona sessile, hemispherical, c.2 × 3.5 mm diam., yellowish with red-brown mottling or uniformly reddish brown, glabrous throughout; outer corona lobes joined laterally across bases of inner lobes, deeply incised with U-shaped sinuses, teeth deltoid-subulate, c.0.6 mm long, divergent, erect, sometimes apparently absent; inner lobes incumbent on anthers and equaling or longer than them; linear-oblong, 1–1.5 mm long, c.0.4 mm wide, apices rounded to truncate. Pollinia dull yellow. Follicles thickly fusiform to narrowly ovoid, c.5 × 1.5 cm thick at middle, glabrous. Seed c.10 × 7 mm.

Zambia. N: Lundazi Dist., Nyika Plateau, 2100 m, 3.i.1959, *Richards* 10425 (K).

Also in Tanzania. Upland pasture; (1500)2100 m.

Conservation status: DD: apparently known only from the type collection and the above collection from the Flora area, but likely to be overlooked. Apparently fairly well established in cultivation from the original collection.

63. **Ceropegia mafekingensis** (N.E. Br.) R.A. Dyer in Bothalia **12**: 256 (1977); in Codd *et al.*, Fl. S. Africa **27**(4): 65 (1980). Type: South Africa, Cape, nr. Mafeking, *Green* 1683 (K000305587 holotype, GRA).

Brachystelma mafekingense N.E. Br. in Fl. Cap. **4**(1): 854 (1908). —Peckover in Brit. Cact. Succ. J. **16**: 177–180 (1998). —Meve, Ill. Handb. Succ. Pl. Asclepiadaceae: 35 (2002). — Bruyns in Strelitzia **34**: 79 (2014).

Ceropegia patriciae Rauh & Buchloh in Kakteen And. Sukk. **15**: 151 (1964). Type: South Africa, Transvaal, near Pretoria, *Rauh* 13269 (HEID holotype).

Tuber ± depressed globose, to 8 cm diam. Stems 1–several, erect to spreading, branching from near base, 1.5–5 cm tall, puberulous. Leaves herbaceous; petiole 1–3 mm long; leaf blade oblong-lanceolate (to ovate), 20–30 × 3–7 mm, ± folded lengthwise, base cuneate to rounded, margins sometimes undulate, apex subacute, abaxial surface puberulous, adaxially glabrous. Inflorescence extra-axillary, peduncle short; cyme umbel-like, 10–20-flowered, flowers mostly opening together. Pedicel 3–8 mm long, puberulous. Sepals narrowly lanceolate, 2–4 mm long, spreading, abaxially pilose, adaxially glabrous. Bud ± cylindrical, slightly beaked. Corolla 10–12 mm long; tubular, tube flask-shaped without distinct basal chamber, obtusely 5-angled, 4–6 × 3–3.5 mm diam., slightly narrower at mouth, exterior purple spotted, minutely puberulous, upper part sometimes papillate, interior yellow and purple punctate; glabrous; sinuses narrowly auriculate; lobes obovate-oblong, 6–7 × 2–4 mm wide, replicate, rotately spreading, margins minutely puberulous, apices acute, adaxially verrucose and dark purple. Corona cupular; outer lobes joined to bases of inner lobes, deeply divided with teeth at bases of inner lobes, about as high as gynostegium, dark purple; inner lobes converging over gynostegium, linear, c.2 mm long, apices obtuse, minutely papillate. Follicles and seeds not recorded.

Botswana. SE: Molepolole, fl. xii.1995, fide Peckover in British Cact. Succ. J. **16**: 177–180 (1998).

Also in Namibia, South Africa. Rocky ridge with scattered bushes and short grass subject to grazing; 1100–1200 m.

Conservation status: LC?: rarely collected but recorded over a wide range, just extending into the Flora area.

64. **Ceropegia pygmaea** Schinz in Verh. Bot. Vereins Prov. Brandenburg **30**: 265 (1888). —Brown in F.T.A. **4**: 439 (1903). —Dinter, Neue Pfl. Sudw.-Afr.: 22 (1914). —Huber in Mem. Soc. Brot. **12**: 137 (1958); in Merxmüller, Prodr. Fl. Sudwestafr. **114**: 24 (1967). —Dyer in Codd *et al.*, Fl. S. Africa **27**(4): 64–65 (1980). —Bruyns in Strelitzia **34**: 79 (1914). Type: Namibia, Olukonda, *Schinz* s.n. (Z holotype, K, ZT). FIGURE 7.3.**151**.

Ceropegia gymnopoda Schltr. in Bull. Herb. Boissier **4**: 450 (1896). Type: Namibia, Uukuambi, 22.ii.1894, *Rautanen* 82 (Z000046714 lectotype, Z000046715, K000305792) designated here.
Ceropegia pumila N.E. Br. in Bull. Misc. Inform. Kew **1898**: 309 (1898). Type: Angola, near Lopollo, 2500 ft, *Welwitsch* 4267 (BM000645903).
Ceropegia pygmaea var. *pumila* (N.E. Br.) H. Huber in Mem. Soc. Brot. 12: 137 (1958).
Brachystelma gymnopodum (Schltr.) Bruyns in Bothalia **25**: 161 (1995); in Strelitzia **34**: 79 (2014). —Meve, Ill. Handb. Succ. Pl. Asclepiadaceae: 29 (2002).

Tuber ovoid to subglobose, 2.5–5 cm diam. Stems 1–3, erect, sparingly branched above, 10–15 cm long, glabrous to pubescent. Leaves herbaceous; petiole absent or to 10 mm long; leaf blade linear to spathulate-obovate, 12–100 × 3–4 mm, apex obtuse to acute, both surfaces pubescent. Inflorescence extra-axillary, peduncle to 8 mm long; cyme 1–3-flowered. Pedicel 3–15 mm long. Sepals 2–5 × c.1 mm, hirsute. Corolla 12–50 mm long, tubular, sharply bent above basal chamber; tube subcylindric, 8–46 mm long, slightly inflated near base, slightly widened towards mouth, exterior pubescent, interior dark purple, glabrous; lobes 2–4 mm long, bases triangular, incurved, replicate, apices free, adaxially dark red, glabrous. Corona 2–3 mm high, outer series cup-shaped, shortly 5-toothed, with sparse, short, purplish hairs within; inner lobes incumbent on anthers, linear-subulate, not exceeding anthers, slightly gibbous at base. Follicles erect-spreading.

Botswana. N: Tsessebe, *Venter et al.* s.n. (UNIN). SE: Lobatsi, fl. xii.1924, *Tapscott* s.n. sub BOL50594 (BOL). **Zimbabwe**. C: Harare, Hatfield, fl. xii., *Whellan* s.n. sub SRGH34864 (SRGH).

Also in Angola and Namibia. Sandveldt, under bushes in loamy sand; 1200–1500 m
Conservation status: LC: known from a wide area, frequent in northern Namibia, uncommon in the Flora area.

Bruyns (1914) gives the type of *Ceropegia pygmaea* as 'Schinz 147' but no collection number is given either in the protologue or on the herbarium sheets. The holotype of *C. gymnopoda* is given as '*Rautanen* 82 (Z)' but there are two sheets in that herbarium, both annotated as type by Schlechter, so a second step lectotypification is needed.

65. **Ceropegia daverichardsii** Bruyns in S. African J. Bot. **112**: 433 (2017), new name, not *Ceropegia richardsiae* Masinde (2012). Type: Zimbabwe C, Kadoma, 7.xii.1995, *Peckover* 257 (PRE0770753-0 holotype).

    Brachystelma richardsii Peckover in Excelsa **17**: 33 (1996). —Meve, Ill. Handb. Succ. Pl. Asclepiadaceae: 31 (2002).

Tuber depressed globose, to 7.5 cm diam., 2 cm deep. Stem solitary, erect, unbranched, herbaceous, 15–30 cm high, c.3 mm thick at base, finely pilose. Leaves herbaceous, spreading; petiole to 2 mm long; leaf blade narrowly lanceolate, 30–100 × 6–18 mm, base cuneate, margin entire, sometimes wavy, apex subacute, both surfaces pilose. Inflorescence extra-axillary, subsessile; cyme clustered, up to 8-flowered, no noticeable scent. Pedicel 2–3 mm long, c.1 mm thick, sparingly pilose. Sepals 1.5–2 × c.0.75 mm wide at base, pilose. Corolla 15–30 mm diam.; tube campanulate, c.3 × 4 mm diam. at mouth, exterior green, glabrous, interior purple, with fine white hairs; lobes linear-triangular, 8–15 mm long, rotate, abaxially green, glabrous, adaxially red to yellowish with red mottling, with fine white hairs near the base. Corona cupular, c.1 × 3 mm diam.; purplish red, outer lobes pouches, yellow with red dots on the inside; inner lobes incumbent on anthers, narrowly triangular, c.1 mm long. Follicles diverging at 30°, linear, 4.5–7.5 × c.0.4 cm, seeds 20–30, dark brown with slightly paler margin, c.10 × 4 mm; coma 15–20 mm.

Zimbabwe. N: Gokwe, 130 km SW Chinhoyi, 4.i.1998, *Bruyns* 7450 (BOL).
Possibly also in Tanzania. In flat swampy mopane bushland, growing in open or under small bushes; 1200–1300 m.
Conservation status: DD: known from very few, widely dispersed records.
Meve (2002) believes that the material from Tanzania named as *C. lancasteri* by Newton in Bradleya **14**: 94–98 (1996) is better placed within this species.

66. **Ceropegia simplex** (Schltr.) Bruyns in S. African J. Bot. **112**: 435 (2017). Type: Mozambique, Inhambane Dist., near Machisugu, fl. xii.1898, *Schlechter* 12121 (B† holotype); Bot. Jahrb. Syst. **38**: 41, fig.5 A-G (1905), illustration designated as lectotype by Masinde in Kew Bull. **62**: 78 (2007); Mozambique, 11 km SW of Pomene, 100 m, 10.i.2004, *Bruyns* 9755 (BOL epitype). FIGURE 7.3.**154**.

    Brachystelma simplex Schltr. in Bot. Jahrb. Syst. **38**: 40 (1905).

Tuber discoid, 2–3 cm diam. Stem solitary, erect, unbranched, densely leafy, 20–40 cm high, uniformly hispidulous. Leaves herbaceous, ascending or spreading, subsessile, petiole shortly hispid; leaf blade linear to linear-ligulate, 30–80 × 2–7 mm, apex acute or subacute, both surfaces hispidulous, adaxially glabrescent. Inflorescence extra-axillary, sessile; cyme fasciculate, 3–8-flowered, flowers opening in succession. Pedicel shorter than calyx, hispidulous. Sepals lanceolate, c.2.5 × 0.3 mm at base, apex acuminate, abaxially hispidulous, margins ciliate. Corolla c.6 mm long, divided to below middle, tube campanulate, shorter than corona, exterior minutely puberulous, interior yellowish, glabrous; sinuses flat; lobes triangular-ovate, subrotate, margins slightly recurved, ciliolate, adaxially dark maroon, glabrous. Corona cupular, yellow with maroon margin; outer lobes broadly oblong, joined laterally for 1/3–2/3 length and then seperated by narrow slits, margin truncate or slightly undulate, slightly incurved, higher than gynostegium, glabrous; inner lobes on inner wall of outer corona at bases of anthers, reduced to cushionlike swellings, sparsely puberulous. Follicles and seeds not recorded.

Mozambique. N: Niassa, 77 km north of Macaloge, 710 m, fl. 23.xi.2000, *Bruyns* 8626 (BOL); c.20 km E of Namina, fl. 24.xii.1998, *Bruyns* 7712 (BOL). Z: 5 km N of Lioma, fl. 26.xii.1998, *Bruyns* 7728 (BOL). GI: Inhambane, 11 km SW of Pomena, 100 m, fl. 10.i.2004, *Bruyns* 9755 (BOL).

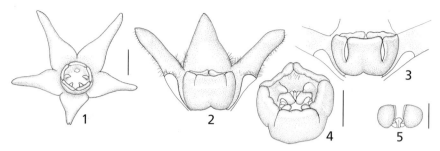

Fig. 7.3.154. —66. CEROPEGIA SIMPLEX. 1, oblique view of flower; 2, side view of dissected flower; 3, side view of centre of dissected flower; 4, oblique view of corona; 5, pollinarium. Scale bars: 1, 2 = 2 mm; 3, 4 = 1 mm; 5 = 0.25 mm. 1, 3 from *Bruyns* 8626; 2, 4, 5 from *Bruyns* 7712. Drawn by Peter Bruyns. Reproduced from Phytotaxa (2018).

Endemic. On both white and pale orange sand under trees; 100–700 m.

Conservation status: LC: rather widely dispersed within Mozambique and clearly under collected.

Bruyns (2017) designated his collection *Bruyns* 9755 as a neotype but Masinde's choice of lectotype, based as it is on an original figure, must stand and Bruyn's collection is better regarded as an epitype.

The Kenyan collections treated as *Brachystelma simplex* by Masinde in F.T.E.A., Apocynaceae (part 2): 309 (2012) differ by the much larger tubers, much branched stems, narrower, sessile leaves, smaller inflorescences (1–4-flowered), glabrous corolla, and emarginate outer corona lobes. Similarly, the description of *B. simplex* by Meve in Ill. Handb. Succ. Pl. Asclepiadaceae: 43 (2002) differs very markedly in the indumentum of the corolla ('exterior hispid to tomentose, interior densely papillose to shortly hairy') and 'emarginate U-shaped' outer corona lobes. The species has also been recorded from West Africa (Benin, Burkina Faso, Ivory Coast, Nigeria) where it was described as a distinct subspecies: *B. simplex* subsp. *banforae* J.-P. Lebrun & Stork (1989). This taxon differs most obviously from Schlechter's plant by the tufted stems, small purple flowers and deeply divided outer corona lobes and Bruyns in Phytotaxa **364**: 130 (2018) has elevated it to a full species, *C. banforae* (J.-P. Lebrun & Stork) Bruyns.

67. **Ceropegia discoidea** (R.A. Dyer) Bruyns in S. African J. Bot. **112**: 433 (2017). Type: South Africa, Transvaal, Zoutpan, 40 km N of Pretoria, *Hardy* 2440 (PRE0469066-0 holotype).

> *Brachystelma discoideum* R.A. Dyer in Fl. Pl. Africa **42**: t.1668 (1973); in Codd *et al.*, Fl. S. Africa **27**(4): 27 (1980). —Meve, Ill. Handb. Succ. Pl., Asclepiadaceae: 28 (2002). —Bruyns in Strelitzia **34**: 78 (2014).

Tuber 7–9 cm diam., upper surface depressed and concave, lower surface rounded or flattish. Stems 1–2(4), somewhat branched, to 10 cm tall, minutely decurved pubescent. Leaves herbaceous, lowest smaller than upper ones; petiole 2–8 mm long, adaxially grooved; leaf blade broadly ovate to sub-circular, 5–15(25) × 5–15(20) mm, abaxial surface minutely pubescent, adaxial glabrous. Inflorescence extra-axillary, cyme 1- or 2-flowered. Pedicel 8–12(15) mm long, slightly decurved, minutely pubescent. Sepals linear-lanceolate, c.2 mm long, minutely pubescent. Corolla 18–22 mm diam., united at the base for 1–5 mm, tube slightly raised into a small cushion; lobes linear-lanceolate, c.10 mm long, spreading, margins slightly recurved, abaxially puberulous, adaxially light yellow, glabrous. Corona sessile, disc-like, dark purple, outer lobes united into annulus, grooved opposite the pollinia, glabrous; inner lobes on inner wall of outer corona at bases of anthers, reduced to cushionlike swellings. Follicles and seeds not recorded.

Zimbabwe. W: Luveve Cemetary road - Western Suburbs, Bulawayo, fl. 8.xii.1988. *Percy-Lancaster* 2265 (SRGH). E: Vumba circular drive, *Venter et al.* 341 (UNIN).
Also in Namibia and South Africa. Among trees and bushes on Kalahari sand; 1200 m.
Conservation status: DD: known from just a few, widely dispersed records.

68. **Ceropegia brevipedicellata** (Turrill) Bruyns in S. African J. Bot. **112**: 433 (2017). Type: South Africa, no locality, ex Pretoria, cult. Kew (K000305832 holotype). FIGURE 7.3.**151**.

    Brachystelma brevipedicellatum Turrill in Bull. Misc. Inform. Kew **1922**: 29 (1922).

    Brachystelma dinteri Schltr. in Bot. Jahrb. Syst. **51**: 144 (1913), not *Ceropegia dinteri* Schltr. (1913). —Dinter, Neue Pfl. Sudw.-Afr.: 15–17, fig.7 (1914). —Turrill in Bull. Misc. Inform. Kew **1922**: 197 (1922). —Huber in Merxmüller, Prodr. Fl. Sudwestafr. **114**: 12 (1967). —Dyer in Bothalia **10**: 375 (1971); in Codd *et al.*, Fl. S. Africa **27**(4): 26–27 (1980). —Percy-Lancaster in Excelsa **13**: 67, 69 (1988). —Boele in Excelsa **14**: 46–47 (1989). —Meve, Ill. Handb. Succ. Pl. Asclepiadaceae: 27–28 (2002). —Bruyns in Strelitzia **34**: 77–78 (2014). Type: Namibia, foothills of Auasberge, *Dinter* 1890 (SAM0007453-0 lectotype, SAM), second step lectotype chosen here.

    Brachystelma ringens E.A. Bruce in Fl. Pl. Africa **28**: t.1096B (1951). Type: South Africa, Cape, near Vryburg, *Phillips* 119 (PRE0469055-0 holotype).

    Ceropegia dinteriana Bruyns in S. African J. Bot. **112**: 433 (2017), new name, based on *Brachystelma dinteri* Schltr., not *Ceropegia dinteri* Schltr.

Tuber subglobose, 3–4 cm diam., upper surface ± depressed. Stem erect, unbranched, 14–30 cm long, shortly hispidulous. Leaves herbaceous, erect-spreading, shortly petiolate; leaf blade elliptical to lanceolate-elliptic, 27–45 × 3–12 mm, apex obtuse to subacute, both surfaces shortly hispidulous. Inflorescences extra-axillary, sessile; cyme fasciculate, small, 3–12-flowered. Pedicel c.3 mm long. Sepals lanceolate, 2–2.5 mm long, densely puberulous. Corolla subrotate, 8–10 mm diam., tube flat, exterior densely puberulous, interior glabrous; lobes triangular-ovate, spreading, apices subacute, adaxially light green or yellowish green with dense brown blotches, less often maroon, transversely rugose, glabrous. Corona cupular, c.1 mm high; outer lobes rounded or emarginate, higher than gynostegium, glabrous; inner lobes incumbent on anthers, incurved, apices triangular obtuse. Anther connective green. Follicles often solitary, slenderly fusiform, c.6 cm long, longitudinally striped or mottled.

Zimbabwe. W: 13km N of Bulawayo, fl. 13.xii.1989. *Percy-Lancaster* 2683 (SRGH). N: Mrewa, *Percy-Lancaster* 1634 (SRGH, UNIN). C: Harare, Waterfalls, 17.xii.1977, *Percy-Lancaster* 188 (SRGH, PRE). E: Inyanga, fl. 24.xi.1955, *Chase* 5549 (SRGH). S: 32km S of Runde River, Mwenezi, 5.xii.1988, *Percy-Lancaster* 2238 (SRGH).
Also in Namibia. Most records from *Brachystegia* woodland on sandy soil, also in *Colophospermum mopane* woodland on darker soils; 1100–1600 m.
Conservation status: LC: one of the members of the group more frequently collected by a few specialist collectors.
Bruyns in Strelitzia **34**: 77 (2014) lectotypified *Brachystelma dinteri* on 2 sheets of *Dinter* 1890 in SAM. A second step lectotypification is therefore required. The chosen sheet has well preserved leafy flowering stems whilst the other sheet is of precociously flowering stems lacking normal leaves.

69. **Ceropegia malawiensis** (Peckover) M.G. Gilbert, comb. nov. Type: Malawi (N), Chitipa Dist., 12.i.2000, *Peckover* 247 (PRU holotype).

    Brachystelma malawiensis Peckover in CactusWorld **37**(2): 114, figs.1,3,4,6,7 (2019).

Tuber discoid, to 4 cm diam., 2 cm deep. Stem often solitary, erect, unbranched, herbaceous, to 10 cm high, very densely puberulous. Leaves herbaceous, spreading, subsessile; leaf blade elliptical, to 40 × 25 mm, base cuneate, margin entire, undulate, apex subacute, both surfaces finely puberulous. Inflorescence extra-axillary, sessile; cymes fasciculate, up to 9-flowered. Pedicel c.1 mm long, puberulous. Sepals lanceolate, densely puberulous. Corolla c.6 mm diam., divided nearly to base; tube very shallow, interior glabrous; lobes triangular-ovate, c.2 × 1.5 mm wide, slightly revolute, spreading, apices subacute, adaxially black, with subapical cluster of ribbonlike vibratile hairs, sometimes absent. Corona cupular, c.3 mm diam., mostly whitish; outer lobes divided c.½ way, tipped with black, glabrous; inner lobes incumbent on anthers, incurved, green. Follicles erect, c.5 × 0.5 cm, green flecked with red, glabrous; seeds up to 14, brown with paler margin, c.13 × 4 mm.

Malawi. N: Chitipa Dist., *Peckover* 247 (PRU).
Known only from type collection. Rock outcrops in Miombo woodland; 835 m.
Conservation status: DD, known only from the type.

This species has been represented on the internet by images labelled as 'Brachystelma sp. aff. lancasteri' but Peckover thought it best compared with the previous species, *Ceropegia brevipedicellata*. It has a very characteristic black corolla usually, but not always, with long versatile hairs near the tips of the lobes, similar to those seen in *C. tavalla* but that species has significantly longer corolla lobes.

70. **Ceropegia arnotii** (Baker) Bruyns in S. African J. Bot. **112**: 433 (2017). Type: South Africa, Cape, ?Colesberg, *Arnot* s.n. (K000305840 holotype).

> *Brachystelma arnotii* Baker in Saunders, Refug. Bot.: t.9 (1869), as 'arnottii'. —Dyer in Bothalia **10**: 373 (1971); in Codd *et al.*, Fl. S. Africa **27**(4): 23–24 (1980). —Meve, Ill. Handb. Succ. Pl. Asclepiadaceae: 21 (2002).
> *Anisotoma arnotii* (Baker) B.D. Jacks., Index Kew. **1**: 139 (1893).
> *Decaceras arnotii* (Baker) Schltr. in Bot. Jahrb. Syst. **18**(Beibl. 45): 26 (1894).
> *Brachystelma grossartii* Dinter, Neue Pfl. Sudw.-Afr.: 16 (1914). Type: Namibia, Okakuja, *Grossart* sub *Dinter* 2698 (SAM0070975-0 lectotype, SAM0004461-0), lectotype designated here.

Tuber napiform, 5–7.5 cm diam., upper surface ± depressed, pale brown. Stems 1–3, erect, simple to somewhat branched, 7–20 cm long, finely grey pubescent with deflexed hairs. Leaves herbaceous, erect-spreading; petiole 1–2 mm long; leaf blade ovate-spathulate to nearly linear, 10–40 × 2–4 mm, longitudinally folded, margins crisped, sometimes slightly undulate, apex acute, abaxial surface densely grey puberulous, adaxial dull green, glabrous. Inflorescence extra-axillary, cyme 1–6-flowered, flowers opening in succession, or sometimes 2–4 together. Pedicel 5–35 mm long, considerably deflexed when flowers expand, grey pubescent. Sepals lanceolate, c.1.5 mm long, acute, finely grey pubescent. Corolla. c.8 mm long, 6–10 mm diam., divided for ²/₃ of length, tube campanulate, 2.5–4 mm deep, exterior puberulous, interior glabrous; lobes ovate to ovate-lanceolate, 5–6 × c.1 mm wide, ascending to widely spreading, subplicate, apices acute, abaxially glabrous, adaxially bright brown to dark purplish-brown with greenish tubercle-like thickening at apices, glabrous. Corona greenish; outer lobes shortly united at base, bifid, segments diverging, c.0.5 mm long, apices spreading-erect, equal or slightly higher than gynostegium; inner lobes on inner wall of outer corona at bases of yellow anthers, reduced and somewhat cushion-like. Follicles and seeds not recorded.

Botswana. N: Nxai Pan, 25.xii.1967, *Lambrecht* 468 (SRGH). SE: c.2 km SE of Takatokwane Camp, fl. 8.xii.1969, *Maguire* 7866 (J).
Also in Namibia, South Africa and Lesotho. Among trees and bushes on Kalahari sand; 1200–1600 m.
Conservation status: LC: fairly frequently collected, particularly in Botswana.

The species was named after the collector, Mr. Arnot of Colesberg, and the spelling 'arnottii' must be treated as a correctable orthographic error. *Decaceras arnotii* (Baker) K. Schum. in Engler, Nat. Pflanzenfam. **4**(2): 267 (1895) is a later isonym; *D. arnoldii* Schltr. in Bot. Jahrb. Syst. **18**(Beibl. 45): 26 (1894), seems to be an orthographic error for this species.

A second stage lectotypification of *Brachystelma grossartii* is needed as there are 2 separate sheets of the type collection: the sheet selected has been quite extensively annotated and includes more material.

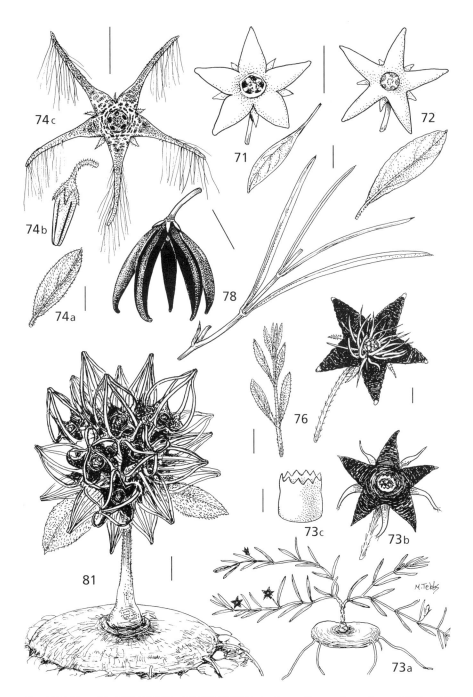

Fig. 7.3.**155**. —**71**. CEROPEGIA VAHRMEIJERI, leaf and flower, from *Bruyns* 8532. —**72**. CEROPEGIA CUPULATA, leaf and flower, from *Verhagen* 13. —**73**. CEROPEGIA PROSTRATA, 73a habit, 73b flower, 73c corona, from *Milne-Redhead* 3242. —**74**. CEROPEGIA TAVALLA, 74a leaf, 74b bud, 74c flower, from *Whellan* in SRGH 34818. —**76**. CEROPEGIA INCANA, leafy shoot and flower, from *Kimberley* 131/180. —**78**. CEROPEGIA PLOCAMOIDES, leafy shoot and flower, from *White* 6893. —**81**. CEROPEGIA BARBERAE, habit, from *Biegel* 5856. Scale bars: habits = 10 mm; flower of 71, 72, 74, 78 = 5 mm; 73b, 73c, 76 = 1 mm. Drawn by Margaret Tebbs.

71. **Ceropegia vahrmeijeri** (R.A. Dyer) Bruyns in S. African J. Bot. **112**: 435 (2017). Types: South Africa, Natal, Zululand, Ubombo / Ingwavuma border, near Lala Nek, *Vahrmeijer* 1050 (PRE0469046-1 lectotype, PRE0469046-2), lectotype designated here. FIGURES 7.3.**155**, 7.3.**156**.

   Brachystelma vahrmeijeri R.A. Dyer in Bothalia **10**(2): 378 (1971); in Codd *et al.*, Fl. S. Africa **27**(4): 26 (1980). —Meve, Ill. Handb. Succ. Pl. Asclepiadaceae: 45 (2002).

Tuber slightly compressed above and below, to 10 cm diam., red. Stems several, rarely branched, to 10 cm tall, minutely pubescent or glabrescent. Leaves herbaceous, sessile or shortly petiolate; leaf blade ± elliptic lanceolate, to 30 × 7 mm, base cuneate to attenuate, margin of young leaves minutely toothed, later ± entire, apex acuminate, both surfaces usually glabrous, margins sometimes ciliolate. Inflorescence extra-axillary, sessile; cyme 2–4-flowered, flowers opening in succession, oldest uppermost. Pedicel 5–10 mm long, glabrous. Sepals ± lanceolate, c.2.5 mm long, glabrous. Bud obovoid, sinuses flat, apex obtuse. Corolla c.8 mm long, divided to below middle; tube bowl-shaped, c.3 × 4 mm diam. at mouth, exterior glabrous, interior greenish yellow or cream, rarely white or maroon; glabrous; lobes triangular-ovate, c.5 mm long, almost flat, slightly spreading, slightly fleshy, same colour as tube, glabrous. Corona sessile, cupular, c.1 mm high, outer lobes joined laterally to form 5-sided, truncate cup, rim narrowly incurved, shallowly 10-notched to almost entire, 2 × as high as gynostegium, with a few long hairs on inside; inner lobes arising ½ way up outer corona, incumbent on anthers, linear-oblong, glabrous. Follicles and seeds not recorded.

Mozambique. M: Maputo 20 km N of Manbica 3.i.2001, *Bruyns* 8522 (E, K); between Ponta Malongane & Ponta do Ouro, 26°S 32°E, xi.2001, *Turner* s.n. (K).

Also in South Africa. In whitish sand among clumps of grass, dwarf and stunted trees of *Commiphora*, *Hyphaene coriacea* and *Phoenix reclinata*; below 100 m.

Conservation status: EN: restricted to a fairly local, rather fragile habitat, treated as endangered in South Africa and only just extending into Mozambique.

A second stage lectotypification of *Brachystelma vahrmeijeri* is needed as there were 2 separate sheets of the type collection in PRE, both labelled as "holotype". Bruyns in Phytotaxa **364**: 123–125 (2018) suggests that this species is not seperable from *Ceropegia cupulata* (R.A. Dyer) Bruyns, and should be treated as a synonym but this author is reluctant to follow because there are differences, most notably in pedicel length which coupled by the considerable differences in distribution and ecology, support their separation.

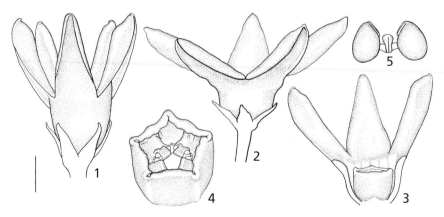

Fig. 7.3.**156**. —**71**. CEROPEGIA VAHRMEIJERI. 1, 2, side view of flower; 3, side view of dissected flower; 4, oblique view of corona; 5, pollinarium. Scale bar: 1–3 = 2 mm; 4 = 1 mm; 5 = 0.25 mm. All from *Bruyns* 8532. Drawn by Peter Bruyns. Reproduced from Phytotaxa (2018).

72. **Ceropegia cupulata** (R.A. Dyer) Bruyns in S. African J. Bot. **112**: 433 (2017). Type: Namibia, Grootfontein, 5 miles W of Aha Mts., *Story* 6400 (PRE0469057-0 holotype). FIGURE 7.3.**155**.

> *Brachystelma cupulatum* R.A. Dyer in Bothalia **10**: 375 (1971); in Codd *et al.*, Fl. S. Africa **27**(4): 26 (1980). —Meve, Ill. Handb. Succ. Pl. Asclepiadaceae: 27 (2002). —Bruyns in Strelitzia **34**: 76–77, fig.124 (2014).

Tuber flattened-discoid, to 10 cm diam. Stem single or branched sparsely above, up to 15 cm tall, sparsely hairy with decurved hairs. Leaves herbaceous, spreading, shortly petiolate; leaf blade linear to ovate, 30–80 × 4–15 mm, base cuneate to attenuate, apex acute, green to grey-green, both surfaces shortly pubescent. Inflorescence from lower nodes, extra-axillary, sessile; cyme to 17-flowered, flowers opening in succession. Pedicel 2–4 mm long, glabrous. Sepals ovate to narrowly lanceolate, 1.5–2 mm long, abaxially sparsely hairy. Bud ellipsoidal, apex rounded. Corolla 6–9 mm long, divided to below middle; tube campanulate, 1.5–2.5(3) mm deep; exterior glabrous, interior green, glabrous; lobes ± oblong, narrowed to apex, 4.5–6 × 2–3 mm wide, semi-erect, margins slightly recurved, apices acute, abaxially sparsely hairy, adaxially green, glabrous. Corona subsessile, cupular, 1.5–3 × c.2.5 mm diam.; outer lobes joined laterally to form 5-sided, truncate cup, rim narrowly incurved, with slits to 0.5 mm deep alternating with anthers, much exceeding gynostegium, yellow?; inner lobes on inner wall of outer corona at bases of anthers, reduced to cushionlike swellings. Follicles diverging, subfusiform, 7–8 cm long. Seeds not recorded.

Botswana. SE: Kweneng Dist., 15 km N of Ngware, 17.xi.1976. *Verhagen* 13 (PRE). Also in Namibia and South Africa. Flats in bushland and woodland on Kalahari sands; 1200–1600 m.

Conservation status: LC: fairly frequently collected over a large area.

73. **Ceropegia prostrata** (E.A. Bruce) Bruyns in S. African J. Bot. **112**: 435 (2017). Type: Zambia, Mwinilunga: SW of Matonchi Farm, fl. 14.xi.1937, *Milne-Redhead* 3242 (K holotype). FIGURE 7.3.**155**.

> *Brachystelma prostratum* E.A. Bruce in Kew Bull. **3**: 462 (1949). —Meve, Ill. Handb. Succ. Pl. Asclepiadaceae: 40 (2002).

Tuber discoid, to 3 cm diam., upper surface ± depressed. Stem prostrate, branched at base, branches 3–5, radiating, 4–6 cm long, internodes 6–10 mm long, to 1 mm thick, minutely pubescent. Leaves herbaceous, erect, subsessile to shortly petiolate, petiole to 2 mm long; leaf blade ± fleshy, linear to narrowly lanceolate, 10–13 × 0.6–2 mm, flat, base narrowly cuneate or attenuate, margin narrowly revolute, apex acute, both surfaces glabrous. Inflorescences from middle and upper nodes, extra-axillary; peduncle 3–4 mm long, 1-flowered. Pedicel 1–3 mm long, green marked with red, pubescent. Sepals narrowly lanceolate, green marked red, 4–5 × c.1 mm, subequal to the corolla lobes, subglabrous. Corolla c.8 mm diam.; tube campanulate, c.2 × 2.5 mm diam. at mouth, exterior glabrous, interior with long white hairs; lobes ovate-triangular, c.3.5 × 2.5 mm at base, rotate, apices acute, adaxially mahogany-red, fleshy with transverse rugose lines. Corona sessile, c.2 mm high; outer lobes joined laterally to form globose-urceolate tube, rim incurved, shortly 10-toothed, exceeding gynostegium; pinkish-red; inner lobes minute, incumbent on anthers, deltoid, c.0.5 mm long. Pollinia orange. Follicles and seeds not recorded.

Zambia. W: Mwinilunga: SW of Matonchi Farm, fl. 14.xi.1937, *Milne-Redhead* 3242 (K). Endemic. In shallow soil overlying laterite, soil full of iron oxide nodules; 1300 m.

Conservation status: DD. Still known only from the type collection made in 1937.

74. **Ceropegia tavalla** (K. Schum.) Bruyns in S. African J. Bot. **112**: 435 (2017). Type: Tanzania, Iringa, Uhehe, near Rugaro, *Goetze* 541 (B† holotype); Zimbabwe, Harare, Hatfield, *Whellan* in SRGH 34949 (SRGH neotype), neotype designated by Boele in Excelsa **14**: 48 (1989). FIGURES 7.3.**155**, 7.3.**157**.

Brachystelma tavalla K. Schum. in Bot. Jahrb. Syst. **28**: 459 (1900). —Percy-Lancaster in Excelsa **13**: 71, 73 (1988). —Boele in Excelsa **13**: 47–48 (1988); in Excelsa **14**: 48 (1989). —Masinde in Kew Bull. **62**: 78–80, fig.17 (2007); in F.T.E.A., Apocynaceae (part 2): 297–298 (2012). —Meve, Ill. Handb. Succ. Pl. Asclepiadaceae: 44 (2002).

Tuber globose, 1–3 cm diam. Stems 1-few, erect, somewhat branched, 22–30 cm long, subtomentose to tomentose. Leaves herbaceous, spreading, sessile; leaf blade linear-lanceolate to lanceolate-elliptic, 12–60 × 3–20 mm both surfaces subtomentose to tomentose. Inflorescence extra-axillary, sessile; cyme 1–4-flowered. Pedicel 4–8 mm long, decurved with buds facing downwards, subtomentose. Sepals lanceolate to linear-lanceolate, 2.5–3 × 0.7–1.7 mm, abaxially tomentose, adaxially glabrous. Bud narrowly ovoid-fusiform, with flat sinuses, subacute. Corolla to 12 mm long, 18–20 mm diam., divided almost to base; tube vestigial, 0.5–1.5 mm deep; lobes lanceolate, 6–12 × c.2 mm wide, rotate, margin ciliate, abaxially sparsely puberulous, adaxially pale yellow to purple, with 6–10 mm long violet vibratile hairs covering outer half to whole of lobe. Corona sessile, cupular, c.1.6 × 3 mm diam., yellow; outer lobes joined laterally to form 5-sided, truncate cup, with deep sinuses alternating with anthers exceeding gynostegium, yellow with darker margin, glabrous; inner lobes on inner wall of outer corona incumbent on anthers, narrowly oblong, 0.75–0.9 mm long, much shorter than anthers, apices rounded to truncate, glabrous. Follicles and seeds not recorded.

Zambia. N: Kalombo Falls, 1200 m, fl. 5.xii.2003, *Bruyns* 9603 (BOL). **Zimbabwe**. N: 10 km N of Shamva, fl. 14.xii.1987, *Percy-Lancaster* 2071 (SRGH). C: Hartley Dist., Poole Farm, fl. 30.xii.1944, *Hornby* 3138 (K, SRGH). E: 15km N of Mutare, fl. 19.xii.1985, *Percy-Lancaster* 1545 (SRGH). **Mozambique**. N: Niassa, 79 km north of Macaloge 710 m, fl. 24.xi.2000, *Bruyns* 8636 (BOL, MO); 6 km south of Marrupa, 750 m, fl. 2.i.2004, *Bruyns* 9712 (E).

Also in Tanzania and D.R. Congo. In *Brachystegia* woodland on sandy soil; 700–1500 m.

Conservation status: LC: one of the members of the group more frequently collected by a few specialist collectors in Zimbabwe.

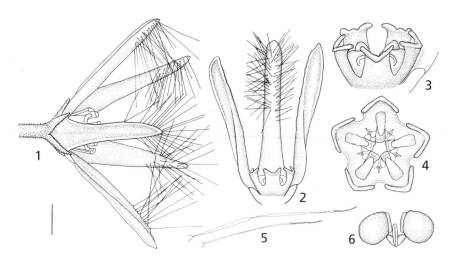

FIG. 7.3.**157**. —**74**. CEROPEGIA TAVALLA. 1, side view of flower; 2, side view of dissected flower; 3, side view of corona; 4, face view of corona; 5, hairs on face of corolla; 6, pollinarium. Scale bar: 1, 2 = 2 mm; 3, 4 = 1 mm; 5 = 0.5 mm; 6 = 0.25 mm. All from *Bruyns* 8636. Drawn by Peter Bruyns. Reproduced from Phytotaxa (2018).

75. **Ceropegia lancasteri** (Boele) Bruyns in S. African J. Bot. **112**: 434 (2017). Type: Zimbabwe, Bulawayo, 200–300 yards from Bulawayo Station on the line to Victoria Falls, *Rogers* 5444, sub BOL 42694 (BOL holotype, S).

    Brachystelma lancasteri Boele in Excelsa **16**: 30 (1994). —Meve, Ill. Handb. Succ. Pl. Asclepiadaceae: 33 (2002).

Tuber present. Stems regularly branched c.8 cm above tuber, herbaceous, 11–16 cm long, puberulous. Leaves herbaceous, subsessile; leaf blade elliptical to slightly obovate, 6–17 × 3–7 mm, apex acute, abaxial surface puberulous, adaxially glabrous. Inflorescence extra-axillary, cyme fasciculate, dark purple, 2- or 3-flowered. Pedicel 2–3.5 mm long, puberulous. Sepals 1.5–2.5 × 1–1.5 mm, puberulous. Bud with globose base and five-angled beak. Corolla 9.5–10.5 mm long, divided for more than ⅔ of length; tube campanulate, (1.5)3 mm deep, c.3 mm diam. at mouth, not completely enclosing the corona, exterior red, white puberulent, interior light red with a few whitish spots, glabrous; lobes with short ovate base, slightly bent and spreading from above gynostegium, lobes then linear-oblong, 6.5–8 × 2–2.5 mm (base) –1 mm (top), strongly replicate, semi-erect or rotately spreading to reflexed, abaxially red, densely white puberulent, adaxially base light red with few whitish spots, rest dark liver red or yellowish-green, (glabrous or) thinly puberulous. Corona with outer corona lobes shortly united at base, c.2 mm long, arched over gynostegium; inner lobes closely incumbent on anthers, linear, straight, apices tapering and not meeting over the centre. Follicles and seeds not recorded.

Zimbabwe. W: 20 km east of Bulawayo, fl. 9.xii.1977, *Percy-Lancaster* 181 (SRGH). Probably endemic. Open grassland; 1300–1400 m.
Conservation status: DD. Known for sure only from two collections.

A record of *Brachystelma lancasteri* from Tanzania by Newton in Bradleya **14**: 94–98 (1996) and included by Masinde in Kew Bull. **62**: 52, 54, fig.3 (2007) and in F.T.E.A., Apocynaceae (part 2): 296–297 (2012) was thought by Meve (2002) to be based on material of *C. daverichardsii*.

76. **Ceropegia incana** (R.A. Dyer) Bruyns in S. African J. Bot. **112**: 434 (2017). Type: South Africa, Natal, 22 km SW of Lichtenburg, 4700 ft, fl. 19.ii.1949, *Acocks* 12476 (PRE0659406-0 holotype). FIGURE 7.3.**155**.

    Brachystelma incanum R.A. Dyer in Bothalia **12**: 54 (1976); in Codd *et al.*, Fl. S. Africa **27**(4): 25 (1980). —Meve, Ill. Handb. Succ. Pl. Asclepiadaceae: 27 (2002).

Tuber c.5 cm diam., slightly compressed. Stem 1(2), spreading-decumbent, sparsely branched from the base, 5–10 cm long, internodes 5–10 mm long, shortly pubescent with curved hairs. Leaves herbaceous, petiole c.5 mm long; leaf blade obovate to ovate-elliptic, 1.5–2 × 1–1.5 cm, flat or folded upwards, abaxial surface pubescent, adaxially pubescent or glabrous. Inflorescence extra-axillary, sessile; cyme (1)2–4-flowered, flowers opening in succession. Pedicel 10–15 mm long, slender, pubescent. Sepals linear-lanceolate, 2–4 mm long, pubescent. Corolla rotate, 9–10 mm diam., divided nearly to base; tube short, closely surrounding base of corona, exterior pubescent, interior with ring of long hairs around corona and sometimes spreading on to base of lobes; lobes triangular, rotate, margins slightly recurved, adaxially dull reddish to dark violet-black with rugose surface and smooth greenish-yellow tips, glabrous. Corona shortly stipitate, campanulate, 1-seriate in appearance, c.1.5 mm diam., outer corona adpressed to and about as high as staminal column, outer margin obscurely 5-lobed; lobes notched at apex, c.1 mm wide; inner lobes on inner wall of outer corona at bases of anthers, reduced and cushion-like. Follicles and seeds not recorded.

Botswana. SW: 21°22.5'S 28°52.5'E, fl. 16.xii.1984, *Clinning* s.n. (PRE). **Zimbabwe**. C: Rusape, 38 km E, 13 km down Baddeley Siding to Nyanga, 1430 m, fl. 6.i.1991, *M. & R. Kimberley* 131 (K, SRGH).
Also in South Africa. Woodland on sandy soil; 1400–1500 m.
Conservation status: LC: moderately frequently collected in South Africa with extensions into Botswana and Zimbabwe.

77. **Ceropegia albipilosa** (Peckover) Bruyns in S. African J. Bot. **112**: 433 (2017). Type: Zimbabwe: Chinhoyi, *Peckover* 256 (PRE0752548-0 holotype, SRGH). FIGURE 7.3.**158**.

> *Brachystelma albipilosum* Peckover in Cact. Succ. J. (Los Angeles) **69**: 155 (1997). —Meve, Ill. Handb. Succ. Pl. Asclepiadaceae: 21 (2002).

Tuber depressed-globose, 2.5–6 cm diam., c.2 cm deep. Stems 1–few, erect or ascending, few-branched, 6–10 cm long, 2–5 mm thick, laxly tomentose. Leaves herbaceous, erect, petiole c.3 mm long, glabrous; leaf blade linear-lanceolate, 60–100 × 2–5 mm; both surfaces glabrous. Inflorescence extra-axillary, sessile, cyme 1-flowered. Pedicel 20–50 mm long, spreading horizontally, pubescent. Sepals narrowly triangular, c.1.5 mm long, minutely pubescent. Corolla rotate, c.7 mm diam., divided for ½–¾ of length; tube shallow, c.1.5 mm deep, 4 mm diam. at mouth, exterior gray green, interior blackish green or purple, glabrous; lobes triangular-ovate, 3–3.5 mm long, semi-erect, margins revolute, adaxially shaggy white tomentose and papillose. Corona ± sessile, c.3 mm diam.; purplish-red, outer lobes quite widely separated, pouched at base, bifid to c.½ length, teeth subulate, recurved, 0.7–1 mm long, papillose-pubescent; inner lobes incumbent on bases of anthers, pillow-shaped, wider than long, apices rounded, glabrous. Follicles linear, 4.5–7.5 × c.0.3 cm wide, green mottled red, glabrous. Seed dark brown with a paler margin, c.8 × 3 mm.

Zimbabwe. N: Chinhoyi, *Peckover* 256 (PRE, SRGH). **Mozambique**. N: Lichinga, 77 km N of Macaloge, 700 m, fl. 23.xi.2000, *Bruyns* 8627 (BOL, E).

Endemic. On quartzite ridge or in white sand, under *Brachystegia* woodland; 700–1100 m.

Conservation status: DD. Known for sure only from two collections.

The protologue also recorded this species from around Chegutu and south of Mutare. It also noted apparent hybrids with *C. tavalla* and with *C. gracilior.*

78. **Ceropegia plocamoides** (Oliv.) Bruyns in S. African J. Bot. **112**: 435 (2017). Type: Tanzania Tabora/ Dodoma boundary, Uyansi, Ngunda-Nukali, Jiwa la Mkoa, *Speke & Grant* s.n. (K000305679 holotype). FIGURE 7.3.**155**.

> *Brachystelma plocamoides* Oliv. in Trans. Linn. Soc. London, Bot. **29**: 112 (1875). —Percy-Lancaster in Excelsa **13**: 71–72 (1988). —Boele in Excelsa **14**: 48 (1989). —Masinde in Kew Bull. **62**: 57, fig.5 (2007); in F.T.E.A., Apocynaceae (part 2): 298 (2012). —Meve, Ill. Handb. Succ. Pl. Asclepiadaceae: 39 (2002).

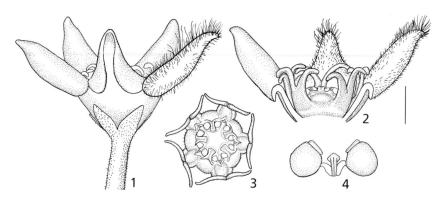

Fig. 7.3.**158**. —**77**. CEROPEGIA ALBIPILOSA. 1, side view of flower; 2, side view of dissected flower; 3, face view of corona; 4, pollinarium. Scale bar: 1–3 = 1 mm; 4 = 0.25 mm. All from *Bruyns* 8627. Drawn by Peter Bruyns. Reproduced from Phytotaxa (2018).

Brachystelma linearifolium Turrill in Bull. Misc. Inform. Kew **1914**: 248 (1914). Type: cultivated at K from material sent from Zimbabwe by Hislop, 7.vii.1914 (K000305672 holotype); Zimbabwe, s. loc., received 2.vi.1920, *Hislop* 81 (K000305671 epitype), designated here.

Tuber globose, 0.5–1.5 cm in diam. Stem erect to ascending, repeatedly trichomotously or dichotomously branched from base, 10–40 cm high, internodes 30–40 mm long, 1–2 mm thick, leafy, glabrous throughout. Leaves herbaceous, suberect, sessile or shortly petiolate, petiole 0–3 mm long; leaf blade ± linear, 35–100 × 1–10 mm, base narrowly cuneate or attenuate, margin entire, apex shortly and abruptly uncinate-mucronate or acute, both surfaces glabrous. Inflorescence subterminal, sessile; cyme 1–3-flowered, flowers opening successively, usually facing downwards below leaves. Pedicel 12–14 mm long, glabrous, occasionally with a linear bract at about middle. Sepals linear-lanceolate, 3–4 × c.0.7 mm at the base, glabrous. Corolla 15–30 mm diam., divided nearly to base; tube very shallow, c.0.5 mm deep, c.3 mm diam. at mouth, exterior probably purplish, glabrous; interior dark purple, glabrous; lobes narrowly linear-lanceolate to spathulate-ligulate, (6)11–20 × 2.5–5 mm wide, often strongly incurved, ± plicate, apices acute, incurved but not connate, abaxially brown, densely minutely puberulent, adaxially dark purple to velvety dark red with orange-grey centre, usually distally prominently keeled, glabrous, sometimes prominently rugulose. Corona sessile, cupular, c.2 mm high, 4 mm diam., pale purple, glabrous throughout; outer lobes united at base into saucer-like structure, deeply bifid teeth narrowly triangular, erect or ascending, c.1.5 × 0.6 mm, strongly pleated at centre and at adjoining points behind the inner lobes where they spread outwards; inner lobes at bases of anthers, rudimentary, to 0.25 mm long, apices obtuse. Follicles narrowly divergent, ellipsoid-fusiform, 4–6 × 1–1.5 cm wide at middle, red striped, glabrous. Seed ovate, brown with a narrow pale margin, c.10 × 6 mm.

Zambia. E: Lusaka Dist., Chisamba, Wardy Farm, 1100 m, fl. 1.ii.1995, *Bingham & Harder* 10354 (K). S: Mazabuka Dist., Nachibanga stream between Choma and Pemba, fl. 9.ii.1960, *White* 6893 (K). **Zimbabwe**. N: Gokwe, Sengwa Research Station, 1000 m, fl. 11.i.1969, *Jacobsen* 4154 (PRE, SRGH). C: Harare, 1650 m, xii.1919, *Eyles* 1954 (K, PRE, SRGH). S: Chiredzi Dist., Ghona-re-Zhou Game Reserve, *Kelly* 515 (SRGH). **Malawi**. N: Vipya, Vernal Pool, ± 70 km south of Mzuzu, fl. 25.ii.1967, *Pawek* 873 (SRGH). C: Dedza, Chongoni Forest, fl. 7.iv.1970, *Salubeni* 1469 (SRGH). S: Zomba, Magomero roadside, *Salubeni* 2906 (SRGH). **Mozambique**. MS: Manica, 69 km towards Machaze, 250 m, 2.i.2001, *Bruyns* 8749 (MO).

Also in Tanzania, Ghana, Ivory Coast, Nigeria. Open *Brachystegia* or *Colophospermum* woodland on sandy soil, sometimes near granitic outcrops; 250–1800 m.

Conservation status: LC: one of the more frequently collected members of the group.

The holotype of *Brachystelma linearifolium* consists of a single flower and detached leaf from a plant cultivated in Kew. The original collector later (1920) sent an excellent collection (K000305671) with 4 flowering stems which fully deserves designation as an epitype.

79. **Ceropegia stenifolia** Bruyns in S. African J. Bot. **112**: 435 (2017), new name, not *Ceropegia stenophylla* C.K. Schneid. (1916). Type: Namibia, near Grootfontein, *Dinter* 2361 (B† holotype, SAM0054280-0 lectotype), designated by Dyer (1980).
   Siphonostelma stenophyllum Schltr. in Bot. Jahrb. Syst. **51**: 148 (1913).
   Brachystelma stenophyllum (Schltr.) R.A. Dyer in Bothalia **10**: 376 (1971); in Codd *et al.*, Fl. S. Africa **27**(4): 34 (1980). —Meve, Ill. Handb. Succ. Pl. Asclepiadaceae: 43 (2002).

Tuber semi-globose, 3–6 cm diam., depressed at apex. Stem 1, erect, spreading-branched, herbaceous, 5–15 cm long, glabrous to sparsely, minutely puberulous. Leaves herbaceous, erect spreading to suberect, subsessile; leaf blade narrowly linear, 30–50 × 1.25–2 mm, slightly plicate, both surfaces glabrous, midrib prominent on lower surface. Inflorescence extra-axillary; sessile cyme 2- or 3-flowered. Pedicel 3–8 mm long. Sepals lanceolate to triangular, 1.5–2 mm long, glabrous. Corolla 12–20 mm long, divided for 60% of length; tube ± narrowly campanulate, 5–8 × 6–7 mm wide at mouth, exterior minutely pubescent, interior maroon, glabrous; lobes with base ovate-lanceolate, linear above, 3.5–14 mm long, revolute, apices connate to form ellipsoid cage. Corona cupular, c.2 mm diam.; yellowish, outer lobes united laterally, rim minutely 2-toothed, 1.5–2 mm high, inner lobes arising from within outer corona, incumbent on anthers, oblong, 0.5–1.0 mm high, apices obtuse. Follicles filiform to sub-fusiform, 5–7 cm long. Seeds not recorded.

Botswana. SE: At Malatswae turnoff from A14 road, 21°47'S 25°57'E, 18.vii.1988, *Venter* s.n. (UNIN).

Also in Namibia and South Africa.

Conservation status: LC?: infrequently collected but from a wide area and very likely to be overlooked by collectors.

80. **Ceropegia nutans** (Bruyns) Bruyns in S. African J. Bot. **112**: 434 (2017). Type: Mozambique, Zambézia, Namuli, 4 km towards Malema, fl. 4.i.2004, *Bruyns* 9729 (BOL150389 lectotype, BOL, MO-2010650), second step lectotype designated here. FIGURE 7.3.**159**.

   Brachystelma nutans Bruyns in Novon **16**: 452(453), fig.1 (2006).

Tuber discoid, to 10 cm diam. Stem prostrate to ascending, 5–20 cm long, 1–2 mm thick, reddish-green, puberulous. Leaves herbaceous, sessile; leaf blade linear, 40–80 × 3–5 mm, abaxial surface glabrous except for puberulous midrib, adaxially glabrous. Inflorescences from all but basal 2 nodes, extra-axillary, sessile, cyme 1-flowered, flower usually facing downwards. Pedicel 20–25 mm, ascending with deflexed apex, puberulous. Sepals brown, c.5 mm long. Corolla 13–16 mm long, divided nearly to base; tube cupular, c.1 mm deep; lobes linear, 12–15 × c.3 mm at base, incurved but not connate, apices acute, abaxially green, puberulous, adaxially green at base, reddish brown towards apex, finely white pubescent. Corona sessile, cupular, c.1 mm high, 3.5 mm diam., yellow-green, glabrous throughout; outer lobes deeply bifid, c.2.5 mm long; erect, inner lobes incumbent on anthers, triangular, slightly longer than anthers, yellow-green with green tips, apices acute, with white hairs mainly near bases. Follicles and seeds not recorded.

Mozambique. Z: Zambézia, 4 km towards Malema, upper slopes of Mt Namuli, 1500 m, 4.i.2004, *Bruyns* 9729 (BOL, MO); Mt Namuli, above waterfall, 1400 m, 5.i.2004, *Bruyns* 9741 (BOL).

Endemic. Bare ground on shallow soil overlying granite; 1400–1500 m.

Conservation status: DD: known only from two collections from a relatively small area but likely to be overlooked.

There are two sheets of the type in BOL both labelled as "holotype". The sheet with the more complete specimen, BOL150389, is here designated as the lectotype.

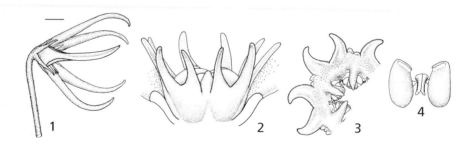

Fig. 7.3.**159**. —**80**. CEROPEGIA NUTANS. 1, side view of flower; 2, side view of dissected flower and corona; 3, face view of part of corona; 4, pollinarium. Scale bar: 1 = 3 mm; 2, 3 = 1 mm; 4 = 0.25 mm. All from *Bruyns* 9729. Drawn by Peter Bruyns. Reproduced from Novon (2006).

81. **Ceropegia barberae** (Hook.f.) Bruyns in S. African J. Bot. **112**: 433 (2017). Type: South Africa, Transkei, valleys of Tsomo River, painting by Mrs. Barber (K iconotype). FIGURE 7.3.**155**.

> *Brachystelma barberae* Hook.f. in Bot. Mag. **92**: t.5607 (1866). —Schlechter in Bot. Jahrb. **18**(Beibl. 45): 25 (1894). —Brown in Fl. Cap. **4**(1): 864 (1908). —Medley-Wood, Natal Pl. **6**: t.587 (1912). —Phillips in Fl. Pl. South Africa **4**: t.354 (1929). —Dyer in Codd *et al.*, Fl. S. Africa **27**(4): 34, 36, fig.7 (1980). —Percy-Lancaster in Excelsa **13**: 63, 65, 67 (1988). — Boele in Excelsa **14**: 46 (1989). —Meve, Ill. Handb. Succ. Pl. Asclepiadaceae: 22 (2002).
> *Dichaelia barberae* (Hook.f.) Bullock in Kew Bull. **7**: 359 (1953).

Tuber 8–20 cm diam., upper surface depressed. Stems 1–few, 5–10 cm tall, ± coarsely pubescent. Leaves herbaceous, spreading, base attenuate into short petiole; leaf blade oblanceolate to linear-oblong, 70–100 × 25–50 mm; base tapering, apex acute to obtuse, both surfaces coarsely pubescent, denser abaxially. Inflorescence apparently terminal, usually of two opposite, sessile, cymes; cyme umbel-like, spherical, 10–15 cm diam., up to 25-flowered, most opening simultaneously, strongly foetid. Pedicel 10–20 mm long, spreading-pubescent. Sepals linear-triangular, 8–11 × c.1 mm, abaxially pubescent. Corolla 20–45 mm long, 20–25 mm diam., divided for c.¾ of length; tube campanulate, 5–7 × c.10 mm wide at mouth, exterior pubescent or glabrous, interior yellow with concentric brownish-maroon lines and spots, pubescent or glabrous; sinuses acute, flat; lobes triangular-ovate at base, tapering to linear above, 25–50 mm long, revolute, apices connate to form ellipsoid cage, apex subacute, abaxially puberulent, adaxially reddish brown or yellow with reddish brown stripes, glabrous. Corona campanulate, c.1 mm high; outer lobes truncate to subtruncate, sometimes with lateral teeth; inner lobes arising from inner wall of outer lobes, incumbent on anthers, obtuse-linear, c.1 mm long. Pollinia brown. Follicles erect-spreading, thickly fusiform, c.5 × 1 cm, thick-walled. Seeds not recorded.

Zimbabwe. W: Bulawayo, 1905, *Munro* s.n. (K). C: Harare ['Salisbury'], Groombridge vlei near Edinburgh Rd., 1480 m, 14.ix.1981, *Biegel* 5856 (K). E: Chipinge Dist., Gungunyana Forest Reserve, x.1963, *Goldsmith* 45/63 (SRGH).

Also in South Africa. Grassland on sandy soil subject to burning; 1500 m.

Conservation status: LC: Uncommon in the Flora area but often collected in South Africa.

82. **Ceropegia rehmannii** (Schltr.) Bruyns in S. African J. Bot. **116**: 141 (2018). Type: South Africa, Transvaal, Houtbochberg, 1880, *Rehmann* 5877 (Z-000046729 holotype).

> *Brachystelma rehmannii* Schltr. in Bull. Herb. Boissier **4**: 449 (1896).
> *Brachystelma foetidum* Schltr. in Bot. Jahrb. Syst. **20**(Beibl. 51): 52 (1895), not *Ceropegia foetida* (E.A. Bruce) Bruyns (2017). —Brown in Fl. Cap. **4**(1): 840 (1908). —Skan in Bot. Mag.: t.8817 (1919). —Dyer in Fl. Pl. South Africa **24**: t.940 (1944); in Codd *et al.* Fl. S. Africa **27**(4): 9–10, fig.7 (1980). —Boele in Excelsa **14**: 47 (1989). —Meve, Ill. Handb. Succ. Pl. Asclepiadaceae: 29–30 (2002). Type: South Africa, Transvaal, Elsberg, 22.x.1893, 5400 ft, *R. Schlechter* 3547 (BOL140266 lectotype, B 10 0153222).
> *Ceropegia foetidissima* Bruyns in S. African J. Bot. **112**: 434 (2017), illegitimate name.

Tuber depressed, 10–18 cm diam., lower surface rounded. Stems 1–several, branched, 7–20 cm long, pubescent with rather coarse hairs. Leaves herbaceous, spreading, shortly petiolate; leaf blade linear-lanceolate to ovate, 5–30 × 3–10 mm, margin ± undulate, abaxial surface pubescent, adaxially pubescent or glabrous. Inflorescence extra-axillary to subterminal; cyme 1- or 2(6)-flowered, flowers with strong putrid smell. Pedicel 6–10 mm long, hairy. Sepals lanceolate, 3–5 mm long, pubescent. Corolla 25–50 mm diam., tube campanulate, 6–10 mm deep; lobes lanceolate-triangular, 10–25 mm long, spreading, margins reflexed, apices attenuate-elongate to bluntly pointed; abaxially finely hairy, adaxially basally greenish-yellow and spotted purple brown, dark purple brown at tips, glabrous to rarely puberulous. Corona subcylindrical; outer lobes erect, bifid, equalling the gynostegium, purplish-brown; inner lobes incumbent on anthers, linear, short, apices obtuse, glabrous. Pollinia subpyriform. Follicles fusiform, 6.5–10 cm long, pubescent. Seeds not recorded.

Botswana. Fide Setshogo, Prelim. Checkl. Pl. Botswana: 25 (2005). **Zimbabwe**. W: Bulawayo, *Percy-Lancaster* s.n. (fide Venter, personal communication, sight record of cultivated plant).

Widespread in South Africa, Lesotho. Grassveldt.

Conservation status: LC: One of the more frequently collected members of the group in South Africa.

In his Flora of Southern Africa account Dyer indicated that the 'Berlin holotype' of *Brachystelma foetidum* had been destroyed and designated a sheet in the Bolus herbarium as the lectotype. In fact a sheet of *Schlechter* 3547 does survive in Berlin but the sheet in the Bolus Herbarium is more complete, with extra information on the label, a section of tuber, absent on the B sheet, and has more flowers, and has been anotated as 'holotype'. It seems best to accept Dyer's lectotypification and treat the B sheet as an isolectotype.

83. **Ceropegia spatuliloba** M.G. Gilbert, sp. nov. Most similar to *Ceropegia praelongum* S. Moore but with shorter pedicels (4–5 mm long, not 15–45 mm), a deeper campanulate corolla tube (not shallowly bowl-shaped) and distinctly spatulate corolla lobes incurved at apices (not linear with recurved margins). Type: Mozambique, Manica e Sofala: Manica, Serra Zuira, planalto de Tsetserra, vertente a c.2 km de vacaria, estrada para Vila Pery, 2100 m, fl. 5.xi.1965, *Torre & Pereira* 12702 (LISC holotype). FIGURE 7.3.**160**.

Tuber discoid, c.2.5 cm diam., 1.2 cm deep. Stems 3 or 4, erect, sparingly branched, to 7 cm long, internodes 0.5–1.5 cm long, c.1.2 mm thick, uniformly minutely puberulous. Leaves possibly not fully developed, herbaceous, ± erect, sessile; leaf blade linear, 20–26 × 1.5–1.8 mm, mostly strongly revolute and narrower, base narrowly cuneate, apex subacute, abaxial surface very sparsely minutely puberulent, adaxial glabrous. Inflorescences several per stem towards apex, extra-axillary, sessile; cyme umbel-like, mostly 2-flowered, flowers opening in succession. Pedicel 4–7 mm long, sparsely minutely puberulous. Sepals triangular-lanceolate, c.2.5 × 0.7 mm, acute, very sparsely minutely puberulent. Bud distinctly beaked, glabrous. Corolla 5.5–6.5 × 3.5–3.8 mm diam., divided to halfway down; tube campanulate, c.2.5 mm deep, exterior and interior pure white, glabrous; sinuses flat; lobes linear-spatulate, c.4 × 1.8 mm wide at base, narrowing to 0.2 mm then widening to 0.4 mm, spreading ± rotately, apices acute, minutely incurved, abaxially glabrous, adaxially dull yellow, slightly greenish near bases, glabrous. Corona sessile, 2-seriate, c.2 mm diam.; outer corona ± bowl-shaped, lobes joined laterally, bluntly triangular, ascending, apices deeply notched, glabrous; inner lobes free from outer corona, at bases of anthers, bluntly triangular, less than half as long as anther, glabrous. Pollinia D-shaped. Follicles and seeds not recorded.

Known only from the type. Grassland on stony reddish-brown soil.

Conservation status: DD

The only material available gives the impression that flowering is precocious and that the mature leaves could prove to be larger than indicated in the description. There are also digital images taken 13.xi.2008 by Stefaan Dondeyne from about the same locality near the border with Zimbabwe.

84. **Ceropegia chlorozona** (E.A. Bruce) Bruyns in S. African J. Bot. **112**: 433 (2017). Type: South Africa, Transvaal, Barberton, *Thorncroft* s.n. (K spirit 1004 prob. holotype).

    Brachystelma chlorozonum E.A. Bruce in Hooker's Icon. Pl. **34**: t.3370 (1938). —Dyer in Fl. Pl. Africa **39**: t.1659 (1972); in Codd *et al.*, Fl. S. Africa **27**(4): 24–25, fig.7 (1980). —Meve, Ill. Handb. Succ. Pl. Asclepiadaceae: 25 (2002).

Tuber 5–7 cm diam., upper surface somewhat depressed. Stem erect, simple to sparingly branched, 10–14 cm long, internodes 13–20 mm long, white hispid-pubescent. Leaves herbaceous, spreading, shortly petiolate; leaf blade elliptic-lanceolate to ovate-elliptic, 20–55 × 15–25 mm, base somewhat cuneate, apex acute to subacute, both surfaces hispid-pubescent to pilose. Inflorescence extra-axillary, cyme 1-flowered. Pedicel 4–5 mm long, decurved. Sepals triangular-lanceolate, c.3.5 × 1.5 mm at base, hispid-pubescent. Corolla to 30 mm diam., divided to ½ of length; tube campanulate, 4–6 mm deep, 8–13 mm diam. at mouth, exterior shortly hispid-pubescent, interior with well-defined concentric bands, often with 1–1.5 mm broad green zone; glabrous around corona, minutely hairy above ring-markings; sinuses developed into small, acuminate, intermediate lobes; lobes ovate-deltoid, 4–5 mm long and wide, ± rotate to slightly recurved, margins slightly reflexed, apices cuspidate, adaxially uniformly green or brown, minutely hairy. Corona cupular, c.1.5 mm deep, purplish-brown, outer lobes forming shallow pouches, notched at apex, incurved, apices and margin minutely white pubescent; inner lobes incumbent on anthers, oblong, c.1.5 mm long, shorter than anthers, dark brown or tinted green, apices rounded, glabrous. Follicles thickly fusiform, c.4 × 1.2 cm, thick-walled. Seed broadly ovate with prominent winged margin.

Mozambique. M: Ponta do Ouro c.300 m north of the South African Border, 26.86°S 32.86°E, 14.viii.1988, *Venter* 13013a. (UNIN, not seen).

Also in South Africa (eastern Transvaal, northern Natal). No habitat data; below 30 m.

Conservation status: VU: Only just extending into the Flora area and regarded as 'near threatened' in South Africa.

The species was originally described from a plant cultivated at Kew from a tuber sent from Barberton and presumed to have been collected locally. There is no dried herbarium material but there is a note that material was preserved in spirit. The only collection assigned to this taxon in the Kew spirit collection, #1004, lacks any provenance but is almost certainly the type material.

85. **Ceropegia schultzei** (Schltr.) Bruyns in S. African J. Bot. **112**: 435 (2017). Type: Namibia, Okuja, *Dinter* 2528 (SAM lectotype), designated by Bruyns (1995).

    Kinepetalum schultzei Schltr. in Bot. Jahrb. Syst. **51**: 149 (1913).

    Tenaris schultzei (Schltr.) E. Phillips in Bothalia **4**: 41 (1941).

    Brachystelma schultzei (Schltr.) Bruyns in Bothalia **25**: 162 (1995); in Strelitzia **34**: 81–81 (2014). —Meve, Ill. Handb. Succ. Pl. Asclepiadaceae: 42 (2002).

Tuber semi-globose to napiform, c.5 cm diam. Stem single, erect, somewhat branched above, 5–20(55) cm long, c.1.5 mm thick, minutely puberulous. Leaves herbaceous, spreading to erect-spreading, subsessile; leaf blade narrowly linear, 25–50(70) × 1–3(20) mm, margins entire, greyish-green, both surfaces minutely puberulous. Inflorescence extra-axillary, cymes 2–4-flowered, flowers opening in succession. Pedicel very slender, 5–12 mm long, c.0.5 mm thick, spreading, minutely adpressed pubescent. Sepals lanceolate, very small, acute, reflexed in the apical part, minutely puberulent. Corolla 60–100 mm diam., divided nearly to base; tube almost flat, <0.5(0.75) mm deep; lobes linear, 18–50 × (0.4)1–2 mm wide, totally reflexed from the base and ± apressed to pedicel at anthesis, apices slightly thickened, abaxially yellowish-green to purplish, finely pubescent, adaxially pale green with darker spots towards centre, pubescent with clavate hairs at base. Corona sessile, c.3.5 × 2.5 mm diam., outer lobes triangular, ascending, apices deeply notched, c.0.5 mm long; inner lobes incumbent on anthers, linear, 1.5–2 mm long, apices connivent over gynostegium, exserted from corolla, papillate. Follicles erect to slightly diverging, linear-fusiform, 6–7.5 × 0.4 cm, glabrous. Seeds not recorded.

Botswana. SW: Bokspits, 1 km north of airstrip, 16.iv.1989, *Venter* s.n. (UNIN); 22°22.5'S 25°07.5'E, 14.xii.1969, *Maguire* in PRE0345813-0 (PRE); 22°22.5'S 25°37.5'E, 31.i.1982, *Snyman & Noailles* in PRE0638876-0 (PRE).

Also found in Namibia and South Africa. Flats between trees and bushes on Kalahari sand; c.900 m.

Conservation status: LC?: infrequently collected in SW Botswana, 'widespread but rare in Namibia'.

86. **Ceropegia punctifera** Bruyns in S. African J. Bot. **112**: 435 (2017), new name, not *Ceropegia punctata* (Masson) Bruyns (2017). Type: Zimbabwe, C: Hartly Dist., Poole Farm, 27.ii.1954, *Hornby* 3334 (SRGH holotype, K000305669). FIGURE 7.3.**160**.

Brachystelma punctatum Boele in Excelsa **16**: 31 (1994). —Meve, Ill. Handb. Succ. Pl. Asclepiadaceae: 40 (2002).

Tuber spherical, to 12 cm diam., lower surface flattened. Stems straight, irregularly branched, to 30 cm long, puberulous. Leaves herbaceous, spreading to erect; petiole 4–6 mm long; leaf blade obovate-oblong, 23.5–36 × 10–15.5 mm, base attenuate, margins entire, apex acute, abaxial surface minutely hispid, adaxial sparsely pubescent. Inflorescence extra-axillary, sessile; cyme 2–5-flowered, flowers opening in succession. Pedicel 1.5–2 mm long, puberulous. Sepals lanceolate, c.2.5 × 0.7 mm, acute, hispid-pubescent. Bud ovoid, apex subacute, pubescent. Corolla to c.7 × 11–12 mm diam., divided for more than 2/3 of length; tube campanulate, 2–2.5 mm deep, exterior glabrous, interior greenish yellow with dark red dots, glabrous; lobes oblong-lanceolate, 3.5–4 × c.1.5 mm wide at base, ± spreading, margins slightly revolute, apices subacute to obtuse, slightly thickened, abaxially puberulous, adaxially yellow green or maroon with yellow dots, glabrous at bases, thickened apices sometimes hairy. Corona sessile, cupular, c.1.5 mm diam., outer lobes separate, ascending, apices shallowly notched to obscurely 3-lobed, c.0.8 mm long, exceeding gynostegium, adaxially thinly puberulous; inner lobes on inner wall of outer corona at bases of anthers, reduced to small dark brown cushions. Follicles and seeds not recorded.

Zimbabwe. C: Hartly Dist., Poole Farm, 27.ii.1954, *Hornby* 3334 (SRGH, K). W: Hwange, *Ellert* sight record mentioned in protolog.

Endemic. *Julbernardia* woodland; c.1200 m.

Conservation status: DD: known only from two collections but likely to be overlooked.

The only two records are separated by c.300 km. The K isotype has distinctively hairy tips to the corolla lobes not mentioned in the protologue and not apparent in the Hwange images.

87. **Ceropegia schinziata** Bruyns in S. African J. Bot. **112**: 435 (2017), new name, not *Ceropegia schinzii* (A. Berger & Schltr.) Bruyns. Type: Namibia, Ovamboland, near Olukunda, *Schinz* s.n. (Z-000001585 holotype, K000305831, ZT-00009990).

Craterostemma schinzii K. Schum. in Bot. Jahrb. Syst. **17**: 154 (1893)
Brachystelma schinzii (K. Schum.) N.E. Br. in F.T.A. 4(1): 471 (1903).

Tuber discoid. Stem usually solitary, sparingly branched, 5–10 cm tall, c.1.5 mm in diam., finely pubescent. Leaves herbaceous, ± sessile; leaf blade linear, 10–50 × 2–5 mm, both surfaces finely pubescent. Inflorescence extra-axillary, sessile; cyme several-flowered, flowers opening in succession. Pedicel 12–16 mm long, c.0.5 mm thick, ascending, finely pubescent. Sepals lanceolate, to 3 mm long. Corolla 8–10 mm diam. deeply divided; tube shallowly cupular, c.0.5 mm deep; lobes ovate to lanceolate, 4–6 × 2–4 mm wide, not recurved, abaxially green, finely pubescent, margins with vibratile, subclavate hairs, adaxially purple brown with white or pale yellow centre, finely pubescent to glabrous. Corona sessile, white with purple-brown margins, glabrous throughout; outer lobes truncate, annular or cup-shaped, margins notched to 1 mm long inner lobes incumbent on anthers. cushionlike, < 0.5 mm long. Follicles and seeds not recorded.

Zimbabwe. W: Bulawayo, Kamativi. Records based on 2 images published by Boele as "Brachystelma species nova" in Excelsa **14**: 50 (1989).

Also in Namibia. Flat areas on white sand between tree; c.1100 m.

Conservation status: DD: known only from two collections but likely to be overlooked.

No collection number was given in either the protologue or on either of the Zurich collections, #163 is only given on the K sheet.

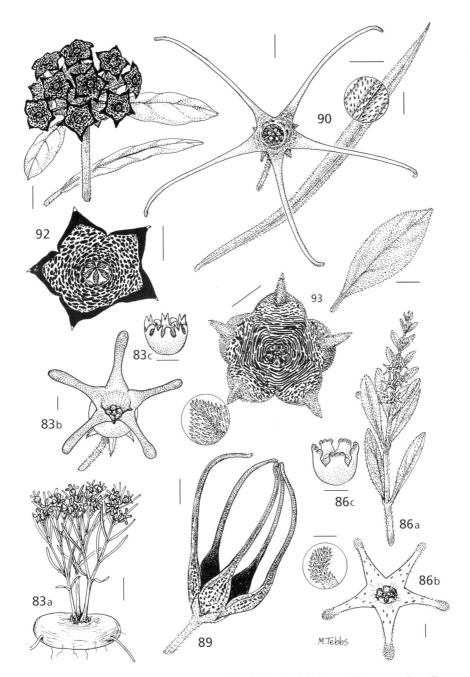

Fig. 7.3.**160**. —**83**. CEROPEGIA SPATULILOBA, 83a habit, 83b flower, 83c corona, from *Torre & Pereira* 12702. —**86**. CEROPEGIA PUNCTIFERA, 86a flowering stem, 86b corolla (inset showing tip of lobe), 86c corona, from *Hornby* 3334. —**89**. CEROPEGIA FLORIBUNDIOR, flower, from *Whellan* 576. —**90**. CEROPEGIA GRACILIDENS, leaf (inset showing indumentum) and flower, from *Bruyns* 7727. —**92**. CEROPEGIA BUCHANANII, inflorescence with leaf and flower, from *Wright* 318. —**93**. CEROPEGIA TOGOENSIS, leaf and flower (inset showing tip of lobe), from *Fanshaw* 4021. Scale bars: all habits, inflorescences and leaves = 10 mm; flower of 89, 90, 92 and 93 = 5 mm; flower and corona of 83 and 86 = 1mm. Drawn by Margaret Tebbs.

88. **Ceropegia neofurcata** Bruyns in S. African J. Bot. **112**: 434 (2017), new name, not *Ceropegia furcata* Werderm. (1939). Type: Zimbabwe, W: Matobo Dist., Longsdale, Matopos Research Station, *Darbyshire* s.n. in MRSH 3070 (SRGH holotype).

Brachystelma furcatum Boele in Excelsa **16**: 29 (1994). —Meve, Ill. Handb. Succ. Pl. Asclepiadaceae: 30 (2002).

Tuber spherical, c.4.5 cm diam., lower surface flattened. Stems erect-spreading, irregularly branched, c.12 cm long, puberulous. Leaves herbaceous, sessile; leaf blade lanceolate to linear, 9–17 × 2–3 mm, apex acute. Inflorescence extra-axillary; cyme 1-flowered. Pedicel 2–3 mm long, puberulous. Sepals elliptic, c.1.4 × 0.9 mm, puberulous. Corolla c.12 mm long, divided for more than ²/₃ of length; tube campanulate, c.3 mm deep; both surfaces glabrous; lobe bases triangular, c.1.7 mm wide, then linear, c.9 mm long, c.0.3 mm wide, inflexed but not connate, abaxially glabrous, adaxially yellowish, dark purple towards tips, pubescent at base with white hairs up to 0.8 mm long. Outer corona lobes free with apices bifid, c.0.8 mm long, puberulous; inner lobes incumbent on anthers, linear, c.0.6 mm long, as high as the gynostegium, apices truncate. Follicles and seeds not recorded.

Known only from the type collection. Area of very saline soil ('Isiquaqua patches') in *Colophospermum mopane* woodland; c.1350 m.

Conservation status: DD.

89. **Ceropegia floribundior** Bruyns in S. African J. Bot. **112**: 433 (2017), new name, not *Ceropegia floribunda* N.E. Br. (1903). Type: Zimbabwe, without locality, *Hislop* s.n. (K000305674 holotype). FIGURE 7.3.**160**.

Brachystelma floribundum Turrill in Bull. Misc. Inform. Kew **1922**: 197 (1922). —Percy-Lancaster in Excelsa **13**: 67, 70 (1988). —Boele in Excelsa **14**: 47 (1989). —Meve, Ill. Handb. Succ. Pl. Asclepiadaceae: 29 (2002). —Masinde in Kew Bull. **62**: 62, fig.8 (2007); in F.T.E.A., Apocynaceae (part 2): 300–301 (2012).

Brachystelma floribundum var. *mlimakito* Masinde in Kew Bull. **62**(1): 64, fig.9 (2007). Type: Tanzania, Ufipa Dist., foot of Mt. Kito, *Richards* 10224 (K000197084 holotype).

Tuber subglobose, 1–15 cm diam., 2–2.7 cm deep, upper surface concave, whitish. Stems 2 or 3, 4–6 cm long, purplish, puberulous. Leaves herbaceous, sessile; leaf blade linear to linear-lanceolate, to 65 × 9 mm, apex acute, abaxial surface puberulous on the midrib, adaxial glabrous. Inflorescence extra-axillary; cyme to 12-flowered. Pedicel 12–15 mm long, puberulous. Sepals lanceolate, c.4 × 1.5 mm wide at base, acute, puberulous. Corolla to 35 mm long, divided for c.⁴/₅ of length; tube campanulate, c.7 × 10 mm diam. at mouth, exterior light green with purple spots, interior light green with concentric purple lines and spots; lobes triangular and folded lengthwise for c.5 mm, then linear, to 28 mm long, usually spreading, apices occasionally remaining connate to form cage, adaxially blackish-purple, glabrous. Corona sessile or subsessile, cupular, 3.2–5 × 3.2–3.5 mm diam., outer lobes deeply bifid, teeth linear, divergent, to 1 mm higher than gynostegium, purple spotted, finely hairy; inner lobes incumbent on anthers, overlapping in centre, sometimes almost absent, glabrous. Follicles and seeds not recorded.

Zambia. N: Lundazi Dist., Nyika Plateau, *Richards* 10425 (K?). **Zimbabwe**. W: Bulawayo, xii.1942, *Brain* 5038 (SRGH). N: Lomagundi Dist., 48 km NE of Mhangura, *Percy-Lancaster* 138 (SRGH). C: Harare ['Salisbury'], Hatfield, xi.1951, *Whelan* 576 (K, SRGH). E: Chimanimani ['Melsetter'] Dist., plot next to Chimanimani Hotel, xii.1966, *Plowes* 2603 (K, SRGH).

Also found in Tanzania. Grassy areas in *Brachystegia* or *Julbernardia* woodland on sandy soils, sometimes near rock outcrops, less often in grassland or *Colophospermum mopane* woodland on sandy clay; 1100–1500 m.

Conservation status: LC. Frequently collected, particularly in Zimbabwe.

90. **Ceropegia gracilidens** Bruyns in Phytotaxa **364**: 113, figs.1,2 (2018). Type: Mozambique, Nampula, 43 km east of Ribáuè (1438DC), ± 600 m, 25.xi.2000, *Bruyns* 8567 (BOL holotype, MO). FIGURES 7.3.**160**, 7.3.**161**.

Tuber discoid, 5–8 cm diam. Stem solitary, erect, usually unbranched, 12–30 cm long, 3–4 mm thick, brownish green, finely pubescent. Leaves in 4–6 pairs, herbaceous, ± sessile; leaf blade linear, 100–150 × 4–16 mm, apex gradually acute, both surfaces finely pubescent. Inflorescences several per stem towards apex, extra-axillary, sessile; cyme umbel-like, 6–8-flowered, flowers opening ± simultaneously, strongly evil-smelling. Pedicel c.10 × 1 mm, sparsely puberulous. Sepals lanceolate, 3–4 × c.1 mm, finely pubescent. Bud to 40 mm long. Corolla campanulate to ± rotate, 50–70 mm diam., deeply divided; tube very shallowly bowl-shaped, 2–3 mm deep, interior transversely ringed with purple-brown on white; glabrous; lobes narrowly lanceolate-caudate, 25–35 × 8–10 mm broad at base, ascending to widely spreading, margins slightly reflexed, abaxially pale green, finely pubescent, adaxially bright yellow-green, glabrous. Corona 2-seriate; outer corona obscurely cup-like, lobes laterally fused to inner lobes, each divided nearly to base, teeth triangular-linear, erect or slightly spreading, obtuse, 2–2.5 mm long, exceeding gynostegium, yellow-brown with darker purple-brown spots, glabrous; inner lobes incumbent on anthers and ± equaling them, linear obtuse, c.1 mm long, glabrous. Follicles and seeds not recorded.

Mozambique. N: Cabo Delgado, 13 km north of Nantulo, 450 m, 1.i.2004, *Bruyns* 9703 (BOL); Nampula, 43 km east of Ribáuè, ± 600 m, 17.xi.2000, *Bruyns* 8567 (BOL, MO). Z: Zambézia, 5 km north of Lioma, 26.xii.1998, *Bruyns* 7727 (K).

Endemic to Mozambique. Patches of shallow soil within open *Brachystegia* woodland, often growing with other succulents; 450–600 m.

Conservation status: LC?. Known from several sites in northern Mozambique.

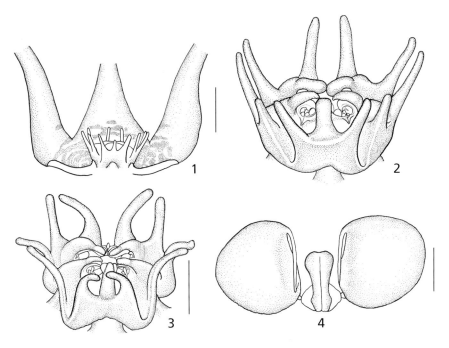

Fig. 7.3.**161**. —**90**. CEROPEGIA GRACILIDENS. 1, side view of centre of dissected flower; 2, 3, side view of corona; 4, pollinarium. Scale bars: 1 = 5 mm; 2, 3 = 2 mm; 4 = 0.25 mm. 1, 2, 4 from *Bruyns* 8567; 3 from *Bruyns* 9703. Drawn by Peter Bruyns. Reproduced from Phytotaxa (2018).

Probably most closely related to the preceding species, *Ceropegia floribundior* Bruyns, from which it differs by the usually solitary erect stem, shallower corolla-tube and the glabrous outer coronal lobes. Superficially similar to *C. glenense* (R.A. Dyer) Bruyns but with much shorter inner corona lobes and distinctive outer corona lobes with very erect, long (c.2.5 mm), slender teeth, also much longer leaves (to 16 cm long, not 3 cm).

91. **Ceropegia cyperifolia** Bruyns in Phytotaxa **364**: 116, figs.3,4 (2018). Type: Mozambique: Niassa Province, 40 km south of Lichinga, 1230 m, 25.xi.2000, *Bruyns* 8642 (BOL holotype, E, MO). FIGURE 7.3.**162**.

Tuber discoid, 4–8 cm diam. Plant spreading close to ground, usually branched near base, to 15 cm wide, c.10 cm tall. Stems slender, terete, green with reddish patches, sparsely puberulous. Leaves herbaceous, ± sessile; leaf blade filiform, 40–80 × c.1 mm, canaliculate, apex acute, both surfaces glabrous. Inflorescences several per plant, extra-axillary or terminal on side branches, sessile, sparsely puberulous; cyme umbel-like, 8- to 11-flowered in dense clusters alongside nodes and terminating branches, flowers opening in quick succession to ± simultaneously, mushroomy ammonia-like smell; pedicel 3–4 × c.1 mm thick. Sepals lanceolate, c.4.5 × 1.5 mm broad at base, attenuate, sparsely puberulous along midrib. Corolla 35–40 mm diam., deeply divided; tube shallowly bowl-shaped; c.3 × 8 mm wide, interior white with concentric rings of purple-brown; lobe bases triangular, c.4 × 5 mm, then lobe slender and ± filiform, c.15 × 1–1.5 mm wide, usually spreading, abaxially pale green spotted with purple-red, glabrous and papillate except towards base of tube, adaxially green to greenish brown on lobes, glabrous. Corona sessile, 2-seriate, c.2.5 × 4.5 mm broad, outer corona ± cup-like, lobes laterally fused to inner lobes, each divided nearly to base, teeth triangular, erect, slightly converging c.1.5 mm tall less than height of anthers, red becoming pale yellow towards base, glabrous outside and with dense tuft of straight white hairs inside near apices of teeth; inner lobes incumbent on anthers and slightly exceeding them, linear-clavate obtuse, c.1 mm long, dorsally slightly swollen and wrinkled towards bases, pale yellow suffused with red towards base, glabrous. Follicles and seeds not recorded.

Mozambique. N: Niassa, 40 km south of Lichinga, 1230 m, 25.xi.2000, *Bruyns* 8642 (BOL, E, MO); ± 12 km south of Maua, 650 m, 3.i.2004, *Bruyns* 9721 (BOL, E); 60 km north of Mandimba, 28.xii.1998, *Bruyns* 7741 (BOL); 30 km NE of Metarica, 600 m, 3.i.2004, *Bruyns* 9722 (BOL).

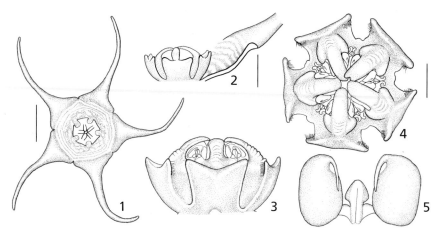

Fig. 7.3.**162**. —**91**. CEROPEGIA CYPERIFOLIA. 1, face view of flower; 2, side view of part of dissected flower; 3, side view of corona; 4, face view of corona; 5, pollinarium. Scale bars: 1 = 5 mm; 2 = 2 mm; 3, 4 = 1 mm; 5 = 0.25. All from *Bruyns* 8642. Drawn by Peter Bruyns. Reproduced from Phytotaxa (2018).

Endemic to Mozambique. Locally common in level areas of shallow, gravelly soil on low granitic domes with scattered *Xerophyta*, species of Commelinaceae and Cyperaceae, and small grasses; 600–1230 m.

Conservation status: LC?. Known from four localities within Niassa separated by nearly 200 km.

92. **Ceropegia buchananii** (N.E. Br.) Bruyns in S. African J. Bot. **112**: 433 (2017). Type: Malawi, S: Shire Highlands, 1881, *Buchanan* 116 (K000305676 holotype). FIGURE 7.3.**160**.

    Brachystelma buchananii N.E. Br. in Bull. Misc. Inform. Kew **1895**: 263 (1895). —Boele in Excelsa **14**: 46 (1989). —Lauchs in Kakteen Sukk. (Berlin) **53**: 236 (2002); in Asklepios **86**: 20 (2002). —Meve, Ill. Handb. Succ. Pl. Asclepiadaceae: 23 (2002). —Masinde in Kew Bull. **62**: 60, fig.7 (2007); in F.T.E.A., Apocynaceae (part 2): 300 (2012).

    Brachystelma magicum N.E. Br. in Bull. Misc. Inform. Kew **1895**: 263 (1895). Type: Tanzania, ?Ulanga Dist., near Ujiji, 1884, *Belgian consul at Zanzibar* s.n. (K000305684 holotype).

    Brachystelma shirense Schltr. in J. Bot. **33**: 339 (1895). Type: Malawi, S: Mt. Sochi, *Scott Elliot* 8666 (B† holotype, K000305678 lectotype), designated by Bullock in Kew Bull. **17**: 188 (1963).

    Brachystelma nauseosum De Wild. in Ann. Mus. Congo Belge Bot., sér. 5 **1**: 191 (1904). Type: D.R. Congo, Ufuru River valley, 25.x.1901, *Cabra & Michel* 63 (BR0000008948007 holotype).

Tuber 5–10 cm diam., upper surface concave. Stem erect, 15–35 cm tall, puberulous. Leaves herbaceous, horizontally spreading; petiole to 5 mm long; leaf blade elliptic-ovate, 40–125 × 18–70 mm, base cuneate, apex obtuse with or without short triangular point, both surfaces glabrous to pubescent especially on veins. Inflorescence terminal, sessile; cyme umbel-like, 5–30-flowered, flowers opening ± simultaneously, with strong foetid smell. Pedicel 25–40 mm long, shortly pubescent. Sepals lanceolate, 3–6 mm long, pubescent. Corolla 18–25 mm diam., divided to halfway down; tube saucer-shaped, interior cream to pale yellow with dense concentric blackish purple lines and blotches and dark margin, glabrous or minutely papillose-pubescent; sinuses slightly raised in bud, usually projecting at anthesis, sometimes produced into prominent rounded lobes; lobes proper triangular, spreading-recurved, apices acute. Corona sessile, c.2 × 4 mm diam., mostly dark brown; outer corona lobes laterally fused to inner lobes, deeply divided, erect or ascending, teeth deltoid-subulate, c.1 mm long, usually each with a prominent subapical dense tuft of short white hairs; inner lobes incumbent on anthers, linear-oblong, meeting or almost meeting in centre, glabrous. Pollinia obliquely oblong. Follicles narrowly divergent, thickly fusiform. 6.5–7 × c.0.9 cm, glabrous. Seeds not recorded.

Botswana. SE: Kanye, 14.i.1993, *Cole* 248 (K). **Zambia**. E: Petauke Dam, 850 m, 6.xii.1958, *Robson* 859 (K). **Zimbabwe**. E: Mutare Dist., Umtali, Butler North, Left Camp no. 2, 4900 ft, 14.xi.1957, *Crook* s.n. in SRGH 83579 (K, SRGH). N: Lomagundi Dist. Mukamba Farm, 48 km NE of Mhangura, 13.xii.1975, *Percy-Lancaster* 139 (SRGH). **Malawi**. N: Karonga Dist., 16 mi W of Karonga on Stevenson road, 620 m, 3.i.1974, *Pawek* 7749 (K). S: Shire Highlands, 4500 ft, xii.1893, *Scott Elliot* 8520 (K). **Mozambique**. Z: Zambézia, 4 km towards Malema, upper slopes of Mt Namuli, 1500 m, 4.i.2004, *Bruyns* 9728 (BOL, E).

Also in D.R. Congo, Tanzania, Uganda. Grassy areas in *Brachystegia* woodland; 600–1700 m. Conservation status: LC, widely distributed.

93. **Ceropegia togoensis** (Schltr.) M.G. Gilbert, comb. nov. Type: Togo: On grassy steppes at the foot of the Agome mountains, iii.1899, *Schlechter* 12961 (B† holotype); Bot. Jahrb. Syst. **38**: 41, fig.5 H–O (1905), illustration designated here as lectotype. FIGURE 7.3.**160**.

    Brachystelma togoense Schltr. in Bot. Jahrb. Syst. **38**: 40 (1905).

Tuber orbicular to discoid, 4–15 cm diam., to 5cm deep, upper surface concave. Stem usually solitary, erect, simple or few-branched, 15–30 cm long, densely shortly hispidulous. Leaves herbaceous, erect or spreading, shortly petiolate; leaf blade lanceolate-oblong to oblong-ligulate or oblanceolate, 40–60(100) × 8–13(30) mm, base cuneate, apex ± acute, both surfaces softly puberulous. Inflorescence terminal, sessile; cyme umbel-like, 3–10-flowered, flowers opening ± simultaneously. Pedicel 5–8(14) mm long, densely puberulous. Sepals lanceolate, c.4 mm long, abaxially densely puberulous. Corolla to 18 mm diam., divided to halfway down; tube shallowly bowl-shaped, c.8 mm diam. at mouth, exterior densely puberulous, interior yellow densely concentrically barred and mottled dull purple, glabrous or sparsely pilose; sinuses minutely toothed; lobes broadly triangular ovate, 6–7 × c.4 mm wide at base, ± spreading, apices acute, adaxially uniformly dull purple, densely finely white pubescent. Corona sessile, c.3 mm diam., outer corona lobes each forming deep pouch, laterally fused to inner lobes, ± deeply divided, erect or ascending, teeth deltoid-subulate, erect or recurved, yellow or dark brown, usually with adaxial line of stiff reflexed, white hairs along middle, otherwise glabrous; inner lobes incumbent on anthers, linear-oblong, slightly longer than anthers, meeting in centre but not overlapping, apices obtuse or very obscurely lobed, glabrous. Pollinia obliquely oblong, dark amber. Follicles and seeds not recorded.

Zambia. N: Lumangwe, 14.xi.1957, *Fanshaw* 4021 (K). **Malawi**. S: Blantyre Dist., Ndirande Mt., *Moriarty* 478 (K).

Widespread in West Africa: Congo (Brazzaville), Benin, Cameroon, Dahomey, Ivory Coast, Ghana, Nigeria, Togo. Habitat not recorded in Flora area; c.1200 m

Conservation status: LC: widespread in West Africa, just extending into Flora area.

49. **HOODIA** Decne.[21]

Hoodia Decne. in Candolle, Prodr. **8**: 664 (1844), nom. cons. prop. —Brown in F.T.A. **4**(1): 490–492 (1903); in Fl. Cap. **4**(1): 896–897 (1909). —Bruyns in Bot. Jahrb. Syst. **115**: 145–270 (1993); in Stapeliads S. Africa Madagascar **1**: 92–129 (2005).
   Monothylaceum G. Don, Gen. Hist. **4**: 116 (1837), nom. rej. prop.
   Scytanthus Hook., Icon. Pl. **7**: t.605–606 (1844).
   Ceropegia sect. *Hoodia* (Decne.) Bruyns in S. African J. Bot. **112**: 420 (2017).

Clump-forming stem succulents; latex clear or at most cloudy. Stems mostly erect, cylindrical, with 11–34 rows of spine-tipped tubercles along the stem. Inflorescences arising mostly towards the apex of the stems, glabrous. Corollas small and deeply lobes to large, flat and somewhat plate-like. Corolline corona absent or forming a weakly developed swelling around the mouth of the corolla tube. Gynostegial corona in 2 series; outer lobes fused basally to the inner lobes and forming small pouches; inner lobes adpressed to the back of the anthers, dorsiventrally flattened. Pollinia D-shaped, attached basally to the translator arms, and with a germination crest on the inner face. Follicles paired, erect, fusiform, glabrous.

Twelve species in south-western Africa with one extending into dry western Angola, and a single species in the Flora Zambesiaca region. Records of two other species of *Hoodia* by Setshogo (2005) in the Preliminary Checklist of the Plants of Botswana appear to be erroneous – *H. gordonii* is native to Namibia and South Africa, while *H. parviflora* is restricted to Angola and Namibia.

Hoodia currorii Decne. in Candolle, Prodr. **8**: 665 (1844). Type: Angola, Elephant Bay, 1840, *Curror* s.n. (K spirit collection holotype, K000306204 epitype), epitype designated by Figueiredo & Smith in Phytotaxa **423**: 299 (2019).
   Scytanthus currorii Hook., Icon. Pl. **7**: t.605–606 (1844), generic name illegitimate.
   Ceropegia currorii (Decne.) Bruyns in S. African J. Bot. **112**: 420 (2017).

---

[21] by D.J. Goyder

Stems succulent, erect to ascending, 10–100 cm tall and 4–6(8) cm thick. Tubercles prominent, fused in lower half of stem into 11–16(24) angles along the stem, each tubercle tipped with a sharp spine 6–10 mm long. Inflorescence with 1–4 flowers opening successively. Pedicels 3–50(60) mm long. Sepals 4–8 mm long, ovate-lanceolate. Corolla 4–18 cm diam., rotate or concave-rotate, circular to broadly 5-lobed, brick-red to flesh-pink or yellowish pink with darker veins, bristles 0.5–3.5 mm long arising from dome-shaped papillae; corolla tube 3–6 mm long, 5–9 mm across, pentagonal; corolla lobes narrowing abruptly into subulate tips 6–20 mm long. Gynostegium enclosed within corolla tube. Corona 2–3 mm tall, deep red-purple or red-brown; outer lobes bifid and forming a 5-lobed cup; inner lobes 0.5 mm long, rectangular, obtuse.

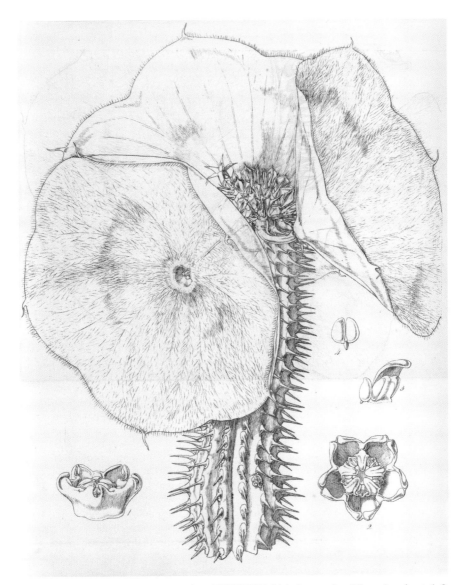

Fig. 7.3.**163**. HOODIA CURRORII subsp. LUGARDII. Main image, tip of flowering shoot; 1, 2, gynostegium and surrounding corona from the side, and from above; 3, stamen; 4, pollinarium. Reproduced from Hooker's Icones Plantarum (1844).

Subsp. **lugardii** (N.E. Br.) Bruyns in Bot. Jahrb. Syst. **115**: 205 (1993). Type: Botswana,
Chukutsa Salt Pan, fl. 21.x.1899, *Lugard* 303 (K holotype). FIGURE 7.3.**163**.

Hoodia lugardii N.E. Br. in F.T.A. **4**(1): 491 (1903).
Ceropegia currorii subsp. *lugardii* (N.E. Br.) Bruyns in S. African J. Bot. **112**: 420 (2017).

Pedicels 3–7 mm long. Corolla 4–7.5 cm diam., concave-rotate.

Botswana. SW: 120 km E of Ghanzi, *Plowes* 2589 (SRGH). SE. N of Lake Xau
[Lake Dow], 22.iii.1965, *Wild & Drummond* 7319 (K, PRE). **Zimbabwe**. S: 50 km W of
Beitbridge, fl. 15.viii.1973, *Leach & Knott* 15085 (BOL, K, LISC, M, PRE, SRGH, WIND).
Also known from northern Limpopo Province in South Africa. Open mopane or
acacia bushland on calcareous soils; 500–1000 m.

50. **STAPELIA** L.[22]

Stapelia L., Sp. Pl.: 217 (1753), conserved name. —Leach in Excelsa Taxon. Ser. **3**:
1–157 (1985). —Bruyns, Stapeliads S. Africa Madagascar **2**: 418–489 (2005).

Gonostemon Haw., Syn. Pl. Succ.: 27 (1812).
Ceropegia sect. *Stapelia* (L.) Bruyns in S. African J. Bot. **112**: 421 (2017).

Clump-forming stem succulents; latex clear or at most cloudy. Stems erect or procumbent-erect,
cylindrical, rounded or sharply 4-angled, glabrous or densely pubescent. Inflorescences basal,
1–10-flowered, pedunculate. Flowers strongly foetid, not nectariferous. Corolla rotate, shallowly
cupular or rarely campanulate, fused basally to half of length, fleshy, rugose or unsculptured,
glabrous or pubescent; corolla lobes ovate or triangular, reflexed or spreading, often ciliate.
Corolline corona absent. Gynostegial corona in 2 series, with 5 staminal and 5 interstaminal
parts fused only at the very base; "outer" interstaminal corona lobes ovate, rectangular or deltate,
spreading, "inner" staminal corona lobes subulate, often bilobed, erect, occasionally with
inflexed subulate or falcate adaxial appendage shorter than staminal parts. Gynostegium (sub-)
sessile or atop a column. Pollinia ovoid or reniform. Follicles paired, fusiform, erect, glabrous
or pubescent.

Stapelia comprises around forty species in southern Africa and several hybrids have
been recorded. In addition to the four species from the Flora region listed below,
in 1886 and 1896 respectively, Schinz and Fleck made collections of *S. schinzii* A.
Berger & Schltr., allegedly from Lake Ngami in northern Botswana. The species has
not been collected there since, despite the area having been extensively collected by
the Lugards.

1. Corolla less that 5 cm in diameter, glabrous; flowers arising at any point along the
   stem . . . . **4.** *kwebensis*
– Corolla at least 8 cm in diameter, pubescent or bearded; flowers arising near the
   base of the stem only. 2
2. Stems procumbent-ascending, leaf rudiments 6–11 mm long; corolla lobes with
   an acute rather than long-attenuate apex . . . . **3.** *gettliffei*
– Stems firmly erect, leaf rudiments 2–3 mm long; corolla lobes long-attenuate at
   tip 3
3. Inner corona lobes with dorsal horn united to the main lobe only at the base;
   corolla 12.5–40 cm in diameter; stems green suffused with purple. . . . **1.** *gigantea*
– Inner corona lobes with dorsal horn united to the main lobe for most of its
   length; corolla 8–13 cm in diameter; stems pale green . . . . **2.** *unicornis*

[22] by D.J. Goyder

1. **Stapelia gigantea** N.E. Br. in Gard. Chron., ser. nov. **7**: 684 (1877). —Leach in Excelsa Taxon. Ser. **3**: 10 (1985). —Bruyns, Stapeliads S. Africa Madagascar **2**: 472 (2005). Type: South Africa, Natal, *Gerrard* 717 (K holotype). FIGURE 7.3.**164**.

   Stapelia nobilis N.E. Br. in Bot. Mag. **127**: t.7771 (1901). Type: South Africa, Port Elizabeth, 1897, cultivated at Kew, v.1900, *Griffiths* s.n. (K holotype).

   Stapelia marlothii N.E. Br. in Bull. Misc. Inform. Kew **1908**: 436 (1908). Type: Zimbabwe, Matopo Hills, *Marloth* 3414 (K holotype, PRE, STE).

   Stapelia gigantea var. *pallida* E. Phillips in Fl. Pl. South Africa **5**: t.181 (1925). Type: South Africa, hort. PRE.

Fig. 7.3.**164**. STAPELIA GIGANTEA. Flowering shoot. Reproduced from The Gardeners' Chronicle (1888).

Stapelia youngii N.E. Br. in Bull. Misc. Inform. Kew **1931**: 43 (1931). Type: Zimbabwe, near Harare [Salisbury], *Young* in Herb. Moss 17301 (K holotype).
Stapelia cylista C.A. Lückh. in S. African Gard. **23**: 139 (1933). Type: White & Sloane, Stapelieae ed. 2, **2**: pl. 15, opposite p. 524 (1937), iconotype.
Gonostemon giganteus (N.E. Br.) P.V. Heath in Calyx **1**: 17 (1992).
Gonostemon giganteus var. *nobilis* (N.E. Br.) P.V. Heath in Calyx **3**: 7 (1993).
Gonostemon giganteus var. *marlothii* (N.E. Br.) P.V. Heath in Calyx **3**: 7 (1993).
Gonostemon giganteus var. *pallidus* (E. Phillips) P.V. Heath in Calyx **3**: 7 (1993).
Gonostemon giganteus var. *youngii* (N.E. Br.) P.V. Heath in Calyx **3**: 7 (1993).
Ceropegia gigantea (N.E. Br.) Bruyns in S. African J. Bot. **112**: 422 (2017).

Compact, dense clump-forming stem-succulent. Stems erect, 10–30 cm long, 1–3.5 cm wide, 4-angled with broadly concave faces. Tubercles inconspicuous. Leaf rudiments 2–3 mm long. Inflorescences 1–4-flowered arising near the base of the stems; peduncle stout. Pedicels 3–6 cm long. Sepals ovate, 10–16 mm long. Corolla rotate, 12.5–40 cm in diam., yellowish to pale reddish, transversely rugulose, wrinkles red-brown or purple, covered by fine, whitish to reddish hairs up to 12 mm long; corolla tube flat to slightly campanulate, 3–10 cm in diam.; corolla lobes spreading, ovate to triangular, 8–15 cm long, attenuate. Gynostegial corona purple, usually atop a short, yellow stipe; interstaminal corona lobes rectangular, occasionally spathulate, 5–6 mm long, 1.5–2.5 mm wide, erect-diverging, apically crenate, with or without mucro; staminal corona lobes 8–12 mm long, 1–2 mm wide, erect, with central subulate and dorsal broadly winged appendages.

Botswana. N: 50 km W of Francistown [cult. SRGH], *Drummond* 8492 (SRGH). SE: Kanye [cult. Mutare 11.i.1970], *Campbell* s.n. sub *Plowes* 3050 (K, MO, PRE, SRGH). **Zambia**. C: Shimabala Cave c.23 km S of Lusaka, fl. 5.ii.1995, *Harder & Bingham* 2589 (K, MO). E: Luangwa Valley c.30 km S of Kakumbi [cult. SRGH 7.i.1966 & Mutare 14.ii.1971], *Bainbridge* 1049 (NBG, PRE, SRGH). S: Livingstone Dist., Victoria Falls, 4th gorge [cult. Mutare 10.iv.1980], *Mitchell* s.n. sub *Plowes* 4154 (K, PRE, SRGH). **Zimbabwe**. N: Darwin Dist., Chesa Purchase Area, Nora Dam [cult. Mutare 12.ii.1978], *Blake* 125 sub *Plowes* 4058A (K, SRGH). W: 50 km SSE of Bulowayo, *Bullock* 11 sub *Leach* 12397 (K, PRE, SRGH). C: Harare [Salisbury], fl. 17.iii.1925, *Eyles* 4487 (K). E: Chipinga Dist., 20 km SE of Chisumbanje [cult. Mutare 7.ii.1977], *Plowes* 4767 (K, PRE, SRGH). S: Zimbabwe Ruins, fl. 4.iii.1962, *Leach* 11382 (K, PRE, SRGH). **Malawi**. S: Cape Maclear, 22.vii.1984, *Salubeni & Patel* 3808 (K, MAL). **Mozambique**. Z: 20 km N of Gurué [Vila Junqueiro, cult. Mutare 6.xii.1979], *Blake* s.n. sub *Plowes* 2633 (LISC, PRE, SRGH). MS: Garuso Mt. [cult. Mutare 18.ii.1980], *Bey* s.n. sub *Plowes* 4057 (K, LISC, SRGH). GI: Massangena, fr. 1.viii.1973, *Correia & Marques* 3134 (K, LMU). M: Between Catuane and Lagoa Mandjene, fl. 14.iv.1949, *Myre & Balsinhas* 603 (K, LMA, PRE).
Also occurs in South Africa, Swaziland and Lesotho. Occasionally naturalised elsewhere. Frequently on rock, but also recorded from sandy or alluvial soils. 0–1200 m.

2. **Stapelia unicornis** C.A. Lückh. in S. African Gard. **28**: 228 (1938). —Leach in Excelsa Taxon. Ser. **3**: 21 (1985). —Bruyns, Stapeliads S. Africa Madagascar **2**: 475 (2005). Type: Swaziland, *Postma* s.n. sub *Lückhoff* 258 (not traced); [cult. Cape Town ii.1938], *Lückhoff* s.n. (BOL neotype), designated by Leach in Excelsa Taxon. Ser. **3**: 21 (1985).
Gonostemon unicornis (C.A. Lückh.) P.V. Heath in Calyx **1**: 17 (1992).
Ceropegia unicornis (C.A. Lückh.) Bruyns in S. African J. Bot. **112**: 422 (2017).

Loosely-packed clump-forming stem-succulent. Stems erect except for the horizontal base, 6–16 cm long, 1–1.5 cm wide, 4-angled with concave faces. Tubercles inconspicuous. Leaf rudiments 2–3 mm long. Inflorescences 1–5-flowered arising near the base of the stems; peduncle stout. Pedicels 0.8–3 cm long. Sepals ovate, 5–10 mm long. Corolla rotate to shallowly campanulate, 8–13 cm in diam., yellowish to pale reddish, transversely rugulose, wrinkles red-brown, covered by fine, purple to red hairs up to 10 mm long; corolla tube shallowly campanulate, 5 mm deep; corolla lobes spreading or reflexed, ovate to triangular, 4.5–6 cm long, attenuate. Gynostegial corona dark purple, usually atop a short, pinkish stipe; interstaminal corona lobes linear, 5–6 mm long, 1 mm wide, ascending and recurved near apex; staminal corona lobes 7–9 mm long, 3–4 mm wide, erect, with dorsal and broadly winged appendages fused for most of their length.

Mozambique. M: Goba Fronteira [cult. Maputo 2001], *Bolnick* s.n. (K).
Restricted to the Lebombo mountains of Swaziland, southern Mozambique and northern KwaZulu-Natal. Growing on rocky slopes, c.300 m.

3. **Stapelia gettliffei** R. Pott in Ann. Transvaal Mus. **3**: 226 (1913), as "*gettleffii*". — Leach in Excelsa Taxon. Ser. **2**: 70 (1980) & **3**: 30 (1985). —Bruyns in Stapeliads S. Africa Madagascar **2**: 470 (2005). Type: South Africa, Limpopo Province, Makhado [Louis Trichardt], Soutpansberg, *Gettliffe* s.n. sub *Transvaal Mus. Herb.* 9343 (PRE holotype).

    Gonostemon gettliffei (R. Pott) P.V. Heath in Calyx **1**: 17 (1992).
    Ceropegia gettliffei (R. Pott) Bruyns in S. African J. Bot. **112**: 422 (2017).

Mat-like clump-forming stem-succulent. Stems decumbent-ascending and often limp and trailing, 5–15 cm long, 1–1.5 cm wide, 4-angled with concave faces. Tubercles inconspicuous. Leaf rudiments 6–11 mm long. Inflorescences 1–3-flowered arising near the base of the stems; peduncle short. Pedicels 5–8 cm long. Sepals lanceolate, 9–15 mm long. Corolla rotate, 9–12 cm in diam., reddish purple on margins and between the wrinkles, transversely rugulose, wrinkles cream, with fine, purplish hairs up to 10 mm long towards the margins and in the fused portion of the corolla; corolla tube very shallow, to 2 mm deep; corolla lobes spreading, narrowly ovate, 4–6 cm long, acute. Gynostegial corona dark purple-brown to red; interstaminal corona lobes linear and channelled, 5 mm long, ascending and recurved near apex; staminal corona lobes 10–12 mm long, 1–1.5 mm wide, erect, with deltoid dorsal and broadly winged appendages fused for most of their length.

Botswana. SE: Mahalapye River, fl. 25.iii.1965, *Wild & Drummond* 7317 (K, SRGH). **Zimbabwe**. E: Chipinga Dist., Mutema Pan [cult. Mutare 1965-1967], *Plowes* 2490 (K, LISC, PRE, SRGH). S: Nuanetsi Dist., Malipate, fl. 3.v.1961, *Drummond & Rutherford-Smith* 7688 (K, SRGH). **Mozambique**. GI: 20 km SE of Pafuri, Limpopo River, fl. 11.vii.1964, *Leach & Mockford* 12302 (K, SRGH).
Also known from the Limpopo and Mpumalanga provinces of South Africa. Growing in shade of acacia or mopane; 250–1100 m.

4. **Stapelia kwebensis** N.E. Br. in F.T.A. **4**(1): 501 (1903). —Leach in Excelsa Taxon. Ser. **3**: 99 (1985). —Bruyns in Stapeliads S. Africa Madagascar **2**: 427–430 (2005). Types: Botswana, Kwebe Hills, fl. 3.i.1898, *Mrs. Lugard* 29 (K lectotype), designated here, see note below; fr. 18.x.1897, *Mrs. Lugard* 29 (K paralectotype); fl. i.1897, *E.J. Lugard* 112 (K syntype).

    Stapelia kwebensis N.E. Br. var. *longipedicellata* A. Berger in Stapel. & Klein.: 318 (1910). Type: Namibia, near Olukonda, ix.1995, *Schinz* s.n. (Z lectotype) designated by Bruyns, Stapeliads S. Africa Madagascar **2**: 427 (2005).
    Stapelia longipedicellata (A. Berger) N.E. Br. in Bull. Misc. Inform. Kew **1913**: 303 (1913). —Leach in Excelsa Taxon. Ser. **3**: 102 (1985).
    Gonostemon kwebensis (N.E. Br.) P.V. Heath in Calyx **1**: 19 (1992).
    Gonostemon longipedicellatus (A. Berger) P.V. Heath in Calyx **1**: 19 (1992).
    Ceropegia longipedicellata (A. Berger) Bruyns in S. African J. Bot. **112**: 422 (2017).

Compact to diffuse clump-forming stem-succulent. Stems erect to spreading, 6–20 cm long, 0.8–1.5 cm wide, obscurely 4-angled with concave faces. Tubercles inconspicuous. Leaf rudiments 2–7 mm long. Inflorescences 1–6-flowered arising at any point along the stems; peduncle stout.

Pedicels 1–5 cm long. Sepals ovate, 3.5–5.5 mm long. Corolla rotate to slightly campanulate, 1.5–4.5 cm in diam., purple to reddish brown or yellow-brown, transversely rugolose, glabrous but minutely papillate; corolla tube shallowly campanulate, to 3–5 mm deep; corolla lobes spreading to slightly reflexed, ovate, 0.5–1.5 cm long, somewhat attenuate. Gynostegial corona purple-red to red; interstaminal corona lobes rectangular, truncate, 0.5 mm long, spreading; staminal corona lobes 0.5–0.75 mm long, dorsiventrally flattened, erect and adpressed to backs of the anthers, the tips incurved over them.

Botswana. N: Moremi Game Reserve, fl. 20.xi.2007, *A. & R. Heath* 1593 (GAB, K, PSUB). SW: Kuke, near fence, fl. 21.ii.1970, *Brown* 8733 (K, PRE, SRGH). SE: Palapye [cult. Greendale 22.ii.1961], *Leach & Noel* 263 (SRGH). **Zimbabwe**. E: Mutema Pan [cult. SRGH ii. 1966], *Chase* s.n. (K, PRE, SRGH). S: Sentinel Ranch 50 km W of Beitbridge [cult. Greendale 11.i.1961], *Leach* 10695 (K, LISC, M, SRGH, WIND). **Mozambique**. GI: near Pafuri 2.vii.1964 [cult. Nelspruit 1965], *Leach & Mockford* 12307A (K, LISC, M, NBG, PRE, SRGH, Z).

Also recorded from Namibia, SW Angola and the Limpopo province of South Africa; 250–1100 m.

Nomenclatural note: Leach in Excelsa Taxon. Ser. **3**: 99 (1985) chose *Lugard* 29 as the lectotype of *S. kwebensis*, but as this consists of two gatherings, it is necessary to propose a second stage lectotypification. I here select the flowering material gathered on the 3rd January 1898 as lectotype of this name.

51. **DUVALIA** Haw.[23]

Duvalia Haw., Syn. Pl. Succ.: 44 (1812). —Brown in F.T.A. **4**(1): 502–503 (1903); in Fl. Cap. **4**(1): 1024–1036 (1909). —Meve in Pl. Syst. Evol. Suppl. **10**: 1–132 (1997). —Bruyns in Stapeliads S. Africa Madagascar **1**: 68–91 (2005).
      Ceropegia sect. *Duvalia* (Haw.) Bruyns in S. African J. Bot. **112**: 425 (2017).

Mat-forming stem succulents; latex clear or at most cloudy. Stems decumbent to erect, sometimes rhizomatous, cylindrical, somewhat clavate or sphaeroidal, 4–6-angled with usually conical tubercles, glabrous. Inflorescences single, mostly arising in lower half of stem, 1–10(20)-flowered but flowers opening in succession, subsessile. Flowers generally held facing upwards on or close to the ground. Corolla rotate, deeply lobed, glabrous or pubescent; corolla lobes ovate or triangular, spreading, plane or strongly relicate, margins ciliate or not. Corolline corona forming a prominent circular or pentagonal fleshy annulus in the fused portion of the corolla, glabrous to bearded. Gynostegial corona in 2 series; "outer" lobes fused to form a pentagonal or circular disk atop the annulus or surrounded by it; "inner" staminal corona lobes dorsiventrally flattened and adnate to the back of the anthers, somewhat obovate, spreading to ascending. Gynostegium sessile or subsessile. Pollinia ovoid. Follicles paired, fusiform, erect, glabrous.

Around 13 species recognised in southern Africa, NE tropical Africa and Arabia.

Duvalia polita N.E. Br. in Gard. Chron., ser. nov. **6**: 130 (1876); in Fl. Cap. **4**(1): 1026 (1909). —Meve in Pl. Syst. Evol., Suppl. **10**: 92 (1997). —Bruyns in Stapeliads S. Africa Madagascar **1**: 71 (2005). Type: cult. Kew Gardens, vii.1876 (K holotype). FIGURE 7.3.**165**.
      Duvalia dentata N.E. Br. in Bull. Misc. Inform. Kew 1895: 265 (1895). Type: Botswana, 30 miles NW of Kobis [Koobie], 18.i.1862, *Baines* s.n. (K holotype).
      Duvalia transvaalensis Schltr. in Bot. Jahrb. Syst. **20**(Beibl. 51): 54 (1895). Type: South Africa, Klipdam, 14.ii.1894, *Schlechter* 4498 (BOL holotype).
      Duvalia transvaalensis var. *parviflora* L. Bolus in Ann. Bol. Herb. **1**: 194 (1915). Type: South

---

[23] by D.J. Goyder

Fig. 7.3.**165**. DUVALIA POLITA. Main image, flowering stems; 1, 2, stem tubercles showing leaf rudiments; 3, flower; 4, pollinarium. Drawn by Walter Fitch. Reproduced from Curtis's Botanical Magazine (1876).

Africa, cult. Queenstown 1.iii.1914, ex Seringa Farm near Naboomspruit, xii.1913, *Galpin* 8467 (BOL holotype).

Duvalia polita var. *transvaalensis* (Schltr.) A.C. White & B. Sloane, Stapelieae, ed. 2, **2**: 754 (1937).

Duvalia polita f. *intermedia* A.C. White & B. Sloane, Stapelieae, ed. 2, **3**: 1144 (1937). Type: White & Sloane, Stapelieae, ed. 2, **2**: 756, fig.751 lower flower (1937), iconotype (photograph) lectotypified by Meve in Pl. Syst. Evol., Suppl. **10**: 92 (1997).

Duvalia polita var. *parviflora* (L. Bolus) A.C. White & B. Sloane in Cact. Succ. J. (Los Angeles) **14**: 159 (1942), illegitimate combination.

Ceropegia polita (N.E. Br.) Bruyns in S. African J. Bot. **112**: 426 (2017).

Compact to diffuse clump-forming stem-succulent. Stems decumbent, 2–10 cm long, 0.7–1.5 cm wide, 6–angled, often with slender horizontal underground runners extending to 30 cm forming new clumps on surfacing. Tubercles 5–10 mm long, conical. Leaf rudiments 4–6 mm long. Inflorescences 1–4-flowered arising near base of stems and opening in succession. Pedicels 1.5–2.5 cm long. Sepals triangular, 4–5 mm long. Corolla rotate to slightly campanulate, 2–3.5 cm in diam., shiny to dull purple to reddish brown or yellow-brown, sometimes paler towards the centre; fused portion of corolla with raised circular annulus 2.5–5 mm high and 8–12 mm across, minutely papillate; corolla lobes spreading to ascending, ovate, 1–1.5 cm long, attenuate, margins ciliate or not, otherwise glabrous. Gynostegial corona reddish brown; outer corona forming a circular or pentagonal disc; staminal corona lobes somewhat paler than outer lobes, c.1 mm long.

Caprivi. Lake Liambezi, *Bruyns* 2294 (WIND). **Botswana**. N: Selinda Reserve, Maun Road, fl. 4.iv.2005, *A. & R. Heath* 1008 (GAB, K, PSUB). SE: 5 km E of Dodi, N of Gabarone, *Plowes* 6702 (SRGH). **Zambia**. S: Kalomo, *Fanshawe* 9229 (SRGH). **Zimbabwe**. N: 1 km S of Sengwa, *Guy* s.n. sub *Plowes* 5027 (SRGH). W: Hwange [Wankie] Dist., 3 km SE of Inyantue River, 29.iii.1963, *Leach* 11614 (K, SRGH). E: 16 km N of Birchenough Bridge, *Plowes* 3014 (SRGH). S. 70 km S of Masvingo, *Plowes* 3915 (SRGH). **Malawi**. S: Shire River valley S of Liwonde, 5 km west of North Shire, 2.i.1999, *Bruyns* 7755 (MO). **Mozambique**. M: 5 km S of Umbeluzi, 20.vi.1969, *Leach* 14248 (LMA).

Also known from Swaziland and northern provinces of South Africa. Often in mopane woodland or hot dry valleys rich in succulents; 300–1100 m.

52. TRIDENTEA Haw.[24]

Tridentea Haw., Syn. Pl. Succ.: 34 (1812). —Leach in Excelsa Taxon. Ser. **2**: 1–68 (1980), in part. —Bruyns in S. African J. Bot. **61**: 180–208 (1995); in Stapeliads S. Africa Madagascar **2**: 531–550 (2005).

Ceropegia sect. *Tridentea* (Haw.) Bruyns in S. African J. Bot. **112**: 423 (2017).

Clump-forming stem succulents; latex clear or at most cloudy. Stems decumbent, cylindrical, obscurely 4(6)-angled with rectangular tubercles, glabrous. Inflorescences single, mostly arising near base of stem, 1–5-flowered, the flowers opening in succession on a stout, gradually lengthening peduncle 0.5–2(4) cm long. Corolla rotate to campanulate, lobed to about halfway, smooth to rugose or papillate; corolla lobes ovate or triangular, spreading or recurved, margins generally ciliate; tubular portion thickened but not forming a prominent annulus. Gynostegial corona in 2 series; "outer" interstaminal lobes ascending to spreading, simple to trifid; "inner" staminal corona lobes dorsiventrally flattened and adnate to the back of the anthers below and ususally much exceeding them as a terete horn. Gynostegium sessile or subsessile. Pollinia D-shaped. Follicles paired, fusiform, erect, glabrous.

Eight species in South Africa and Namibia, with one just edging into the Flora Zambesiaca region in the extreme south west of Botswana.

---

[24] by D.J. Goyder

Tridentea marientalensis (Nel) L.C. Leach in Trans. Rhodesia Sci. Assoc. **59**: 3 (1978); in Excelsa Taxon. Ser. **2**: 18 (1980). Bruyns in Stapeliads S. Africa Madagascar **2**: 547 (2005). Type: Namibia, Haruchas, near Mariental, *Berger* s.n. sub STE 7044 (not traced); Farm Haruchas on Auob River, 13.iv.1960, *Leistner* 1819 (KMG neotype, PRE), designated by Leach in Excelsa Taxon. Ser. **2**: 18 (1980). FIGURE 7.3.**166**.

Stapelia marientalensis Nel in Kakteenk. **1935**: 118 (1935).

Stapelia auobensis Nel in White & Sloane, Stapelieae ed. 2, **2**: 472 (1937). Type: Namibia, Auob River [cult. Windhoek], *Triebner* s.n. sub STE 2106 (not traced); White & Sloane, Stapelieae ed. 2, **2**: fig.411 (1937), illustration chosen as lectotype by Bruyns in Stapeliads S. Africa Madagascar **2**: 548 (2005).

Ceropegia marientalensis (Nel) Bruyns in S. African J. Bot. **112**: 423 (2017).

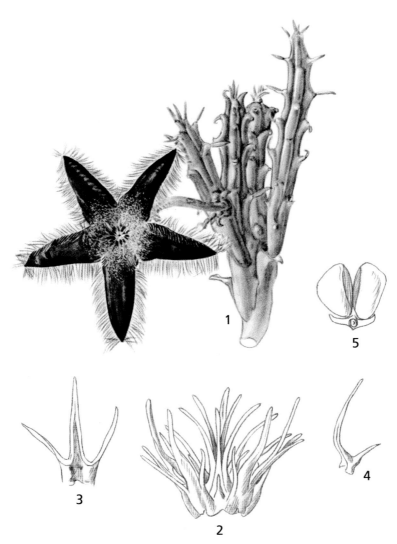

Fig. 7.3.**166**. TRIDENTEA MARIENTALENSIS subsp. MARIENTALENSIS. 1, flowering stem; 2, corona; 3, outer corona lobe; 4, inner corona lobe; 5, pollinarium. Drawn by Cynthia Letty. Reproduced with permission from Flowering Plants of Africa (1960).

Diffuse clump-forming stem-succulent. Stems decumbent, 5–15 cm long, 1–1.5 cm wide, 4-angled. Tubercles with leaf rudiments 5–12 mm long. Inflorescences few-flowered arising near base of stems, flowers opening in succession. Peduncles spreading, 0.5–10 cm long. Pedicels (5)7–10(13) cm long, spreading along the ground, with the flowers somewhat inclined. Sepals ovate lanceolate, 6–7 mm long. Corolla rotate to slightly campanulate, 5–7.5 cm in diam., cream to pale yellow often speckled brown or maroon, speckles becoming denser in lower half of lobes, the upper half of the lobes dark brown or maroon; corolla lobes spreading to reflexed, ovate, 1.8–2.5 cm long, acute, margins with white or purple vibratile cilia. Gynostegial corona pale yellow with brown spots; outer corona 6 mm long, ascending, trifid; inner staminal corona lobes 3–5 mm long, erect and connivent, then diverging above, with laterally flattened dorsal appendages 1–2.5 mm long arising near their base.

Subsp. **marientalensis**.

Corolla more or less flat, yellowish towards the centre.

Botswana. SW: 6 km E of Tsabong [cult. Mutare 19.ii.1970], *Liversedge* s.n. sub *Plowes* 3616A (PRE, SRGH).

Also known from central and southern Namibia, and the Northern Cape province of South Africa. On flat areas of reddish sand or calcrete, often in association with *Rhigozum trichotomum* or acacia scrub; 900 m.

53. **PIARANTHUS** R. Br.[25]

Piaranthus R. Br., On the Asclepiadeae: 12 (1810); in Mem. Wern. Nat. Hist. Soc. **1**: 23 (1811); in Fl. Cap. **4**(1): 1015–1024 (1909). —Meve in Bradleya **12**: 57–102 (1994). —Bruyns in Stapeliads S. Africa Madagascar **2**: 345–368 (2005).
  Huerniopsis N.E. Br. in J. Linn. Soc., Bot. **17**: 171 (1878); in F.T.A. **4**(1): 499–500 (1903); in Fl. Cap. **4**(1): 922–923 (1909).
    Ceropegia sect. *Piaranthus* (R. Br.) Bruyns in S. African J. Bot. **112**: 425 (2017).

Mat-forming stem succulents; latex clear or at most cloudy. Stems decumbent, cylindrical, obscurely 4–5 angled with rounded tubercles, glabrous. Inflorescences 1–3 per stem, mostly arising mostly towards the apex of the stem, 1–5(10)-flowered, the flowers opening in rapid succession on a gradually lengthening knobbly peduncle 0.2–1.5 cm long. Corolla rotate to campanulate, deeply lobed, papillate; corolla lobes spreading, margins rarely ciliate. Gynostegial corona in 2 series; "outer" lobes very weakly developed; "inner" staminal corona lobes dorsiventrally flattened and adnate to the back of the anthers below, with or without an apical appendage. Gynostegium sessile or subsessile. Pollinia ellipsoid. Follicles paired, fusiform, erect, glabrous.

Seven species in southern Africa.

Corolla 4–4.5 cm in diameter; inner corona lobes 8–10 mm long, base orange-yellow, narrowed abruptly for most of their length into a slender cream or white apical appendage . . . . **1.** *atrosanguineus*
Corolla 2–3 cm in diameter; inner corona lobes 3–6 mm long, reddish or brownish, narrowing gradually towards the apex and lacking a slender apical appendage . . . .**2.** *decipiens*

1. **Piaranthus atrosanguineus** (N.E. Br.) Bruyns in Syst. Bot. **24**: 396 (1999). Type: Botswana, northern Kalahari Desert, i.1899, *Lugard* 263, cult. Brown, ix.1901 (K holotype).
    Stapelia atrosanguinea N.E. Br. in Gard. Chron., ser. 3 **30**: 425 (1901).
    Caralluma atrosanguinea (N.E. Br.) N.E. Br. in F.T.A. **4**(1): 485 (1903).

---

[25] by D.J. Goyder

Huerniopsis atrosanguineus (N.E. Br.) A.C. White & B. Sloane, Stapelieae ed. 2, **3**: 972 (1937).

Huerniopsis gibbosa Nel in White & Sloane, Stapelieae ed. 2, **3**: 1174 (1937). Type: Botswana, Lobatsi, *Nel* s.n. sub SUG7358 (BOL holotype).

Huerniopsis papillata Nel in White & Sloane, Stapelieae ed. 2, **3**: 972 (1937). Type: Botswana, Debeeti, *Nel* s.n. sub SUG 7347 (not traced); Bruyns, Stapelieae ed. 2, **3**: fig.1041 (1937), iconotype lectotypified by Bruyns, Stapeliads S. Africa Madagascar **2**: 350 (2005).

Ceropegia atrosanguinea (N.E. Br.) Bruyns in S. African J. Bot. **112**: 425 (2017).

Fig. 7.3.**167**. PIARANTHUS DECIPIENS. Main image, flowering shoots; 1, calyx; 2, gynostegium surrounded by corona lobes; 3, corona lobe. Drawn by Cythia Letty. Reproduced with permission from Flowering Plants of South Africa (1932).

Mat-forming stem-succulent. Stems decumbent, 2–15 cm long, 0.8–2 cm wide, 4(5)-angled. Tubercles with leaf rudiments 3–6 mm long. Inflorescences 1–3-flowered arising in upper half of the stem, flowers opening in succession. Peduncles to 0.5 cm long. Pedicels 5–7 mm long, ascending, with the flowers upward facing. Sepals lanceolate, 6–9 mm long. Corolla rotate, 4–4.5 cm in diam., red or brown, papillate, thickened around the mouth of the corolla tube; corolla lobes spreading to slightly reflexed, ovate, 1.5–1.8 cm long, acute, margins lacking cilia. Inner staminal corona lobes 8–10 mm long, base orange-yellow, narrowed abruptly for most of their length into a slender cream or white apical appendage.

Botswana. SE: Mochudi Dist., fl. 14.iv.1963, *Schlieben* 9606 (K, PRE).

Also known from South Africa (North West and Limpopo Provinces).

In partial shade under *Acacia*, *Grewia* or *Euphorbia* species on shallow stony ground; c.1000 m.

2. **Piaranthus decipiens** (N.E. Br.) Bruyns in Syst. Bot. **24**: 396 (1999). Type: South Africa, *MacOwan* 2246 (K holotype). FIGURE 7.3.**167**.

   Huerniopsis decipiens N.E. Br. in J. Linn. Soc., Bot. **17**: 171 (1878); in F.T.A. **4**(1): 499 (1903); in Fl. Cap. **4**(1): 922 (1909).
   Piaranthus grivanus N.E. Br. in Hooker's Icon. Pl. **20**: t.1924 (1890). Type: South Africa, Griqualand West, Griva, *Barkly* 11 (K holotype).
   Caralluma grivana (N.E. Br.) Schltr. in J. Bot. **36**: 479 (1898).
   Ceropegia decipiens (N.E. Br.) Bruyns in S. African J. Bot. **112**: 425 (2017).

Mat-forming stem-succulent. Stems decumbent, 2–15 cm long, 0.8–2 cm wide, 4(5)-angled. Tubercles with leaf rudiments 3–6 mm long. Inflorescences 1–3-flowered arising in upper half of the stem, flowers opening in succession. Peduncles to 0.5 cm long. Pedicels 2–6 mm long, ascending, with the flowers upward facing. Sepals lanceolate, 6–8 mm long. Corolla rotate, 2–3 cm in diam., red or brown mottled with yellow, papillate, thickened around the mouth of the corolla tube; corolla lobes spreading or ascending, ovate, 7–12 mm long, acute, margins with clavate purplish cilia. Inner staminal corona lobes 3–6 mm long, reddish or brownish, narrowing gradually towards the apex and lacking a slender apical appendage.

Botswana. SE: 6 km NW of Molepolole [cult. Pretoria 6.iii.1955], *Codd* 8917 (K, PRE).

Bruyns suggests this species should also occur in northern Botswana. It is known from central Namibia and South Africa.

In shade under shrubs or small bushes on loam or calcrete, rarely on sand; c.1150 m.

54. **HUERNIA** R. Br.[26]

Huernia R. Br., On the Asclepiadeae: 11 (1810). —Brown in F.T.A. **4**(1): 495–499 (1903); in Fl. Cap. **4**(1): 902–922 (1909). —Leach in Excelsa Taxon. Ser. **4**: 1–197 (1988). —Müller & Albers in Albers & Meve (eds.), Ill. Handb. Succ. Pl. Asclepiadaceae: 159–174 (2002). —Bruyns, Stapeliads S. Africa Madagascar **1**: 130–211 (2005). —Meve in F.T.E.A., Apocynaceae (part 2): 336–343 (2012).
   Ceropegia sect. *Huernia* (R. Br.) Bruyns in S. African J. Bot. **112**: 423 (2017).

Clump-forming or creeping stem succulents; latex clear or at most cloudy. Stems roundly or sharply 4–6-angled; leaf rudiments sessile, 1–5 mm long, horizontally spreading to ascending, caducous, sometimes spinescent. Inflorescences, 1–6-flowered, mostly in lower half of stem, pedunculate. Corolla 1–4 cm long, often bulged or tipped in the sinuses, rotate, campanulate, cyathiform, elongated-conical, urceolate or globose, considerably fused, tube often with central annulus (corolline corona), corolla lobes triangular, often warty or rugose, glabrous, often sculptured with massive conical to cylindrical emergences, often tipped by a papilla. Gynostegial

---

[26] by D.J. Goyder

corona in 2 series: "outer" corona with 5 free staminal and 5 interstaminal parts fused to form a basal ring or disc adpressed to corolla tube, free interstaminal corona lobes, if differentiated, lingulate or rectangular; "inner" corona of staminal lobes atop gynostegium, dolabriform, triangular or clavate, often with a humped back, erect, inflexed or reflexed. Gynostegium sessile. Pollinia oblong-ellipsoid; corpusculum with triangular to oblong basal projections. Follicles paired, fusiform, stout.

Around sixty-seven spp. in Africa and Arabia, with a centre of distribution in southern Africa.

1. Inner corona lobes much exceeding the anthers 2
– Inner corona lobes not or only shortly exceeding the anthers 9
2. Inner corona lobes tapering to a fine point . . . . **5.** *erectiloba*
– Inner corona lobes not tapering to a fine point 3
3. Inner corona lobes distinctly swollen towards apex 4
– Inner corona lobes not or only slightly swollen at apex 7
4. Corolla tube longer than broad 5
– Corolla tube broader than long 6
5. Corolla bicampanulate; corolla tube inside dark maroon; papillae dark maroon
   . . . . **8.** *kirkii*
– Corolla tubular-campanulate; corolla tube inside concentrically dark-lined on pale ground; papillae spotted and banded . . . . **9.** *longituba*
6. Swollen apex of inner corona lobes obconic and truncate, smooth; pedicels at least 15 mm long . . . . **6.** *hystrix*
– Swollen apex of inner corona lobes knob-like and conspicuously papillate; pedicels less than 12 mm long. . . . . **12.** *volkartii*
7. Inner corona lobes ± cylindrical and obtuse . . . . **11.** *levyi*
– Inner corona lobes narrowing towards the apex. 8
8. Corolla campanulate with relatively long tube merging into the limb; stems decumbent to erect . . . . **7.** *hislopii*
– Corolla bicampanulate with a relatively short tube and an abruptly spreading broad limb; stems decumbent to ± prostrate. . . . . **10.** *occulta*
9. Corolla with a prominent raised and thickened annulus around mouth of tube
   10
– Corolla without a prominent annulus 11
10. Stems prostrate; corolla lobes narrowly triangular-lanceolate, at least twice as long as broad . . . . **2.** *procumbens*
– Stems decumbent to erect; corolla lobes broadly triangular, about as broad as long. . . . . **4.** *zebrina*
11. Corolla lobes broadly triangular, as broad as long; tube and lobes with concentric maroon markings on a cream ground . . . . **3.** *leachii*
– Corolla lobes narrowly triangular, at least twice as long as broad; lobes uniformly cream or yellow, tube with a maroon band surrounding a white base. . . **1.** *verekeri*

1. **Huernia verekeri** Stent in Bull. Misc. Inform. Kew **1933**: 145 (1933). Type: Zimbabwe, Sabi valley, *Vereker* s.n. sub GH 5427 (SRGH holotype).

   Huernia verekeri var. *stevensonii* A.C. White & B. Sloane, Stapelieae ed. 2, **3**: 1145 (1937). Type: Zimbabwe, Nyamandhlovu distr., near Sawmills [specimen not preserved according to Leach (1988: 146)].

   Ceropegia verekeri (Stent) Bruyns in S. African J. Bot. **112**: 425 (2017).

   Clump-forming stem succulent. Stems decumbent to prostrate, 3–10 cm long, 0.6–1.2 cm thick, 5–7-angled. Tubercles conical above, tapering to a soft tooth. Inflorescence of 1–5 flowers developing successively; peduncle to 10 mm long; pedicels 10–16 mm long, spreading, holding the flower laterally. Sepals 5–9 mm long, slender, acuminate, mostly reaching well beyond corolla lobe sinus. Corolla rotate, 3.5–4.5 cm diam.; corolla tube white at base, maroon above, 2–3 mm deep and 8–10 mm across; lobes cream or yellow, 12–16 mm long and 5 mm wide at base, narrowly triangular, attenuate. Outer corona lobes fused to form a pentagonal or circular disc; inner corona lobes cream to maroon, adpressed to back of anthers with enlarged obtuse dorsal gibbosity.

Stems mostly 6-angled, decumbent; tubercles 8–15 mm long . . . . a) subsp. *verekeri*
Stems 5-angled, prostrate; tubercles 3–6 mm long . . . . .b) subsp. *pauciflora*

a) Subsp. **verekeri**.

Botswana. N: between Nokareng and Aha Mts, ii.1966, *Wild & Drummond* 6921 (K, PRE, SRGH). **Zambia**. S: near Feira, *Fanshawe* 9424 (SRGH). **Zimbabwe**. N: Mt Darwin, *Bingham* s.n. sub SRGH 3176 (K, LISC, M, SRGH). W: Plumtree, *Davies* s.n. in SRGH 23215 (SRGH). E: Sabi Valley, 40 km N of Rupisi Host Springs [cult. Harare 28.v.1960], *Leach* 9972 (K, SRGH). S: Bikita, fl. 17.xii.1953, *Wild* 4422 (K, PRE, SRGH). **Malawi**. S: 6 km downstream from Mpatamanga gorge, *Campbell-Barker* s.n. sub *Leach* 15134 (K). **Mozambique**. T: Mesuva, *Chase* 2820A (SRGH).
Also known from NE Namibia. Stony places in shade of trees or shrubs; 150–1200 m.

b) Subsp. **pauciflora** (L.C. Leach) Bruyns, Stapeliads S. Africa Madagascar **1**: 138 (2005). Type: Mozambique, near the mouth of the Save River 16 km S of Mambone [cult. Nelspruit iii.1967], *Leach & Bayliss* 11889 (SRGH holotype, K, LISC, PRE).
    Huernia verekeri Stent var. *pauciflora* L.C. Leach in Bothalia **10**: 49 (1969).
    Ceropegia verekeri subsp. *pauciflora* (L.C. Leach) Bruyns in S. African J. Bot. **112**: 425 (2017).

Mozambique. GI: 16 km S of Mambone [cult. Nelspruit iii.1967], *Leach & Bayliss* 11889 (K, LISC, PRE, SRGH).
Grows in shade of *Androstachys johnsonii* thickets; 2–5 m above sea level.

2. **Huernia procumbens** (R.A. Dyer) L.C. Leach in Bothalia **10**: 54 (1969). Type: South Africa, Limpopo Prov., Pafuri, Kruger National Park, *van der Schijff* 3618 (PRE holotype, K).
    Duvalia procumbens R.A. Dyer in Fl. Pl. Africa **31**: t.1218 (1956).
    Ceropegia procumbentior Bruyns in S. African J. Bot. **112**: 424 (2017).

Mat-forming stem succulent. Stems prostrate, 6–50 cm long, 0.7–1.2 cm thick, 5-angled. Tubercles obtuse, barely rising from the stem above. Inflorescence of 1–5 flowers developing successively; peduncle to 15 mm long; pedicels 10–15 mm long, spreading, holding the flower facing upwards. Sepals 8–11 mm long, slender, attenuate. Corolla rotate, 3–5.5 cm diam.; corolla tube pink to maroon, 1 mm deep and forming a raised annulus above; lobes cream with red margins, 13–24 mm long and 5 mm wide at base, narrowly triangular, attenuate. Outer corona lobes broadly obtuse; inner corona lobes pink to pale maroon, adpressed to back of anthers with an erect or ascending obtuse dorsal gibbosity.

Zimbabwe. S: Pesu river gorge, 20 km W of Pafuri [cult. Nelspruit 10.vii.1964], *Leach et al.* 12286A (K, LISC, PRE, SRGH, ZSS).
Also occurs in neighbouring parts of South Africa. Rocky cliffs or outcrops, in shade of *Androstachys johnsonii*; 200–300 m.

3. **Huernia leachii** Lavranos in J. S. African Bot. **25**: 311 (1959). Type: Mozambique, 16 km S of Chimoio [Vila Pery] on Mutare – Beira road [cult. Lavranos], *Leach* 5641 (PRE holotype).
    Ceropegia leachiana Bruyns in S. African J. Bot. **112**: 424 (2017).

Mat-forming stem succulent. Stems procumbent, 3–30(150) cm long, 0.3–0.8 cm thick, 4(5)-angled. Tubercles obscure. Inflorescence of 1–6 flowers developing successively; peduncle to 10 mm long; pedicels 5–40 mm long, flower held facing slightly downwards. Sepals 5–7 mm long, slender, attenuate. Corolla broadly campanulate, 2–2.5 cm diam.; corolla tube cream with concentric maroon stripes and with cylindrical obtuse papillae to 1 mm long in mouth of tube, tube 6 mm deep and 12 mm across; lobes cream with maroon stripes, 6 mm long and 6 mm wide at base, broadly triangular, acute. Outer corona lobes dark maroon, rounded or emarginate and spreading onto base of corolla tube; inner corona lobes yellow with maroon margins, adpressed to back of anthers, apical portions inflexed to meet above gynostegium, linear from a slight dorsal swelling around top of anthers, apices bristly, obtuse.

Mozambique. MS: 16 km S of Chimoio [Vila Pery], 23.iii.1960, *Leach & Wild* 9819 (K, LISC, PRE, SRGH). **Malawi.** S: Mangoche Distr., rock ledge near Lake Malawi [cult. Blantyre], *Campbell-Barker* s.n. (NBG photograph and sketch only, Leach (1988: 122)).

Granite domes and weathered granite, growing with *Coleochloa setifera*, *Xerophyta* and other succulents; 400–800 m.

4. **Huernia zebrina** N.E. Br. in Fl. Cap. **4**(1): 921 (1909). Type: South Africa, KwaZulu-Natal, Eshowe, *Saunders* s.n. (K holotype). FIGURE 7.3.**168**.

    Huernia blackbeardiae H. Jacobsen, Sukkulenten: 75 (1933). Type: Botswana, Serowe, *Blackbeard* s.n. (GRA holotype, PRE).

    Huernia zebrina N.E. Br. var. *magniflora* E. Phillips in Fl. Pl. South Africa **16**: t.613 (1936). Type: South Africa, Limpopo Prov., Potgeitersrust, *Ralston* s.n. sub PRE 20568 (PRE holotype).

    Huernia zebrina N.E. Br. subsp. *magniflora* (E. Phillips) L.C. Leach in Excelsa Taxon. Ser. **4**: 139 (1988).

    Ceropegia zebrina (N.E. Br.) Bruyns in S. African J. Bot. **112**: 425 (2017).

Clump-forming stem succulent. Stems erect to decumbent, 1.5–12 cm long, 0.8–2 cm thick, 4–5(6)-angled. Tubercles deltoid, laterally flattened, narrowing abruptly into a slender tooth. Inflorescence of 1–3 flowers developing successively; peduncle to 5 mm long; pedicels 15–20 mm long, spreading, holding the flower laterally. Sepals 8–10 mm long, ovate, acuminate. Corolla rotate, (2.5)3.5–4.5(5) cm diam.; corolla tube maroon flecked with cream, 6–7 mm deep and 6–8 mm wide enclosed by a broad shiny swollen raised annulus; lobes cream with transverse maroon lines, 10–15 mm long and 15–20 mm wide at base, broadly triangular, attenuate. Outer corona lobes cream with red or maroon margins, deeply to shallowly bilobed and spreading onto base of corolla tube; inner corona lobes red or maroon with yellow dorsal gibbosity, adpressed to back of anthers and inflexed, apices narrowly obtuse, smooth.

Subsp. **zebrina**.

Botswana. SE: Serowe [cult. Nelspruit], *Leach & Bayliss* 12517 (BM, K, MO, NBG, PRE, SRGH, Z). **Zimbabwe.** S: 16 km N of Beitbridge [cult. SRGH], *Mavi* 272 (K, NBG, PRE, SRGH). **Mozambique.** M: 16 km N of Moamba [cult. Nelspruit], *Leach & Bayliss* 11738 (LISC, MO, NBG, SRGH).

Also known from Namibia, South Africa (Limpopo and KwaZulu-Natal), and Swaziland. Stony areas often with calcrete, in shade of small bushes or trees; 50–1200 m.

5. **Huernia erectiloba** L.C. Leach & Lavranos in Kirkia **3**: 38 (1963). Type: Mozambique, Nampula Prov., 45 km E of Ribáuè, 17.v.1961, *Leach & Rutherford-Smith* 10914 (SRGH holotype, PRE).

    Ceropegia erectiloba (L.C. Leach & Lavranos) Bruyns in S. African J. Bot. **112**: 424 (2017).

Clump-forming stem succulent. Stems decumbent, 5–10(20) cm long, 0.4–1.2 cm thick, 4–5-angled. Tubercles deltoid, slightly flattened laterally, tapering abruptly into a slender tooth. Inflorescence of 1–3 flowers developing successively; peduncle very short; pedicels 10–

20 mm long, spreading, holding the flower facing upwards. Sepals 5–7 mm long, attenuate. Corolla bicampanulate, 2.2–3 cm diam.; corolla tube maroon, 10 mm deep and 7–10 mm across then opening abruptly to form a maroon-spotted cream rotate limb, throat of corolla tube papillate with inward-pointing bristles to 2 mm long; lobes cream with maroon spots, erect or somewhat inflexed, 6–8 mm long and 11 mm wide at base, broadly triangular, acuminate. Outer corona lobes cream or yellow with maroon markings, quadrate to shortly bilobed and spreading onto the base of the corolla tube; inner corona lobes cream mottled with red, adpressed to back of anthers below, then connivent-ascending above, reaching about

Fig. 7.3.**168**. HUERNIA ZEBRINA subsp. ZEBRINA. Main image, flowering shoots; 1, flower bud; 2, calyx; 3, half-flower minus the gynostegium; 4, gynostegium and corona from above; 5, stamen; 6, cross-section of stem. Drawn by E. Niemeyer. Reproduced with permission from Flowering Plants of South Africa (1932).

twice the height of the gynostegium.

Mozambique. N: Nampula Prov., 60 km E of Nampula, 21.v.1962, *Leach & Rutherford-Smith* 10955 (K, LISC, SRGH). Z: Monte Ile [cult. Sintra 3.iii.1966], *Torre & Correia* 14994 (LISC).

On granite domes and outcrops in association with *Xerophyta* and other succulents; 300–700 m.

6. **Huernia hystrix** (Hook.f.) N.E. Br. in Gard. Chron., ser. nov. **5**: 795 (1876). Type: South Africa [cult. Kew], *McKen* s.n. (specimen not preserved). Lectotype: Curtis's Botanical Magazine **95**: t.5751, designated by Bruyns in Stapeliads S. Africa Madagascar **1**:187 (2005). FIGURE 7.3.**169**.

  Stapelia hystrix Hook.f. in Bot. Mag. **95**: t.5751 (1869).
  Ceropegia hystrix (Hook.f.) Bruyns in S. African J. Bot. **112**: 424 (2017).

Clump-forming stem succulent. Stems decumbent-ascending, 2–7(10) cm long, 0.8–1 cm thick, 5-angled. Tubercles flattened laterally, deltoid, frequently terminating in a yellow-tipped tooth 2–3 mm long. Inflorescence of 1–3 flowers developing successively; peduncle to 5 mm long; pedicels 15–60 mm long, spreading, holding the flower on the ground facing upwards. Sepals 7–12 mm long. Corolla with campanulate tube and spreading or recurved lobes, 3–5 cm diam.; corolla tube cream with concentric maroon lines, 5–6 mm deep and 10–12 mm across; lobes cream with coarser and less continuous maroon lines, 12–15 mm long and 12–15 mm wide at base, broadly triangular, attenuate, lobes and mouth of tube covered with obtuse maroon-blotched papillae 3–5.5 mm long. Outer corona lobes cream, short and rectangular, often fused to form disc; inner corona lobes white to greenish cream below, cream speckled with maroon above, dorsiventrally flattened from a slightly gibbous base, apex expanded abruptly into an inverted foot-like appendage.

Subsp. **hystrix**.

Zimbabwe. S: Chefu river, Gona-re-Zhou [cult. Greendale sub *Leach* 14233], *Blake* 119 (K, SRGH). **Mozambique.** M: 16 km N of Moamba [cult. Nelspruit i.1965], *Leach & Bayliss* 12254 (BOL, K, LISC, MO, NBG, PRE, SRGH).

Also known from South Africa (Limpopo, Gauteng, Mpumalanga, KwaZulu-Natal) and Swaziland. In leaf litter under trees or shrubs, on sandy or alluvial soils; 50–500 m.

7. **Huernia hislopii** Turrill in Bull. Misc. Inform. Kew **1922**: 30 (1922). Type: Zimbabwe [cult. Kew 3.ix.1920], *Hislop* s.n. (K holotype).

  Ceropegia hislopii (Turrill) Bruyns in S. African J. Bot. **112**: 424 (2017).

Clump-forming stem succulent. Stems decumbent, 3–15 cm long, 0.8–1.2 cm thick, 5-angled. Tubercles deltoid, laterally flattened, tapering abruptly into a short tooth. Inflorescence of 1–5 flowers developing successively; peduncle to 5 mm long; pedicels 6–25 mm long, spreading, holding the flower facing upwards. Sepals 6–10 mm long, narrowly ovate-acuminate. Corolla with campanulate tube and rotate limb, 4–6 cm diam., cream or white irregularly spotted with maroon and papillate except for basal portion of tube; corolla tube 10–15 mm deep and 8–10 mm across at the mouth, basal chamber subglobose marked with concentric maroon lines and lacking papillae, narrowing abruptly into the papillate upper portion of tube; lobes 12–22 mm long and 15–18 mm wide at base, broadly triangular, attenuate, corolla lobe sinuses with small teeth at the mid-point. Outer corona lobes dark truncate or emarginate, partially fused and spreading onto the corolla tube, dark maroon; inner corona lobes cream spotted with maroon, the tips somewhat darker, adpressed to back of anthers below then connivent-divergent above tapering to a finely tuberculate rounded apex reaching around twice the height of the gynostegium.

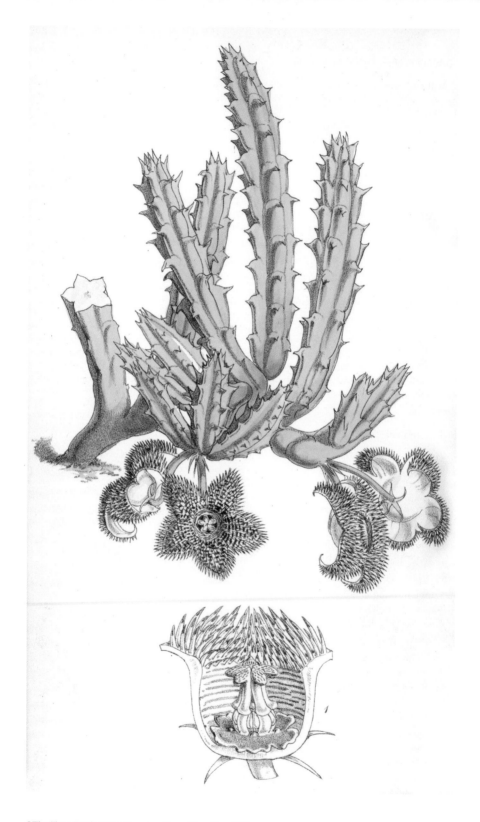

1. Upper portion of corolla tube ± equal in length to the swollen basal portion; corolla lobes mostly longer than broad; inner corona lobes without apical swelling . . . . a) subsp. *hislopii*
- Upper portion of corolla tube longer than the swollen basal portion; corolla lobes at most as long as broad; inner corona lobes swollen apically 2
2. Corolla papillae with apical bristle much shorter than papilla. . . c) subsp. *robusta*
- Corolla papillae with apical bristle longer than above and up to as long as the papilla. . . . . b) subsp. *cashelensis*

a) Subsp. **hislopii**.

Zimbabwe. N: Paradise Pools, Bindura, *West* s.n. sub *Plowes* 2567 (SRGH). W: Old Gwanda Road, *Plowes & Bullock* 2489 (BOL, COI, K, SRGH). C: Domboshawa, *Leach* 5734 (SRGH). E: Vumba, Norseland, iii.1949, *Wild* 2805 (K, SRGH). S: 6 km S of Lundi River on Beitbridge road [cult. Nelspruit i.1965], *Leach* 11642 (K, SRGH). **Mozambique**. MS: Manica, *Garcia* s.n. in *Mendonça* 563 (LISC).
Found in shallow soils on granitic domes; 600–1400 m.

b) Subsp. **cashelensis** (L.C. Leach & Plowes) Bruyns, Stapeliads S. Africa Madagascar **1**: 193 (2005). Type: Zimbabwe, Chimanimani [Melsetter] Distr., 10 km W of Cashel [cult. Greendale], *Leach* 5404 (PRE holotype, K, SRGH).

> *Huernia longituba* N.E. Br. subsp. *cashelensis* L.C. Leach & Plowes in J. S. African Bot. **32**: 49 (1966).
> *Ceropegia hislopii* subsp. *cashelensis* (L.C. Leach & Plowes) Bruyns in S. African J. Bot. **112**: 424 (2017).

Zimbabwe. E: Chitora Farm c.15 km E of Banti Forest [cult. SRGH], *Mavi* 605 (K, PRE, SRGH).
Open *Brachystegia* woodland on dry western slopes of the Chimanimani Mts; 1100–1600 m.

c) Subsp. **robusta** L.C. Leach & Plowes in J. S. African Bot. **32**: 53 (1966). Type: Zimbabwe, Lupani Dist., Mabikwa [cult. Nelspruit], *Leach* 11628 (SRGH holotype, BM, G, K, LISC).

> *Ceropegia hislopii* subsp. *robusta* (L.C. Leach & Plowes) Bruyns in S. African J. Bot. **112**: 424 (2017).

Zimbabwe. N: Gokwe Distr., Charama road, *Bingham* 874 (NBG, SRGH). W: Lupani Dist., 20 km SE of Halfway Hotel on Victoria Falls road [cult. Greendale], *Leach* 9835 (K, SRGH). C: Quetanga, 24 km E of Kwekwe [Que Que], [cult. Mutare i.1972 sub *Plowes* 3294], *Crow* s.n. (K, SRGH).
Sand or gravel flats under mopane or acacia; 900–1200 m.

8. **Huernia kirkii** N.E. Br. in Fl. Cap. 4(1): 920 (1909). Type: South Africa, Mpumalanga, Komatipoort, 28.vii.1902, *Kirk* 76 (K holotype).

> *Huernia bicampanulata* I. Verd. in Fl. Pl. South Africa **12**: t.449 (1932). Type: South Africa, Limpopo Province, Naauwpoort Farm c.50 km S of Polokwane [Pietersburg], *Van Son* s.n. in PRE 10136 (PRE holotype).
> *Ceropegia kirkii* (N.E. Br.) Bruyns in S. African J. Bot. **112**: 424 (2017).

Fig. 7.3.**169**. HUERNIA HYSTRIX subsp. HYSTRIX. Main image, flowering stems; 1, half-flower exposing gynostegium and corona. Drawn by Walter Fitch. Reproduced from Curtis's Botanical Magazine (1869).

Clump-forming stem succulent. Stems decumbent to prostrate, 3–15 cm long, 0.8–1.2 cm thick, 5-angled. Tubercles laterally flattened, deltoid, tapering abruptly into a small acute tooth. Inflorescence of 1–5 flowers developing successively; peduncle to 5 mm long; pedicels 6–15 mm long, spreading, holding the flower facing upwards. Sepals 6–16 mm long, narrowly ovate-acuminate. Corolla 3–5 cm diam., bicampanulate with a subglobose-campanulate tube opening abruptly into a saucer-shaped limb, papillate except in tube; corolla tube dark maroon, to 20 mm deep; limb and lobes cream spotted with maroon; lobes 8–10 mm long and 11–15 mm wide at base, erect or spreading slightly, broadly triangular, acuminate. Outer corona lobes dark maroon, truncate-emarginate and spreading onto the base of the corolla; inner corona lobes cream spotted with maroon, adpressed to back of anthers below then connivent-divergent above, the apex distinctly swollen, bristly.

Zimbabwe. S: Clarendon Cliffs, Lower Lundi valley [cult. Mutare sub *Plowes* 2623], *Liversedge* s.n. (BOL, K, PRE, SRGH). **Mozambique**. GI: Mabalane, Limpopo River [cult. Nelspruit i.1965], *Leach & Bayliss* 11775 (BM, COI, K, PRE, SRGH).

Also recorded from South Africa (Limpopo, Mpumalanga & KwaZulu-Natal). In sandy or stony soil under trees or bushes; 50–500 m.

9. **Huernia longituba** N.E. Br. in Fl. Cap. **4**(1): 912 (1909). Type: South Africa, Northern Cape Province, near Douglas, Griqualand West, iv.1906, *E. Pillans* sub *N.S. Pillans* 609 (BOL lectotype, GRA).

    Ceropegia longituba (N.E. Br.) Bruyns in S. African J. Bot. **112**: 424 (2017).

Clump-forming stem succulent. Stems decumbent, 3–12 cm long, 1–2 cm thick, 4–5-angled. Tubercles deltoid, laterally flattened, tapering abruptly into a small acute tooth. Inflorescence of 1–3 flowers developing successively; peduncle very short; pedicels 8–10(18) mm long, spreading, holding the flower laterally, resting on the ground. Sepals 8–10 mm long, slender. Corolla tubular-campanulate, 2–4 cm diam., papillate; corolla tube up to 20 mm deep, cream with concentric maroon markings in basal chamber, cream with fine red or maroon spots above and on lobes; lobes erect or spreading, 8–10 mm long and 8–10 mm wide at base, broadly triangular, attenuate. Outer corona lobes dark maroon, truncate-emarginate and spreading onto the base of the corolla; inner corona lobes cream spotted with maroon, adpressed to back of anthers below then connivent-divergent above, the apex distinctly swollen, bristly.

Botswana. SW: Near Tsane, *Paterson* s.n. in Bulawayo Museum 427 (SRGH). SE: 16 km S of Lobatsi, *Leach et al.* 12451 (BM, BOL, G, K, LISC, PRE, SRGH).

Also occurs in South Africa (Northern Cape and Free State). Sandy flat areas under trees and bushes; 1100–1200 m.

10. **Huernia occulta** L.C. Leach & Plowes in J. S. African Bot. **32**: 57 (1966). Type: Zimbabwe, near Masvingo [Zimbabwe], *Leach & Plowes* 11661 (SRGH holotype, K, PRE).

    Ceropegia occultiflora Bruyns in S. African J. Bot. **112**: 424 (2017).

Clump-forming stem succulent. Stems decumbent to nearly prostrate, 3–15 cm long, 0.3–0.8 cm thick, (4)5-angled. Tubercles deltoid, weakly flattened laterally, tapering abruptly into a small spreading tooth. Inflorescence of 1(3) flowers developing successively; peduncle to 5 mm long; pedicels 8–20(25) mm long, spreading, holding the flower facing upwards. Sepals 6–9 mm long. Corolla bicampanulate with basal tube opening abruptly to a saucer-shaped limb, 4–6 cm diam., papillate; corolla tube 7–10 mm deep and 7–10 mm wide at throat, dark maroon; lobes erect or spreading, 10–12 mm long and 14–18 mm wide at base, broadly triangular, acuminate, middle of sinus marked by a short tooth. Outer corona lobes dark maroon, truncate-emarginate and spreading onto the base of the corolla; inner corona lobes cream spotted with maroon, adpressed to back of anthers below then connivent-divergent above, tapering to the finely tuberculate apex.

Zimbabwe. S: Matibi Mission SW of Fort Victoria [cult. Mutare sub *Plowes* 3109], *Bey* 105 (K, SRGH).

Known only from southern Zimbabwe. Growing in grassy tufts on granite slabs; c.1000 m.

11. **Huernia levyi** Oberm. in Fl. Pl. South Africa **16**: t.616 (1936). Type: Zimbabwe, Hwange [Wankie], iv.1932, *Levy* s.n. in Herb. Transvaal Mus. 31142 (PRE).

   Ceropegia levyi (Oberm.) Bruyns in S. African J. Bot. **112**: 424 (2017).

Clump-forming stem succulent. Stems decumbent, 4–10 cm long, 0.8–1.5 cm thick, 4–5-angled. Tubercles deltoid, laterally flattened, tapering abruptly into a short tooth. Inflorescence of 1–3 flowers developing successively; peduncle to 5 mm long; pedicels 7–11 mm long, spreading, holding the flower laterally, resting on the ground or somewhat ascending from it. Sepals 5–6 mm long. Corolla tubular campanulate, 2–2.5 cm diam., papillate; corolla tube 22–35 mm deep, dark maroon in lower half, cream with maroon spots above and on lobes; lobes spreading, 5–8 mm long and 12–15 mm wide at base, broadly triangular, attenuate, sinus marked with short tooth. Outer corona lobes dark maroon, forming a disc and spreading onto the base of the corolla; inner corona lobes reddish brown, becoming darker towards tips, adpressed to back of anthers below then connivent-erect and subcylindrical above, the apex somewhat clavate, finely tuberculate.

Caprivi. Mpilila Island, *Killick & Leistner* 3403 (PRE). **Botswana**. N: Pandamatenga, *P.A. Smith* 4308 (PSUB, SRGH) & *P.A. Smith* 4308, cult. KGW sub *Leach* 16868 (K, NBG, PRE). **Zambia**. C: 60 km from Kabwe [Broken Hill], *Tapscott* s.n. in BOL 31304 (BOL). E: Great East Road c.215 km from Lusaka [cult. Harare sub *Leach* 14801], *Anton-Smith* s.n. (BOL, BR, K, MO, PRE, SRGH). S: Livingstone Dist., Songwe River gorge [cult. Harare sub SRGH 3394], *Mitchell* 3013 (G, K, NBG, PRE, SRGH, Z). **Zimbabwe**. N: E of Mhangura [Mangula], 15.xii.1975, *Percy-Lancaster* 141 (K, SRGH). W: 8 km SE of Inyantui River [cult. Greendale, v.1960], *Leach* 9974 (BOL, K, SRGH).

Most collections are from the Zambezi valley. Generally found in hot stony places amongst small bushes or under mopane; 500–1000 m.

12. **Huernia volkartii** Werderm. & Peitscher in Gartenflora **85**: 78 (1936). Type: Angola, Cuanza Sul, Uku [Vila Nova do Seles], *Gossweiler* s.n., cult. Jena (specimen not preserved). Type: White & Sloane, Stapelieae ed. 2, **3**: fig.958 (1937), neotype designated by Bruyns (2005: 201).

   Huernia montana Kers in Bot. Not. **122**: 179 (1969). Type: Angola, Huíla Province, Chela Mountains 25 km NW of Lubango [Sa da Bandeira] on road to Tundavala, 30.iv.1968, *Kers* 3460 (S holotype, K).

   Ceropegia volkartii (Werderm. & Peitscher) Bruyns in S. African J. Bot. **112**: 425 (2017).

Mat- or clump-forming stem succulent. Stems prostrate, decumbent or erect, 2.5–30(50) cm long, 0.5–0.8 cm thick, (4)5-angled. Tubercles deltoid, slightly flattened laterally, tapering abruptly into a small tooth. Inflorescence of 1–3 flowers developing successively; peduncle very short; pedicels 6–12 mm long, spreading, holding the flower slightly upwards. Sepals 5–8 mm long. Corolla campanulate, 2–3 cm diam., conspicuously papillate; corolla tube 6–9 mm deep, 10–15 mm wide at mouth, cream with concentric maroon markings near base, cream with maroon spots or blotches above and on lobes; lobes spreading to recurved, 4–9 mm long and 7–10 mm wide at base, broadly triangular, attenuate, sinus mid-point marked by well-developed tooth. Outer corona lobes dark maroon, truncate-emarginate or bifid, spreading onto the base of the corolla; inner corona lobes cream spotted with maroon, adpressed to back of anthers below then connivent-divergent above, the apex distinctly swollen and sometimes spreading horizontally, finely tuberculate.

Stems decumbent to erect . . . . a) var. *volkartii*
Stems prostrate . . . . b) var. *repens*

a) Var. **volkartii**.

Zimbabwe. E: Chimanimani Mts, Bundi valley [cult. Mutare 1968], *Plowes* 2598A (K, SRGH). S: Belingwe Dist., 3 km S of Mnene Mission [cult. Nelspruit ii.1966], *Leach & Bullock* 12864 (K, SRGH).

Also recorded from western Angola and northern Namibia, although Bruyns (2005: 202) notes some vegetative differences compared with material from the Flora region. On granite domes at 1000–1600 m.

b) Var. **repens** (Lavranos) Lavranos in J. S. African Bot. **38**: 43 (1972). Type: Mozambique, Jaegerslust Farm, Garuso Dist. [cult. Bryanston 29.xii.1959], *Schweickerdt* 3469 (PRE holotype, SRGH).

   Huernia repens Lavranos in J. S. African Bot. 27: 11 (1961).
   Ceropegia volkartii var. repens (Lavranos) Bruyns in S. African J. Bot. **112**: 425 (2017).

Zimbabwe. E: 27 km E of Chipinga, *Percy-Lancaster* 112 (SRGH). S: Mutirikwi [Kyle] Dam near Masvingo, *Plowes* 2483 (SRGH). **Mozambique**. MS: Espungabera Mt, *Percy-Lancaster* 74 (SRGH). 900–1100 m.

55. ORBEA Haw.[27]

Orbea Haw., Syn. Pl. Succ.: 37 (1812), conserved name. —Leach in Excelsa Taxon. Ser. **1**: 1–75 (1978). —Bruyns in Aloe **37**: 72–76 (2001); in Syst. Bot. Monogr. **63**: 1–196 (2002); in Stapeliads S. Africa Madagascar **1**: 240–329 (2005). —Meve in F.T.E.A., Apocynaceae (part 2): 343–356 (2012).
   Podanthes Haw. in Syn. Pl. Succ: 32 (1812), rejected name.
   Diplocyatha N.E. Br. in J. Linn. Soc., Bot. **17**: 167 (1878).
   Stapeliopsis E. Phillips in Fl. Pl. South Africa **12**: t.445 (1932), illegitimate name.
   Stultitia E. Phillips in Fl. Pl. S. Afr. **13**: t.520 (1933).
   Orbeopsis L.C. Leach in Excelsa Taxon. Ser. **1**: 61 (1978).
   Orbeanthus L.C. Leach in Excelsa Taxon. Ser. **1**: 71 (1978).
   Pachycymbium L.C. Leach in Excelsa Taxon. Ser. **1**: 69 (1978).
   Angolluma R. Munster in Cact. Succ. J. New S. Wales **17**: 63 (1990).
   Ballyanthus Bruyns in Aloe **23**: 76 (2000).
   Ceropegia sect. *Orbea* (Haw.) Bruyns in S. African J. Bot. **112**: 426 (2017).

Decumbent to prostrate, often clump-forming stem succulents; latex clear or at most cloudy. Stems cylindrical, conical or club-shaped, 1–25 cm long, 10–30 mm wide, roundly 4-angled to irregularly 4–5-angled, glabrous, occasionally rhizomatous. Tubercles in 4 or 4–5 rows, borderline to leaves not always visible. Leaf rudiments sessile, triangular-deltate to conical-subulate, acute, more or less caducous; stipules, if present, pointed, ovoid or globoid, glandular. Inflorescences lateral, with 1–40 flowers developing in succession on short peduncles. Flowers mostly with scent of faeces. Corolla rotate to campanulate, fused to half of length, occasionally with a central annulus (corolline corona), adaxially usually warty or rugose, glabrous, papillate or with hairs; corolla lobes lanceolate, ovate or triangular, spreading to reflexed. Gynostegial corona in 2 series: "outer" corona of 5 staminal and 5 interstaminal parts fused just basally to nearly completely to form a rotate or cyathiform structure; free interstaminal corona lobes deltoid-triangular, saccate, lingulate or bilobed; "inner" staminal corona lobes subulate, lingulate, ovoid, triangular or clavate, occasionally with hump. Gynostegium sub-sessile or atop a column. Guide rails normally vertical, basally widened. Mericarps two, obclavate-fusiform.

---

[27] by D.J. Goyder

Nearly 60 species in Africa and Arabia.

In addition to the species native to the region, *Orbea variegata* (L.) Haw. is commonly encountered in cultivation.

1. Inflorescences with 1–3 flowers opening in gradual succession 2
- Inflorescences with at least 3 flowers opening simultaneously 11
2. Corolla with a conspicuous ring or dome-like annulus surrounding or supporting the gynostegium; inflorescences 1 per stem, near base 3
- Corolla lacking a conspicuous annulus around the gynostegium; inflorescences several along the stem, especially towards apex. 8
3. Annulus surrounding the gynostegium . . . . **13.** *paradoxa*
- Annulus supporting the gynostegium 4
4. Outer corona disc-like around the gynostegium . . . . **12.** *maculata*
- Outer corona of 5 discrete lobes 5
5. Inner corona lobes dorsiventrally flattened and inflexed over head of gynostegium . . . . **11.** *longidens*
- Inner corona lobes terete, erect, much exceeding anthers 6
6. Corolla conspicuously rugulose . . . .**14.** *tapscottii*
- Corolla smooth or only weakly rugulose 7
7. Corolla rotate, lobes spreading; inner corona lobes not obviously clavate not tuberculate . . . . **10.** *halipedicola*
- Corolla lobes strongly reflexed; inner corona lobes clavate and tuberculate . . . .
 . . . . **15.** *umbracula*
8. Corolla at least 25 mm in diameter 9
- Corolla 20 mm or less in diameter 10
9. Corolla lobes lacking marginal cilia; inner corona lobes 4–5 mm long, slender, entire, connivent-erect . . . . **2.** *lugardii*
- Corolla lobes with clavate marginal cilia near base; corona lobes 7–9 mm long, filiform, bifid, contorted . . . . **16.** *rogersii*
10. Corolla rotate; outer corona forming a pentagonal disc . . . .**1.** *schweinfurthii*
- Corolla campanulate; outer corona urceolate. . . . .**3.** *carnosa*
11. Inner corona lobes lacking a dorsal horn . . . .**4.** *caudata*
- Inner corona lobes with prominent dorsal horn. 12
12. Corolla white or cream with maroon to brown spots . . . .**6.** *knobelii*
- Corolla red or yellow, not spotted 13
13. Inner corona lobes (not including dorsals horn) at most equalling anthers . . 14
- Inner corona lobes much exceeding anthers 15
14. Corolla lobes at least 3× as long as broad . . . .**7.** *lutea*
- Corolla lobes only slightly longer than broad . . . . **5.** *melanantha*
15. Corolla 35–45 mm in diameter. . . . .**9.** *valida*
- Corolla 55–85 mm in diameter. . . . . **8.** *huillensis*

1. **Orbea schweinfurthii** (A. Berger) Bruyns in Aloe **37**: 76 (2000). Type: D.R. Congo, Rutihuru plains at Nswiwa, Itande, 9.iv.1891, *Stuhlmann* 2208 (NY holotype).

     Caralluma schweinfurthii A. Berger in Stapel. & Klein: 103 (1910).

     Caralluma piaranthoides Oberm. in Fl. Pl. South Africa **15**: t.599 (1935). Type: Zimbabwe, Hwange [Wankie], *Levy* 8444 (PRE holotype, K).

     Pachycymbium schweinfurthii (A. Berger) M.G. Gilbert in Bradleya **8**: 23 (1990).

     Angolluma schweinfurthii (A. Berger) Plowes in Excelsa **16**: 118 (1994).

     Ceropegia schweinfurthii (A. Berger) Bruyns in S. Afr. J. Bot. **112**: 428 (2017).

Mat-forming non-rhizomatous stem succulent. Stems decumbent or occasionally erect, 3–15 cm long, 0.4–1.2 cm thick, 4-angled. Tubercles tapering into a conical spreading tooth. Inflorescence 1–4 per stem near the apex; 1–3(8) flowers developing successively; peduncle very short; pedicels 2–4 mm long, erect. Sepals 2–3 mm long. Corolla rotate, 1–1.5 cm diam., densely papillate; corolla tube 0.5 mm deep, forming a somewhat pentagonal annulus beneath the corona, yellow; lobes spreading and slightly recurved, 3–3.5 mm long and 3 mm wide at base, triangular-ovate, acute, yellow with maroon spots coalescing towards the tips. Corona yellow to cream speckled with purple; outer lobes spreading and fused laterally to form a pentagonal disc with an entire or toothed margin; inner corona lobes dorsiventrally flattened and adpressed to back of anthers, deltoid, usually with 1–4 apical teeth.

Botswana. N: Selinda Spillway, Motswiri camp, fl. 18.iii.2003, *A. & R. Heath* 421 (GAB, K, PSUB). **Zambia**. B: Machili, 10.ii.1961, *Fanshawe* 6234 (K, NDO). **Zimbabwe**. W: 14 km from Hwange [Wankie] on W bank of Lukosi river [cult. Nelspruit ii/iii.1965], *Leach* 12123 (K). **Malawi**. S: Liwonde National Park, between Chiungune Hill and Shire River, fl. 20.iii.1977, *Brummitt et al.* 14873 (K, MAL).

Also known from D.R. Congo, Rwanda, Tanzania and Uganda. Seasonally wet areas in mopane or *Combretum* woodland; 450–1000 m.

2. **Orbea lugardii** (N.E. Br.) Bruyns in Aloe **37**: 75 (2000). Type: Botswana, Toteng [Totin] near Lake Ngami, xii.1896, *Lugard* 74 (K holotype).

Caralluma lugardii N.E .Br. in F.T.A. **4**(1): 487 (1903).
Caralluma longicuspis N.E. Br. in Fl. Cap. **4**(1): 884 (1909). Type: Namibia, cult. *Pillans* 14 (BOL holotype).
Pachycymbium lugardii (N.E. Br.) M.G. Gilbert in Bradleya **8**: 28 (1990).
Ceropegia lugardiana Bruyns in S. African J. Bot. **112**: 427 (2017).

Clump-forming rhizomatous stem succulent. Stems erect, 4–15 cm long, 0.6–1.2 cm thick, 4-angled. Tubercles tapering into a conical spreading acuminate tooth. Inflorescences 2–6 per stem in the upper half or near the apex; 1–3(7) flowers developing successively or occasionally together; peduncle short; pedicels 3–6 mm long, ascending and generally holding the flower facing upwards. Sepals 3–5 mm long. Corolla rotate to campanulate, 3–4.5 cm diam., finely papillate all over; corolla tube 3–6 mm deep and 5–8 mm across, more or less enclosing the gynostegium, red or brown; lobes spreading to ascending, 18–25 mm long and 3–4 mm wide at base then narrowing abruptly into slender recurved limb, narrow portion yellow, yellow-green, red or brown, wider basal portion mostly same colour as the tube. Outer corona lobes red to purple-brown, ascending and deeply bifid with diverging teeth, 1–2 mm long; inner corona lobes yellow above from a reddish base, dorsiventrally flattened and adpressed to back of anthers then connivent-erect, 4–5 mm long.

Botswana. N: Toteng [Totin] near Lake Ngami, xii.1896, *Lugard* 74 (K). SE: Matsitamma, 5.xii.1966, *McClintock* K17 (K). 900–1100 m.

3. **Orbea carnosa** (Stent) Bruyns in Aloe **37**: 73 (2000). Type: South Africa, NW Province, Zilkaat's Nek, Magaliesberg, 13.xii.1913 [cult. Pretoria, 14.iv.1914 (K) & 22.ii.1915 (PRE)], *Pole Evans* s.n. in U.D.A. Herb. 11020 (PRE lectotype, K), designated by Leach & Plowes in J. S. African Bot. **33**: 100.

Caralluma carnosa Stent in Bull. Misc. Inform. Kew **1916**: 42 (1916).
Pachycymbium carnosum (Stent) L.C. Leach in Excelsa Taxon. Ser. **1**: 71 (1978).
Ceropegia keithii (R.A. Dyer) Bruyns subsp. *carnosa* (Stent) Bruyns in S. African J. Bot. **112**: 427 (2017).

Clump-forming rhizomatous stem succulent. Stems erect to spreading, 4–15 cm long, 1–2 cm thick, 4-angled. Tubercles laterally flattened and broadly deltoid, usually with a minute pair of denticles near tip. Inflorescences 1–6(10) per stem, mostly in the upper half with 1–2(3) flowers developing successively; peduncle very short; pedicels (1)2–6(12) mm long, spreading to descending so flowers often nodding. Sepals 2.5–5 mm long. Corolla campanulate, 0.6–2 cm diam., irregularly rugulose with scattered erect papillae, uniformly yellow to maroon, or pale brown or maroon with irregular cream markings; corolla tube 2.5–4 mm deep and 4–7 mm across, the mouth constricted by an inward-facing annulus, more or less enclosing the gynostegium; lobes spreading to ascending, 3–5 mm long and 2–8 mm wide at base, broadly triangular. Corona cream with red-purple markings; outer corona lobes erect then spreading, truncate-emarginate to deeply bifid, fused laterally to form an urceolate cup exceeding the anthers; inner corona lobes 1 mm long, adpressed to back of anthers and slightly exceeding them.

Subsp. **keithii** (R.A. Dyer) Bruyns in Syst. Bot. Monogr. **63**: 89 (2002). Type: Swaziland, Lebombo [Ubombo] Mts, 30 km from Stegi, iii.1935, *Keith* s.n. (PRE holotype).

> *Caralluma keithii* R.A. Dyer in Fl. Pl. South Africa 15: t.600 (1935).
> *Caralluma fosteri* Pillans in White & Sloane, Stapelieae ed. 2, **1**: 292 (1937). Type: South Africa, Mpumalanga, Lydenburg Dist. [cult. Rosedale iii.1934], *Pillans* s.n. (BOL holotype).
> *Caralluma schweickerdtii* Oberm. in Bothalia **3**: 250 (1937); in White & Sloane, Stapelieae ed. 2, **1**: 294 (1937). Type: South Africa, Limpopo province, Soutpansberg Dist., Chapudi Farm near Waterpoort, *Obermeyer et al.* 411 (PRE holotype).
> *Pachycymbium keithii* (R.A. Dyer) L.C. Leach in Excelsa Taxon. Ser. **1**: 71 (1978).
> *Pachycymbium lancasteri* Lavranos in Cact. Succ. J. (Los Angeles) **56**: 196 (1984). Type: South Africa, Limpopo Province, Giyani, N of Letaba River, *Percy-Lancaster* s.n. (NBG holotype).
> *Ceropegia keithii* (R.A. Dyer) Bruyns in S. African J. Bot. **112**: 427 (2017).
> *Caralluma carnosa* sensu Oberm. in Fl. Pl. South Africa **15**: t.592 (1935), non Stent (1916).

Botswana. SE: 10 km W of Gobojango, ii.1989, *Hargreaves* 6389 (GAB). **Zimbabwe**. C: near Lake MacIlwaine, W of Harare [Salisbury], *Leach* 15108 (SRGH). E: Nyanga Dist., Manyika TTL, 4.i.1967, *Plowes* 2608 (BM, K, LISC, MO, PRE, ZSS). S: near Beitbridge, *Rushworth* 479 (SRGH). **Mozambique**. M: Lebombo mountains close to border with Swaziland, 20 km N of South African border, fl. 6.xii.2001, *Goyder* 5031 (K, LMU).

Also known from Swaziland and northern provinces of South Africa. Stony ground or on rock or occasionally sand. Flowers in subsp. *keithii* are larger than in subsp. *carnosa*, and measure 10–20 mm in diameter, with corolla lobes 6–8 mm broad; 150–1400 m.

4. **Orbea caudata** (N.E. Br.) Bruyns in Aloe **37**: 73 (2000). —Meve in F.T.E.A., Apocynaceae (part 2): 354 (2012). Type: Malawi, Namadzi [Namasi], iv.1899, *Cameron* 25 (K holotype).

> *Caralluma caudata* N.E. Br. in F.T.A. **4**(1): 485 (1903).
> *Caralluma longecornuta* Gomes e Sousa in Moçambique Doc. Trimestral **4**: 44 (1935) & **6**: 20 (1936), nomen nudum. Type: Mozambique, Niassa, Mandimba, *Torre* 4 (COI, PRE).
> *Caralluma caudata* var. *fusca* C.A. Lückh. in White & Sloane, Stapelieae ed. 2, **3**: 1144 (1937). Type: White & Sloane, Stapelieae ed. 2, **1**: fig.287 (1937), lectotypified by Bruyns in Syst. Bot. Monogr. **63**: 105 (2002).
> *Caralluma praegracilis* Oberm. in White & Sloane, Stapelieae ed. 2, **3**: 1161 (1937). Type: South Africa, KwaZulu-Natal, Nongoma, *Gerstner* 752 (PRE, not found); White & Sloane, Stapelieae ed. 2, **1**: fig.1212 (1937), lectotypified by Bruyns in Syst. Bot. Monogr. **63**: 105 (2002).
> *Orbeopsis caudata* (N.E. Br.) L.C. Leach in Excelsa Taxon. Ser. **1**: 68 (1978).
> *Ceropegia caudata* (N.E. Br.) Bruyns in S. African J. Bot. **112**: 427 (2017).

Clump-forming non-rhizomatous stem succulent. Stems erect to decumbent, 4–15 cm long, 0.6–1.1 cm thick, 4-angled. Tubercles tapering into a conical spreading to ascending attenuate tooth. Inflorescence usually 1 per stem in the upper half; (2)3–7 flowers developing together; peduncle short; pedicels 9–20 mm long, horizontal to ascending. Sepals 6–7 mm long. Corolla rotate, 3.5–6(9.5) cm diam., rugulose, finely papillate with coarser papillae in the tube, yellow to greenish, dotted with purple-brown to maroon or brick red, the markings coalescing or becoming darker in the tube; corolla tube 1.5–2 mm deep, enclosing lower half of the gynostegium only;

lobes spreading, 18–25(35) mm long and 5–7(11) mm wide at base then narrowing gradually into slender acute tip. Outer corona lobes purple-red with cream tips, quadrate and slightly bifid, ascending, 1.5–2 mm long; inner corona lobes cream to white with reddish margins, dorsiventrally flattened and adpressed to back of anthers, the tips overlapping in the centre of the flower, 1–1.5 mm long.

Tubercles 5–15 mm long, stems olive-green flecked with purple; Zambia, Malawi and
   Mozambique . . . . a) subsp. *caudata*
Tubercles 12–25 mm long, stems grey-green flecked with purple; Caprivi, Botswana
   and Zimbabwe . . . . b) subsp. *rhodesiaca*

a) Subsp. **caudata**.

Zambia. N: Nchalanga [Nachalanga] Hill near Mbala [cult. Nelspruit sub *Leach* 13296], *Richards* s.n. (K, SRGH). **Malawi**. C. Lilongwe [cult. Pretoria iv.1954], *Reynecke* 30 (K, SRGH). S. Malawi, Namadzi [Namasi], iv.1899, *Cameron* 25 (K). **Mozambique**. N: Niassa, Mandimba, fl. i.1937, *Torre* 4 (COI, PRE).

Also known from SW Tanzania. Bruyns (2005: 257) suggests the record from KwaZulu-Natal may be an error as the species has never been recollected in this region; 500–1600 m.

b) Subsp. **rhodesiaca** (L.C. Leach) Bruyns in Aloe **37**: 73 (2000). Type: Zimbabwe, Belingwe Dist., S of Mnene Mission, *Leach & Bullock* 13145 (SRGH holotype, BM, BOL, BR, K, LISC, PRE, ZSS).

   Caralluma chibensis C.A. Lückh. in S. African Gard. **25**: 56 (1935). Type: Zimbabwe, Chibi, *Jackson* s.n. in *Luckhoff* 182 (not traced).
   Caralluma caudata var. *chibensis* (C.A. Lückh.) C.A. Lückh. in White & Sloane, Stapelieae ed. 2, **1**: 352 (1937).
   Caralluma caudata var. *stevensonii* Oberm. in White & Sloane, Stapelieae ed. 2, **3**: 1156 (1937). Type: Zimbabwe, near Harare, *Stevenson* s.n. sub Transvaal Museum 34947 (PRE holotype, not traced).
   Caralluma caudata var. *milleri* Nel in White & Sloane, Stapelieae ed. 2, **3**: 1158 (1937). Type: Angola, Cuando Cubango, Cubango [Okavango] River E of Rundu, *Miller* 7390 (STE holotype, not traced).
   Caralluma caudata subsp. *rhodesiaca* L.C. Leach in Bothalia **11**: 134 (1973).
   Orbeopsis caudata subsp. *rhodesiaca* (L.C. Leach) L.C. Leach in Excelsa Taxon. Ser. **1**: 68 (1978).
   Orbeopsis chibensis (C.A. Lückh.) Plowes in Asklepios **118**: 5 (2014).
   Ceropegia caudata subsp. *rhodesiaca* (L.C. Leach) Bruyns in S. African J. Bot. **112**: 427 (2017).

Caprivi. Near Katima Mulilo [cult. Pretoria iii.1953 & ii.1954], *Codd* 7595 (BM, PRE, SRGH). **Botswana**. N: Selinda Reserve, Wild Dog Pan Camp, 6.ii.2003, *A. & R. Heath* 232 (GAB, K). **Zimbabwe**. N: Mrewa Dist., Shawanoe R. [cult. Nelspruit i.1965], *Leach & Muller* 12166 (COI, K, M, NBG, SRGH, WIND). W: Bulowayo, Burnside, fl. 15.i.1966, *Bullock* 77 (K, SRGH). C: Rusape Dist., Chiduku, 5.ii.1965, *Plowes* 2465 (K, SRGH). E: Chipinga Dist., Sabi Gorge [cult Mutare 6.ii.1965], *Plowes* 2404 (K, SRGH). S: Moodies Pass [cult. Greendale 6.ii.1960], *Leach* 9761 (K, LISC, SRGH).

Also known from a single collection is southern Angola. Generally found on the sandy floor of mopane or *Brachystegia* woodland, sometimes on granite; 300–1300 m.

5. **Orbea melanantha** (Schltr.) Bruyns in Aloe **37**: 76 (2000). Type: South Africa, Limpopo Province, Sandloop, *Schlechter* 4694 (B† holotype); Bandolierkop [cult. Greendale 3.ii.1960], *Leach* 9757 (K neotype, KIEL, LMA, PRE, SRGH), neotype designated by Bruyns (2005: 260).

Stapelia melanantha Schltr. in Bot. Jahrb. Syst. **38**: 50 (1905).

Caralluma melanantha (Schltr.) N.E. Br. in Fl. Cap. **4**(1): 885 (1909).

Stapelia furcata N.E. Br. in Fl. Cap. **4**(1): 973 (1909). Type: South Africa, "Transvaal", *Todd* s.n. (K holotype).

Caralluma leendertziae N.E. Br. in Ann. Transvaal Mus. **2**: 47 (1909). Type: South Africa, Limpopo Province, Potgietersrust, *Leendertz* 1279 (K holotype).

Caralluma rubiginosa Werderm. in Repert. Spec. Nov. Regni Veg. **30**: 54 (1932). Type: cult. Berlin (not traced).

Caralluma melanantha var. *sousae* Gomes e Sousa in Moçambique Doc. Trimestral **4**: 46 (1935). Type: Mozambique, Maputo Dist., Mangulane, xi.1934, *Gomes e Sousa* s.n. (not traced); White & Sloane, Stapelieae ed 2, **1**: 356, fig.292 (1937), lectotype designated by Plowes in Asklepios **118**: 17 (2014).

Caralluma australis Nel in White & Sloane, Stapelieae ed 2, **3**: 1153 (1937). Type: South Africa, Limpopo Province, Pietersburg, *Kirsten* s.n. sub STE 5881 (NBG holotype).

Orbeopsis melanatha (Schltr.) L.C. Leach in Excelsa Taxon. Ser. **1**: 66 (1978).

Orbeopsis sousae (Gomes e Sousa) Plowes in Asklepios **118**: 17 (2014).

Orbeopsis rubra Plowes in Asklepios **118**: 16 (2014). Type: Mozambique, Maputo Province, 16 km N of Moamba, *Leach* 12261 (SRGH holotype, K).

Orbeopsis karibensis Plowes in Asklepios **118**: 17 (2014). Type: Zimbabwe, Zambezi Valley, upper end of Kariba Dam between Binga and Deka, collector unknown, cult. sub *Plowes* 4088 (SRGH holotype).

Ceropegia melanantha (Schltr.) Bruyns in S. African J. Bot. **112**: 427 (2017).

Mat-forming non-rhizomatous stem succulent. Stems decumbent, 3–10 cm long, (1)1.5–4 cm thick, 4-angled. Tubercles laterally flattened, tapering into a stout deltoid apical tooth with minute denticles near tip. Inflorescence 1(2) per stem near or above the middle; 3–13 flowers opening simultaneously; peduncle stout, to 15 mm long; pedicels 15–45 mm long, spreading to ascending. Sepals 6–10 mm long. Corolla rotate, 4–6.5 cm diam., deep red-brown to maroon- or purple-black, sometimes finely spotted with yellow, rugulose and finely papillate; corolla tube 2.5–4 mm deep, cupular; lobes spreading with tips often reflexed, 14–18 mm long and 9–14 mm wide at base, triangular-ovate, acute, with vibratile purple clavate cilia along the margins. Corona red-brown to purple-black; outer lobes spreading-ascending, ± square, 3–4 mm long, with 2–3 tuberculate ridges and 2 or more apical teeth; inner corona lobes 2–3 mm long, dorsiventrally flattened and adpressed to back of anthers below, then laterally flattened and raised above them with one or two dorsal horns.

Zimbabwe. N: Zambezi Valley, upper end of Kariba Dam between Binga and Deka, collector unknown, cult. sub *Plowes* 4088 (SRGH). **Mozambique**. M: 16 km N of Moamba, *Leach* 12261 (K, SRGH).

Also occurs in northern provinces of South Africa. Sandy soils in acacia or *Brachystegia* woodland, sometimes in stony ground on in shallow soil over granite; 50–500 m.

6. **Orbea knobelii** (E. Phillips) Bruyns in Aloe **37**: 75 (2000). Type: Botswana, near Molepolole, *Knobel* s.n. sub PRE 8308 (PRE holotype).

Stapelia knobelii E. Phillips in Fl. Pl. South Africa **10**: t.363 (1930).

Caralluma langii White & Sloane, Stapelieae: 61 (1933). Type: Botswana, near Gaborone, *Van Son* s.n. sub *White & Sloane* 113 (not traced); White & Sloane, Stapelieae: fig.40 (1933), lectotype designated by Bruyns, Stapeliads S. Africa Madagascar (2005: 264).

Caralluma knobelii (E. Phillips) E. Phillips in Fl. Pl. South Africa **15**: t.593 (1935).

Caralluma knobelii var. *langii* (A.C. White & B. Sloane) A.C. White & B. Sloane, Stapelieae ed. 2, **1**: 368 (1937).

Caralluma kalaharica Nel in White & Sloane, Stapelieae ed. 2, **3**: 1165 (1937). Type: Botswana, NW of Lake Ngami, Tsau, *Nel* s.n. (not traced); White & Sloane, Stapelieae ed 2, **3**: fig.1216 (1937), lectotype designated by Bruyns, Stapeliads S. Africa Madagascar (2005: 264).

Orbeopsis knobelii (E. Phillips) L.C. Leach in Excelsa Taxon. Ser. **1**: 65 (1978).

Ceropegia knobelii (E. Phillips) Bruyns in S. African J. Bot. **112**: 427 (2017).

Clump-forming rhizomatous stem succulent. Stems stout, decumbent, 3–10 cm long, 1–2.5 cm thick, 4-angled. Tubercles tapering into a conical spreading to ascending tooth. Inflorescence 1 per stem in lower half; 3–10 flowers developing successively; peduncle less than 10 mm long; pedicels 1–15 mm long, spreading or ascending. Sepals 3–5 mm long. Corolla rotate, 2.5–3.5 cm diam., white to cream, irregularly blotched with purple-brown, irregularly rugulose-papillate; corolla tube 1.5–2 mm deep, shallow-cupular; lobes spreading and slightly recurved, 10–14 mm long and 7–10 mm wide at base, triangular-ovate, acute, margins with scattered clavate cilia. Corona brown or purple; outer lobes spreading-ascending 1.5–3 mm long, ± square with a notched truncate apex, upper face with 2 radial ridges; inner corona lobes 3–4 mm long, dorsiventrally flattened and adpressed to back of anthers below then erect and laterally flattened above, with a backward-pointing tooth 1 mm long near the base.

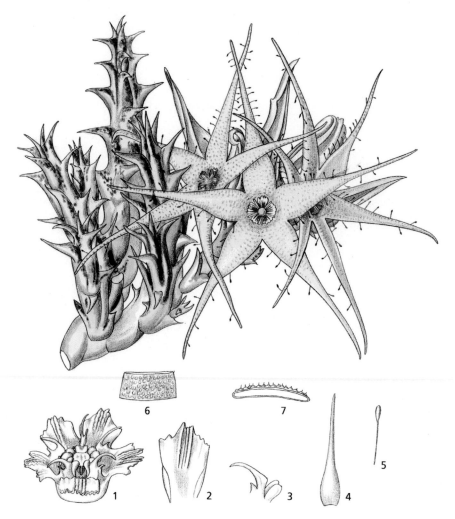

Fig. 7.3.**170**. ORBEA LUTEA subsp. LUTEA. Main image, flowering stems; 1, gynostegium and corona with one corona lobe removed; 2, outer corona lobe; 3, inner corona lobe; 4, calyx lobe; 5, vibratile hair from corolla; 6, 7, corolla surface from above and in section. Drawn by Cynthia Letty. Reproduced with permission from Flowering Plants of South Africa (1936).

Botswana. N: Reported by Nel in White & Sloane (1937: 1165). SE: near Molepolole, *Knobel* s.n. sub PRE 8308 (PRE).

Also found in the Northern Cape Province of South Africa. In shade, on deep fine sand or on calcrete around pans; 900–1000 m.

7. **Orbea lutea** (N.E. Br.) Bruyns in Aloe **37**: 75 (2000). Type: South Africa, Klip Drift, *Tuck* s.n. in *MacOwan* 2240 (K lectotype), designated by Leach in J. S. African Bot. **36**: 163 (1970). FIGURE 7.3.**170**.

    Caralluma lutea N.E. Br. in Hooker's Icon. Pl. **20**: t.1901 (1890).
    Caralluma lateritia N.E. Br. in F.T.A. **4**(1): 486 (1903). Type: Botswana, Botletle Flats near Tame's and Rakops villages, iv.1899, *Lugard* 307 (K holotype).
    Caralluma vansonii Bremek. & Oberm. in Ann. Transvaal Mus. **16**: 429 (1935). Type: Botswana, N'Kate, Nata River, viii.1930 [cult. Pretoria iii.1931], *van Son* s.n. in Herb. Transvaal Mus. 28760 (PRE holotype, F, K, PRE).
    Caralluma lutea var. *lateritia* (N.E. Br.) Nel in White & Sloane, Stapelieae ed. 2, **1**: 373 (1937).
    Caralluma lateritia var. *stevensonii* A.C. White & B. Sloane, Stapelieae ed. 2, **1**: 371 (1937). Type: Zimbabwe, Matetsi, 50 km SE of Victoria Falls, *Stevenson* s.n. (?K holotype).
    Caralluma lutea var. *vansonii* (Bremek. & Oberm.) C.A. Lückh., Stapelieae: 62 (1952).
    Orbeopsis lutea (N.E. Br.) L.C. Leach in Excelsa Taxon. Ser. **1**: 64 (1978).
    Orbeopsis lateritia (N.E. Br.) Plowes in Asklepios **118**: 8 (2014).
    Orbeopsis matabelensis Plowes in Asklepios **118**: 9 (2014). Type: Zimbabwe, Alicedale, 20 km SW of Nyamandhlovu, *Plowes* 1877 (SRGH holotype).
    Ceropegia lutea (N.E. Br.) Bruyns in S. African J. Bot. **112**: 427 (2017).

Mat-forming occasionally rhizomatous stem succulent. Stems decumbent, 3–20 cm long, 1–2.5 cm thick, 4-angled. Tubercles laterally flattened, tapering into a conical spreading tooth. Inflorescences 1–2 per stem in the upper half; 3–30 flowers opening together; peduncle short; pedicels 12–30 mm long. Sepals 5–8 mm long. Corolla rotate, 3.5–6.5 cm diam., irregularly rugulose-papillate except in tube, uniformly yellow to red-brown, or red-brown to almost black and blotched with yellow; corolla tube 2 mm deep, enclosing lower half of the gynostegium only; lobes spreading, 18–35 mm long and 6–15 mm wide at base then narrowing gradually into slender attenuate tip. Corona red- to purple-brown or blackish, outer lobes sometimes with yellow margins; outer corona lobes sub-quadrate, 2 mm long and 3 mm broad, margins of the 5 lobes ± contiguous to form a disc; inner corona lobes 0.8–1 mm long dorsiventrally flattened and adpressed to back of anthers, with a laterally flattened and somewhat recurved dorsal horn 2.5–3 mm long.

Subsp. **lutea**.

Botswana. N: Near Jack's Camp, Makgadikgadi Pans, fl. 2.iii.2017, *A. & R. Heath* 2803 (GAB, K). SE: 60 km N of Serowe [cult. Harare sub SRGH 2694, i.1966], *Wild & Drummond* 7280 (K, SRGH). **Zimbabwe**. W: Nyamandhlovu Pasture Research Station, fl. xi.1952, *Plowes* 1529 (K, SRGH).

Also found in northern provinces of South Africa. On calcrete or stony outcrops, in shade or out in the open; 900–1200 m.

Subsp. *lutea* has corolla lobes 3–5 times as long as broad. To the west of this zone, principally in Namibia, it is replaced by subsp. *vaga*, whose corolla lobes are only 1.5–2.5 times as long as broad.

8. **Orbea huillensis** (Hiern) Bruyns in Aloe **37**: 74 (2000). Type: Angola, Huíla, xii.1859, *Welwitsch* 4266 (BM lectotype, LISU), designated by Bruyns (2005: 273), as·holotype.

    Caralluma huillensis Hiern, Cat. Afr. Pl. **1**: 697 (1898).
    Caralluma gossweileri S. Moore in J. Bot. **50**: 367 (1912). Type: Angola, Huíla Province, Forte Princeza Amelia near Kuvango, x.1905, *Gossweiler* 2098 (BM holotype, COI, K).

Caralluma tsumebensis Oberm. in White & Sloane, Stapelieae ed. 2, **3**: 1163 (1937). Type: Namibia, Tsumeb, *Nägelsbach* s.n. sub Herb Transvaal Mus 32820 (PRE holotype).
Orbeopsis gossweileri (S. Moore) L.C. Leach in Excelsa Taxon. Ser. **1**: 67 (1978).
Orbeopsis huillensis (Hiern) L.C. Leach in Excelsa Taxon. Ser. **1**: 68 (1978).
Orbeopsis tsumebensis (Oberm.) L.C. Leach in Excelsa Taxon. Ser. **1**: 68 (1978).
Ceropegia huillensis (Hiern) Bruyns in S. African J. Bot. **112**: 427 (2017).

Clump-forming mostly non-rhizomatous stem succulent. Stems stout, decumbent, 4–30(50) cm long, 1.5–3 cm thick excluding teeth, 4(5)-angled. Tubercles laterally flattened, tapering into a large deltoid spreading tooth with a pair of denticles towards the tip. Inflorescences 1–3 per stem from base to near apex; 5–40 flowers opening together; peduncle stout, 15 mm long; pedicels 10–45 mm long, horizontal to ascending. Sepals 5–9 mm long. Corolla rotate, 5.5–8.5 cm diam., transversely rugulose-papillate, dark brown, maroon, dark purple, or yellow in subsp. *flava*; corolla tube 5–7 mm deep, completely enclosing the gynostegium; lobes spreading, 20–45 mm long and 5–9 mm wide at base then narrowing gradually into attenuate tip, margins somewhat reflexed with or without scattered white or maroon vibratile hairs. Corona deep purple-brown, or orange in subsp. *flava*; outer corona lobes spreading to ascending, 1.5–3 mm long, oblong-acute to truncate-dentate; inner corona lobes 1.5–3.5 mm long, dorsiventrally flattened and adpressed to back of anthers then ascending into a slender entire or bifid tongue.

Subsp. **huillensis**.

Botswana. N: Selinda Reserve, Crane Pan, 13.xi.2004, *A. & R. Heath* 675 (GAB, K). **Zambia**. S: Kalumbi, Kafue National Park [cult. Harare i.1967], *Mitchell* 3012 (K, SRGH). **Zimbabwe**. W: Nyamandhlovu Pasture Research Station, xi.1953, *Plowes* 1884 (K, SRGH).
Also known from southern Angola and northern Namibia. Growing in fine sandy soil, generally in shade; 900–1200 m.

9. **Orbea valida** (N.E. Br.) Bruyns in Aloe **37**: 76 (2000). Type: Botswana?, *Holub* s.n. (K holotype).
Caralluma valida N.E. Br. in Bull. Misc. Inform. Kew **1895**: 264 (1895).
Ceropegia valida (N.E. Br.) Bruyns in S. African J. Bot. **112**: 428 (2017).

Clump-forming rhizomatous or non-rhizomatous stem succulent. Stems stout, decumbent, 4–20 cm long, 1.5–2.5 cm thick excluding teeth, 4(5)-angled. Tubercles laterally flattened, tapering into a large deltoid spreading tooth with a pair of denticles towards the tip. Inflorescences 1–3 per stem from base to near apex; 5–40 flowers opening together; peduncle stout, 15 mm long; pedicels 10–45 mm long, horizontal to ascending. Sepals 5–9 mm long. Corolla rotate, 3.5–4.5(5.5) cm diam., transversely rugulose-papillate, deep maroon to pinkish red; corolla tube 2–6 mm deep, partially or completely enclosing the gynostegium; lobes spreading, 15–20(25) mm long and 5–9 mm wide at base then narrowing gradually into attenuate tip, margins somewhat reflexed with scattered white or maroon vibratile hairs. Corona deep purple-brown; outer corona lobes spreading to ascending, 1.5–3 mm long, oblong-acute to truncate-dentate; inner corona lobes 1.5–3.5 mm long, dorsiventrally flattened and adpressed to back of anthers then ascending into a slender entire or bifid tongue.

Corolla tube 5–6 mm deep, enclosing the gynostegium; stems lacking underground rhizomes . . . . .a) subsp. *valida*
Corolla tube 2–3 mm deep, gynostegium partially exserted; stems with underground rhizomes . . . . .b) subsp. *occidentalis*

a) Subsp. **valida**.

Caprivi. Reported by Bruyns (2005: 278). **Botswana**. N: No locality, *Holub* s.n. (K). **Zambia**. B: 5 km S of Shangombo [cult. Pretoria], 26.ii.1954, *Codd* 7485 (PRE). S:

Katombora Forest Extn. [cult. SRGH1964, 1965], *Bainbridge* 1017 (K, SRGH).
Zimbabwe. W: Bembesi River, Lupane Dist. [cult. SRGH 1965], *Walter* 10 (K, SRGH).
Known only from this region. Open dry mopane woodland; 900–1200 m.

b) Subsp. **occidentalis** Bruyns in Aloe **37**: 76 (2000). Type: Botswana, north-east of
D'Kar, *Bruyns* 6465 (BOL holotype).
   Orbeopsis occidentalis (Bruyns) Plowes in Asklepios **118**: 10 (2014).
   Ceropegia valida subsp. *occidentalis* (Bruyns) Bruyns in S. African J. Bot. **112**: 428 (2017).

Botswana. N: north-east of D'Kar, *Bruyns* 6465 (BOL).
Also known from northern Namibia. Among *Terminalia sericea* and *Commiphora* on
red to orange sands; 1100–1200 m.

10. **Orbea halipedicola** L.C. Leach in Excelsa Taxon. Ser. **1**: 40 (1978). Type:
   Mozambique, between Buzi and Gorongosa Rivers, *Ambrose* s.n. sub *Leach* 12396
   (SRGH holotype, LISC, PRE).
   Stapelia halipedicola (L.C .Leach) P.V. Heath in Calyx **1**: 16 (1992).
   Orbea halipedicola subsp. *septentrionalis* L.C. Leach in Excelsa Taxon. Ser. **1**: 43 (1978).
   Type: Mozambique, Gorongosa Park, 14.v.1972, *Tinley* s.n. (SRGH holotype).
   Stapelia halipedicola var. *borealis* P.V. Heath in Calyx **1**: 16 (1992). Type as for *Orbea
   halipedicola* subsp. *septentrionalis*.
   Ceropegia halipedicola (L.C. Leach) Bruyns in S. African J. Bot. **112**: 427 (2017).

Clump-forming often rhizomatous stem succulent. Stems slender, decumbent, 5–15 cm
long, 0.4–0.7 cm thick, 4-angled. Tubercles tapering into a slender conical ascending to slightly
incurved tooth with a pair of denticles towards the tip. Inflorescence 1 per stem near the base; 1–3
flowers opening in gradual succession; peduncle short, less than 5 mm long; pedicels 20–30 mm
long, spreading with apex upturned so flower faces upwards. Sepals 4–6 mm long. Corolla rotate,
3–4.2 cm diam., smooth and not rugulose, but with small acute papillae on annulus and towards
apex of lobes, deep red to purple-brown with irregular yellow to whitish markings; corolla tube
1.5–2.5 mm deep, formed by a convex raised annulus 8–9 mm across; lobes spreading, 14–16 mm
long and 7–10 mm wide at base, ovate and slightly attenuate, margins with white or red-purple
vibratile hairs at least in the lower half. Outer corona lobes dark purple-brown, spreading and
slightly recurved onto the rim of the annulus, 2 mm long, subquadrate; inner corona lobes 3–4
mm long, dorsiventrally flattened and adpressed to back of anthers then erect or connivant
above, tapering to a slightly clavate and obscurely rugulose apex.

Mozambique. MS: Cherinda [cult. Greendale], *Leach* 11243 (K, SRGH).
Only known from the coastal flats between the Save and Buzi Rivers, and in
Gorongosa National Park. Under mopane, or on bare white sandy flats sometimes
seasonally inundated; 0–50 m.

11. **Orbea longidens** (N.E. Br.) L.C. Leach in Kirkia **10**: 290 (1975); in Excelsa Taxon.
   Ser. **1**: 36 (1978). Type: Mozambique, Delagoa Bay, recd. ix.1883, *Monteiro* s.n. (K
   syntype); cult. Norwich sub *Tillett* s.n., several dates (K syntype).
   Stapelia longidens N.E. Br. in Gard. Chron., ser. 3 **18**: 324 (1895).
   Ceropegia longidens (N.E. Br.) Bruyns in S. African J. Bot. **112**: 427 (2017).

Clump-forming stem succulent, rhizomatous or not. Stems slender, erect or decumbent, 5–12
cm long, 0.5 cm thick, 4-angled. Tubercles tapering into a slender conical ascending tooth with
a pair of denticles 2/3 of the distance towards the tip. Inflorescence 1 per stem near the base;
1–3 flowers opening in gradual succession; peduncle short, less than 5 mm long; pedicels 20 mm
long, spreading sometimes with ascending apex. Sepals 6–8 mm long. Corolla rotate, 3–5 cm
diam., smooth, glabrous or minutely pubescent towards tips of lobes, cream or yellowish with
small maroon to purple-brown spots in tube which become larger on lobes and more confluent
distally; corolla tube 5 mm deep, broadly campanulate and steep-sided and with a raised annulus

surrounding the base of the gynostegium 4–6 mm across; lobes spreading to reflexed, 16–20 mm long and 6–12 mm wide at base, ovate to ovate-lanceolate, acute, margins with or without spathulate cilia at least in the lower half. Outer corona lobes cream with a broad purple patch down the middle, spreading and slightly recurved onto the rim of the annulus, 1.2–1.5 mm long, subquadrate; inner corona lobes 1.2–1.5 mm long, dorsiventrally flattened and adpressed to back of anthers.

Mozambique. M: Delagoa Bay [cult. Tillett, 25.x.1892] (K).

Maputaland endemic occurring in northern KwaZulu-Natal and southern Mozambique; growing in grey or white sand within 50 km of the coast; 10–100 m.

Bruyns (2005: 291) reports that this species is now very rare around Maputo.

12. **Orbea maculata** (N.E. Br.) L.C. Leach in Excelsa Taxon. Ser. **1**: 49 (1978). Type: Botswana, near T'Klakane Pits, northern Kalahari Desert, iv.1899, *Lugard* 297 (K holotype). FIGURE 7.3.**171**.

Caralluma maculata N.E. Br. in F.T.A. **4**(1): 487 (1903).

Ceropegia rangeana (Dinter & A. Berger) Bruyns subsp. *maculata* (N.E. Br.) Bruyns in S. African J. Bot. **112**: 427 (2017).

Clump-forming rhizomatous stem succulent. Stems slender to stout, erect, 2–10 cm long, 0.6–1.5 cm thick, 4-angled. Tubercles laterally flattened, deltoid, tapering to a conical tooth with a pair of denticles towards the tip. Inflorescence 1 per stem near the base; 1–5 flowers opening in gradual succession; peduncle short, less than 5 mm long; pedicels 25–50 mm long, spreading with apex upturned so flower faces upwards. Sepals 4–6 mm long. Corolla rotate, 3–7.5 cm diam., smooth to slightly papillate, pale greenish yellow to white, with red-purple or maroon dots, the dots generally coalescing into more solid blocks towards the tips of the lobes; corolla tube 0.5–1 mm deep, formed by a small thickened annulus around the base of the gynostegium; lobes spreading and slightly recurved towards the apices, 12–32 mm long and 6–15 mm wide at base, oblong, acute, margins with white or purple vibratile hairs in the lower two-thirds. Corona yellow to orange or mink or maroon; outer corona lobes spreading and forming a disc around the gynostegium 2 mm wide; inner corona lobes 1 mm long at most, dorsiventrally flattened and adpressed to back of anthers.

Subsp. **maculata**.

Botswana. SE: Boatlename, 110 km N of Molepolole, fl. 18.v.1934, *Knobel* 3 (PRE). **Zimbabwe**. W: Mweme River near Kezi, *Bullock* 116 (SRGH). E: Mutema Pan, Sabi Valley [cult. Mutare 3.vi.1969], *Plowes* 3289 (K, SRGH). S: 7 km W of Birchenough Bridge, 25.v.1959, *Leach & Noel* 9031 (K, SRGH).

Also known from South Africa (Limpopo Province and northern KwaZulu-Natal). Stony ground generally in open mopane woodland; 500–1100 m.

13. **Orbea paradoxa** (I. Verd.) L.C. Leach in Excelsa Taxon. Ser. **1**: 55 (1978). Type: Mozambique, Ressano Garcia, *Blignaut & Van der Merwe* 403 (PRE lectotype) designated by Leach in Excelsa Taxon. Ser. **1**: 55 (1978) as holotype.

Stultitia paradoxa I. Verd. in Fl. Pl. South Africa **17**: t.677 (1937).

Stapelia paradoxa (I. Verd.) P.V. Heath in Calyx **1**: 16 (1992).

Ceropegia paradoxa (I. Verd.) Bruyns in S. African J. Bot. **112**: 427 (2017).

Clump-forming rhizomatous stem succulent. Stems slender, apparently prostrate, 3–6 cm long, 0.4–0.8 cm thick, 4-angled. Tubercles tapering into a conical tooth which is somewhat flattened above towards the apex and with 1–3 pairs of denticles towards the tip. Inflorescence 1 per stem near the base; 1–3 flowers opening in gradual succession; peduncle short, less than 5 mm long; pedicels 8–10 mm long, spreading with apex upturned so flower faces upwards. Sepals 4–6 mm long. Corolla campanulate, 1.8–3.5 cm diam., smooth and somewhat shiny, pale greenish white distally on the lobes, elsewhere cream or white with red-purple or pink transverse markings,

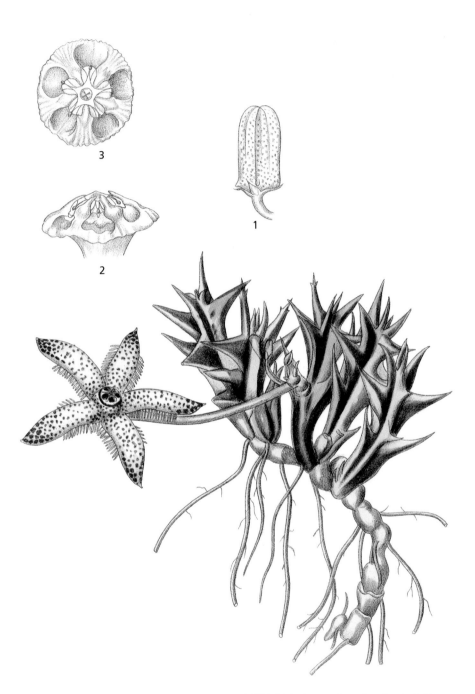

Fig. 7.3.**171**. ORBEA MACULATA subsp. MACULATA. Main image, flowering shoots; 1, flower bud; 2, 3, gynostegium and corona from the side, and from above. Drawn by Cynthia Letty. Reproduced with permission from Flowering Plants of South Africa (1933).

coalescing into more continuous bands towards the base and around the annulus, sometimes with white ring surrounding the mouth of the tube; corolla tube 6–8 mm deep, with a circular to pentagonal annulus 2–3 mm tall enclosing the gynostegium and with stiff erect bristles towards the base; lobes spreading and somewhat convex, 6–10 mm long and 6–10 mm wide at base, ovate deltate, acute, margins with clavate purple hairs in the lower half. Outer corona lobes ovate or oblong, concave, spreading, 1.5–2 mm long, 1.8–2 mm wide, maroon or yellow suffused with brown or red; inner corona lobes 1 mm long, dorsiventrally flattened, adpressed to back of anthers, the tips meeting in the middle, yellow.

Mozambique. M: Catuane [cult. Nelspruit], *Leach & Bayliss* 11938 (SRGH).
Restricted to the foot of the Lebombo Mts in southern Mozambique and neighbouring Swaziland and South Africa (Mpumalanga and KwaZulu-Natal). Found in seasonally waterlogged soils below about 300 m.

14. **Orbea tapscottii** (I. Verd.) L.C. Leach in Kirkia **10**: 291 (1975); in Excelsa Taxon. Ser. 1: 31 (1978). Type: Botswana, Lobatsi, *Tapscott* s.n. (K holotype). FIGURE 7.3.**172**.
   Stapelia tapscottii I. Verd. in Bull. Misc. Inform. Kew **1927**: 357 (1927).
   Stultitia tapscottii (I. Verd.) E. Phillips in Fl. Pl. South Africa **13**: t.520 (1933).
   Ceropegia tapscottii (I. Verd.) Bruyns in S. African J. Bot. **112**: 428 (2017).

Clump-forming non-rhizomatous stem succulent. Stems erect, 5–12 cm long, 0.5–1 cm thick, 4-angled. Tubercles tapering to an ascending to spreading conical tooth with a pair of minute denticles towards the tip. Inflorescence 1 per stem near the base; 1–3(10) flowers opening in gradual succession; peduncle absent or up to 15 mm long; pedicels 6–12 mm long, ascending. Sepals 3–5 mm long. Corolla rotate, 2.2–6 cm diam., rugulose, pale brown to brown-purple or red, with cream ridges; corolla tube 1–1.5 mm deep and 4–5 mm across, surrounded by a raised thickened annulus; lobes spreading to slightly recurved, 10–20 mm long and 8–13 mm wide at base, ovate, acute, margins with vibratile hairs in the lower half of the lobe. Corona red to deep maroon or pale yellow flecked with maroon; outer corona lobes 1 mm long; inner corona lobes 4–5 mm long, dorsiventrally flattened and adpressed to back of anthers then connivant-erect then diverging towards the clavate-tuberculate tips, often with a laterally compressed dorsal horn arising near the base.

Botswana. SE. 16 km S of Moshupa [cult. Nelspruit 1965], 27.ix.1964, *Leach et al.* 12467 (BOL, BR, GRA, K, MO, SRGH).
Also occurs in northern provinces of South Africa (NW Province, Limpopo, Gauteng, Mpumalanga). Flat sandy areas often under *Vachellia tortilis*; 1100–1300 m.

15. **Orbea umbracula** (M.D. Hend.) L.C. Leach in Kirkia **10**: 291 (1975). Type: Zimbabwe, Bikita Distr., Moodies Pass, *Leach & Pienaar* 5584 (PRE holotype).
   Stultitia umbracula M.D. Hend. in Fl. Pl. Africa **35**: t.1374 (1962).
   Stapelia umbracula (M.D. Hend.) P.V. Heath in Calyx **1**: 16 (1992).
   Ceropegia umbracula (M.D. Hend.) Bruyns in S. African J. Bot. **112**: 428 (2017).

Clump-forming rarely rhizomatous stem succulent. Stems slender, erect, 4–10 cm long, 0.6–1 cm thick, 4-angled. Tubercles tapering into a slender spreading-ascending conical tooth with a pair of denticles 5–8 mm from the tip. Inflorescence 1 per stem in lower half; 1–8 flowers opening in succession; peduncle up to 10 mm long; pedicels 15–40 mm long, erect to spreading holding the flower facing upwards or outwards. Sepals 6 mm long. Corolla with strongly reflexed lobes, 3–4.5 cm diam. when flattened out, minutely papillate, dull maroon to brown, becoming shiny with yellow markings on lobes; corolla tube 2 mm deep, surrounded by a raised thickened annulus; lobes strongly reflexed and somewhat convex, 16–20 mm long and 7–9 mm wide at base, ovate and somewhat rugulose distally, acute, margins with fine white cilia. Outer corona lobes spreading, subquadrate with 2 raised radial ridges and an apical tooth, dull maroon to brown; inner corona lobes 4–5 mm long, dorsiventrally flattened and adpressed to back of anthers, then extending as a terete connivant-erect tooth with a recurved clavate-tuberculate apex.

Zimbabwe. E: Maranke T.T.L. 30 km S of Odzi, 25.ii.1971, *Plowes* 3824 (K, NBG, PRE, SRGH). S: Moodies Pass [cult. Harare iv.1860], *Leach* 9874 (K, SRGH). **Mozambique**. GI: Planicie Davetave, Mambone, 14.xii.1998, *Bruyns* 7680 (BOL, K).

Only known from this area, associated with the Save River drainage system. Found in sandy soils under *Brachystegia* in Zimbabwe, and in leaf litter under *Baikaea* or *Androstachys* in Mozambique; 900–1100 m.

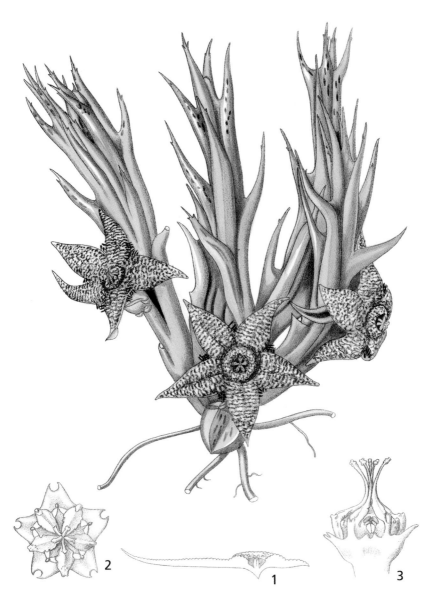

Fig. 7.3.**172**. ORBEA TAPSCOTTII. Main image, flowering shoots; 1, cross-section of flower; 2, 3, gynostegium and corona from above, and from the side. Drawn by Cynthia Letty. Reproduced with permission from Flowering Plants of South Africa (1933).

Fig. 7.3.**173**. ORBEA ROGERSII. Main image, flowering stem; 1, corolla lobe; 2, gynostegium and corona; 3, inner corona lobe. Drawn by Cynthia Letty. Reproduced with permission from Flowering Plants of South Africa (1935).

16. **Orbea rogersii** (L. Bolus) Bruyns in Aloe **37**: 76 (2000). Type: Botswana, Mahalapye, *Rogers* 6298 (BOL holotype). FIGURE 7.3.**173**.

> *Stapelia rogersii* L. Bolus in Ann. Bolus Herb. **1**: 194 (1915).
> *Caralluma rogersii* (L. Bolus) E.A. Bruce & R.A. Dyer in Bull. Misc. Inform. Kew **1934**: 303 (1934).
> *Pachycymbium rogersii* (L. Bolus) M.G. Gilbert in Bradleya **8**: 28 (1990).
> *Angolluma rogersii* (L. Bolus) Plowes in Excelsa **16**: 120 (1994).
> *Ceropegia rogersii* (L. Bolus) Bruyns in S. African J. Bot. **112**: 428 (2017).

Clump-forming non-rhizomatous stem succulent. Stems slender, erect, 3–10 cm long, 0.8–1 cm thick, 4-angled. Tubercles tapering to a slender spreading to ascending conical tooth with a pair of denticles towards the tip. Inflorescences 1–8 per stem mainly in the upper half; 1–3(8) flowers opening in succession from a swollen peduncular patch; pedicels 8–15 mm long, ascending. Sepals 8–15 mm long. Corolla rotate-campanulate with upcurved lobes, 2.5–3.5 cm diam., finely papillate on lobes and with cylindrical papillae on annulus, pale greenish yellow, white on the united basal portion; corolla tube 0.5–1 mm deep, surrounded by a small thickened annulus around the base of the gynostegium; lobes 18–22 mm long and 3–5 mm wide at base, narrowly oblong, acute, replicate for most of their length, basal margins with transparent clavate-globose non-vibratile cilia. Corona cream; outer corona lobes spreading, 1.5–3 mm long, rectangular to lanceolate with 1–3 apical teeth; inner corona lobes 8–10 mm long, dorsiventrally flattened and adpressed to back of anthers at base then forming bifid or occasionally trifid slender filiform appendages which become contorted apically, a dorsal appendage is similarly filiform and contorted.

Botswana. N: Selinda Reserve, fl. 28.iii.2005, *A. & R. Heath* 980 (GAB, K, PSUB). SE: Mahalapye, *Rogers* 6298 (BOL). **Zimbabwe**. S: 11 km E of Beitbridge–Masvingo [Fort Victoria] road, on Chituripasi road, fl. 10.i.1961, *Leach* 10680 (K, SRGH).

Also known from scattered localities in northern South Africa and Swaziland. Generally found in flat sandy areas in shade of shrubs or surrounding base of acacia or mopane trees; 500–1000 m.

56. TAVARESIA Welw.[28]

Tavaresia Welw. in Bol. Ann. Cons. Ultramar. Lisb. **7**: 79 (1854). —Brown in F.T.A. **4**(1): 493–494 (1903); in Fl. Cap. **4**(1): 901–902 (1909). —Bruyns, Stapeliads S. Africa Madagascar **2**: 526–530 (2005).

> *Decabelone* Decne. in Ann. Sci. Nat. Bot., sér. 5 **13**: 404 (1871).
> *Ceropegia* sect. *Tavaresia* (Welw.) Bruyns in S. African J. Bot. **112**: 421 (2017).

Clump-forming stem succulents; latex clear or at most cloudy. Stems cylindrical, 6–14-angled; tubercles tipped by 3 fine bristles. Inflorescences solitary, subsessile, arising near the base of the stems. Corolla tubular with relatively short spreading lobes, papillate. Gynostegial corona in 2 series, fused to each other: "outer" corona fused basally to form a somewhat pouched tube with 10 erect filiform appendages topped with spherical knobs; "inner" staminal corona lobes arched over backs of anthers. Gynostegium on a short column. Pollinia D-shaped, longer than broad. Follicles paired, erect, terete, glabrous.

Two species are recognised in southern Africa and Angola – *Tavaresia angolensis* Welw. and *T. barklyi* (Dyer) N.E. Br. *Tavaresia meintjesii* R.A. Dyer and *T. thompsoniorum* van Jaarsv. & R. Nagel are believed to be naturally occurring hybrids, the first between *T. barklyi* and *Stapelia gettliffei*, the second between *T. barklyi* and *Huernia urceolata* L.C. Leach.

[28] by D.J. Goyder

Tavaresia barklyi (Dyer) N.E. Br. in F.T.A. **4**(1): 494 (1903); in Fl. Cap. **4**(1): 901 (1909). Type: South Africa, Northern Cape Province, Du Plooy's Dam near the Orange River, 1871, *Barkly* s.n. [comm. 3/1874] (K lectotype, K), designated by Bruyns, Stapeliads S. Africa Madagascar **2**: 528 (2005). FIGURE 7.3.**174**.

Decabelone barklyi Dyer in Bot. Mag. **101**: t.6203 (1875).

Decabelone grandiflora K .Schum. in Engler & Prantl, Nat. Pflanzenfam. **4**(2): 276 (1895). Type not designated.

Euphorbia antunesii Pax in Bot. Jahrb. Syst. **34**: 79 (1904). Type: Angola, Huila, *Antunes* 72 (B† holotype).

Tavaresia grandiflora (K. Schum.) A. Berger, Stapel. & Klein.: 45 (1910).

Tavaresia grandiflora var. *recta* Van Son in White & Sloane, Stapelieae, ed. 2 **3**: 1145 (1937). Type not designated.

Ceropegia barklyana Bruyns in S. African J. Bot. **112**: 421 (2017).

Clump-forming stem-succulent. Stems ±erect from a shortly decumbent base, 1.5–30 cm long, 1–2.5 cm wide, (8)10–14-angled; tubercles conical, tipped with 3 bristles, the central one longest, erect, the two laterals deflexed. Inflorescences with few flowers opening in succession. Pedicels 5–15 mm long, spreading. Sepals 5–15 mm long, lanceolate, attenuate. Corolla deeply tubular to funnel-shaped, usually ascending initially, then spreading or curved downwards towards the tip, yellow flecked with round or longitudinally elongated maroon spots becoming denser and coalescing towards the base of the tube, papillate, each papilla tipped with a short bristle; corolla tube 2.5–15 cm long; lobes spreading, broadly triangular, 1.2–2.5 cm long, margins lacking cilia. Outer corona 9–18 mm long, lower parts white, upper filiform parts orange or white streaked with maroon; inner corona lobes 1.5 mm long, white streaked with maroon.

Botswana. SE: Palapye Road, iii.1910, *Kensit* s.n. (K). **Zimbabwe**. W: Hwange [Wankie], fl. 8.i.1936, *Eyles* 8544 (SRGH). S: cult. Harare, 6.iii.1967 ex Intentengwe River, 17 miles NW of Beitbridge, *Leach* 13654 (K, SRGH).

Also known from Angola, Namibia and South Africa. Dry bush with mopane, *Catophractes* and *Sesamothamnus*; 500–1000 m.

Fig. 7.3.**174**. TAVARESIA BARKLYI. Main image, flowering stem; 1, section through corona, and gynostegium from above; 2, portion of corona from within; 3, pollinarium (inverted). Drawn by Walter Fitch. Reproduced from Curtis's Botanical Magazine (1875).

57. AUSTRALLUMA Plowes[29]

Australluma Plowes in Haseltonia **3**: 54 (1995). —Bruyns, Stapeliads S. Africa
Madagascar **1**: 61–65 (2005).
Ceropegia sect. *Australluma* (Plowes) Bruyns in S. African J. Bot. **112**: 425 (2017).

Rhizomatous clump-forming stem succulents; latex clear or at most cloudy. Stems 4-angled;
tubercles rectangular, leaf rudiments ascending to spreading, 0.5–2 mm long. Inflorescences
1–10 per stem, subsessile, arising in the upper half of the stem. Corolla rotate, deeply lobed,
smooth to rugulose, papillate. Gynostegial corona in 2 series: "outer" corona of 5 fused lobes
forming a disc; "inner" staminal corona lobes arched over backs of anthers. Gynostegium
subsessile. Pollinia D-shaped, longer than broad. Follicles paired, erect, terete, glabrous.

Two species in southern Africa.

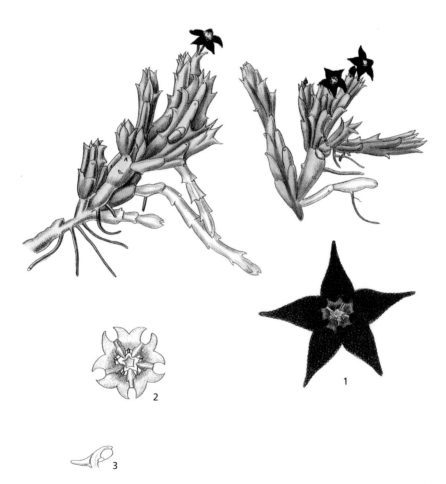

Fig. 7.3.**175**. AUSTRALLUMA UBOMBOENSIS. Main image, flowering shoots; 1, flower from
above; 2, gynostegium and corona from above; 3, stamen. Drawn by Cynthia Letty. Reproduced
with permission from Flowering Plants of South Africa (1932).

---

[29] by D.J. Goyder

Australluma ubomboensis (I. Verd.) Bruyns, Stapeliads S. Africa Madagascar **1**: 63 (2005). Type: South Africa, KwaZulu-Natal, Lebombo [Ubombo] Mts, iii.1930, *Pole Evans* s.n. in Nat. Herb. 8764 (PRE). FIGURE 7.3.**175**.

Caralluma ubomboensis I. Verd. in Fl. Pl. South Africa **12**: t.443 (1932).
Pachycymbium ubomboensis (I. Verd.) M.G. Gilbert in Bradleya **8**: 25 (1990).
Angolluma ubomboensis (I. Verd.) Plowes in Excelsa **16**: 119 (1994).
Orbea ubomboensis (I. Verd.) Bruyns in Aloe **37**: 76 (2000).
Ceropegia ubomboensis (I. Verd.) Bruyns in S. African J. Bot. **112**: 425 (2017).

Clump-forming stem-succulent. Stems erect from underground rhizomes or nearly prostrate and lacking rhizomes, 1.5–8 cm long, 4–10 mm wide, 4-angled; tubercles obscure with short spreading deltoid tooth. Inflorescences 1–3 near apex, shortly pedunculate. Pedicels 1–7 mm long, erect. Sepals 1.5–2 mm long, ovate-lanceolate. Corolla rotate, deeply lobed, red, maroon or dark purple brown, transversely rugulose and papillate; corolla tube very shallow; lobes spreading to recurved, broadly triangular to ovate, 3–7 cm long, margins lacking cilia. Outer corona lobes 1–1.5 mm long, spreading onto corolla, bifid, dark purple-brown, red or maroon; inner corona lobes 0.5–1 mm long, white with reddish margins to wholly maroon.

Zimbabwe. S: 6 km S of Gutu, fl. 13.i.1964, *Leach* 12062 (K, SRGH). **Mozambique**. MS: 38 km N of Save, 2.i.2001, *Bruyns* 8750 (BOL, K).

Also known from South Africa (Limpopo & KwaZulu-Natal Provinces) and Swaziland. Dry bush, frequently on rock; 50–1400 m.

58. **EMICOCARPUS** K. Schum. & Schltr.[30]

Emicocarpus K. Schum. & Schltr. in Bot. Jahrb. Syst. **29**(Beibl. 66): 21 (1900). — Brown in Fl. Cap. **4**(1): 518 (1907).
Lobostephanus N.E. Br. in Hooker's Icon. Pl. **27**: t.2692 (1901).

Prostrate herb; latex colour and rootstock not recorded. Leaves opposite, subentire to deeply dissected. Inflorescences extra-axillary, pedunculate, umbelliform. Corolla lobed almost to the base. Corona in two series. Outer corona corolline, arising from the fused basal portion of the corolla, with a minute tubular basal portion, and 20 erect lobes in two whorls, the outer consisting of 5 simple segments opposite the corolla-lobes, and the inner of 5 tripartite segments alternating with them, with their lateral lobes infolded so as to stand in front of the simple outer lobe, as if they were appendages arising from the base of it. Gynostegial corona of 5 lobes inflexed over the backs of the anthers. Pollinaria with pendant pollinia attached by long abruptly bent translator arms. Stylar head produced into a long rostrate appendage exserted beyond the anthers. Follicles indehiscent, only one of the pair developing, obtriangular-pyramidal, the points developed into 3 spreading spines; seeds single, curved, subterete, lacking a tuft of hairs at either end.

A monotypic genus with assumed affinities to the South African genus *Eustegia* R. Br.

Emicocarpus fissifolius K. Schum. & Schltr. in Bot. Jahrb. Syst. **29**(Beibl. 66): 21 (1900). Type: Mozambique, Maputo [Lourenço Marques], Delagoa Bay, 30.xi.1897, *Schlechter* 11535 (K000305354 lectotype, B†, BR, E, GRA, HBG, K000305352, K000305353, PRE, NU, S, SAM, WAG), lectotype designated here. FIGURE 7.3.**176**.

Lobostephanus palmatus N.E. Br. in Hooker's Icon. Pl. **27**: t.2692 (1901). Type: Mozambique, Delagoa Bay, 1893, *Junod* 502 (K holotype, Z).

Prostrate herb; stems to 1.3 m, minutely pubescent along one line. Leaves with petiole to c.5 mm long, minutely pubescent; lamina of upper leaves subentire, linear with short, laterally directed basal teeth; lamina of lower leaves more deeply dissected, palmate, with variable

[30] by D.J. Goyder

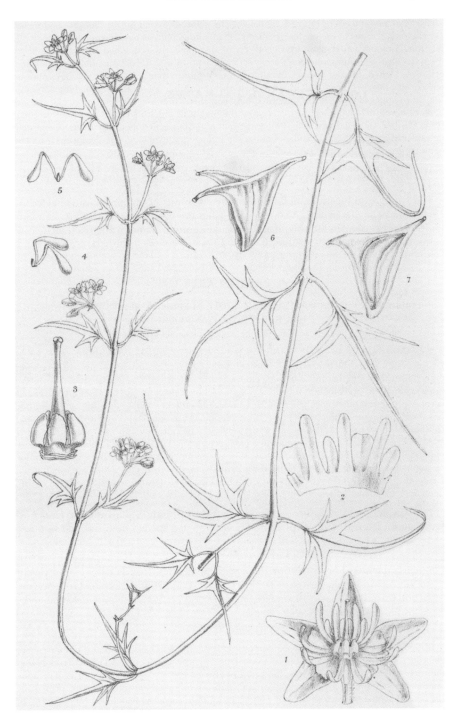

Fig. **7.3.176**. EMICOCARPUS FISSIFOLIUS. Main image, flowering shoot; 1, flower; 2, corona; 3, gynostegium; 4, 5, pollinaria; 6, 7, indehiscent fruits. Reproduced from Hooker's Icones Plantarum (1901).

numbers of lateral teeth or lobes, 2–4 × 0.5–3.5 cm, glabrous except for the margins. Peduncles 1–2 cm long, pubescent on one line; pedicels 2–3 mm long, pubescent along one line. Sepals c.1 mm long, oblong, acute, glabrous. Corolla cream or tinged with red, lobed almost to the base; fused basally for c.1 mm; lobes apparently reflexed at anthesis, c.2 × 0.7 mm, oblong, subacute, glabrous. United, tubular portion of outer corona no more than 0.5 mm long; lobes opposite teh corolla lobes shorter and broader than the remaining lobes, 0.6–1 mm long, oblong; remaining lobes of outer series somewhat narrower, 1–1.5 mm long. Inner, gynostegial corona lobes dorsiventrally flattened, just exceeding the height of the anthers and appressed to their backs. Anther wings 0.3 mm long. Pollinaria minute. Stylar head appendage rostrate, exserted beyond anthers for c.1.5 mm. Follicles solitary, indehiscent, c.1 cm long and measuring 1–1.3 cm between the triangular distal projections, glabrous. Seeds solitary, ecomose.

Mozambique. M: Marracuene, Maxaquene, fl. & fr. 19.iv.1961, *Macuácua* 95 (K, LMA).

The five known localities all fall within the city limits of Maputo. Last collected in 1966 from near the airport. Also recorded from the experimental grounds between Universidade Eduardo Mondlane and Avenida Kenneth Kaunda, an area now heavily cultivated. Sandy ground; c.30 m.

59. **PERGULARIA** L.[31]

Pergularia L., Syst. Nat. ed. 12, **2**: 191 (1767); in Mant. Pl.: 8 (1767). —Goyder in Kew Bull. **61**: 245–256 (2006).
    Doemia R. Br., On the Asclepiadeae: 39 (1810).

Tomentose, pubescent or occasionally glabrous twiners or twining shrubs with white latex. Leaves ovate to suborbicular, strongly cordate. Inflorescences extra-axillary, initially appearing umbelliform but lengthening into a raceme. Corolla with a short tube and spreading lobes. Corolla adnate to gynostegium for whole length of tube. Corolline corona absent. Gynostegial corona in 2 series; outer corona ring with 5 interstaminal lobes arising from the column at or just above the mouth of the corolla tube; inner corona of 5 staminal lobes adnate to the staminal column about half way up the anther wings, semisagittate in outline with a free subulate apex arched over the head of the column and a basal projection. Pollinaria with pendant, subsessile pollinia with translucent outer margins. Follicles generally paired, reflexed on the pedicel or occasionally spreading, fusiform to narrowly ovoid, curved slightly into a long attenuate beak, ± smooth to strongly echinate, glabrous or pubescent. Seeds flattened, ovate, margins entire, dentate or crenulate, pubescent on both sides.

Two species in Africa, Arabia and eastwards through India and Sri Lanka to Bangladesh. Records from China and SE Asia refer to species now assigned to other genera.

Pergularia daemia (Forssk.) Chiov., Res. Sci. Somalia Ital. **1**: 115 (1916). —Goyder in F.T.E.A., Apocynaceae (part 2): 364 (2012). Type: Yemen, Zabid, 1763, *Forsskål* (not traced); Sana'a, 10.xi.1937, *Rathjens* 37/7 (BM neotype), designated by Goyder in Kew Bull. **61**: 249 (2006).
    Asclepias daemia Forssk., Fl. Aegypt.-Arab.: 51 (1775).

Scrambling herbaceous twiner, sometimes woody towards the base; stems and inflorescences glabrous or pubescent with stiff, spreading hairs. Leaves with petiole (1)2–5(12) cm long, minutely pubescent; lamina thin, 2–11 × 1.5–11 cm, broadly ovate to suborbicular, apex attenuate to acuminate, base strongly cordate, glabrous to pubescent with stiff hairs principally restricted to veins beneath to densely pubescent on both surfaces. Peduncles (2)4–11(20) cm long, glabrous or pubescent; pedicels 1.5–4 cm long. Sepals 1.5–4 × 0.6–1 mm, narrowly ovate

---

[31] by D.J. Goyder

to oblong, acute, glabrous or pubescent. Corolla greenish white or yellow-green, sometimes marked with pink or brown outside; tube 2–6 mm long; lobes 5–10(16) × 1.8–5 mm, ovate or oblong to elliptic, subacute or acute, glabrous abaxially, upper surface bearded towards the margins and frequently at the base of the lobes, otherwise glabrous. United, tubular portion of outer corona barely visible beyond mouth of corolla tube or exserted for 0.5–1 mm, lobes arising from the interstaminal portions of the tube c.0.5–1 × 0.5–1.5 mm, oblong, truncate, entire or shallowly toothed. Inner corona lobes slender to stout, 2.5–7(9) mm long, the apical projection reaching to the top of the column, often projecting much further, the basal tails spreading outwards from the column. Filament tube extending 2–4 mm beyond mouth of corolla tube but frequently obscured by staminal corona lobes; anther wings 0.8–2 mm long. Pollinaria with laterally compressed corpusculum c.0.3 × 0.1 mm in dorsal view, with a keel c.0.2 mm deep viewed laterally; translator arms extremely short, c.0.2 mm long, flattened and curved; pollinia flattened, c.0.6–0.9 × 0.3 mm, oblanceolate, the outer margin of each pollinium translucent for most of its length. Follicles solitary or paired, 5–8 × 0.8–2 cm, narrowly ovoid to fusiform with a long, curved beak, subglabrous to densely pubescent, with or without soft, pubescent processes. Seeds 4–7 × 2–4 mm, flattened, ovate, entire or sometimes with a somewhat crenulate or dentate margin, densely pubescent on one or both faces.

Corolla tube much shorter than the lobes, 2–5 mm long, campanulate; united tubular
    portion of outer corona generally exserted from mouth of corolla tube by less
    than 0.5 mm; gynostegial stipe largely obscured by staminal corona lobes; anther
    wings (1.1)1.3–2 mm long. . . . . a) subsp. *daemia*
Corolla tube ± equal in length to the lobes, 4–6 mm long, narrowly cylindrical; united
    tubular portion of outer corona exserted for c.1 mm from mouth of corolla tube;
    gynostegial stipe not obscured by staminal corona lobes; anther wings 0.8–1 mm
    long. . . . . .b) subsp. *barbata*

a) Subsp. **daemia**. FIGURE 7.3.**177**.

> *Cynanchum cordifolium* Retz., Observ. Bot. **2**: 15 (1781). Type not traced.
> *Cynanchum extensum* Jacq., Misc. Austriac. **2**: 353 (1781); in Icon. Pl. Rar. **1**: 6, t.54 (1782). Type: Jacquin, Icon. Pl. Rar. **1**: t.54 (1782), lectotype designated by Liede in Asklepios **51**: 65 (1990).
> *Cynanchum bicolor* Andrews in Bot. Repos. **9**: t.562 (1809). Type: Bot. Repos. **9**: t.562 (1809), iconotype.
> *Doemia extensa* (Jacq.) W.T. Aiton, Hort. Kew. ed. 2, **2**: 76 (1811).
> *Cynanchum echinatum* Thunb., Observ. Cynanch.: 8 (1821). Type: *Herb. Thunberg* 6298 (UPS holotype).
> *Doemia bicolor* Sweet, Hort. Brit. ed. 2: 361 (1830). Type as for *Cynanchum bicolor.*
> *Doemia scandens* (P. Beauv.) Loud., Hort. Brit.: 94 (1830).
> *Doemia angolensis* Decne. in Ann. Sci. Nat. sér. 2 **9**: 337 (1838). Type: Angola (P holotype).
> *Raphistemma ciliatum* Hook. f. in Bot. Mag. **94**: t.5704 (1868). Type: "*Raphistemma* ?sp. nov. Hort. Kew 10/67" (K holotype).
> *Doemia cordifolia* (Retz.) K. Schum. in Abh. Königl. Akad. Wiss. Berlin **1894**: 17 (1894); in Engler, Pflanzenw. Ost-Afrikas **C**: 324 (1895).
> *Doemia barbata* Schltr. in Bot. Jahrb. Syst. **20**(Beibl. 51): 43 (1895), illegitimate name, not Klotzsch (1861). Type: South Africa, near Ramakopa, 4000', 15.ii.1894, *Schlechter* 4507 (B† holotype, BOL).
> *Pergularia extensa* (Jacq.) N.E. Br. in Fl. Cap. **4**(1): 758 (1908).

Stems and inflorescences pubescent. Leaf lamina 2.5–11 × 2.5–11 cm, pubescent. Corolla tube 2–4(5) mm long, campanulate, c.1/2–1/3 the length of the lobes; lobes (5)6–10(16) × 3–5 mm, ovate to elliptic. United portion of outer corona generally barely visible beyond mouth of corolla tube but occasionally exserted for 0.5–1 mm Inner corona lobes slender to stout, 3.5–7(9) mm long, the apical projection reaching at least 1 mm beyond top of column, often much more. Anther wings (1.1)1.3–2 mm long. Follicles fully or partially recurved alongside the pedicel, pubescent, sometimes covered with soft, pubescent processes to c.1 cm in length, processes sometimes reduced or absent. Seeds with an entire or somewhat dentate margin.

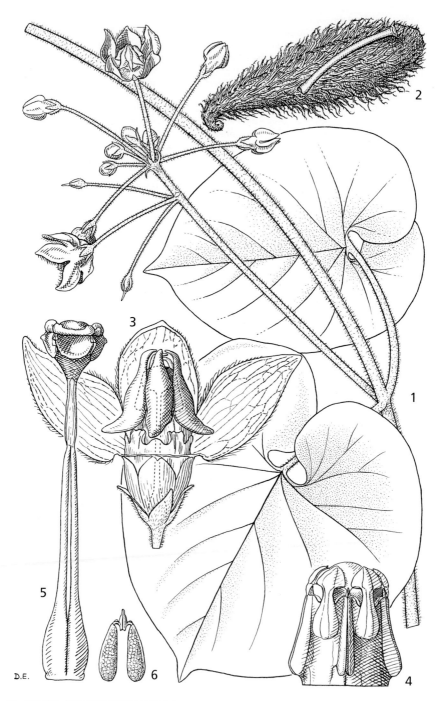

Fig. 7.3.**177**. PERGULARIA DAEMIA subsp. DAEMIA. 1, habit (× 1); 2, single follicle (× 1); 3, flower (× 4); 4, upper part of gynostegium with corona removed (× 8); 5, paired ovaries and stylar head (× 8); 6, pollinarium (× 12). All from *Faulkner* 2317. Drawn by D. Erasmus. First published in Flora of Ethiopia and Eritrea (2003).

Caprivi. About 60 km from Katima on road to Linyanti, 27.xii.1958, *Killick & Leistner* 3132 (K, PRE). **Botswana**. N: Botletle River, 9 km E of Makalamabedi, fl. 21.iii.1965, *Wild & Drummond* 7212 (K). SW: Deception Pan, Central Kalahari Game reserve, fr. iv.1975, *Owens* 24 (K, PSUB, SRGH). SE: Ilalamabele–Mosu area, near Soa Pan, fl. 9.i.1974, *Ngoni* 292 (K, SRGH). **Zambia**. N: Mwengo, Abercorn [Mbala] Dist., fl. 9.vi.1951, *Bullock* 3947 (K). **Zimbabwe**. E: Chipinga Dist., Gungunyana Forest Reserve, fl. iii.1964, *Goldsmith* 8/64 (K, SRGH). S: Gwanda Dist., Special Native Area "G", fl. 17.xii.1956, *Davies* 2326 (K, SRGH). **Malawi**. S: Mchesi Mt., Nabwato valley stream, Mulanje Massif, fl. 3.vi.1988, *Chapman* 9127 (K, MO). **Mozambique**. Z: Estrada Milange–Quelimane, 5 km do 1ª cruzamento para Mocuba, fl. 20.v.1949, *Barbosa & Carvalho* 2761 (K, LMA). MS: 20 km SW of Dombe, just N of Lucite River, fl. 22.iv.1974, *Pope & Müller* 1238 (K, SRGH). GI: Bazaruto Island, N extremity, near lighthouse, fl. & fr. viii.1937, *Gomes e Sousa* 1994 (K). M: 4 km W of bridge over Tembe River, c.30 km W of Bela Vista, fl. 4.xii.2001, *Goyder* 5028 (K, LMU).

Widespread across drier tropical or subtropical regions of subsaharan Africa, the Arabian peninsula and the Indian subcontinent. Scrambling over shrubs in dry bushland or occurring near seasonal watercourses in more arid areas; 50–1750 m.

b) Subsp. **barbata** (Klotzsch) Goyder in Kew Bull. **61**: 251 (2006). Type: Mozambique, 'Inhambane' *Peters* s.n. (K lectotype, B†), designated by Goyder in Kew Bull. **61**: 251 (2006).

  Doemia barbata Klotzsch in Peters, Naturw. Reise Mossambique **6**(1): 274 (1861).
  Pergularia barbata (Klotzsch) Brenan in Mem. New York Bot. Gard. **8**: 505 (1954), non *Doemia barbata* Schltr. in Bot. Jahrb. **51**: 43 (1895).

Stems and inflorescences pubescent. Leaf lamina 2.5–8(11) × 2.5–8(11) cm, pubescent. Corolla tube 4–6 mm long, cylindrical, ± equal to the lobes; lobes 5–7 × 1.8–3 mm, oblong to elliptic. United portion of outer corona extending for 0.5–1 mm beyond the mouth of the corolla tube. Inner corona lobes very slender, 2.5–5 mm long, the apical projection reaching ± to top of column. Anther wings 0.8–1 mm long. Follicles spreading at 180° to each other and at right angles to the pedicel, but frequently somewhat recurved distally, densely pubescent, frequently with scattered low warty processes. Seeds not seen.

Zambia. B: Machili, fl. 10.i.1961, *Fanshawe* 6113 (K, NDO). S: Gwembe Valley, bank of Zongwe River, fl. & fr. 1.vii.1961, *Angus* 2949 (FHO, K). **Zimbabwe**. N: Old Chiswiti, Darwin Dist., fl. & fr. 5.vi.1965, *Bingham* 1481 (K, SRGH). W: Deka River, Hwange [Wankie], fl. 24.vi.1934, *Eyles* 8050 (K). E: Chimanimani Hot Springs, fl. 21.x.1948, *Chase* 1190 (K, SRGH). S: Chitsa's Kraal, fl. 12.vi.1950, *Wild* 3498 (K, SRGH). **Malawi**. N: 15 km S of Karonga on Chilumba road, fl. 26.v.1989, *Goyder et al.* 3286 (K, MAL). Lower Mwanza River, Chikwawa Dist., fl. 6.x.1946, *Brass* 18019 (K). **Mozambique**. T: Mazoe River 5 km from Zimbabwe [S. Rhodesia] border, fl. & fr. 22.ix.1948, *Wild* 2588 (K, SRGH). MS: Boka, Lower Buzi, fl. xii.1906, *Swynnerton* 1916 (K). GI: 25 km S of Mambone, Save River, fl. 10.x.1963, *Leach & Bayliss* 11898 (K, SRGH).

Zambezi valley and tributaries. Thickets and riverine vegetation; 100–600 m.

Zambian material has a glabrous calyx, whereas in other regions the calyx is pubescent. Some material from northern Botswana (e.g. *P.A. Smith* 1325 (K) – Botswana N, Dindinga Island) is intermediate between subsp. *barbata* and subsp. *daemia* in that fruit characters suggest placement in subsp. *barbata*, but the short corolla tube is more characteristic of subsp. *daemia*.

60. **CALOTROPIS** R. Br.[32]

Calotropis R. Br., On the Asclepiadeae: 28 (1810).
Madorius Kuntze, Rev. Gen. Pl. **2**: 421 (1891).

Erect shrubs with large succulent leaves. Inflorescences terminal and extra-axillary, pedunculate, with several clusters of umbelliform cymes. Corolla c.3 cm in diam., lobed for c.²/₃ of its length. Corolline corona absent. Gynostegial corona lobes laterally compressed with an upturned spur at the base, radiating out from the gynostegium to which they are adnate for their entire length. Pollinaria with oblong, slightly winged corpuscula and short, cylindrical translator arms; pollinia flattened, pear-shaped. Style head pentagonal, depressed or raised slightly above the top of the anthers. Follicles large, ovate or subglobose, smooth, usually only one of the pair developing, with spongy fibrous mesocarp and a papery endocarp. Seeds ovate, planoconvex, pubescent.

Three species in Africa, Arabia and southern and SE Asia, introduced elsewhere in the tropics.

Corona lobes oblong, truncate, broadest at the top; corolla campanulate; follicles inflated, subglobose. . . . . **1.** *procera*
Corona lobes ovate, rounded, broadest around the middle; corolla rotate or slightly reflexed; follicles not inflated, ovate in outline. . . . . **2.** *gigantea*

1. **Calotropis procera** (Aiton) W.T. Aiton, Hort. Kew. ed. 2, **2**: 78 (1811). —Ali in Notes Roy. Bot. Gard. Edinburgh **38**: 290 (1980). —Goyder in F.T.E.A., Apocynaceae (part 2): 368 (2012). Type: Jacq., Observ. Bot. **3**: t.69 (1768) as *Asclepias gigantea*, lectotype designated by Ali in Notes Roy. Bot. Gard. Edinburgh **38**: 290 (1980); topotype from Jamaica. FIGURE 7.3.**178**.

?*Apocynum syriacum* S.G. Gmel., Reise Russland **2**: 198 (1774), nomen nudum.
Asclepias procera Aiton, Hort. Kew. **1**: 305 (1789).
Madorius procerus (Aiton) Kuntze, Revis. Gen. Pl. **2**: 421 (1891).
Calotropis syriaca (S.G. Gmel.) Woodson in Ann. Missouri Bot. Gard. **17**: 148 (1930), illegitimate name.
Calotropis gigantea var. *procera* (Aiton) P.T. Li in J. S. China Agric. Univ. **12**(3): 39 (1991).

Stout, weakly to strongly branched succulent shrub, 2–5 m tall; young parts shortly tomentose, often appearing farinose, glabrescent. Leaves subsessile or shortly petiolate, the petiole up to 5 mm long; lamina 7–26 cm long, 4–15.5 cm wide, ovate, oblong-ovate, elliptic or more usually broadly obovate, apex rounded or obtuse with a short acuminate tip, base cordate, often obscuring the petiole, slightly succulent, with conspicuous midrib and arched lateral veins, glaucous. Inflorescences many-flowered, usually with at least 2 major branches each with successive subumbelliform clusters of flowers. Peduncles robust, 3–8 cm long, very occasionally absent; bracts caducous, 7–12 mm long, ovate-lanceolate, acute. Pedicels slender, 10–35 mm long. Calyx lobes 4–5mm long, 3–4 mm wide, ovate, acute or subacute. Corolla campanulate, united for c.¹/₃ of its length, white with purple tips to the lobes, glabrous; lobes erect or spreading, 7–10 mm long, 6–7 mm wide, ovate, acute. Corona white or purple; lobes 5–6 mm long, c.3 mm wide, oblong, obliquely truncate or rounded, cleft radially in the upper half, minutely scabrid along the margins, basal spur c.2 mm long. Gynostegium c.7 mm high, the head of the column forming a drum c.5 mm in diam., 2 mm deep. Corpusculum c.0.4 mm long, slightly winged; translator arms 0.2–0.3 mm long; pollinia 1.2–1.5 mm long, 0.6–0.7 mm wide. Follicles 8–13 cm long, ovoid to subglobose, with one side somewhat flattened, obtuse or depressed at the apex. Seeds c.7 mm long, 5 mm wide, ovate, with a narrow rim, minutely pubescent.

Zambia. N: Lake Tanganyika, Mpulungu, fl. 6.iii.1952, *Richards* 1371 (K). C: Katondwe, fl. 13.iv.1988, *Phiri* 2425 (K, UZL). **Zimbabwe**. N: Chimanda Tribal Trust Land near Mahutwe, fl. & fr. 18.xi.1971, *Patterson* s.n. in GH214207 (K, SRGH).

[32] by D.J. Goyder

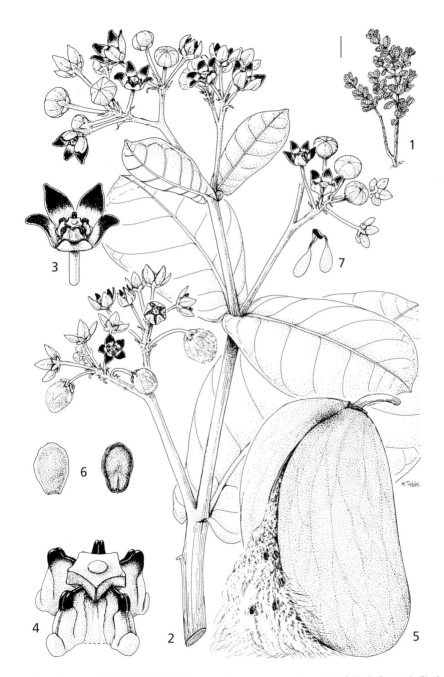

Fig. 7.3.**178**. CALOTROPIS PROCERA. 1, habit; 2, flowering branch (× ²/₃); 3, flower (× 2); 4, gynostegium with corona (× 4); 5, follicle (× ²/₃); 6, seeds (without coma) (× 4); 7, pollinarium (× 8). 1, 7 from *Goyder et al.* 3212; 2–6 from *Greenway & Kirrika* 10969. Scale bar in 1 = 0.5 m. Drawn by Margaret Tebbs. First published in Flora of Ethiopia and Eritrea (2003).

Malawi. N: Chilumba to Karonga, fl. & fr. 16.ix.1975, *Pawek* 10106 (K, MAL, MO, SRGH, UC). C: N of Chitala on Kasache road, fl. & fr. 12.ii.1959, *Robson* 1579 (K). S: 10 km S of Chikwawa near Lengwe Game Reserve, fl. 5.v.1989, *Goyder & Radcliffe-Smith* 3212 (K, MAL, PRE). **Mozambique**. Z: Quelimane Dist., Mocuba, fl. 2.vii.1949, Faulkner Kew 440 (K). T: 3 km S of Tete, fl. & fr. 6.v.1960, *Leach & Brunton* 9933 (K, SRGH). MS: Chemba, no cruzamento da estrada para a Tambara, fl. 4.iv.1962, *Balsinhas & Macuácua* 550 (K, LMA).

Drier parts of tropical Africa, Arabia and south Asia; naturalised elsewhere in the tropics.

Often found at the edge of villages on disturbed or degraded land (often an indicator of overgrazing), usually on sand; 100–900 m

2. **Calotropis gigantea** (L.) W.T. Aiton, Hort. Kew., ed. 2 **2**: 78 (1811). —Goyder in F.T.E.A., Apocynaceae (part 2): 368 (2012). Type: Sri Lanka, *Hb. Hermann* **2**: p.74, no.112 (BM lectotype), designated by Huber in Revis. Handb. Fl. Ceylon **1**(1): 35 (1973); see also Jarvis, Order out of Chaos: 321 (2007).

   Asclepias gigantea L., Sp. Pl.: 214 (1753).
   Madorius giganteus (L.) Kuntze, Revis. Gen. Pl. **2**: 421 (1891).

Like *Calotropis procera* but corolla ovoid-truncate in bud, the lobes rotate or somewhat reflexed, margins revolute. Corona lobes 8–11 mm high, ovate, narrowing gradually towards the rounded upper margin, not split apically, but with an auricle, c.1 mm long, either side of the scabrid margin; basal spur 2–4 mm long. Corpusculum 0.6–0.8 mm long; translator arms c.0.4 mm long; pollinia c.1.6 mm long, 0.6 mm wide. Follicles 7–10 cm long, 2.5–4 cm wide, ovate, acute and somewhat beaked. Seeds 5–7 mm long, 3–4 mm wide.

Mozambique. N: Wimbi Beach, Pemba, fl. & fr. 13.xii.2008, *Goyder & Crawford* 5099 (K, LMA, LMU, P). M: right bank of River Maputo near Salamanga, fl. 29.ix.1947, *Gomes e Sousa* 3619 (K).

Almost certainly introduced in Africa – often associated with Indian settlements where it is used medicinally. Native to southern and SE Asia from India to Indonesia; introduced and becoming naturalised in New World tropics.

Sandy soil on coral rock or bare and degraded land, often at the edge of a village; 1–100 m.

61. KANAHIA R. Br.[33]

Kanahia R. Br., On the Asclepiadeae: 28 (1810). —Field *et al.* in Nordic J. Bot. **6**: 790 (1986).

Erect, multistemmed riverine shrubs branching mainly near the base, glabrous in all parts except the corolla. Leaves opposite with stipules or colleters distributed along interpetiolar line or restricted solely to the leaf axils. Inflorescences extra-axillary; flowers arranged in an indeterminate condensed spiral. Corolla ± rotate to campanulate with a shaggy indumentum of white hairs towards the apex and margins of the lobes. Corolline corona absent. Gynostegium stipitate. Gynostegial corona lobes arising near the base of the anther wings, subglobose or somewhat compressed laterally. Anther wings vertical, frequently somewhat flared at base; anther appendages membranous, inflexed over apex of stylar head. Pollinaria with an ovoid corpusculum, straight or slightly articulated translator arms and pollinia circular or only slightly flattened in section. Stylar head flat. Follicles single or paired, inflated or not; seeds generally inflated, not differentiated into disc and rim.

Two species in tropical Africa and Arabia.

---

[33] by D.J. Goyder

Kanahia laniflora (Forssk.) R. Br. in Salt, Voy. Abyss., App.: 64 (1814), as *Kannahia laniflora*. —Brown in F.T.A. **4**(1): 296 (1902). —Goyder in F.T.E.A., Apocynaceae (part 2): 370 (2012). Type: Yemen, Djöbla, *Forsskål* s.n. (C holotype). FIGURE 7.3.**179**.

Asclepias laniflora Forssk., Fl. Aegypt.-Arab.: 51 (1775).

Kanahia kannah Schult. in Roemer & Schultes, Syst. Veg. **6**: 94 (1820). Type as for *Kanahia laniflora*.

Asclepias laniflora Delile, Cent. Pl. d'Afr. Voy. Caillaud: 49, t.64 (1826), illegitimate name, non Forssk. (1775). Type: Sudan, Mt. Aqarô, *Caillaud* s.n. (MPD holotype).

Kanahia delilei Decne. in Ann. Sci. Nat., sér. 2 **9**: 330 (1838); in Candolle, Prodr. **8**: 537 (1844). Type as for *Asclepias laniflora* Delile, non Forssk.

Kanahia forsskalii Decne. in Candolle, Prodr. **8**: 537 (1844). Type as for *Kanahia laniflora*.

Gomphocarpus glaberrimus Oliv. in Trans. Linn. Soc. London **29**: 110, t.120 (1875). Type: Tanzania, Marenge M'Khali, *Speke & Grant* s.n. (K holotype).

Asclepias glaberrima (Oliv.) Schltr. in J. Bot. **33**: 335 (1895), illegitimate name, not Sessé & Moc. (1888).

Kanahia glaberrima (Oliv.) N.E. Br. in F.T.A. **4**(1): 297 (1902).

Asclepias coarctata S. Moore in J. Bot. **46**: 297 (1908). Type: Mozambique, lower Umswirizwi River, *Swynnerton* 248 (BM lectotype, K), designated here; paralectotype: lower Buzi River, *Swynnerton* 1895 (BM, K).

Kanahia monroi S. Moore in J. Bot. **49**: 156 (1911). Type: Zimbabwe, Victoria, *Monro* s.n. (BM holotype).

Asclepias rivalis S. Moore in J. Bot. **52**: 337 (1914). Type: Angola, Lucalla River, *Gossweiler* 5771 (BM holotype).

Erect, multistemmed shrub to 2.5 m tall; stems branching mainly near the base and arising from a sandy coloured non-tuberous rootstock. Leaves 6–15(20) × 0.3–1.5(2.5) cm, linear-lanceolate, tapering gradually into the attenuate apex and the channeled petiole, margins smooth, glabrous. Inflorescences extra-axillary; peduncles 1.5–9 cm long; rachis up to 1.5(3) cm long; bracts deciduous, basal bract in each inflorescence to c.15(25) × 2(4) mm, lanceolate, acute, other bracts smaller and more filiform; flowers spreading or erect on slender pedicels 1–2.5(3) cm long. Sepals 3–10 × 1–3 mm, ovate to narrowly lanceolate, acute. Corolla ± rotate to campanulate, divided ± to the base or occasionally lobes united for about half their length, cream or white, sometimes tinged with green; lobes 7–10(13) × 2.5–5 mm, ovate to elliptic, subacute, glabrous on outer face, inner face minutely pubescent with a shaggy indumentum of white hairs towards the apex and margins. Gynostegium with stipe 1–2.5 mm long. Corona lobes white, arising near the base of the anther wings and reaching the top of the column or to about halfway, 2–4 × 1–2 mm, subglobose or somewhat compressed laterally, truncate and with a deep or shallow groove distally, the margins sometimes produced into a pair of teeth arching towards or over the head of the staminal column. Anther wings 2–3 mm long, vertical, frequently somewhat flared at base. Pollinaria with corpusculum 0.4–0.5 × 0.2 mm, ovoid, dark brown; translator arms 0.2–0.3 mm long, slender, straight or slightly articulated; pollinia 0.8–1.5 × 0.2–0.4 mm, ovoid-subcylindrical or linear-oblanceolate in outline and slightly flattened in section. Stylar head flat, level with the top of the anthers. Follicles single or paired, inflated or not, (3)3.5–6 × (0.5)1–2 cm, ovoid with a rounded apex to subcylindrical with a slender beak; fruiting pedicel not contorted; seeds generally inflated, not differentiated into disc and rim, 3–4 × 1–2 mm, ovoid, tapered towards the coma, smooth; coma 6–10 mm long.

Zambia. C: Kabwe [Broken Hill], fl. 9.x.1963, *Fanshawe* 8071 (K, NDO). E: Mwangazi River, fl. 26.xi.1958, *Robson* 711 (K). **Zimbabwe**. N: River Chingombe, 48 km from Makuti on Kariba road, fl. 17.ii.1981, *Philcox & Leppard* 8679 (K). W: E of Mbala Lodge on Lukozi River, fl. 21.x.1968, *Rushworth* 1208 (K, SRGH). C: Ruzawi River, Romsley Estate, fl., *Hopkins* s.n. in GH22652 (K, SRGH). E: Rusape Hot Springs, fl. i.1947, *Chase* s.n. in GH16607 (K, SRGH). S: Nuanetsi River Bridge, fl. 3.ii.1973, *Ngoni* 187 (K, SRGH). **Mozambique**. N: Cabo Delgado, Metoro, fl. & fr. 31.i.1984, *Groenendijk et al.* 897 (K, LMU). T: Rio Mucangadzi para Heitor Dias, fl. 8.iv.1972, *Macêdo* 5172 (K, LMA). MS: Gorongosa Dist., Rio Nhandugué near Vunduzi, fl. 23.vii.2006, *Goyder, Massingue & Timberlake* 4090 (K, LMA).

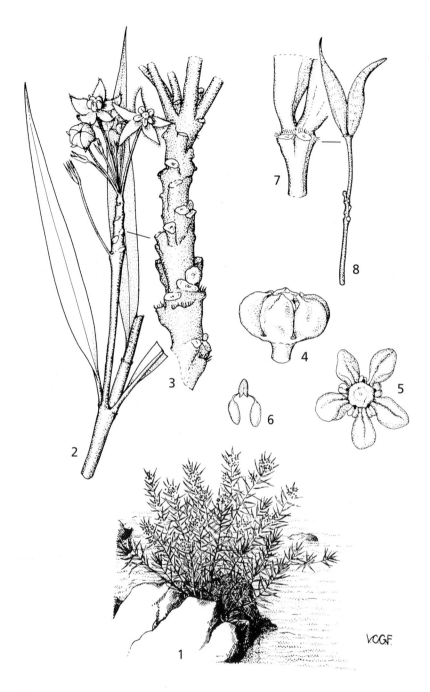

Fig. 7.3.**179**. KANAHIA LANIFLORA. 1, habit; 2, part of flowering branch (× 2/3); 3, axis of inflorescence; 4, 5, lateral and apical view of gynostegium and corona (× 5); 6, pollinarium (× 10); 7, base of paired follicles; 8, infructescence (× 2/3). 2–6 from *Mooney* 8580; 7, 8 from *Mesfin & Kagnew* 1815. Drawn by Victoria C. Friis. Adapted from Nordic Journal of Botany (1986).

Widely distributed in tropical Africa from Limpopo Province of South Africa northwards to Ivory Coast in the west and Somalia in the east, also known from Yemen and Saudi Arabia.

Growing in sand or among rocks in seasonally inundated watercourses; 100–1000 m.

62. PERIGLOSSUM Decne.[34]

Periglossum Decne. in Candolle, Prodr. **8**: 520 (1844). —Bester & Nicholas in Phytotaxa **282**: 28–36 (2016).

Perennial herbs with erect annual stems arising from a vertical napiform tuber; latex white. Leaves opposite, linear. Inflorescences extra-axillary at upper nodes, formed of globose, pedunculate clusters of flowers. Corolla lobed almost to the base. Corolline corona absent. Gynostegium narrowed basally and apically to form a barrel-shaped structure. Gynostegial corona of 5 erect and somewhat fleshy, dorsiventrally flattened staminal lobes; interstaminal lobes rudimentary or filiform. Anther wings minute and broadly triangular, positioned at the top of the highly convex and inflated anthers; anther appendages conspicuous, inflexed over the stylar head. Pollinaria with a minute corpusculum, and slender sigmoid curved translator arms which are at least twice as long as the pollinia. Stylar head completely obscured by the anther appendages. Follicles apparently single by abortion, slender, fusiform, erect; seeds flattened, with a terminal plume of hairs.

Three species in Southern Africa. Formerly confused with the morphologically similar *Cordylogyne*, the two genera appear in separate well supported clades according to Bester & Nicholas in Phytotaxa **282**: 28–36 (2016), citing studies by Goyder *et al.* in Ann. Missouri Bot. Gard. **94**: 423–434 (2007) and Bester (unpublished data).

Periglossum mackenii Harv., Thes. Cap. **2**: 7, t.111 (1863). —Brown in Fl. Cap. **4**(1): 584 (1907). Type: South Africa, near Durban, *Gerrard & McKen* 664 (TCD holotype, K, NH). FIGURE 7.3.**180**.

> *Periglossum kassnerianum* Schltr. in Bot. Jahrb. Syst. **20**(Beibl. 51): 40 (1895). —Brown in Fl. Cap. **4**(1): 584 (1907). Type: South Africa, Kl. Olifant Rivier, 22.xii.1893, *Schlechter* 4043 (AMD, B†, BOL, GRA, K, MEL, NH, PRE).
> *Periglossum mossambicense* Schltr. in Bot. Jahrb. Syst. **38**: 33 (1905). —Brown in Fl. Cap. **4**(1): 583 (1907). Type: Mozambique, Dondo [25-mile station] near Beira, fl. iv.1898, *Schlechter* 12284 (B† holotype); Bot. Jahrb. Syst. **38**: 33, fig.2 (1905), lectotype designated here.
> *Cordylogyne kassnerianum* (Schltr.) Eyles in Trans. Roy. Soc. South Africa **5**: 448 (1916).
> *Cordylogyne mossambicense* (Schltr.) Eyles in Trans. Roy. Soc. South Africa **5**: 448 (1916).

Perennial herb with slender vertical, tuber. Stems (10)30–50 cm tall, erect, simple or occasionally branched at base, glabrous except for two lines along the stem that are minutely pubescent. Leaves often reduced towards the base of the stem; upper leaves 40–140 × 0.5–3 mm, spreading or gently ascending, narrowly linear, acute, with revolute margins, glabrous except for the midrib below. Inflorescences 1 or more at upper nodes; peduncles 1–3(13) cm long, pubescence more more conspicuous than on the stems, flowers 6–20 or occasionally more in a subglobose head; pedicels 1–1.5 mm long, white-pubescent. Calyx lobes 2–3.5 × 1 mm, ovate to lanceolate, acute, pubescent, often reddish. Corolla lobed almost to the base, frequently tinged reddish outside, green within; lobes held erect, 3–5 × 1–1.5 mm, oblong or lanceolate, glabrous. Corona lobes 3–4 mm long, c.1 mm wide, slightly taller thatn the staminal column, dorsiventrally flattened and somewhat fleshy, erect, lanceolate to oblanceolate, rounded or subacute apically, tapering gradually or more abruptly towards the base, inner face somewhat swollen above with transverse and/or vertical lines or ridges. Gynostegium 2–3 mm high. Follicles 80–90 × 3–4 mm, narrowly fusiform with an attenuate beak, minutely pubescent; seeds c.4 mm long, 3 mm wide, ovate and slightly verrucose, with a narrow inflated rim.

---

[34] by D.J. Goyder

Botswana. N: Moremi Game Reserve, Khwai River floodplain, fl. & fr. 27.xi.2007, *A. & R. Heath* 1465 (GAB, K, PSUB). **Zambia**. W: Mashi/Caprivi, fl. 4.xi.1962, *Fanshawe* 7131 (K, NDO). **Zimbabwe**. N: Sasame River source, 2 km from Gokwe on Kwekwe [Que-Que] road, fl. 7.i.1963, *Bingham* 377 (K, SRGH). W: Matopos, Besna Kobila Farm, fl. xi.1956, *Miller* 3916 (K, SRGH). C: Harare [Salisbury], between Avondale West and Mabelreign, fl. 23.x.1955, *Drummond* 4922 (K, SRGH). E: Mutare [Umtali], fl. 4.xi.1953, *Chase* 5146 (K, SRGH). **Mozambique**. Z: Namagoa, 200 km inland from Quelimane, fl. iii.1943, *Faulkner* 145 (K, LMA, PRE). GI: Maua, Tuane, Rio Uagunumbo, fl. 10.iii.1970, *Balsinhas* 1616 (LMA). M: Costa do Sol, fl. 4.xi.1963, *Balsinhas* 657 (K, LMA).

Widely distributed in Southern Africa. Seasonally flooded or moist grassland. Sea level to 1500 m.

I have retained a broader concept of this species than that of Bester & Nicholas in Phytotaxa **282**: 28–36 (2016). There is subtle variation in both the shape and the ventral ornamentation of the staminal corona lobes, but these are often impossible to observe in herbarium material and I have failed to apply them taxonomically with any degree of confidence. The type of *Periglossum mossambicense* is no longer extant, and it appears that N.E. Brown, and subsequently Bester and Nicholas, misinterpreted a specimen at Kew collected by Junod. Schlechter made no mention of the filiform interstaminal lobes 2 mm long that N.E. Brown regarded to be diagnostic, and the illustration accompanying the diagnosis, which I propose to designate as the lectotype of *P. mossambicense*, does not depict this structure either. On manipulation of the one flower of *Junod* 189 that appeared to have a filiform interstaminal corona lobe, it turned out to be another staminal corona lobe seen edge-on. Any interstaminal structures are at most vestigial.

South African *Periglossum angustifolium* Decne. is clearly distinct with its sagittate corona lobes, and the newly described *P. podoptyches* Nicholas & Bester from KwaZulu-Natal has distinctive basal structures on the ventral face of the corona lobe not seen in elsewhere in the genus.

Fig. 7.3.**180**. PERIGLOSSUM MACKENII. 1, flowering shoot; 2, inflorescence. Photographed by B. Wursten.

63. **STENOSTELMA** Schltr.[35]

Stenostelma Schltr. in Bot. Jahrb. Syst. **18**(Beibl. 45): 6 (1894). —Bullock in Kew Bull. **7**: 417 (1952). —Bester & Nicholas in Phytotaxa **361**: 41–55 (2018).
Krebsia Harv. in Gen. S. Afr. Pl., ed. 2: 233 (1868), illegitimate name, non Eckl. & Zeyh., Enum. Pl. Afric. Austral.: 179 (1836).

Perennial herbs, annual stems simple or branching near the base from a napiform tuber. Latex white. Inflorescences extra-axillary, sessile or pedunculate, subglobose. Corolla lobed almost to the base, the lobes somewhat pouched below, convex and reflexed above. Corolline corona absent. Gynostegial corona lobes [in the Flora region] well exserted from the flower, fleshy, subulate or slightly winged and with a quadrate swelling near the base on the outer surface. Anther appendages erect, ovate, scarious, completely obscuring the truncate or conical stylar head. Pollinia attached apically to the rather long, winged translator arms. Follicles lanceolate, erect.

Eight species in southern and south tropical Africa. Bester & Nicholas (2018) indicate that *Stenostelma* is an early-divergent lineage of the Asclepiadinae.

Inflorescences pedunculate; calyx 1.5–2 mm long; corolla lobes 3 mm long; follicles smooth . . . .**1.** *capense*
Inflorescences sessile or subsessile; calyx 4–5 mm long; corolla lobes 5–6 mm long; follicles ornamented with soft protruberances . . . . **2.** *corniculatum*

1. **Stenostelma capense** Schltr. in Bot. Jahrb. Syst. **18**(Beibl. 45): 6 (1894). —Bullock in Kew Bull. **7**: 417 (1952). Type: South Africa, near Kimberley, xii.1892, *Flanagan 1693* (PRE 0345553-0 lectotype, BM, BOL, GRA, K), lectotype designated by Bester & Nicholas in Phytotaxa **361**: 43 (2018). FIGURE 7.3.**181**.
    Schizoglossum aciculare N.E. Br. in F.T.A. **4**(1): 363 (1902); in Fl. Cap. **4**(1): 620 (1907). Type: Botswana, Ngamiland, near Kwebe, xii.1896, *Lugard 82* (K holotype).

Perennial herb with vertical, tuberous rootstock. Stems 4–45 cm, branched near the base, suberect or spreading initially then becoming erect, pubescent, with short white hairs restricted mainly to two lines along the stem. Leaves often reduced towards the base of the stem; upper leaves 20–70(90) × 0.6–2.5(3.0) mm, erect or ascending, linear, acute, with revolute margins, pubescent on upper surface and on the midrib below. Umbels 1 to many, extra-axillary; peduncles 5–16 mm, bifariously pubescent, with 5–10 flowers in a subglobose head; pedicels c.3 mm long, pubescent, becoming contorted in fruit. Calyx lobes 1.5–2 × 0.5–1 mm, ovate to lanceolate, acute, pubescent. Corolla lobed almost to the base, purplish green, lobes c.3 × 1.5 mm, lanceolate, somewhat attenuate with a subacute apex, the lower part concave giving the corolla the appearance of being subglobose, the upper half convex, reflexed somewhat swollen towards the apex and minutely papillate. Corona lobes fleshy, c.3.5 mm long, cream, attached immediately below the anthers, the lower portion a subrhomboid plate c.1 mm long, included within the concave portion of the corolla 'tube' and about equalling the gynostegium, the inner face slightly pouched at the base and with two basal auricles forming minute inflexed teeth, the upper portion an erect or slightly inward-pointing, subulate horn 2–2.5 × 0.25 mm. Follicles 30–55 × 3–4 mm, narrowly lanceolate with an attenuate, occasionally bifid apex, somewhat contorted at the base, pubescent.

Botswana. N: Ngamiland, near Kwebe, fl. xii.1896, *Lugard 92* (K). **Zambia**. C: Kabwe [Broken Hill], fl. 9.x.1963, *Fanshawe 8041* (K, NDO). **Zimbabwe**. W: Plumtree, fl. 17.ii.1922, *Eyles 3290* (K). C: Strathavon, Harare [Salisbury], fl. 23.x.1955, *Drummond 4917* (K, SRGH).
Also in South Africa and Namibia. Drier ground in dambos, flowers after burning; 900–1500 m.

---

[35] by D.J. Goyder

2. **Stenostelma corniculatum** (E. Mey.) Bullock in Kew Bull. **7**: 417 (1952). Type: South Africa [Eastern Cape Province], near Table Mountain, Queenstown Div., 7.xii.1832, *Drège* 3423 (P lectotype, K), lectotype designated by Bester & Nicholas in Phytotaxa **361**: 43 (2018).

> *Lagarinthus corniculatus* E. Mey., Comm. Pl. Afr. Austr.: 208 (1838). —Brown in Fl. Cap. **4**(1): 587 (1907).
>
> *Gomphocarpus stenoglossus* Schltr. in J. Bot. **32**: 257 (1894). Type: South Africa [Eastern Cape Province], Kreili's country, *Barber* 293 (K lectotype, GRA), lectotype designated by Bester & Nicholas in Phytotaxa **361**: 43 (2018).
>
> *Krebsia corniculata* (E. Mey.) Schltr. in Bot. Jahrb. Syst. **20**(Beibl. 51): 41 (1895).
>
> *Krebsia stenoglossa* (Schltr.) Schltr. in J. Bot. **33**: 270 (1895).

Perennial herb with vertical, tuberous rootstock. Stems 20–50 cm, simple or branched near the base, suberect or spreading initially then becoming erect, pubescent, with short white hairs restricted mainly to two lines along the stem. Leaves often reduced towards the base of the stem; upper leaves 80–130 × 2–6 mm, spreading or ascending, linear, acute, with revolute margins, glabrous. Umbels 1 to many, extra-axillary; peduncles 0–3 mm long, with 5–10 flowers in a subglobose head; pedicels 2–3 mm long, pubescent, contorted in fruit. Calyx lobes 4–5 × 1 mm, linear to narrowly triangular, acute, pubescent. Corolla lobed almost to the base, greenish white within, lobes 5–6 × 2–2.5 mm, lanceolate-oblong, somewhat attenuate with a subacute apex, the lower part concave giving the corolla the appearance of being subglobose, the apex reflexed, minutely papillate. Corona lobes c.4 mm long, cream with a purple midline, attached immediately below the anthers, slightly pouched at the base, then erect, narrowly triangular and somewhat keeled outside towards the base, tapering gradually to an attenuate apex. Follicles (immature) lanceolate, ornamented with soft spines in 6 longitudinal rows, pubescent; seeds not seen.

Mozambique. M: Lebombo mountains, ± on border with Swaziland c.20 km N of South African border, fl. 6.xii.2001, *Goyder* 5032 (K, LMU).

Also known from Swaziland and South Africa. Growing in peaty soil in grassland; 200 m.

Fig. 7.3.**181**. STENOSTELMA CAPENSE.1, flowering shoot; 2, flower from above. Photographed by T. Benn.

7. APOCYNACEAE

64. SCHIZOGLOSSUM E. Mey.[36]

Schizoglossum E. Mey., Comm. Pl. Afr. Austr.: 218 (1838). —Kupicha in Kew Bull. **38**: 599–672 (1984).

Perennial herbs with simple or little-branched erect annual stems arising from a napiform tuber. Latex white. Leaves opposite. Inflorescences extra-axillary, pedunculate. Corolla lobed almost to the base. Corolline corona absent. Gynostegial corona of dorsiventrally flattened lobes, generally with one or two appendages on the ventral face. Anther appendages inflexed over the stylar head. Pollinia with translucent marginal germination zone, attached apically or laterally to the short slender, slightly winged translator arms. Follicles single by abortion, held erect from a contorted fruiting pedicel, ornamented with soft appressed bristles; seeds with a coma of white hairs.

Schizoglossum barbatum Britten & Rendle in Trans. Linn. Soc., Bot.: **4**: 27 (1894). Type: Malawi, Mt. Mulanje [Milanji], x.1891, *Whyte* 33, 82, 103, 116 (BM syntypes), *Whyte* s.n. (K isosyntype).

Perennial herb with vertical napiform tuber. Stems c.40 cm long, erect, simple or branched, minutely pubescent. Leaves opposite, sessile, erect, 3–6.5 × 0.1–0.2 cm, linear, margins strongly revolute, glabrous except on the midrib beneath. Inflorescences extra-axillary, pedunculate. Peduncles longer at lower nodes, 0.2–4 cm long, minutely pubescent; pedicels 2–5 mm long minutely but densely pubescent. Calyx lobes c.2 mm long, narrowly triangular, pilose. Corolla maroon; lobes reflexed, 3–4 × 1.5 mm, lanceolate, acute, outer face sparsely pubescent with soft white hairs, inner face densely pilose. Gynostegial corona of five dorsiventrally flattened lobes; basal portion of lobes 0.8–1 × 0.8–1 mm, slightly fleshy and more or less rectangular in outline, the upper margin ± level with top of staminal column, with a thickened midline extending into a filiform apical tooth 0.5–2.5 mm long. Follicles and seeds not known.

Malawi. S: Mt. Mulanje, close to the CCAP cottage, fl. 2.xi.1986, *J.D. & E.G. Chapman* 8176 (K, MO).

Known only from Mt Mulanje in southern Malawi. Montane grassland at c.2000 m.

This species was excluded from *Schizoglossum* by Kupicha (1984) but remained unplaced in her treatment. Despite similar pollinarium morphology, she also excluded it from *Aspidoglossum* due to differences in corona lobe shape, or the possession of pedunculate inflorescences.

65. ASPIDOGLOSSUM E. Mey.[37]

Aspidoglossum E. Mey., Comm. Pl. Afr. Austr.: 200 (1838). —Kupicha in Kew Bull. **38**: 599–672 (1984).
Rhinolobium Arn. in Mag. Zool. Bot. **2**: 420 (1838).

Perennial herb with flowering shoots produced annually and dying back or burnt to ground level each year, all parts exuding a milky latex. Stems erect, branched or not. Leaves opposite or rarely irregularly inserted, verticillate (sect. *Verticillus*), sessile or subsessile, narrowly linear or rarely broader (sect. *Verticillus*), margins revolute. Flowers in sessile or subsessile fascicles at upper nodes. Pedicels pilose. Calyx lobes triangular, pilose. Corolla campanulate, spreading or rarely reflexed, rarely tips united (*A. connatum*). Corolline corona absent. Gynostegial corona lobes mostly of two types; type A (Fig. 7.3.**182A**) comprising an erect, dorsally compressed, ± quadrate basal portion with peaked, square or sloping shoulders, the apex produced into an attenuate tooth, ventral face with two faint vertical wings or ridges running into an appendage similar to the apical tooth of the main lobe; type B (Fig. 7.3.**182B**) with basal portion of lobe ± as in type A but with truncate apex, apical tooth absent, ventral surface with two often well-developed vertical ridged running into an inflexed appendage. Pollinarium with subcylindrical

---

[36, 37] by D.J. Goyder

corpusculum, rarely broader and with lateral flanges (*A. hirundo*); translator arms short, slender or flattened and ribbon like, attached subapically to the pollinia; pollinia subcylindrical or pyriform and dorsally compressed, with or without translucent germination zone near point of attachment to translator arms. Follicles single, pubescent, with or without soft appressed bristles; fruiting pedicel contorted to hold follicle erect.

A genus of 35 species from tropical and southern Africa.

1. Leaves whorled, at least in the inflorescence (sect. *Verticillus*) 2
– Leaves opposite throughout 3
2. Plant at least 35 cm tall; stems simple or branched. . . . . **8.** *nyasae*
– Plant 7–30 cm tall; stems never branched . . . .**7.** *ovalifolium*
3. Corona lobes of type A (sect. *Aspidoglossum*; Fig. 7.3.**182A**) 4
– Corona lobes not as above 9
4. Corolla lobes completely glabrous . . . . **6.** *glabellum*
– Corolla lobes with some indumentum. 5
5. Corona lobes 1–2 mm long. 6
– Corona lobes 3–5.5 mm long 8
6. Corolla lobes glabrous on outer surface . . . . **3.** *crebrum*
– Corolla lobes pilose on outer surface. 7
7. Stems to 35 cm long; translator arms contorting to invert pollinia when released from anthers. . . . .**5.** *breve*
– Stems at least 40 cm long; translator arms not contorting to invert pollinia when released from anthers . . . .**2.** *erubescens*
8. Pedicels 1–3 mm long; ventral surface of corolla lobes densely pilose at base; inner and outer lobes of corona ± equal in length; corpusculum subcylindrical, without flanges. . . . .**1.** *biflorum*
– Pedicels 3–9 mm long; ventral surface of corolla lobes with short white hairs distributed mainly towards the apex and margins; inner lobe of corona about half the length of the outer lobe; corpusculum rhomboid with linear flanges projecting out and back from the widest point. **4.** *hirundo*
9. Plants very slender, usually at least 75 cm long; corolla usually pubescent on ventral surface, the hairs sometimes very long, those at the tip often stouter than the rest; corona lobes of type B (Fig. 7.3.**182B**) or 'penguin'-shaped (*A. elliotii*; sect. *Virga*) 10
– Plants always less than 75 cm tall; corolla glabrous on ventral surface; corona lobes not as above (sect. *Latibrachium*). 16
10. Corolla lobes 5.5–9 mm long, often united at their tips to form a lantern . . . . . .
. . . .**12.** *connatum*
– Corolla lobes 2–4.5 mm long, never united at tip. 11
11. Corona lobes 'penguin'-shaped; corpusculum laterally compressed and somewhat arcuate viewed from side. . . . .**15.** *elliotii*
– Corona lobes of type B (Fig. 7.3.**182B**); corpusculum not curved. 12
12. Corona pilose. . . . .**10.** *lanatum*
– Corona glabrous 13
13. Ventral surface of corolla lobes long-pilose, particularly towards apex and margins. 14
– Ventral surface of corolla lobes glabrous or shortly pubescent, sometimes with a few stout hairs at tip. 15
14. Corolla lobes 2–3 mm long; upper margin of corona lobes produced into 3 teeth, the middle one much longer than outer pair . . . . **14.** *interruptum*
– Corolla lobes (2)3–4.5 mm long; upper margin of corona lobes not as above. . .
. . . .**11.** *angustissimum*

15. Ventral surface of corolla completely glabrous, or glabrous in the centre and pubescent around the edge or only at the tip; basal portion of corona lobes 1.5–2 × 1–1.5 mm. . . . . **9.** *eylesii*
– Ventral surface of corolla glabrous or often with stout white hairs at tip; basal portion of corona lobes c.1 × 0.5 mm . . . . **13.** *masaicum*
16. Corolla lobes reflexed or spreading; corona lobes with pair of narrowly triangular teeth on upper margin and a third tooth on inner face. . . . . **16.** *araneiferum*
– Corolla lobes spreading or campanulate; corona lobes not as above 17
17. Leaves pubescent on upper surface and on midrib below; corona with lobules c.0.5 mm long alternating with main lobes; pollinia 0.3–0.4 mm long . . . . **17.** *rhodesicum*
– Leaves glabrous; corona with minute pairs of rounded teeth alternating with the lobes; pollinia 0.9–1 mm long. . . . .**18.** *delagoense*

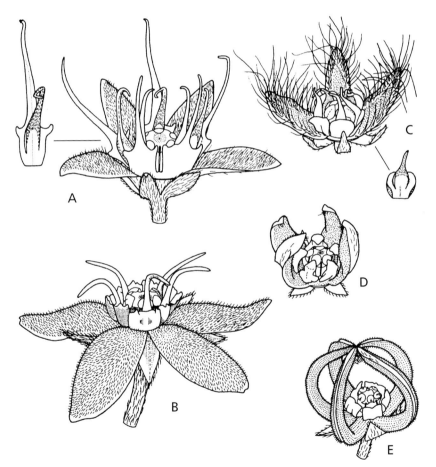

Fig. 7.3.**182**. ASPIDOGLOSSUM flowers. A, flower with corona of type A, from *Corby* 553. B, flower with corona of type B, from *Devenish* 1088. C, flower and corona lobe of ASPIDOGLOSSUM ANGUSTISSIMUM, from *De Wilde* 510. D, flower of ASPIDOGLOSSUM ELLIOTII, from *Smeds* 1426. E, flower of ASPIDOGLOSSUM CONNATUM, from *Williamson et al.* 574A. Drawn by Frances Kristina Kupicha. Adapted from Kew Bulletin (1984).

Sect. **Aspidoglossum**.

1. **Aspidoglossum biflorum** E. Mey., Comm. Pl. Afr. Austr.: 201 (1838). Type: South
   Africa [Eastern Cape Province], Windvogelberg, *Drège* s.n. (LUB† holotype);
   Drège 3427 (K fragment, lectotype).

   Schizoglossum excisum Schltr. in J. Bot. **32**: 259 (1894). —Brown in Fl. Cap. **4**(1): 638
   (1907). Type: South Africa [Eastern Cape Province], near Tsomo, *Barber* s.n. (K lectotype),
   designated by Kupicha in Kew Bull. **38**: 638 (1984).

   ?*Schizoglossum strictum* Schltr. in Bot. Jahrb. Syst. **20**(Beibl. 51): 22 (1895). —Brown in Fl.
   Cap. **4**(1): 642 (1907). Type: South Africa [ZwaZulu-Natal], near Ingagane, *Schlechter* 3405
   (B† holotype).

   ?*Schizoglossum tubulosum* Schltr. in Bot. Jahrb. Syst. **20**(Beibl. 51): 23 (1895). —Brown in
   Fl. Cap. **4**(1): 636 (1907). Type: South Africa [ZwaZulu-Natal], near Newcastle, *Schlechter*
   3410 (B† holotype).

   Schizoglossum venustum Schltr. in Bot. Jahrb. Syst. **20**(Beibl. 51): 24 (1895). Type South
   Africa [Mpumalanga], near Klein Oliphants River, fl. 26.xi.1893, *Schlechter* 2794 (K
   lectotype), designated by Kupicha in Kew Bull. **38**: 638 (1984).

   ?*Schizoglossum venustum* var. *concinnum* Schltr. in Bot. Jahrb. Syst. **20**(Beibl. 51): 25 (1895).
   Type: South Africa [ZwaZulu-Natal], near Cato's Ridge, *Schlechter* 3262 (B† holotype).

   Schizoglossum biflorum (E. Mey.) Schltr. in Bot. Jahrb. Syst. **20**(Beibl. 51): 25 (1895); in J.
   Bot. **34**: 449 (1896). —Brown in Fl. Cap. **4**(1): 641 (1907).

   Schizoglossum shirense N.E. Br. in Bull. Misc. Inform. Kew **1895**: 253 (1895); in F.T.A. **4**(1):
   361 (1902). Type: Malawi, Shire valley, *Waller* s.n. (K lectotype), designated by Kupicha in
   Kew Bull. **38**: 638 (1984).

   Schizoglossum gwelense N.E. Br. in F.T.A. **4**(1): 360 (1902). Type: Zimbabwe, near Gwele, xi-
   xii.1899, *Cecil* 131 (K lectotype), designated by Kupicha in Kew Bull. **38**: 638 (1984); Harare,
   Rand 190 (BM paralectotype).

   Schizoglossum conrathii Schltr. in Bot. Jahrb. Syst. **38**: 27 (1905). —Brown in Fl. Cap.
   4(1):637 (1907). Type: South Africa [Gauteng], near Pretoria, fl. x.1898, *Conrath* 989 (K
   lectotype), designated by Kupicha in Kew Bull. **38**: 640 (1984).

   ?*Schizoglossum biflorum* var. *concinnum* (Schltr.) N.E. Br. in Fl. Cap. **4**(1): 641 (1907).

   Schizoglossum biflorum var. *integrum* N.E. Br. in Fl .Cap. **4**(1): 642 (1907). Type: South
   Africa [Eastern Cape Province], Linch's Post near Kowie River, *Bowie* s.n. (K holotype, BM).

   Schizoglossum biflorum var. *gwelense* (N.E. Br.) N.E. Br. in Fl. Cap. **4**(1): 642 (1907).

Perennial herb with slender vertical napiform tuber. Stems single, 25–90 cm long, erect, simple
or occasionally with a single ascending to erect branch, pubescent with short curled hairs. Leaves
opposite and paired even in the inflorescence, sessile, erect, (2)3–6(8) × 0.1–0.2(0.6) cm, linear,
margins strongly revolute, sparsely to densely pubescent on upper surface and on midrib below
with short, soft or stiff white hairs. Inflorescences extra-axillary, sessile, (1)2–7-flowered at each of
the upper nodes. Pedicels 1–3 mm long, densely pubescent. Calyx lobes 2–3 × 1–1.5 mm, ovate-
lanceolate, attenuate, pilose. Corolla rotate or campanulate, green, maroon or white, divided ±
to the base or lobes united for c.1/4 of their length; lobes 3–5 × 1.5–2.5 mm, ovate, acute, outer
face sparsely pubescent with soft white hairs, inner face densely pilose at base, often becoming
sparser apically. Corona lobes of type A, white, yellowish or purple, adjacent lobes meeting at
margins to form a pentagonal drum surrounding the staminal column; basal portion of lobes
c.1.5 × 2 mm, slightly fleshy, the upper margin ± level with top of staminal column, shoulders
sloping, square or with a short curled horn, apical tooth arising from middle of upper margin,
3–5 mm long, linear, erect or spreading slightly; appendage on inner face erect, almost as long
as tooth on main lobe, linear. Staminal column c.1 mm long. Follicles (immature) pubescent.

Zimbabwe. C: Makoni Dist., 1.5 km. from Eagles Nest, fl. 29.xi.1955, *Drummond* 5067
(K, SRGH). E: S of Penhalonga, 1934, *Gilliland* Q929 (K). **Malawi**. S: Shire valley, *Waller*
s.n. (K lectotype). **Mozambique**. N: Cabo Delgado, cut-line S of Olumbi, fl. 9.xii.2008,
Goyder et al. 5096 (K, LMA). Z: Mocuba Dist., Namagoa plantations, fl. xii.1943,
Faulkner P342 (K, PRE). MS: Chupanga, fl. 8.i.1863, *Kirk* s.n. (K paralectotype).

   Also known from South Africa, Lesotho and Namibia. In grassland or open
woodland; 100–1500 m.

2. **Aspidoglossum erubescens** (Schltr.) Bullock in Kew Bull. **9**: 589 (1954). Type: Malawi, Shire Highlands, Mulanje, xii.1893, *Scott Elliot* 8671 (BM holotype, K).

  Schizoglossum erubescens Schltr. in J. Bot. **33**: 306 (1895). —Brown in F.T.A. **4**(1): 359 (1902).

  Schizoglossum strictissimum S. Moore in J. Bot. **40**: 254 (1902). —Brown in F.T.A. **4**(1): 358 (1902). Type: Zimbabwe, Bulowayo, xii. 1897, *Rand* 195 (BM holotype, K).

  ?*Schizoglossum pentheri* Schltr. in Ann. Nat. Hofmus. Wien. **18**: 397, t.5 (1903). Type: Zimbabwe, Matabeleland, Ligombwe, 19.xii.1895, *Penther* 2414 (B† holotype, K photograph).

Perennial herb with vertical napiform tuber. Stems single, 40–100 cm long, simple or rarely sparsely branched. Leaves sessile, 2–7 × 0.1–0.25 cm, linear, margins revolute, densely pubescent with short curled hairs on upper surface and on midrib below. Inflorescences sessile, 2–6-flowered at each of the upper nodes. Pedicels to 1.5 mm long, densely pubescent. Calyx lobes 1.5–2 × 0.7–1 mm, ovate, acute, pilose. Corolla spreading to slightly reflexed, lobed ± to the base, green tinged with brown or purple; lobes 2.5 × 2 mm, broadly ovate, acute, upper surface densely pubescent or completely glabrous, lower surface pubescent. Corona lobes of type A but outer tooth not always developed, deep dingy purple; basal portion of lobes 0.5–1 × 0.8–1.5 mm with square or sloping shoulders, apical tooth c.0.5 mm long, spreading or reduced to a dorsal swelling, ventral tooth 0.5–1 mm long, entire or bifid and inflexed over staminal column. Staminal column c.0.5 mm long. Follicles 6–8 cm long, lanceolate in outline, attenuate, puberulent; fruiting pedicel contorted to hold follicle erect. Seeds c.5 × 3 mm, ovate, with a convoluted margin c.0.5 mm wide, central disc verrucose; coma c.2.5 cm long.

Zimbabwe. W: Bulowayo, fl. & fr. xii.1897, *Rand* 195 (BM, K). C: Arcturus, Ewanrigg, fl. 26.xii.1947, *Christian* in GH18346 (K, SRGH). **Malawi**. S: Zomba, fl. & fr. xii.1896, *Whyte* s.n. (K). **Mozambique**. N: Murrupula, Cabo Mucarre, fl. 10.i.1961, *Carvalho* 403 (K, LMA, LMU).

Not known elsewhere. Grassland; 750–1500 m.

3. **Aspidoglossum crebrum** Kupicha in Kew Bull. **38**: 642 (1984). Type: Zimbabwe, Mutare [Umtali] Dist., NE corner of Mutare golf course, 1080 m, 10.xii.1951, *Chase* 4228 (LISC holotype, BM, BR, COI, K, SRGH).

Perennial herb with slender vertical napiform tuber. Stems single, 30–70 cm long, erect, unbranched, pubescent with short curled white hairs. Leaves opposite and paired, even in the inflorescence, sessile, erect, 2–7 × 0.1–0.3 cm, linear, pubescent with soft white hairs on upper surface and on midrib below. Inflorescences sessile, 2–10-flowered at each of the upper nodes. Pedicels 1–3 mm long, pubescent. Calyx lobes 2–2.5 × 1 mm, ovate to lanceolate, attenuate, pilose. Corolla campanulate, lobed ± to the base, dark purple or brown; lobes scarcely opening at anthesis, 3–4 × 1.5–2 mm, ovate, acute, glabrous on the outer surface, inner surface densely pubescent in upper half, glabrous below. Corona lobes of type A, basal portion c.0.5–1.5 × 1.5 mm with square to acute shoulders, longer than staminal column, apical tooth c.1 mm long, narrowly triangular; ventral tooth c.2 mm long, conspicuously longer than the outer tooth, contorted distally. Staminal column c.1 mm long; anther wings broadly triangular. Follicles (immature) densely pubescent.

Zimbabwe. C: Between Riverside and Nyazura [Inyazura], fl. 25.xii.1956, *Pole-Evans* 5049 (K). E: Mutare [Umtali] golf course, fl. 12.xii.1956, *Chase* 6263 (K, SRGH).

Known only from eastern Zimbabwe. Grassland; 1000–1100 m.

4. **Aspidoglossum hirundo** Kupicha in Kew Bull. **38**: 642 (1984). Type: Mozambique, Ribáuè, 78 km from Altomolócuè, 4.xii.1967, *Torre & Correia* 16350 (LISC holotype).

Perennial herb with vertical napiform tuber. Stems single, 40–80 cm long, erect, unbranched, slender, bifariously pubescent. Leaves opposite and paired even in inflorescence, sessile, erect, 4.5–9 × 0.1–0.15 cm, linear, margins revolute, glabrous or sparsely pubescent on upper surface. Inflorescences sessile, 1–5-flowered at each of the upper nodes. Pedicels 3–9 mm long, pubescent.

Calyx lobes c.1.5 × 0.5 cm, ovate-lanceolate, acute, sparsely pilose. Corolla spreading, lobed ± to the base, yellow-green; lobes 3.5–5 × 1.5–2 mm, ovate, acute, outer face glabrous, inner face pubescent with very short white hairs distributed mainly towards the apex and margins. Corona lobes of type A, c.2.5 mm long, basal portion tapering gradually into the erect, linear apical tooth; inner tooth about half length of outer tooth, slightly inflexed. Staminal column c.1.5 mm long. Follicles not seen.

Mozambique. Z: Quelimane, Gilé Dist., 14 km from Namarrua, fl. 11.xii.1997, *Dungo* 138 (LMU). N: Mozambique, Ribáuè, 78 km. from Altomolócuè, fl. 4.xii.1967, *Torre & Correia* 16350 (LISC).

Known only from Nampula and Zambesia provinces of Mozambique. Open *Brachystegia* woodland and seasonally inundated grassland; 30–500 m.

A GBIF record from the Eastern highlands of Zimbabwe was determined as this species by Pieter Bester (PRE), but I have not been able to examine this material myself.

5. **Aspidoglossum breve** Kupicha in Kew Bull. **38**: 643 (1984). —Goyder in F.T.E.A., Apocynaceae (part 2): 374 (2012). Type: Malawi, Nyika Plateau, near Chelinda camp, 10.i.1967, *Hilliard & Burtt* 4399 (E holotype, K).

Perennial herb with slender vertical napiform tuber. Stems single or occasionally paired, 10–35 cm long, erect, slender and unbranched, pubescent with spreading white hairs. Leaves opposite and paired even in inflorescence, sessile, erect, (1)2–6 × 0.1–0.2 cm, linear, margins strongly revolute, pubescent with spreading white hairs on upper surface and on midrib below. Inflorescences sessile or with peduncle to 0.6 cm, 3–8-flowered at each of the upper nodes. Pedicels 3–8 mm long, sparsely to densely pubescent. Calyx lobes 1.5–2.5 × 0.5–0.8 mm, narrowly triangular to ovate, attenuate, sparsely to densely pilose in upper half of outer face, lower half with shorter reddish hairs. Corolla campanulate, lobed ± to the base, yellow-green or reddish brown outside, paler within and on margins; lobes 2.5–4 × 1.5–2 mm, ovate, acute, weakly replicate, outer surface with white hairs towards the apex, commonly with reddish ones below; inner face glabrous. Corona lobes of type A, green or yellow, sometimes tinged purple on the horns, as tall as or longer than the staminal column; basal portion spreading, slightly fleshy, 0.5–0.7 × 1 mm, with square shoulders, apical tooth erect, narrowly triangular, 0.7–1 mm long; inner tooth erect or inflexed over apex of stigma head, about as long as outer tooth. Staminal column c.1 mm long. Follicles not seen.

Malawi. N: Nyika National Park, Nganda peak 15 km NNE of Chelinda, fl. 23.i.1992, *Goyder et al.* 3557 (BR, K, MAL, PRE).

Known only from the Nyika Plateau in Malawi and the Kitulo Plateau of southern Tanzania. Broken turf among rocks in montane grassland; 2400–2600 m.

6. **Aspidoglossum glabellum** Kupicha in Kew Bull. **38**: 643 (1984). Type: Zimbabwe, Glencoe forest reserve, slopes of Mt. Pene, 22.xi.1955, *Drummond* 4952 (PRE holotype, BR, K, SRGH).

Perennial herb with slender vertical napiform tuber. Stems single, 30–60 cm long, erect and unbranched, bifariously pubescent with short curled white hairs. Leaves opposite, even in inflorescence, sessile, erect, 1–3.5(5) × 0.1–0.3 cm, sparsely pubescent on upper surface and on midrib below, or glabrescent. Inflorescences sessile, 1–3(4)-flowered at each of the upper nodes. Pedicels 3–5 mm long, pubescent. Calyx lobes 1–2 × 0.5–0.7 mm, ovate to triangular, acute or attenuate, pilose or subglabrous. Corolla spreading, yellow-green; lobes 3.5–4 × 1.5–2 mm, ovate, acute, glabrous on both surfaces. Corona lobes of type A; basal portion c.0.5–1 × 1.5 mm, with square to acute shoulders, ± as long as the staminal column; apical tooth 1.5–2.5 mm long, narrowly triangular, contorted at the tip; ventral tooth ± as long as outer tooth. Staminal column 0.5 mm long. Follicles (immature) to c.6 cm long, slender, lanceolate in outline with pronounced beak, densely pubescent; fruiting pedicel contorted to hold follicle erect.

Zimbabwe. E: Melsetter Dist., Glencoe Forest Reserve, fl. 27.xi.1967, *Simon & Ngoni* 1327 (K, SRGH). **Mozambique**. MS: Eastern Chimanimani Mts, Mevumodzi Valley

near Eastern Lakes, 28.x.2014, *Wursten* 1054 (BR, K, LMA, SRGH).
Known only from Eastern Zimbabwe and the adjacent Manica Province of Mozambique. Montane grassland; 900–1950 m.

Sect. **Verticillus** Kupicha in Kew Bull. **38**: 645 (1984).

7. **Aspidoglossum ovalifolium** (Schltr.) Kupicha in Kew Bull. **38**: 649 (1984). Type: South Africa [Eastern Cape Province], Komgha, fl. xi.1892, *Flanagan* 1307 (BOL lectotype, GRA, K, NBG, NH, PRE), designated by Kupicha in Kew Bull. **38**: 649 (1984).

> *Schizoglossum ovalifolium* Schltr. in Bot. Jahrb. Syst. **18**(Beibl. 45): 5 (1894).
> *Schizoglossum striatum* Schltr. in J. Bot. **32**: 356 (1894). —Brown in Fl. Cap. **4**(1): 631 (1907). Type: South Africa, KwaZulu-Natal ["Natal"], *Saunders* s.n. (K ?holotype).
> *Schizoglossum pumilum* Schltr. in Bot. Jahrb. Syst. **20**(Beibl. 51): 21 (1895). —Brown in Fl. Cap. **4**(1): 628 (1907). Type: South Africa [Mpumalanga], between Waterval River and Zuikerboschrand, fl. 18.x.1893, *Schlechter* 3496 (BOL lectotype, K, Z), designated by Kupicha in Kew Bull. **38**: 649 (1984).
> ?*Schizoglossum robustum* Schltr., in J. Bot. **33**: 267 (1895). —Brown in Fl. Cap. **4**(1): 631 (1907). Type: South Africa [KwaZulu-Natal], near Ixopo, i.1895, *Schlechter* 6659 (B† holotype).
> *Schizoglossum contracurvum* N.E. Br. in Fl. Cap. **4**(1): 628 (1907). Type: South Africa [KwaZulu-Natal], Greenwich Farm, Riet Vlei, fl. xi-xii.1899, *Fry* in *Galpin* 2747 (PRE lectotype, GRA, K), designated by Kupicha in Kew Bull. **38**: 649 (1984).
> *Schizoglossum robustum* var. *inandense* N.E. Br. in Fl. Cap. **4**(1): 632 (1907). Type: South Africa [KwaZulu-Natal], Inanda, *Wood* 316 "partly" (BOL lectotype, K, NH, PRE, SAM), designated by Kupicha in Kew Bull. **38**: 649 (1984).
> *Schizoglossum robustum* var. *pubiflorum* N.E. Br. in Fl. Cap. **4**(1): 632 (1907), pro parte. Type: South Africa [Gauteng], Jeppes Town Ridges, Johannesburg, fl. i-ii.1899, *Gilfillan* in *Galpin* 6230 (K lectotype), designated by Kupicha in Kew Bull. **38**: 649 (1984).

Perennial herb with slender vertical napiform tuber. Stems single, 7–30 cm long, erect, simple, densely pubescent. Leaves irregular or opposite below, verticillate in inflorescence, sessile, spreading or ascending, 1.4–3.5 × 0.05–1.2 cm, acicular to oblong to broadly elliptic, very rarely suborbicular, upper surface glabrous or pubescent with stiff white hairs, lower surface glabrous or pubescent. Inflorescences sessile or shortly pedunculate, 2–13-flowered at each of the upper nodes. Pedicels 5–10 mm long, pubescent. Calyx lobes 2–6 × 0.5–2 mm, pubescent with red or white hairs. Corolla spreading to reflexed, green striped with brown or purple; lobes 4–7 × 2–4 mm, ovate, lower surface pubescent, upper surface glabrous or densely pubescent with short white hairs. Corona lobes of type A, white marked with pink, basal portion 1–3 mm high, with acute or rounded shoulders, apical tooth to c.5 mm long, erect; inner tooth entire or bifid, much shorter than main tooth, inflexed. Follicles c.4 cm long, with soft spreading or appressed processes c.4 mm long.

Kupicha in Kew Bull. **38**: 649 (1984) reports this species from **Mozambique** (M). Also known from Swaziland and South Africa (Gauteng, Mpumalanga, Free State, KwaZulu-Natal, Eastern Cape). Grassland.

8. **Aspidoglossum nyasae** (Britten & Rendle) Kupicha in Kew Bull. **38**: 650 (1984). Type: Malawi, Mulanje, x.1891, *Whyte* s.n. (BM holotype, K).

> *Schizoglossum nyasae* Britten & Rendle in Trans. Linn. Soc. London, Bot. **2**(4): 26 (1894). —Brown in F.T.A. **4**(1): 363 (1902).
> *Schizoglossum multifolium* N.E. Br. in Bull. Misc. Inform. Kew **1895**: 253 (1895). Type: Malawi, *Buchanan* 965 (K holotype).
> *Schizoglossum leptoglossum* Weim. in Bot. Not. **1935**: 387 (1935). Type: Zimbabwe, near Inyanga village, 20.i.1931, *Norlindh & Weimarck* 4455 (LD holotype, BM, BR, LISU, PRE, SRGH).

Perennial herb with slender vertical napiform tuber. Stems single, (35)45–100 cm long, erect, simple or branched, densely pubescent with short curled white hairs. Leaves irregular, opposite or more usually verticillate below, verticillate in inflorescence, sessile, spreading or ascending, (2)3–5(10) × (0.2)0.4–1.5(2.5) cm, linear, oblanceolate or occasionally broadly elliptic, upper surface glabrous or pubescent with stiff white hairs especially towards the margins, lower surface glabrous except for midrib or occasionally sparsely pubescent. Inflorescences sessile, 4–11-flowered at each of the upper nodes. Pedicels 7–16 mm long, minutely pubescent with white or rusty hairs. Calyx lobes 4–6 × 1–1.5 mm, lanceolate or narrowly triangular, attenuate, pubescent with red or white hairs. Corolla spreading to reflexed, green or brownish; lobes 5–6.5(9.5) × 2.5(4) mm, ovate to lanceolate, acute, often somewhat replicate, lower surface glabrous or rusty pubescent, upper surface densely pubescent with short white hairs. Corona lobes of type A, white with pink tinge or spots, basal portion 2–3 × 2–2.5 mm, with acute shoulders, slightly longer than the column, glabrous or occasionally pubescent on inner face; apical tooth 4–6 mm long, tapering gradually and coiled once at the tip; inner tooth entire, bifid or quadrifid, erect, half to as long as main tooth; corona lobes united briefly at the base by a minute lobule obscuring the mouth of the anther wing fissure. Staminal column c.1 cm long; corpusculum c.0.5 × 0.2 mm, oblong-ovate in outline, black. Follicles 7.9–9 cm long, lanceolate in outline with prominent beak, minutely puberulent and with soft spreading or appressed processes c.3–6 mm long; fruiting pedicel contorted to hold follicle erect.

Zimbabwe. N: Umvukwe Range (Great Dyke), Vanad Pass, fl. 5.ii.1997, *Goyder & Paton* 4101 (EA, K, PRE, SRGH). W: Matobo Dist., Besna Kobila farm, fl. i.1935, *Miller* 4990 (K, SRGH). C: Marandellas Dist., fl. 7.ii.1946, *Greatrex* s.n. in GH14433 (K, SRGH). E: Mutare [Umtali] Dist., Odzani Heights, fr. 24.ii.1957, *Chase* 6368 (K, SRGH). **Malawi**. S: Mangochi, Namizimu forest reserve, fl. 21.ii.1982, *Hepper* 7330 (K). **Mozambique**. N: Niassa, Vila Cabral, serra de Massangulo, 25.ii.1964, *Torre & Paiva* 10815 (LISC). Z: Ribáuè, serra de Ribáuè, Mapaluè, 23.iii.1964, *Torre & Paiva* 11372 (LISC).

Montane grassland; 1100–1700 m.

Two collections from eastern Zimbabwe (*Norlindh & Weimarck* 4455; *Pope* 1181) have hairs on the corona lobes. One of these (*Norlindh & Weimarck* 4455) also has broader pollinia and flattened translator arms – a combination of characters paralleled in *Aspidoglossum delagoense*.

Sect. **Virga** Kupicha in Kew Bull. **38**: 651 (1984).

9. **Aspidoglossum eylesii** (S. Moore) Kupicha in Kew Bull. **38**: 652 (1984). Type: Zimbabwe, Mazowe Dist., i.1907, *Eyles* 500 (BM holotype, NBG/SAM, SRGH).

 Schizoglossum eylesii S. Moore in J. Bot. **52**: 149 (1914).

Perennial herb with slender vertical napiform tuber. Stems single, 55–140 cm long, simple or with a single erect branch, pubescent with short, curled white hairs. Leaves opposite and paired even in inflorescence, sessile, erect, (3)5–10 × 0.1–0.3 cm, linear, margins strongly revolute, minutely appressed-pubescent on upper surface and on midrib below. Inflorescences sessile, 3–7(13)-flowered at each of the upper nodes. Pedicels 2–4 mm long, pubescent. Calyx lobes 1–2 × 0.5–1 mm, ovate to triangular, acute, pubescent. Corolla campanulate, lobed ± to the base, yellow-green within, sometimes tinged with brown or maroon outside; lobes 2.5–3.5 × 1–1.5 mm, ovate, acute, flat or replicate, outer face pilose, inner face glabrous or glabrous in the centre and pubescent around the edge or only at the tip. Corona lobes of type B; basal portion of lobe 1.5–2 × 1–1.5 mm, slightly fleshy, ± as long as staminal column, shoulders sloping or acute, tooth erect or inflexed, 0.5–1 mm long, narrowly triangular, delimited from basal portion by transverse dorsal ridge. Staminal column c.1 mm long Follicles 5–7 cm long, lanceolate in outline, beaked, densely pubescent; fruiting pedicel contorted to hold follicle erect.

Zimbabwe. N: Miami, K.34 Experimental Farm, fl. 7.iii.1947, *Wild* 1859 (K, SRGH). E: Mutare [Umtali] Dist., W of Cross Hill, East Commonage, fl. 22.iii.1960, *Chase* 7314 (K, SRGH). S: Victoria Dist., Kyle National Park Game Reserve, near base of Mtunumushava Hill, fl. 21.v.1971, *Grosvenor* 512 (K, SRGH). **Malawi**. C: Ntchisi Forest

Reserve, fl. 26.iii.1970, *Brummitt* 9416 (K, MAL).

Not known elsewhere. Open deciduous woodland; 900–1800 m.

Aspidoglossum woodii (Schltr.) Kupicha from central KwaZulu-Natal is closely allied to this taxon, differing principally in the longer apical and marginal hairs on the upper surface of the corolla. *A. eylesii* should perhaps be treated as a northern subspecies of *A. woodii*.

10. **Aspidoglossum lanatum** (Weim.) Kupicha in Kew Bull. **38**: 652 (1984). —Goyder in F.T.E.A., Apocynaceae (part 2): 375 (2012). Type: Zimbabwe, near Inyanga village, 11.i.1931, *Norlindh & Weimarck* 4183 (LD holotype, BM).

   Schizoglossum lanatum Weim. in Bot. Not. **1935**: 392 (1935).

Perennial herb with slender vertical napiform tuber. Stems single, 35–75 cm long, simple or with a single erect branch, pubescent with short curled white hairs. Leaves opposite and paired even in inflorescence, sessile, erect, 2.5–4 × 0.1–0.15 cm, minutely appressed-pubescent on upper surface and on midrib below. Inflorescence 3–9(11)-flowered at each of the upper nodes. Pedicels 2–5 mm long, pubescent. Calyx lobes 1–2.5 × c.0.5–1 mm, ovate to triangular, acute, pubescent. Corolla campanulate, lobed ± to the base, yellow-green, sometimes tinged purple; lobes 2.5–3.5 × 1–1.5 mm, ovate, acute, flat or replicate, outer face pilose, inner face evenly pubescent. Corona lobes of type B; basal portion 1–1.5 × c.1 mm, ± as long as the staminal column, shoulders sloping or acute, moderately to densely pilose along upper margin; appendage c.0.5 mm long, narrowly triangular, erect or inflexed and densely pilose. Follicles (very immature) densely pubescent, apparently with soft linear processes.

Zimbabwe. E: Inyanga Dist., fl. 12.i.1951, *Chase* 3568 (K, SRGH).

Also recorded from SE Tanzania. Grassland and open woodland; 1400–1800 m.

11. **Aspidoglossum angustissimum** (K. Schum.) Bullock in Kew Bull. **7**: 418 (1952); —Goyder in F.T.E.A., Apocynaceae (part 2): 375 (2012). Type: D.R. Congo, Niamniamland, Gumango Hill, *Schweinfurth* 3879 (B† holotype, K lectotype), designated by Kupicha in Kew. Bull. **38**: 653 (1984). FIGURE 7.3.**182C**.

Perennial herb with vertical slender woody napiform tuber. Stems single or paired, (50)80–135 cm long, erect, simple or commonly with several erect branches, pubescent with short curled white hairs. Leaves opposite and paired even in inflorescence, sessile or subsessile, erect, 3–8(11) × 0.1–0.6 cm, linear, margins strongly revolute, appressed-pubescent on upper surface and on midrib below. Inflorescences sessile, 1–13-flowered at each of the upper nodes. Pedicels 1–7 mm long, pubescent. Calyx lobes 1–2 × 0.5 mm, ovate, acute, pilose. Corolla campanulate, divided ± to the base, white, sometimes tinged brown or purple; lobes (2)3–4.5 × 1–2 mm, ovate but strongly replicate so appearing linear, incurved at the tip, outer face pilose, inner face pubescent with conspicuous white hairs to 2 mm long towards the apex and margins. Corona lobes of type B; white or greenish, sometimes spotted with brown or purple; basal portion 0.5–1.5 mm, shorter than or ± as long as the staminal column, shoulders square or rounded; tooth 0.6–1 mm long, linear to broadly triangular, entire or the broader teeth shallowly to deeply bifid, erect or inflexed. Staminal column 0.5–1 mm long. Follicles 5–7 cm long, fusiform with a long attenuate beak, pubescent; fruiting pedicel contorted to hold follicles erect.

Corona lobes with clearly defined basal portion 1–1.5 mm long; tooth projecting from
   inner face linear to narrowly triangular, entire . . . . a) subsp. *angustissimum*
Corona lobes with poorly defined boundary between basal portion and apical tooth;
   basal portion (to widest point) c.0.5 mm long; tooth broadly triangular, often
   deeply bifid . . . . b) subsp. *brevilobum*

a) Subsp. **angustissimum**.

   Schizoglossum angustissimum K. Schum. in Bot. Jahrb. Syst. **17**: 123 (1893). —Brown in F.T.A. **4**(1): 357 (1902).

Schizoglossum elatum K. Schum. in Bot. Jahrb. Syst. **17**: 123 (1893). Type: Kenya, *Fischer* 398 (B† holotype).
Schizoglossum whytei N.E. Br. in F.T.A. **4**(1): 357 (1902). Type: Malawi, Kondowe to Karonga, vii.1896, *Whyte* 353 (K lectotype).
? *Schizoglossum zernyi* Markgraf in Notizbl. Bot. Gart. Berlin **14**: 116 (1938). Type: Tanzania, Matengo Plateau WSW of Songea, above Ugano, *Zerny* 427 (B† holotype).
Aspidoglossum whytei (N.E. Br.) Bullock in Kew Bull. **7**: 418 (1952).

Zambia. N: Nkali (Kali) Dambo, fl. 15.i.1955, *Richards* 4094 (K). **Malawi**. N: Nyika Plateau NE of Nganda, fl. 10.iv.1997, *Patel et al.* 5116 (K, MAL). S: Goche, Kirk Range, fl. 30.i.1959, *Robson* 1362 (K).
Widespread throughout tropical Africa from D.R. Congo, Central African Republic and Sudan to Malawi and Zambia. Seasonally waterlogged grassland; 1200–1800 m.

b) Subsp. **brevilobum** Goyder, subsp. nov. Differs from the type subspecies by the corona lobes with a shorter basal portion (c.0.5 mm rather than 1–1.5 mm long) and broader often bifid appendage less clearly differentiated from the basal portion. Type: Zimbabwe, Mutare (Umtali) Commonage, fl. 10.x.1956, *Chase* 6262 (K holotype, SRGH).

Zambia. S: Kaloma, fl. 10.ii.1965, *Fanshawe* 9167 (K, NDO). **Zimbabwe**. N: Shamva, fl. 19.xii.1921, *Eyles* 3249 (K). E: Mutare (Umtali) Commonage, fl. & fr. 10.xii.1951, *Chase* 4226 (K, SRGH).
Not known elsewhere. Grassland; 900–1300 m.
Material from Zambia and northern Zimbabwe has somewhat shorter corolla lobes than the bulk of the collections which come from eastern Zimbabwe.

12. **Aspidoglossum connatum** (N.E. Br.) Bullock in Kew Bull. **7**: 419 (1952). —Goyder in F.T.E.A., Apocynaceae (part 2): 376 (2012). Type: Zambia, Fwambo, S of Lake Tanganyika, 1894, *Carson* 17 (K holotype). FIGURES 7.3.**182E**, 7.3.**183**.
Schizoglossum connatum N.E. Br. in Bull. Misc. Inform. Kew **1895**: 69 (1895); in F.T.A. **4**(1): 356 (1902).
Schizoglossum vulcanorum Lebrun & Taton in Bull. Jard. Bot. État. Bruxelles **17**: 66 (1943). Type: D.R. Congo, between Kingi and Busogo, *Lebrun* 8647 (BR holotype).

Perennial herb with subglobose to napiform tuber. Stems single, 40–90 cm long, erect, simple or commonly with one or more erect branches, minutely pubescent with short curled white hairs. Leaves opposite and paired even in the inflorescence, 3–9 × 0.05–0.3 cm, linear, margins strongly revolute, appressed pubescent on upper surface and on midrib below. Inflorescences sessile, 3–7-flowered at each of the upper nodes. Pedicels 1–6 mm long, minutely pubescent. Calyx lobes 1.5–2.5 × 0.4–1 mm, triangular to ovate, acute, pilose. Corolla spreading to campanulate, divided ± to the base, white, cream, yellow, pink or purple; lobes 5.5–9 × 1–2 mm, linear-lanceolate, plane or more usually strongly replicate, united at the tips to form a lantern or free and spreading, glabrous or pilose on outer face, inner face glabrous to densely pubescent. Corona lobes of type B, green, cream or purple, adjacent lobes abutting to form a drum around the column; basal portion 1–1.5 × 1–1.5 mm, ± as long as or longer than the staminal column, shoulders rounded, square or toothed; ventral tooth minute. Staminal column 0.5–1 mm long. Follicle c.7 cm long, fusiform with a long attenuate beak, pubescent; fruiting pedicel contorted to hold follicle erect.

Zambia. N: pans near Mbala (Abercorn), fl. 20.i.1955, *Richards* 4180 (K). W: Kitwe, fl. 28.ii.1954, *Fanshawe* 888 (K, NDO). **Malawi**. C: Dedza Dist., Chongoni Forest boundary, fl. 4.i.1968, *Salubeni* 931 (K, MAL, SRGH).
Also known from Kenya, Tanzania, Uganda, D.R. Congo, S. Sudan and Guinea Bissau. Seasonally waterlogged grassland; 900–1700 m.

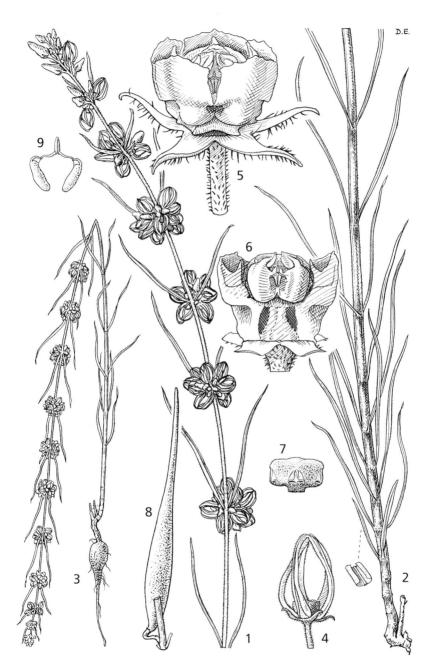

Fig. 7.3.**183**. ASPIDOGLOSSUM CONNATUM. 1, 2, habit (× 1); 3, habit including tuber (× ¹/₃); 4, flower (× 3); 5, gynostegium and corona (× 12); 6, gynostegium with two corona lobes cut away to expose stamens and ovaries (× 12); 7, stylar head (× 12); 8, follicle (× 1); 9, pollinarium (× 20). 1–3 from *Richards* 12012; 4–7, 9 from *Watermeyer* 3; 8 from *Chandler* 1569. Drawn by D. Erasmus. Reproduced from Flora of Tropical East Africa (2012).

13. **Aspidoglossum masaicum** (N.E. Br.) Kupicha in Kew Bull. **38**: 656 (1984). — Goyder in F.T.E.A., Apocynaceae (part 2): 378 (2012). Type: Kenya, Kilimanjaro, *Johnston* s.n. (K holotype, BM).

> *Schizoglossum masaicum* N.E. Br. in Bull. Misc. Inform. Kew **1895**: 252 (1895); in F.T.A. **4**(1): 358 (1902).
> *Schizoglossum fuscopurpureum* Schltr. & Rendle in J. Bot. **34**: 98 (1896). —Brown in F.T.A. **4**(1): 361 (1902). Type: Angola, near Huilla, *Welwitsch* 4177 (BM lectotype, K, LISU), designated by Kupicha in Kew Bull. **38**: 656 (1984) as holotype.
> *Schizoglossum baumii* N.E. Br. in F.T.A. **4**(1): 361 (1902). Type: Angola, near Kavenga on R. Kubango, *Baum* 413 (K holotype, BM, COI, E, Z).
> *Schizoglossum altum* N.E. Br. in Bull. Misc. Inform. Kew **1906**: 250 (1906). Type: Malawi, Thondwe (Ntondwe), *Cameron* 107 (K holotype).
> *Schizoglossum semlikense* S. Moore in J. Bot. **50**: 361 (1912). Type: D.R. Congo, Ruwenzori Dist., Semliki Valley, *Kässner* 3282a (BM holotype).

Perennial herb with globose to napiform tuber. Stems single, 30–110 cm long, simple or branched, usually extremely slender, pubescent. Leaves opposite and paired even in inflorescence, 1–6 × 0.05–0.2 cm, linear, margins strongly revolute, pubescent on upper surface and on midrib below. Inflorescences sessile, 2–10-flowered at each of the upper nodes. Pedicels 1–5 mm long, pilose. Calyx lobes c.1.5 × 0.2–0.5 mm, lanceolate or narrowly ovate, acute, pilose. Corolla spreading to campanulate, divided ± to the base, white or greenish, sometimes flushed pink or purple; lobes 2–4 × 1 mm, ovate, subacute, plane (but commonly replicate in East Africa), pilose on outer surface, glabrous on inner face, often with stout white hairs at apex. Corona lobes of type B; basal portion c.1 × 0.5 mm, with square or rounded shoulders, tooth to c.0.5 mm long, triangular, inflexed. Follicles 3–6 cm long, narrowly fusiform with a pronounced beak, pubescent; fruiting pedicel contorted to hold follicle erect.

Zambia. B: near Luene River, Mankoya, fl. 20.xi.1959, *Drummond & Cookson* 6648 (K, SRGH). C: Lusaka, fl. 30.xi.1964, *Robinson* 6261 (K). **Malawi**. S: Thondwe (Ntondwe), *Cameron* 107 (K).

Widespread in tropical Africa from Namibia and Angola in the southwest to Kenya and Ethiopia in the northeast. Seasonally waterlogged grassland; 1000–1500 m.

14. **Aspidoglossum interruptum** (E. Mey.) Bullock in Kew Bull. **7**: 419 (1952). — Goyder in F.T.E.A., Apocynaceae (part 2): 378 (2012). Type: South Africa, Cape Province, Aliwal North Div., Wittebergen, *Drège* s.n. (K lectotype, BM, CGE, K, MO), designated by Kupicha in Kew Bull. **38**: 658 (1984). FIGURE 7.3.**184**.

> *Lagarinthus interruptus* E. Mey., Comm. Pl. Afr. Austr.: 208 (1838).
> *Lagarinthus abyssinicus* Benth. & Hook.f., Gen. Pl. **2**: 753 (1876), invalid name.
> *Schizoglossum barberae* Schltr. in Bot. Jahrb. Syst. **18**(Beibl. 45): 27 (1894). —Brown in Fl. Cap. **4**(1): 658 (1907). Type: South Africa, Cape Province, Tsomo R., *Barber* 847 (K lectotype), designated by Kupicha in Kew Bull. **38**: 658 (1984).
> *Schizoglossum abyssinicum* K. Schum. in Engler & Prantl, Nat. Pflanzenfam. **4**(2): 233 (1895). Type: ?Ethiopia, *Schimper* 1633 (K).
> *Schizoglossum altissimum* Schltr. in Bot. Jahrb. Syst. **20**(Beibl. 51): 13 (1895). —Brown in Fl. Cap. **4**(1): 660 (1907). Type: South Africa, Mpumalanga, Iuxta R., near Lydenburg, *Schlechter* 3944 (K lectotype, BM, BR, GRA, K, NH, Z), designated by Kupicha in Kew Bull. **38**: 658 (1984).
> *Schizoglossum interruptum* (E. Mey.) Schltr. in J. Bot. **34**: 450 (1896). —Brown in Fl. Cap. **4**(1): 660 (1907).
> ?*Schizoglossum morumbenense* Schltr. in Bot. Jahrb. Syst. **38**: 28 (1905). Type: Mozambique, Inhambane Dist., near Morumben, *Schlechter* 12098 (B† holotype).
> *Schizoglossum lasiopetalum* Schltr. in Bot. Jahrb. Syst. **38**: 29 (1905). Type: Mozambique, Matola near Delagoa Bay, 10.xii.1897, *Schlechter* 11685 (COI lectotype, BM, BR, E, GRA, K, PRE, Z), designated by Kupicha in Kew Bull. **38**: 658 (1984).

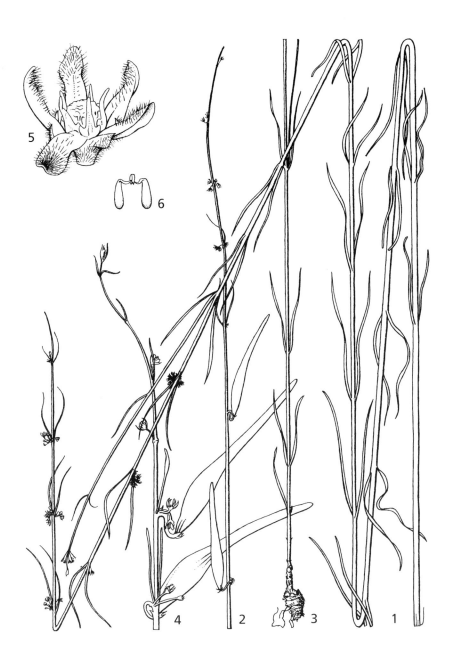

Fig. 7.3.**184**. ASPIDOGLOSSUM INTERRUPTUM. 1–3, habit (× ²/₃); 4, upper part of stem with fruit (× ²/₃); 5, flower (× 10); 6, pollinarium (× 26). 1, 5, 6 from *Bruce* 96; 2 from *Maitland* 1925; 3, 4 from *Wallace* 282. Drawn by E. Papadopoulos. Reproduced from Flora of Ethiopia and Eritrea (2003).

Schizoglossum gracile Weim. in Bot. Not. **1935**: 384 (1935), non *Lagarinthus gracilis* E. Mey. Type: Zimbabwe, Maconi, near Maidstone village, fl. 4.i.1931, *Norlindh & Weimarck* 4093 (LD holotype).

Perennial herb with napiform tuber. Stems single, 30–110 cm long, simple or branched, extremely slender, pubescent. Leaves opposite and paired even in inflorescence, 1.5–4 × c.0.05 cm, linear, margins strongly revolute, pubescent. Inflorescences sessile, 1–11-flowered at each of the upper nodes. Pedicels 2–5 mm long, pilose. Calyx lobes 0.7–1.5 × 0.3–0.6 mm, lanceolate, acute, pilose. Corolla campanulate, divided ± to the base, yellow-green to brown, sometimes pinkish outside; lobes 2–3 × c.1 mm, ovate, acute, plane or more usually replicate, outer surface pilose, inner surface long-pilose particularly towards the apex and margins. Corona lobes with basal portion c.0.5 × 0.8 mm, the upper margin developed into three teeth, the central one c.0.5 mm long, narrowly triangular, much longer than the lateral pair. Gynostegium c.0.5 mm long. Follicles (immature) fusiform with an attenuate beak, densely puberulent.

Zambia. N: North Luangwa National Park, fl. 17.iii.1994, *P.A. Smith* 428 (K). C: Chakwenga Headwaters, fl. 19.i.1964, *Robinson* 6223 (K). **Zimbabwe**. W: Matobo Dist., Besna Kobila farm, fl. & fr. xii.1956, *Miller* 4024 (K, SRGH). C: Harare (Salisbury) Dist., Cleveland Dam, fl. 23.xi.1951, *Wild* 3691 (K, SRGH). E: Mutare (Umtali) Dist., fl. 12.xi.1954, *Chase* 5326 (K, SRGH). **Mozambique**. M: Namaacha, fl. 25.iv.1947, *Pedro & Pedrogão* 811 (LMA).
Widespread but local in tropical and southern Africa. Grassland; 1100–1500 m.

15. **Aspidoglossum elliotii** (Schltr.) Kupicha in Kew Bull. **38**: 659 (1984). —Goyder in F.T.E.A., Apocynaceae (part 2): 380 (2012). Type: without locality, *Scott Elliot* s.n. (BM holotype). FIGURE 7.3.**182D**.
   Schizoglossum elliotii Schltr. in J. Bot. **33**: 305 (1895). —Brown in F.T.A. **4**(1): 359 (1902).
   Schizoglossum debile Schltr. in J. Bot. **33**: 305 (1895). Type: Uganda, Buddu, *Scott Elliot* 7471 (BM holotype, K).

Perennial herb with napiform tuber. Stems single, 80–120 cm long, simple or branched, pubescent. Leaves opposite even in inflorescence, 2–6 × 0.05–0.1 cm, minutely pubescent. Inflorescences sessile, 2–6-flowered at each of the upper nodes. Pedicels 4–6 mm long, pilose. Calyx lobes c.1.5 × 0.7 mm, ovate, acute, pilose. Corolla campanulate, divided ± to the base, green or brown; lobes 3–4 × 1.5–2 mm, ovate, subacute, plane or weakly replicate, pilose on outer surface, long-pilose within especially towards the apex and margins. Corona lobes c.1.5 mm long, fleshy, with lateral flanges to c.0.5 mm from base and an inward-pointing linear or triangular tooth c.0.5 mm long at apex - lobe described by Kupicha (1984) as 'penguin-shaped'. Follicles not seen.

Zambia. E: Katete, fl. 29.i.1957, *Wright* 134 (K). S: Mapanza, Choma, fl. 23.xii.1958, *Robinson* 2953 (K). **Zimbabwe**. N: 1.5 km. N of Gokwe, fl. 2.i.1964, *Bingham* 1023 (K, SRGH).
Also known from D.R. Congo, Rwanda, Burundi, Tanzania, Uganda and Ethiopia. Poorly drained grassland and open *Brachystegia* woodland; 1100–1400 m.
This species is very closely allied to *A. interruptum*, which differs in the dorsally flattened corona and the more conventional, subovoid, form of the corpusculum.

Sect. **Latibrachium** Kupicha in Kew Bull. **38**: 659 (1984).

16. **Aspidoglossum araneiferum** (Schltr.) Kupicha in Kew Bull. **38**: 661 (1984). Type: South Africa, KwaZulu-Natal, Newcastle, fl. 5.x.1893, *Schlechter* 3428 (Z lectotype), designated by Kupicha in Kew Bull. **38**: 661 (1984).
   Schizoglossum araneiferum Schltr. in Bot. Jahrb. Syst. **20**(Beibl. 51): 13 (1895). —Brown in Fl. Cap. **4**(1): 642 (1907).

Schizoglossum polynema Schltr. in Bot. Jahrb. Syst. **38**: 30 (1905). Type: Mozambique, Ressano Garcia, 24.xii.1897, *Schlechter* 11907 (BOL lectotype, BM, GRA, K, Z), designated by Kupicha in Kew Bull. **38**: 661 (1984).

Perennial herb with slender napiform tuber. Stems single, 30–60 cm long, simple or branched, pubescent. Leaves opposite even in inflorescence, 1.5–3.5 × 0.05–0.15 cm, linear, glabrous. Inflorescences sessile, 3–7-flowered at each of the upper nodes. Pedicels 4–7 mm long, pilose. Calyx lobes 1.5–2 × 0.7–1 mm, ovate, acute, pilose with white or rusty hairs. Corolla spreading to reflexed, divided ± to the base, dull green; lobes 2.5–3.5 × 1–2 mm, ovate, subacute, plane or weakly replicate, entirely glabrous or sparsely pilose on outer face only. Corona lobes with basal portion c.0.5 × 1 mm with square or rounded shoulders, a pair of teeth arising from the upper margin and single tooth on the ventral face; apical teeth 1–2 mm long, linear-triangular, united briefly at the base and contorted above; ventral tooth linear, ± as long as apical teeth. Gynostegium c.0.5 mm long. Follicles (immature) pubescent.

Zimbabwe. N: Umvukwe Mt., fl. 5.iii.1961, *Richards* 14569 (K, SRGH). **Mozambique**. M: Maputo, between Changalane and Cataune, fl. 21.xii.1952, *Myre & Carvalho* 1422 (K, LMA).

Also known from South Africa and Lesotho. Grassland; 300–1700 m.

17. **Aspidoglossum rhodesicum** (Weim.) Kupicha in Kew Bull. **38**: 663 (1984). Type: Zimbabwe, 3 km. W. of Mt. Inyangani, 5.xii.1930, *Fries, Norlindh & Weimarck* 3441 (LD holotype, BM, BR, LISU, PRE, SRGH).

Schizoglossum rhodesicum Weim. in Bot. Not. **1935**: 390 (1935).

Perennial herb with subglobose to napiform tuber. Stems single, 20–50 cm long, simple or occasionally with a single erect branch, pubescent. Leaves opposite even in inflorescence, 1.5–5 × 0.1–0.3 cm, linear, pubescent on upper surface and on midrib below. Inflorescences sessile, (2)6–12-flowered at each of the upper nodes. Pedicels 2–6 mm long, pilose. Calyx lobes 1.5–2.5 × 0.5–1 mm, ovate-triangular, acute, pilose. Corolla campanulate, divided ± to the base, green, yellow or whitish; lobes 3–4 × 1.5–2 mm, ovate, subacute, plane or weakly replicate, sparsely pilose outside, glabrous within. Corona lobes with narrow basal stalk c.0.5 mm long; distal portion 1–1.5 × 1–1.5 mm, broadly triangular, the outer tips commonly curled forwards; ventral tooth narrowly triangular, concave, slightly shorter than main lobe; corona lobes linked at the base by short lobules c.0.5 mm long which alternate with the main lobes. Gynostegium 0.5 mm long. Follicles not seen but described by Kupicha as c.2.5 cm long, densely lanate, bearing bristles.

Zimbabwe. E: Mare River, Inyanga, fl. 27.x.1946, *Wild* 1570 (K, SRGH).

Reported also from central Mozambique by Kupicha in Kew Bull. **38**: 663 (1984). Restricted to this area of eastern Zimbabwe and the adjacent part of Mozambique. Open montane grassland; 1600–2200 m.

18. **Aspidoglossum delagoense** (Schltr.) Kupicha in Kew Bull. **38**: 663 (1984). Type: Mozambique, Hangwane, Rikatla, 1893, *Junod* 484 (BR lectotype, Z), designated by Kupicha in Kew Bull **38**: 663 (1984).

Schizoglossum delagoense Schltr. in Bull. Herb. Boissier **4**: 446 (1896). —Brown in Fl. Cap. **4**(1): 645 (1907).

?*Schizoglossum biauriculatum* Schltr. in Bot. Jahrb. Syst. **38**: 29 (1905). —Brown in Fl. Cap. **4**(1): 639 (1907). Type: Mozambique, near Catembe [Katembe], Delagoa Bay area, *Schlechter* 11610 (B† holotype).

Perennial herb with napiform tuber. Stems single or paired, 20–60 cm long, simple, pubescent. Leaves opposite, even in inflorescence, 2.5–7 × 0.02–0.5 cm, linear, glabrous. Inflorescences sessile, 2–7-flowered at each of the upper nodes. Pedicels 5–8 mm long, pilose. Calyx lobes 1.5–2.5 × 0.5–1 mm, ovate-triangular, acute, pilose. Corolla spreading, lobed ± to the base, yellow-green or brown; lobes 4–5 × 1.5–2 mm, ovate, subacute, plane or replicate, pilose on outer face, glabrous or minutely puberulent within. Corona united at the base to form a cup c.0.5

mm long; main lobes with basal stalk; distal portion c.1–1.5 × 1–1.5 mm, broadly ovate with a rounded apex, lateral margins strongly incurved; ventral tooth about half as long as main lobe; intermediate portion of corona developed into pairs of short, rounded teeth. Gynostegium c.1.5 mm long. Follicles not seen but described by Kupicha as 4.5–5 cm long, slender, glabrescent, without bristles.

Mozambique. GI: road to Praia Sepúlveda, fl. 18.i.1965, *Rodrigues et al.* 282 (LMU). M: Maputo Elephant Reserve, c.5 km along track W of Ponta Milibangalala near head of Lake Munde, fl. 30.xi.2001, *Goyder* 5017 (K, LMU).

Restricted to NE KwaZulu-Natal and southern Mozambique–Maputaland; endemic. Open grassland on sand; sea level to 50 m.

66. **ASCLEPIAS** L.[38]

Asclepias L., Sp. Pl.: 214 (1753). —Goyder in Kew Bull. **64**: 369–399 (2009).
Odontostelma Rendle in J. Bot. **32**: 161 (1894).
Trachycalymma (K. Schum.) Bullock in Kew Bull. **8**: 348 (1953). —Goyder in Kew Bull. **56**: 129–161 (2001).
Aidomene Stopp in Bot. Jahrb. Syst. **87**: 21 (1967).

Slender to robust perennial herbs with annual stems arising from a tuber or fleshy taproot; latex white; stems prostrate to erect, simple or branched. Leaves opposite, linear to broadly ovate. Inflorescences terminal or extra-axillary, nodding or erect, umbelliform, sessile or pedunculate. Flowers 5-merous. Corolla divided almost to the base, lobes campanulate, spreading or reflexed. Corolline corona absent. Gynostegial corona of 5 generally cucullate fleshy lobes arising from the staminal column in a staminal position; minute interstaminal lobes sometimes also present. Pollinia pendant in anther cells; translator arms slender and terete or flattened, sometimes clearly geniculate, but never with a massively expanded proximal portion and slender distal portion. Stylar head rarely projecting much beyond top of anthers (but long-rostrate in *A. longirostra*). Follicles mostly single by abortion, generally held erect, smooth, occasionally ribbed or with lines of soft pubescent processes. Seeds ovate, discoid, with a coma of silky hairs.

A genus with two major centres of distribution, one New World with c.120 spp. mostly centred on southern parts of the North American continent, the other Old World, with c.80 spp., 38 of these in tropical Africa with the remainder in southern Africa. In addition, many segregate genera have been recognised in the Old World. In the broad sense the *Asclepias* radiation comprises some 380–400 species. Molecular surveys of this group do not lend strong support to current generic delimitation (see Goyder *et al.* in Ann. Missouri Bot. Gard. **94**: 423–434 (2007) and Chuba *et al.* in Syst. Bot. **42**: 148–159 (2017)). However, they do not suggest a workable alternative, so the morphologically distinctive African segregate genera *Margaretta*, *Stathmostelma*, *Gomphocarpus*, *Pachycarpus*, *Xysmalobium* and *Glossostelma* have been maintained in this treatment. I can only echo N.E. Brown's view on the subject published in his openly artificial treatment of the group for F.T.A 4(1): 299 (1902): "Undoubtedly *Xysmalobium*, *Asclepias* and *Schizoglossum* are but artificial divisions of one natural genus, since they cannot be separated by characters that do not break down at some point…" but he then goes on to say how he has allocated species to each. We have moved on considerably since then in recognising more natural units, but the phylogenetic structure is still inadequate for a stable realignment of the group at generic level.

Note: Only generic synonyms relevant to the Old World have been cited above.

---

[38] by D.J. Goyder

1. Corolla bright red; corona orange or yellow, with prominent tooth arising from the cavity and arching over the stylar head; annual or short-lived perennial from fibrous, non-tuberous rootstock; pan-tropical weed . . . .**1.** *curassavica*
   – Corolla white, yellow, green or brownish; corona variously coloured, corona tooth absent or inconspicuous and included within the cavity of the lobe; perennial producing annual shoots from tuberous rootstock; plants native to tropical Africa 2
2. Inflorescences erect 3
   – Inflorescences nodding. 16
3. Inflorescences solitary and terminal. 4
   – Inflorescences extra-axillary, sometimes initially appearing terminal 9
4. Corona lobes about as tall as the column, less than 3 mm long. 5
   – Corona lobes twice as long as the column, at least 5 mm long. 8
5. All leaves narrowly linear with an attenuate base, glabrous or subglabrous; inflorescence with 5–10 flowers, corolla campanulate . . . . **15.** *nuttii*
   – At least the lower leaves triangular with a truncate to weakly cordate base, pubescent; inflorescence with 10–35 flowers, corolla somewhat reflexed. 6
6. Follicles smooth. . . . . **26.** *albens*
   – Follicles ornamented. 7
7. Peduncle less than 3 cm long, inflorescences with 10–20 flowers; follicles with discrete, soft, spine-like processes. . . . . **24.** *adscendens*
   – Peduncle at least 6 cm long, inflorescences with 10–35 flowers; follicles with irregularly lobed longitudinal wings . . . . **23.** *densiflora*
8. Proximal margins of corona lobes twice as tall as column, central tooth only marginally longer than the proximal margins; anther wings triangular above the conspicuous notch, curved below it. . . . . **14.** *eminens*
   – Proximal margins of corona lobes ± as tall as column, distal margin forming a tooth twice as long; anther wings triangular, without a conspicuous notch, lower margins not curved . . . .**19.** *grandirandii*
9. Corona shorter than, or about as tall as the column. 10
   – Corona clearly longer than the column. 12
10. Outer face of corolla glabrous . . . .**17.** *minor*
    – Outer face of corolla pubescent. 11
11. Corona 2–4 mm long, laterally compressed, the apex drawn out into an erect tooth; leaves narrowly linear, base cuneate; Zimbabwe and Mozambique . . .**21.** *cucullata*
    – Corona c.1.5 mm long, neither laterally compressed nor with an apical tooth – pouched and with 2 rounded auricles apically; leaves linear to oblong, narrowing abruptly at the base; southern Tanzania and northern Malawi . . **16.** *breviantherae*
12. Leaves narrowly to broadly oblong, secondary veins clearly visible 13
    – Leaves narrowly linear, only the mid-vein visible 14
13. Corona lobes with mid-line drawn out into a long-attenuate tip, 5–7 mm long; outer face of corolla pubescent; highlands of SE D.R. Congo, SW Tanzania, NE Zambia and northern Malawi . . . .**19.** *grandirandii*
    – Corona lobes lacking long drawn-out tip, 3–4 mm long; outer face of corolla glabrous; regions of Botswana and Mozambique bordering South Africa and Swaziland . . . . **25.** *meliodora*
14. Peduncles mostly 3–14 cm long, subglabrous or minutely pubescent; corolla yellow, cream or white adaxially; corona lobes lacking papillae in the cavity and along the midline; stems reddish . . . . **20.** *aurea*
    – Peduncles 1–4 cm long, densely pubescent, sometimes minutely so; corolla greenish white tinged pink adaxially; corona lobes papillose in the cavity and/or along the midline; stems green 15

15. Corona lobes 3–3.5 mm long, papillae present in both cavity of lobe and on midline; translator arms of uniform thickness . . . . **22.** *pygmaea*
–   Corona lobes 4–8 mm long, papillae present only along the midline; translator arms broader distally . . . . **18.** *randii*
16. Leaves ovate or broadly oblong, generally with a truncate to cordate base. . . . 17
–   Leaves linear to lanceolate, tapering gradually into the petiole or with a narrowly truncate base 19
17. Leaves herbaceous, pubescent; venation prominent . . . . **9.** *fimbriata*
–   Leaves semisucculent, glabrous or subglabrous; venation usually indistinct . . . 18
18. Corona lobes laterally compressed and with a central cavity . . . . **7.** *fulva*
–   Corona lobes dorsiventrally flattened, at least distally; margins inrolled, but not forming a distinct central cavity . . . . **11.** *buchwaldii*
19. Stylar head extending c.5 mm beyond anthers as a long-rostrate appendage . . .
. . . . **6.** *longirostra*
–   Stylar head not or barely extending beyond anthers, flat or domed 20
20. Corona lobes spreading from the column, subcylindrical with densely papillate inrolled upper margins; endemic to Chimanimani mts (eastern Zimbabwe). . . .
. . . . **10.** *graminifolia*
–   Corona not as above 21
21. Plant generally taller than 50 cm; most leaves more than 10 cm long 22
–   Plant generally shorter than 50 cm; most leaves less than 10 cm long 24
22. Corona lobes 7–10 mm long, twice the height of the column; sepals at least 8 mm long . . . . **2.** *stathmostelmoides*
–   Corona lobes less than 7 mm long, 1–1.5 times height of column; sepals at most 3 mm long 23
23. Corona lobes 4.5–7 mm long; anther wings 2.5–3 mm long . . . . **3.** *longissima*
–   Corona lobes 3–3.5 mm; anther wings no more than 2 mm long. . **4.** *crassicoronata*
24. Corona lobes laterally compressed for their entire length, not drawn out into a tongue distally . . . . **5.** *amabilis*
–   Corona lobes rounded in section or dorsiventrally flattened, if appearing laterally compressed near attachment to column, then distal margins drawn out into a tongue 25
25. Corona lobes subglobose; a tuft of papillae entirely filling the cavity of the corona. . . . . **8.** *palustris*
–   Corona lobes generally longer than broad; papillae, if present, microscopic. . . 26
26. Gynostegium stipitate, the stipe 2–3.5 mm long; free portion of corona lobes forming an erect dorsiventrally flattened tongue 4–5 mm long with a truncate apex and inrolled margins . . . . **11.** *buchwaldii*
–   Gynostegium sessile; corona lobes not as above 27
27. Corolla white, frequently suffused with pink or purple, but never strongly veined; anther wings c.1.5 mm long; proximal margins of corona lobes reaching ± halfway up anther wings . . . . **13.** *foliosa*
–   Corolla white or pink with deeper veins within; anther wings c.2 mm long; corona lobes with proximal margins not reaching base of anther wings . . . . . . **12.** *ameliae*

1. **Asclepias curassavica** L., Sp. Pl.: 215 (1753). —Goyder in F.T.E.A., Apocynaceae (part 2): 384 (2012). Type: Curaçao, Linn. Herb. 310: 19 (LINN lectotype), designated by Woodson in Ann. Missouri Bot. Gard. **41**: 59 (1954); see also Jarvis, Order out of Chaos: 321 (2007).

   Asclepias nivea L. var. *curassavica* (L.) O. Kuntze, Revis. Gen. Pl. **2**: 418 (1891).
   Asclepias bicolor Moench, Methodus: 717 (1794), illegitimate name. Type as for *A. curassavica* L.

Asclepias aurantiaca Salisb., Prodr. Stirp. Chap. Allerton: 150 (1796), illegitimate name. Type as for *A. curassavica* L.

Annual or short-lived perennial herb with a single stem arising from fibrous rootstock; stems 0.5–2 m, erect, simple or branched, young shoots minutely pubescent, becoming glabrous with age, interpetiolar line often prominent. Leaves with petiole 5–15 mm long, glabrous or minutely puberulent; lamina 6–13 × 1–3 cm, narrowly oblong, apex attenuate, base narrowly cuneate, subglabrous except for the minutely puberulous midrib and margins. Inflorescences extra-axillary, umbelliform with (3)5–10 flowers; peduncles (2)3–5 cm long, glabrous or minutely puberulent; bracts 1–2 mm long, filiform, pubescent; pedicels 1–1.5 cm long, glabrous or minutely pubescent. Sepals 2–2.5 × 0.5–1 mm, narrowly oblong to triangular, acute, pubescent. Corolla bright red, strongly reflexed, lobes 6–7 × 2–3 mm, narrowly obovate, acute, glabrous or minutely pubescent. Gynostegial stipe slender, 1.5–2 mm long. Corona lobes orange or yellow, arising at the top of the gynostegial stipe, 3–4 × 1.5 mm, c.1.5 times the height of the column, somewhat fleshy and semi-cylindrical, not strongly compressed, a falcate tooth c.2 mm long arising from the cavity of the lobe and arching over the head of the column. Anther wings 1.8 mm long, the margin slightly convex, and with a minute notch at the base of the guide rail. Stylar head ± flat. Follicle erect on erect pedicel, 6–7 × 1 cm, fusiform and tapering towards both towards the base and apex, smooth, glabrous. Seeds c.5 × 3.5 mm, flattened, broadly ovate, with a marginal rim c.1 mm wide, smooth or verrucose sparsely; coma c.2.5 cm long.

Zimbabwe. W: Hwange [Wankie] Dist., Matetsi Safari area, Namakuwe R, fl. 13.iii.1981, *Gonde* 360 (K, SRGH). **Malawi**. S: Blantyre, on waste ground in township, fl. 6.i.1956, *Jackson* 1774 (K). **Mozambique**. T: Zumbo, on the Zambezi, fl. 22.vi.1900, *Baum* 1004 (K). M: Maputo [Lourenço Marques], Jardim Vasco da Gama, fl. 20.iii.1972, *Balsinhas* 2400 (K).

A pantropical weed. Native and widespread across the neotropics, cultivated and naturalised widely in Old World tropics. Moist areas.

2. **Asclepias stathmostelmoides** Goyder in Kew Bull. **64**: 375 (2009); —Goyder in F.T.E.A., Apocynaceae (part 2): 385 (2012). Type: D.R. Congo, Katanga, ii.1900, *Verdick* 361 (BR holotype, K photograph). FIGURE 7.3.**185A**.

Stathmostelma verdickii De Wild. in Ann. Mus. Congo Belge, Bot. sér. 5 **1**: 188 (1904), non *Asclepias verdickii* De Wild. (1906: 305).

Perennial herb with a single annual stem arising from a vertical tuber; stems 0.3–1.3 m long, unbranched, erect, densely spreading-pubescent, frequently lacking leaves in upper parts. Leaves with petiole 1–8 mm long; lamina 14–28 × 0.7–3.4 cm, narrowly linear to lanceolate, apex attenuate, the base cuneate or narrowly truncate, secondary veins numerous and frequently conspicuous, ± at right angles to the midrib, pubescent with stiff white hairs on both surfaces. Inflorescences extra-axillary with 6–16 flowers in a nodding umbel, peduncles 2–4.5(12) cm long, spreading-pubescent; bracts filiform, pubescent; pedicels (1.5)2–5 cm long, pubescent. Sepals 8–10 × 0.5–1 mm, narrowly triangular, attenuate, densely pubescent. Corolla rotate to weakly reflexed, green outside, tinged brown or purple within, pubescent on outer surface and towards margins within, inner surface papillate elsewhere; lobes 10–15 × 5–9 mm, ovate, acute, plane or weakly replicate. Gynostegium ± sessile. Corona lobes attached ± at the base of the staminal column, 7–10 × 3–4 mm, about twice height of column, laterally compressed, complicate above, solid below line from point of attachment to column to just below distal end of upper margin, quadrate, upper margins rounded, highest towards outer margin, pale green or white, frequently with a purple tinge or band and a white upper margin. Anther wings c.3 mm long, triangular, flared slightly towards the base; anther appendages c.2 mm long. Stylar head raised above top of anthers by c.1.5 mm, flat. Fruiting pedicel curved to hold follicle erect, follicle (immature) c.6 × 1.5 cm, fusiform, not inflated, smooth, densely pubescent or tomentose with short white hairs.

Zambia. N: Mbala Dist., Chilongowelo, close to path to Plain of Death, 4.ii.1952, *Richards* 951 (K). C: Mulungushi, fl. 9.ii.1964, *Fanshawe* 8268 (K, NDO). S: Mazabuka Dist., Mapanza, fl. 15.ii.1958, *Robinson* 2760 (K, SRGH). **Malawi**. N: Mbawa experimental station, Mzimba, fr. 5.iv.1955, *Jackson* 1598 (K). S: Liwonde National Park, Chiunguni

Horseshoe, fl. 29.xii.1986, *Oudley* 1806 (K). **Mozambique**. Z: Gurue, 26 km from Mutuali to Lioma, fl. 10.ii.1964, *Torre & Paiva* 10505 (LISC).

Scattered localities in Burundi, D.R. Congo, Tanzania, Zambia, Malawi and Mozambique. Grassland and open *Brachystegia* or other deciduous woodland; 500–1500 m.

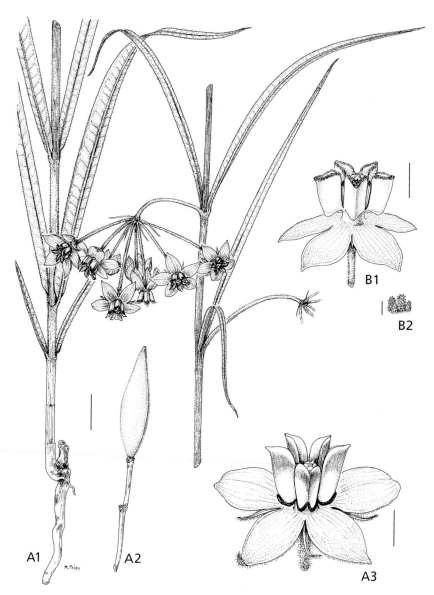

Fig. 7.3.**185**. A. —ASCLEPIAS STATHMOSTELMOIDES. A1, habit; A2, follicle; A3, flower. B. —ASCLEPIAS LONGISSIMA. B1, flower; B2, detail of coronal papillae. Scale bars: A1, A2 = 2 cm; A3, B1 = 5 mm; B2 = 0.2 mm. A1, A3 from *Bidgood et al.* 2647; A2 from *Congdon* s.n.; B1, B2 from *Johnston* 257. Drawn by Margaret Tebbs. Reproduced and adapted from Kew Bulletin (2009).

3. **Asclepias longissima** (K. Schum.) N.E. Br. in F.T.A. **4**(1): 338 (1902). —Goyder in F.T.E.A., Apocynaceae (part 2): 386 (2012). Type: Tanzania, Lake Malawi [Nyasa], *Goetze* s.n. (B† holotype, K). FIGURE 7.3.**185B**.

    Gomphocarpus longissimus K. Schum. in Bot. Jahrb. Syst. **30**: 382 (1901).

Perennial herb with a single annual stem arising from a tuber; stems 0.5–1 m long, unbranched, erect, glabrous or minutely pubescent above, lacking leaves in upper parts. Leaves with petiole 1–5 mm long; lamina 13–27 × 0.5–1.5 cm, narrowly linear to linear-lanceolate, apex attenuate, the base cuneate or narrowly truncate, secondary veins numerous and frequently conspicuous, ± at right angles to the midrib, margins revolute, glabrous. Inflorescences extra-axillary with 2–4 flowers in a nodding umbel, peduncles 1–2.5 cm long, minutely pubescent; bracts filiform, pubescent; pedicels 1.5–2 cm long, minutely pubescent. Sepals 2–3 × 1 mm, triangular, acute, pubescent. Corolla strongly reflexed, purple outside, green tinged red within, pubescent on outer surface and towards margins within, inner surface papillate elsewhere; lobes 7–10 × 4–6 mm, ovate, acute, plane or weakly replicate. Corona lobes attached c.1 mm above the base of the staminal column, 4.5–7 × 3–5 mm, 1–1.5 times height of column, laterally compressed, complicate above, solid in lower half, quadrate, upper margins rounded and densely papillate, highest towards inner margin, pale green, sometimes tinged with purple. Anther wings 2.5–3 mm long, narrowly triangular, flared slightly towards the base; anther appendages c.2 mm long. Stylar head raised above top of anthers by c.1.5 mm, flat. Follicles not seen.

Zambia. N: Mbala Dist., Old road from Chemba village to Cascalawa, fl. 16.ii.1960, *Richards* 12490 (K, SRGH). Mbala [Abercorn] Dist., fl. iv.1932, *Gamwell* 111 (BM). **Mozambique**. N: Marrupa, 20 km from Missor on Lichinga road, fl. 16.ii.1981, *Nuvunga* 534 (K, LMU, LISC).

Also known from from several localities in SW Tanzania. Generally on rocky ground in open *Brachystegia* woodland or grassland; 500–1500 m.

4. **Asclepias crassicoronata** Goyder in Kew Bull. **64**: 377 (2009). —Goyder in F.T.E.A., Apocynaceae (part 2): 388 (2012). Type: Zambia N, 30 km ESE of Kasama, 18.ii.1961, *Robinson* 4390 (K holotype).

Perennial herb with a single annual stem arising from a tuber or fleshy taproot; stems c.1 m long, unbranched, erect, glabrous or minutely pubescent above, lacking leaves in upper parts. Leaves with petiole 1–5 mm long; lamina 11–20 × 0.3–0.4 cm, narrowly linear, apex attenuate, the base cuneate, secondary veins inconspicuous, margins revolute, glabrous except along margins. Inflorescences extra-axillary with 4–6 flowers in a nodding umbel, peduncles 0.5–1.5 cm long, minutely pubescent; pedicels 1.5–2 cm long, slender, minutely pubescent. Sepals 2–3 × 1 mm, triangular, acute, pubescent. Corolla strongly reflexed, green or purple, glabrous or sparsely pubescent on outer surface, minutely papillate within; lobes 4–6 × 2–3 mm, ovate, acute, plane. Corona lobes attached c.1 mm above the base of the staminal column, 3–3.5 × 1.5 mm, slightly taller than column, fleshy, concave above, not laterally compressed, with a pair of acute teeth 0.5–1 mm long on the proximal margins and a fleshy tooth on the distal margin all pointing towards head of column, green. Anther wings 1.5 mm long, curved; anther appendages c.1 mm long. Stylar head raised above top of anthers by c.0.5 mm, flat. Follicles not seen.

Zambia. N: North Luangwa National Park, fl. 16.ii.1994, *P.A. Smith* 244 (K). **Malawi**. N: Mzimba, 12 km S of Eutini, fl. 31.i.1976, *Pawek* 10780 (K, MAL). C: Chipala Hill, 6 km N of Kasungu, fl. 14.i.1959, *Robson & Jackson* 1182 (K).

Also known from southern Tanzania. *Brachystegia* or mixed deciduous woodland on rocky hills; 1000–1900 m.

The flowers of this species are smaller than the similar *Asclepias stathmostelmoides* and *A. longissima*, and the corona is not laterally compressed and cucullate, but fleshy and with an inward-pointing tooth.

5. **Asclepias amabilis** N.E. Br. in Bull. Misc. Inform. Kew **1895**: 70 (1895). —Goyder in F.T.E.A., Apocynaceae (part 2): 380 (2012). Type: Zambia, Fwambo, *Carson* 55 (K lectotype), designated by Goyder in Kew Bull. **64**: 378 (2009).

    Gomphocarpus amabilis (N.E. Br.) Bullock in Kew Bull. **8**: 341 (1953).

Perennial herb with a single annual stem arising from a slender woody taproot or tuber; stems 0.2–0.5 m long, unbranched, erect, subglabrous to minutely rusty-pubescent above. Leaves subsessile; lamina 3–10 × 0.1–0.3 cm, narrowly linear, glabrous or minutely rusty-pubescent. Inflorescences extra-axillary with 5–9 flowers in a nodding umbel, peduncles 2–6 cm long, minutely rusty-pubescent; bracts to c.0.6 cm long, filiform, pubescent; pedicels 1–2.5(3) cm long, minutely rusty-pubescent. Sepals 3–4 × 1 mm, lanceolate to narrowly triangular, attenuate, pubescent, generally reddish brown. Corolla ± rotate to weakly reflexed, lobes 6–9(11) × 3–4.5(6) mm, ovate, subacute, adaxial surface pale green or greenish cream and papillose, occasionally only minutely so, abaxial face dull brown, purple or occasionally pink, pubescent, at least towards to apex. Corona lobes attached 1–1.5 mm above the base of the staminal column, 2.5–3.5(5.5) × 2–2.5(3.5) mm, as tall or taller than the column, laterally compressed, complicate above, quadrate, upper margin extremely variable – from almost entire with teeth or projections reduced to indistinct lobes, to strongly dissected with a pair of erect proximal teeth, the distal upper margins raised and variously undulate, linked to the proximal teeth by a lateral flap or tooth pointing back towards the column, green, cream or yellowish. Anther wings 1.3–1.5 mm long, broadly triangular. Stylar head ± level with top of anthers. Follicles and seed not seen.

Zambia. N: Mbala Dist., Marsh near fish ponds, Kiwimbi, fl. 9.ii.1955, *Richards* 4396 (BR, K). W: Mwinilunga Dist., Kalenda Plain N of Matonchi Farm, fl. 8.xii.1937, *Milne-Redhead* 3556 (K). **Malawi**. N: 8 km SW of Mzuzu, fl. 9.i.1973, *Pawek* 13279 (CAH, K, MAL, MO, UC).

Recorded also from southern Tanzania and the Katanga plateaus of south-eastern D.R. Congo. Seasonally waterlogged grassland (dambo/mbuga); 1200–2300 m.

Robinson 4174, from near Kasama in northern Zambia, is clearly a member of the *A. amabilis* complex – the corona suggests an affinity with the high-montane Tanzanian endemic, *A. edentata.*

6. **Asclepias longirostra** Goyder in Kew Bull. **64**: 380 (2009). Type: Malawi, Fort Manning [Mchinji] Dist., near Tamanda Mission, 8.i.1959, *Robson* 1105 (K holotype).

Perennial herb with one or perhaps more annual stems arising from a slender woody taproot or tuber; stems 0.3–0.5 m long, unbranched, erect, subglabrous to minutely pubescent above. Leaves subsessile; lamina 5–7 × 0.1–0.3 cm, narrowly linear, glabrous. Inflorescences extra-axillary with 4–7 flowers in a nodding umbel, peduncles 2–6 cm long, glabrous; bracts c.0.1 cm long, filiform; pedicels 0.5–1.5 cm long, glabrous or minutely rusty-pubescent. Sepals 3–4 × 1–1.5 mm, lanceolate to narrowly triangular, attenuate, glabrous or sparsely pubescent, reddish or purplish brown. Corolla campanulate, lobes 7–8 × 2–3 mm, ovate, with a long attenuate apex, adaxial surface green, minutely papillose, abaxial face dark purple, glabrous or minutely pubescent towards to apex. Corona lobes attached c.0.5 mm above the base of the staminal column, 2–3 mm long, and slightly taller than the anthers, dorsiventrally compressed and ± oblong in outline, with the inflexed lateral margins produced above into short rounded teeth, somewhat fleshy, probably purple with white margins (see note below). Anther wings c.1.5 mm long, triangular; anther appendages pale, narrowly triangular and drawn out into slender erect projections c.3 mm long lying alongside the stylar head appendage. Stylar head extending beyond top of anthers for c.5 mm as a long, purple, rostrate appendage. Follicles and seed not seen.

Zambia. C: Mutinondo Wilderness Area, W end of Matobo Dambo, fl. 16.xii.2019, *Merrett & Vollesen* in *Bidgood* 9924 (K). **Malawi**. N: Kapopo village, Chulu Native Authority, fl. 15.i.1959, *Jackson* 2295 (K). C: Fort Manning [Mchinji] Dist., near Tamanda Mission, fl. 8.i.1959, *Robson* 1105 (K).

Known only from these three collections. Seasonally waterlogged grassland – described by Jackson as a sandy seasonal swamp.

The rostrate stylar head appendage of this species appears to be unique in African *Asclepias*. The slender tips to the anther appendages look superficially like the teeth on the inner faces of staminal corona lobes seen in *Aspidoglossum*. Their derivation here, however, is quite different. Although highly distinctive, vegetative and floral characters other than the form of the stylar head and the anther appendages suggest a close affinity to *Asclepias amabilis*. Notes on two of the collections state that the 'anther appendages' are 'purple tipped white' or 'dark purple edged white' – these comments almost certainly refer to the corona lobes.

7. **Asclepias fulva** N.E. Br. in Bull. Misc. Inform. Kew **1895**: 254 (1895). —Goyder in F.T.E.A., Apocynaceae (part 2): 391 (2012). Type: Uganda, *Wilson* 112 (K holotype).

Pachycarpus viridiflorus E. Mey., Comm. Pl. Afr. Austr.: 214 (1838). Type: South Africa, Uitenhage Division, N side of the Zuurberg Mountains, 2500-3000', 30.x.1829, *Drège* s.n. (B† holotype, K).

Xysmalobium viridiflorum (E. Mey.) D. Dietr., Syn. Pl. **2**: 903 (1840).

Gomphocarpus viridiflorus (E. Mey.) Decne. in Candolle, Prodr. **8**: 561 (1844).

Asclepias rubicunda Schltr. in J. Bot. **33**: 336 (1895). Type: Uganda, Buddu, February, *Scott Elliot* 7443 (K lectotype, BM), designated by Goyder in Kew Bull. **64**: 381 (2009).

Asclepias dregeana Schltr. in J. Bot. **33**: 337 (1895), new name for *Gomphocarpus marginatus* sensu Schltr., non E. Mey.

Asclepias calceolus S. Moore in J. Bot. **41**: 312, 338 (1903). Type: South Africa, Transvaal, open veld N of Johannesburg, *Rand* 966 (BM holotype, K).

Asclepias dregeana var. *calceolus* (S. Moore) N.E. Br. in Fl. Cap. **4**(1): 697 (1908).

Asclepias dregeana var. *sordida* N.E. Br. in Fl. Cap. **4**(1): 697 (1908). Type: South Africa, Transkei, Kentani, ii.1905, *Pegler* 655 (K holotype, PRE).

Pachycarpus fulvus (N.E. Br.) Bullock in Kew Bull. **8**: 334 (1953).

Asclepias viridiflora (E. Mey.) Goyder in Kew Bull. **52**: 247 (1997), illegitimate name, not Raf., Med. Repos. **5**: 360 (1808).

Perennial herb with annual stems arising from a vertical napiform or fusiform tuber; stems 0.2–0.6 m long, usually single and unbranched, but sometimes branched below, erect, densely puberulent with short rusty hairs. Leaves with petiole 1–6 mm long; lamina 3.5–7.5(10) × 1.2–4 cm, narrowly oblong to ovate-oblong, apex obtuse to acute, base rounded or slightly cordate, margins scabrid, venation prominent with numerous parallel secondary veins at an acute angle to the midrib, indumentum of reddish hairs on both upper and lower surfaces. Inflorescences extra-axillary with 4–10 flowers in a nodding umbel, peduncles 1.5–7(9) cm long, densely rusty-puberulent; bracts 5–10 mm long, filiform, puberulent; pedicels 1–3 cm long, puberulent. Calyx lobes 4–11 × 1.5–3 mm, lanceolate, acute, pubescent, green or brown. Corolla often somewhat reflexed at base, spreading above, rotate or saucer-shaped, outside green or brownish purple, densely pubescent with white or rusty hairs, inner surface pale orange or yellow-brown, glabrous; lobes 8–13 × 4–7.5 mm, ovate, acute. Corona lobes solid except for a shallow sinus along upper and inner margins, laterally compressed, 2–5 × 2–4 mm, quadrate with a truncate base and an obliquely truncate top, the upper margins produced into a pair of teeth extending over the stylar head, yellowish green with a brown or purple projecting rim to 1.5 mm wide around the top and base. Anther wings 2.5 × 1.5 mm, triangular; anther appendages c.1 × 1 mm, broadly ovate, subacute. Stylar head flat or undulate. Follicles c.8–11 × 1.5 cm, fusiform and with 4–6 longitudinal toothed wings or ridges running for all or most of their length, minutely puberulent.

Zambia. N: Mbala [Abercorn] Dist., Ndundu, fl. 16.ii.1957, *Richards* 8201 (K). **Zimbabwe**. N: Lomagundi Dist., Mtorashangu, fl. xii.1973, *Cannell* 570 (K, SRGH). C: Makoni Dist., 1.5 km from Eagles Nest on Rusape road, fl. 29.xi.1955, *Drummond* 5068 (K, PRE, SRGH). E: Mutare, S car park, Murahwa's Hill, fl. 1.i.1965, *Chase* 8220 (EA, K, SRGH).

Distributed from Uganda to eastern regions of South Africa and Lesotho. Open grassland and *Brachystegia* or mixed deciduous woodland; 1000–2100 m.

8. **Asclepias palustris** (K. Schum.) Schltr. in J. Bot. **33**: 336 (1895). —Brown in F.T.A. **4**(1): 349 (1902). —Goyder in F.T.E.A., Apocynaceae (part 2): 392 (2012). Type: Angola, Malange, *Mechow* 401 (K lectotype, B†), designated by Goyder in Kew Bull. **56**: 134 (2001). FIGURE 7.3.**186**.

Gomphocarpus cristatus Decne. in Ann. Sci. Nat., Bot. sér. 2 **9**: 325, t.11D, figs.3, 4 (1838), non *Asclepias cristata* S. Moore in J. Bot. **50**: 343 (1912). Type: Angola, Benguela Plateau, *da Silva* s.n. (P holotype, K photograph).

Gomphocarpus paluster K. Schum. in Bot. Jahrb. Syst. **17**: 127 (1893), as *G. palustris*.

Asclepias cristata S. Moore in J. Bot. **50**: 343 (1912), not *Gomphocarpus cristatus* Decne. (1838). Type: Angola, Kubango, Kapembe, 23.xi.1905, *Gossweiler* 2288 (BM lectotype), designated by Goyder in Kew Bull. **56**: 134 (2001).

Trachycalymma cristatum (Decne.) Bullock in Kew Bull. **8**: 349 (1953). —Goyder in Kew Bull. **56**: 134 (2001).

Asclepias kyimbilae Schltr., invalid name, in sched., *Stolz* 502 (K, WAG).

Perennial herb with annual stems arising from a vertical rootstock with globose to fusiform lateral tubers; stems usually single, 0.1–0.4(0.6) m long, generally unbranched, erect, densely pubescent with spreading white hairs. Leaves subsessile or with petiole to 3 mm long, pubescent; lamina 3–7(8) × (0.1)0.4–1.3 cm, linear to lanceolate, acute, base rounded to truncate, margins stiffly pubescent, venation prominent with secondary veins at c.45° to the midrib, indumentum of spreading white hairs on upper surface and on main veins below. Inflorescences extra-axillary with 4–10 flowers in a nodding umbel, peduncles 1–6(13) cm long, lengthening markedly in fruit, erect, densely spreading-pubescent; bracts 0.2–1 cm long, filiform or occasionally linear-lanceolate, pubescent; pedicels 1–1.5(2) cm long, pubescent. Calyx lobes 3–5(6) × 1–1.5 mm, lanceolate to broadly triangular, acute, densely pubescent, purple. Corolla campanulate or occasionally reflexed, white or greenish, tinged with dull pink especially towards tip outside, densely pubescent towards the apex outside, glabrous or minutely papillate at base within; lobes 5–8 × 3–5 mm, ovate, subacute. Corona lobes cucullate but with a rounded tip making the lobe appear subglobose, c.3 mm long, about half as tall as the column, white with purple tip, the upper margins with a pair of erect triangular teeth proximally, otherwise entire to variously dentate, with a laterally flattened tooth c.0.5 mm long topped with a dense crest of papillae ±

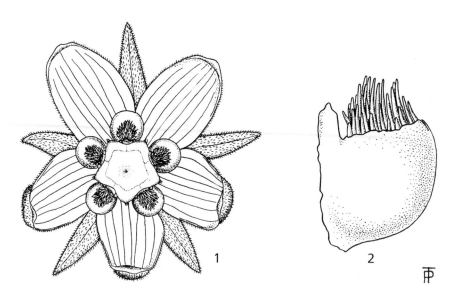

Fig. 7.3.**186**. ASCLEPIAS PALUSTRIS. 1, flower (× 4); 2, corona lobe, lateral view (× 12). From *Milne-Redhead* 3079. Drawn by P. Taylor. Reproduced from Kew Bulletin (2001).

filling the cavity of the lobe, the papillae reaching to or slightly exceeding the upper margins. Anther wings c.1.5 mm long, triangular; anther appendages c.0.5 × 1.5 mm, reniform, inflexed over apex of stylar head. Fruiting pedicel not contorted; follicle erect, c.5 × 0.5 cm, narrowly fusiform with an attenuate apex, not inflated, smooth, densely pubescent. Seeds flattened, c.0.4 × 0.3 cm, ovate, with an inflated and somewhat convoluted rim and a verrucose disc.

Zambia. Chinsali Dist., Great North Road 55 km NE of Mpika, fl. 25.xi.1972, *Kornas* 2704 (K). Mwinilunga Dist., SW of Dobeka Bridge, fl. 4.xi.1937, *Milne-Redhead* 3079 (K). E: Lundazi Dist., Nyika Plateau Rest House, fl. 2.i.1959, *Richards* 10410 (K). **Zimbabwe**. C: Goromonzi, fl. 1.i.1927, *Eyles* 4604 (K, SRGH). **Malawi**. N: Nyika Plateau, Chelunduo Stream at Chelinda Camp, fl. 26.x.1958, *Robson & Angus* 386 (K). S: Mt Mulanje, Lichenya Plateau, fl. 20.x.1971, *Moriarty* 688 (K). **Mozambique**. N: Near Lake Malawi, fl. 1902, *Johnson* 460 (K). Z: Namuli Mountain, Muretha Plateau, fl. 16.xi.2007, *Harris et al.* 320 (K, LMA).

Widely scattered in tropical Africa from Nigeria to Uganda in the north, and Angola to Zimbabwe in the south. Montane or seasonally waterlogged grassland, occasionally in open woodland; 1200–2600 m.

9. **Asclepias fimbriata** Weim. in Bot. Not. **22**: 378 (1935). Type: Zimbabwe, Nyanga, above Pungwe stream, 16.xii.1930, *Fries et al.* 3748 (LD holotype, BM, BR, K photograph, SRGH).

  Trachycalymma fimbriatum (Weim.) Bullock in Kew Bull. **8**: 352 (1953). —Goyder in Kew Bull. **56**: 137 (2001).

Perennial herb with annual stems, rootstock a vertical napiform tuber; stems 0.15–0.5 m long, simple or branched near base only, erect, densely pubescent with spreading white hairs. Leaves with petiole 2–8 mm long, pubescent; lamina (2.5)3–5 × (1.5)2–3 cm, ovate, acute to obtuse, base rounded to truncate or slightly cordate, margins softly pubescent, venation prominent with secondary veins at c.45° to the midrib, indumentum of spreading white hairs on both upper and lower surfaces. Inflorescences extra-axillary with (3)4–9 flowers in a nodding umbel, peduncles 3–7 cm long, erect, densely spreading-pubescent; bracts 0.3–0.6 cm long, filiform, pubescent; pedicels c.1 cm long, densely pubescent. Calyx lobes 3–5 × 1–2 mm, lanceolate to broadly triangular, acute, densely pubescent, purple. Corolla campanulate, purple, densely pubescent towards the apex outside, minutely but densely papillate within; lobes 7–8 × 3–4.5 mm, ovate, subacute. Corona lobes cucullate but with a rounded tip making the lobe appear subglobose, c.3 mm long, about half as tall as the column, white with purple tip, the upper margins densely fimbriate, the hairs c.0.5 mm long, cavity of the lobe without a tooth. Anther wings c.1.5 mm long, margins ± vertical above; anther appendages c.1 × 0.5 mm, ovate, acute and inflexed over apex of stigma head. Fruiting pedicel contorted to hold follicle erect; follicle to c.8 × 1.5 cm, ovoid or narrowly ovoid with an attenuate apex, not inflated, with soft, pubescent processes arranged along weak longitudinal ridges, densely pubescent. Seeds not seen.

Zimbabwe. E. Inyanga Dist., near Mtarazi Falls, fl. & fr. 18.xi.1963, *Chase* 8071 (K, SRGH). **Mozambique**. MS. Serra Zuira, Tsetserra plain, fl. 4.xi.1965, *Torre & Pereira* 12629 (LISC).

Known only from the eastern highlands of Zimbabwe and adjacent parts of Mozambique. Growing in open montane grassland; 1700–2300 m.

10. **Asclepias graminifolia** (Wild) Goyder in Kew Bull. **64**: 383 (2009). Type: Zimbabwe, Chimanimani Mts, slopes of Point 71, *Goodier & Phipps* 199 (SRGH holotype, K photograph).

  Pachycarpus graminifolius Wild in Kirkia **4**: 148 (1964).
  Trachycalymma graminifolium (Wild) Goyder in Kew Bull. **56**: 141 (2001).

Perennial herb with annual stems, rootstock not seen; stems to c.0.6 m long, simple or branched, erect, minutely pubescent with spreading white hairs. Leaves with petiole 1–6 mm long, pubescent; lamina 8–16 × 0.3–1.2 cm, linear, attenuate, base cuneate to truncate, margins softly pubescent, venation prominent with secondary veins at c.45° to the midrib, indumentum of spreading white hairs on upper surface and on midrib below. Inflorescences extra-axillary with 3–6 flowers in a nodding umbel, peduncles 1–2 cm long, erect, densely spreading-pubescent; bracts 0.5–1 cm long, filiform, pubescent; pedicels 1–1.5 cm long, densely pubescent. Calyx lobes 5–6 × 1.5–2.5 mm, lanceolate, acute, densely pubescent, purple. Corolla campanulate, dull purple with paler margins, sparsely pubescent towards the apex outside, glabrous within; lobes 8–10 × 6–7 mm, broadly ovate, subacute. Corona lobes spreading from the column, white or white with purple sides, c.5 × 1.5 mm, canaliculate for most of their length with inrolled upper margins and a slightly upturned hooded tip, about half as tall as the column, the margins densely papillate, cavity of the lobe without a tooth. Anther wings c.2.5 × 1 mm, the margin spreading at c.45° for the top 1 mm, then falling vertically; anther appendages c.2 × 1 mm, subreniform, inflexed over apex of stigma head. Fruiting pedicel contorted; follicle (immature?) c.5.5 × 1 cm, narrowly ovoid with an attenuate apex, not inflated, smooth, densely pubescent. Seeds not seen.

Zimbabwe. E: Chimanimani Mts, E of National Parks Office, fl. 20.iii.1981, *Philcox et al.* 9016 (K).

Endemic to the Chimanimani Mountains. Currently known only from the Zimbabwean side, but it is likely that populations will also be found on the Mozambican side. Growing in rocky quartzite grassland; 1700–2200 m.

11. **Asclepias buchwaldii** (Schltr. & K. Schum.) De Wild. in Ann. Mus. Congo Belge, Bot. sér. 5 **1**: 185 (1904). —Goyder in F.T.E.A., Apocynaceae (part 2): 394 (2012). Type: Tanzania, Usambara, Mombo, *Buchwald* 375 (B† holotype); Iringa Dist., upper slopes of Image Mountain, *Goyder et al.* 3924 (K neotype, DSM, PRE), designated by Goyder in Kew Bull. **56**: 146 (2001).

　　Gomphocarpus buchwaldii Schltr. & K. Schum. in Bot. Jahrb. Syst. **33**: 324 (1903).

　　Asclepias affinis De Wild. in Ann. Mus. Congo Belge, Bot. sér. 5 **1**: 184 (1904), not *Gomphocarpus affinis* Schltr., nec *Asclepias affinis* (Schltr.) Schltr. Type: D.R. Congo, Vieux Kasongo, viii.1896, *Dewèvre* 952 bis (BR holotype).

　　Asclepias buchwaldii var. *angustifolia* De Wild. in Ann. Mus. Congo Belge, Bot. sér. 5 **1**: 185 (1904). Type: D.R. Congo, Vieux Kasongo, viii.1896, *Dewèvre* 952 (BR holotype).

　　Trachycalymma buchwaldii (Schltr. & K. Schum.) Goyder in Kew Bull. **56**: 146 (2001).

Perennial herb with annual stems arising from a slender, vertical, fleshy tuber, occasionally also with fusiform lateral tubers; stems usually single, 0.15–0.6 m long, simple or occasionally branched below, erect, minutely pubescent with spreading white hairs. Leaves sessile or with petiole to 2 mm long; lamina somewhat fleshy, 2.5–7(10) × (0.4)0.8–2.2 cm, ovate to narrowly lanceolate, acute to attenuate, base rounded or rarely subcuneate, margins softly pubescent, venation prominent in dried material with numerous secondary veins at 45°–90° to the midrib, glabrous or with an indumentum of minute white hairs on both surfaces. Inflorescences extra-axillary with 3–6 flowers in a nodding umbel, peduncles 1–4(6) cm long, erect, densely spreading-pubescent; bracts 0.1–0.6 cm long, filiform, pubescent; pedicels 0.7–2 cm long, extending slightly in fruit, densely pubescent. Calyx lobes 2–5 × 0.5–1 mm, lanceolate, acute, densely pubescent, dull purple. Corolla campanulate or occasionally reflexed, outer face dull purple with paler margins, pubescent, green or brownish within and papillose at least towards margins and apex; lobes 5–7 × 2–3 mm, oblong, subacute. Corona lobes violet with darker margins, paler within, slightly fleshy, adnate to column for 2–3.5 mm and reaching the base of the anthers, adnate portion forming a pair of vertical wings with slightly inrolled papillate margins, extending basally into a recurved erect tongue 4–5 mm long and 1–2 mm wide at apex reaching ± to top of gynostegium, margins inrolled, papillate, apex truncate or weakly lobes, minutely but densely papillose, inrolled or not, cavity of the lobe without a tooth. Anther wings 0.7–1 mm, vertical; anther appendages c.1 × 1.5 mm, subreniform, inflexed over apex of stigma head. Fruiting pedicel not contorted, erect; follicles c.15 × 0.5 cm, narrowly fusiform with an attenuate apex and base, not inflated, smooth, densely pubescent. Seeds not seen.

Zambia. N: Mbala Dist., Nkali Dambo, fl. 27.xii.1967, *Simon & Williamson* 1590 (K, SRGH). **Malawi**. N: Nyika National Park, escarpment below Jalawe viewpoint, c.25 km N of Chelinda, fl. 25.i.1992, *Goyder et al.* 3581 (K, MAL).

Known from the southern highlands of Tanzania and adjacent regions of Malawi and Zambia. Also occurs in northern Tanzania, Burundi and D.R. Congo. Growing in grassland or *Brachystegia* woodland, usually on steep rocky hillsides; 900–2400 m.

12. **Asclepias ameliae** S. Moore in J. Bot. **50**: 345 (1912). Type: Angola, between Forte Princeza Amelia and the R. Kubango, 1.xi.1905, *Gossweiler* 2199 (BM holotype), reported in error by Moore as 2176.

> *Gomphocarpus pulchellus* Decne. in Ann. Sci. Nat., Bot. sér. 2 **9**: 325 (1838). Type: Angola, Benguela Plateau, *da Silva* s.n. (P holotype, K).
> *Asclepias pulchella* (Decne.) N.E. Br. in F.T.A. 4(1): 346 (1902), illegitimate name, not *A. pulchella* Salisb. (1796).
> *Trachycalymma pulchellum* (Decne.) Bullock in Kew Bull. **8**: 350 (1953), in part. —Goyder in Kew Bull. **56**: 150 (2001).

Perennial herb with 1(3) annual stems arising from a subglobose or occasionally napiform tuber; stems 0.2–0.4 m long, generally unbranched, erect, densely pubescent with spreading white hairs. Leaves with petiole 1–8 mm long, pubescent; lamina 4–7 × 0.5–2 cm, oblong to elliptic, occasionally lanceolate or oblanceolate, acute, base cuneate, venation prominent with secondary veins at c.45° to the midrib, indumentum of spreading white hairs on both surfaces. Inflorescences terminal or extra-axillary with 4–11 flowers in a ± erect umbel, peduncles (3)5–16 cm long, erect, densely spreading-pubescent; bracts 0.5–1 cm long, filiform or occasionally linear-lanceolate, pubescent; pedicels 1–2.5 cm long, pubescent. Calyx lobes 4–7(9) × 1–2 mm, narrowly lanceolate, acute, pubescent. Corolla broadly campanulate, pinkish mauve outside, white or pale pink frequently with deeper veins within, sparsely pubescent towards the apex outside, glabrous within; lobes 7–10 × 4–6 mm, ovate, subacute. Corona mostly white but pinkish towards base, lobes 4–5 mm long, fleshy and adnate to the column to the base of the anther wings, somewhat cucullate above and tapering gradually into an erect tongue ± as long as the staminal column, proximal margins mostly not reaching the base of the anther wings and with a pair of auricles proximally, with a dense band of papillae across the cavity halfway along the lobe so that the lobe appears to be filled with papillae for the middle third of its length, glabrous at extreme base; interstaminal corona lobes consist of a truncate or emarginate lobule c.0.5 × 0.5 mm situated between the auricles of the principal lobes. Anther wings c.2 mm long, triangular; anther appendages c.1.5 × 1.5 mm, ovate, inflexed over apex of stigma head. Follicles erect, densely pubescent when young, mature fruit not seen.

Zambia. W: Ndola, fl. 18.xii.1954, *Fanshawe* 1724 (K, NDO). C: Mkushi Dist., Fiwila, fl. 7.i.1958, *Robinson* 2664 (K, SRGH).

Recorded from Angola, western and central Zambia, and the Katanga region of D.R. Congo. Open, mixed deciduous woodland, or occasionally in seasonally waterlogged grassland; c.1200–1500 m.

13. **Asclepias foliosa** (K. Schum.) Hiern in Cat. Afr. Pl. **1**: 686 (1898). —Brown in F.T.A. 4(1): 349 (1902). —Goyder in F.T.E.A., Apocynaceae (part 2): 395 (2012). Type: D.R. Congo, Mukenge, *Pogge* 1130 (K lectotype, B†), designated by Goyder in Kew Bull. **56**: 153 (2001).

> *Gomphocarpus foliosus* K. Schum. in Bot. Jahrb. Syst. **17**: 126 (1893).
> *Asclepias modesta* N.E. Br. in F.T.A. 4(1): 348 (1902). Type: Malawi, Namasi, *Cameron* 6 (K lectotype), designated by Goyder in Kew Bull. **56**: 153 (2001).
> *Asclepias modesta* var. *foliosa* N.E. Br. in F.T.A. 4(1): 349 (1902). Type: Angola, Huilla, near Lopollo, *Welwitsch* 4174 (K holotype, BM, LISU).
> *Asclepias lepida* S. Moore in J. Bot. **50**: 344 (1912). Type: Angola, Kubango, near Forte Colui, 18.x.1905, *Gossweiler* 2176 (BM holotype).

Trachycalymma pulchellum sensu Bullock in F.W.T.A. ed. 2 **2**: 92 (1963).
Trachycalymma foliosum (K. Schum.) Goyder in Kew Bull. **56**: 153 (2001).

Perennial herb with 1–3(5) annual stems arising from one or more subglobose or napiform tubers; stems 0.1–0.4 m long, generally unbranched, erect, pubescent with spreading white hairs. Leaves with petiole to 3 mm long, pubescent; lamina (1.5)4–9(10) × 0.2–1.5 cm, linear to elliptic, acute, base cuneate, venation only visible on wider leaves, secondary veins at c.45° to the midrib, glabrous or with indumentum of spreading white hairs on both surfaces. Inflorescences terminal or extra-axillary with 2–4(6) nodding flowers per umbel, peduncles 1–15 cm long, erect, pubescent; bracts 0.2–0.4 cm long, filiform, pubescent; pedicels 0.5–2 cm long, pubescent. Calyx lobes 3–5 × 1(2) mm, ovate to lanceolate, acute, pubescent, purple or occasionally green. Corolla campanulate, white, frequently suffused with pale pink or purple but not strongly veined within, sparsely pubescent outside, glabrous within; lobes 7–10(12) × 3–6(7) mm, ovate-oblong, subacute. Corona purple with white margins (green with white margins in west Africa), lobes 2.5–4 mm long, fleshy and adnate to the column to the base of the anther wings, cucullate above and tapering gradually into a rounded spreading tongue half to as long as the staminal column, proximal margins mostly reaching about midway along the anther wings and with a pair of erect teeth extending further up the column, with a dense band of papillae across the cavity halfway along the lobe and entirely concealed within it, exposed portion of tongue glabrous to minutely papillate; interstaminal corona lobes minute. Anther wings c.1.5 mm long, triangular; anther appendages c.1 × 1 mm, ovate, inflexed over apex of stigma head. Fruiting pedicel not contorted; follicle erect, 20–25 × 0.5 cm, narrowly fusiform with a stipe to 17 cm long and an attenuate apex, not inflated, smooth, minutely puberulent. Seeds not seen.

Zambia. B: Balovale, fl. xii.1953, *Gilges* 310 (K). N: Mpulungu Dist., Munomba R. dambo, fl. 9.xii.2006, *Bingham* 13231 (K). W: Kitwe, fl. 24.xi.1955, *Fanshawe* 2624 (K, NDO). C: Jellis's farm, Lazy J Ranch, 20 km SE of Lusaka, fl. 20.xii.1994, *Bingham* 10210 (K). E: Chadiza, fl. 28.xi.1958, *Robson* 777 (K). S: Livingstone, fl. vi.1955, *Seale* 22 (SRGH). **Malawi**. N: Chitipa Dist., 8 km NE of Chendo, fl. 2.i.1977, *Pawek* 12203 (K, MAL, MO). C: Dedza Dist., Chongoni Forest Reserve, fl. 27.xi.1967, *Salubeni* 905 (K, MAL, SRGH). S: Zomba Dist., Balaka, fl. 6.xii.1956, *Jackson* 2094 (K, SRGH). **Mozambique**. Near Lake Malawi, fl. 1902, *Johnson* 471 (K).

Recorded from the savanna regions of West Africa and the *Brachystegia* belt of southern tropical Africa. Open, mixed deciduous woodland over most of its range, but frequently occurring in grassland in West Africa; (600)1100–2000 m.

14. **Asclepias eminens** (Harv.) Schltr. in J. Bot. **34**: 453 (1896). —Brown in Fl. Cap. **4**(1): 685 (1909). Type: South Africa, KwaZulu-Natal, Zululand, *Gerrard & McKen* 1291 (TCD holotype, BM, K).

> *Gomphocarpus eminens* Harv. in Thes. Cap. **2**: 60, t.195 (1863).
> *Stenostelma eminens* (Harv.) Bullock in Kew Bull. **8**: 342 (1953).

Perennial herb with one to many annual stems arising from a vertical napiform tuber; stems 10–15 cm long, simple or branched below, prostrate to ascending, pubescent in two lines. Leaves with petiole 1–3 mm long; lamina 2–4(5) × 0.1–0.3(0.5) cm, linear, apex acute, base truncate or hastate, margins somewhat revolute, glabrous. Inflorescences solitary and terminal forming umbels of 2–6 flowers; peduncles 1–2 cm long, pubescent; pedicels 1.1–2 cm long, minutely pubescent. Sepals 3–6 × 1 mm, narrowly triangular, acute, sparsely pubescent. Corolla spreading to reflexed, lobes 7–10 × 2.5–3.5 mm, oblong to lanceolate, acute, adaxial surface grey-green or greenish white, minutely but densely papillate, abaxial face green tinged brown or purple, glabrous. Gynostegium stipitate. Corona lobes 5–6 mm long, erect, closely appressed to the staminal column and long-exceeding it, cucullate, green or white with a fleshy midline green below and pinkish towards the short apical tooth. Anther wings 3 mm long, triangular in upper 2/3, ending at a characteristic notch, then rounded below. Stylar head ± flat. Fruiting pedicel not contorted, follicles 5–7 cm long, erect, fusiform, smooth; seeds not seen.

Zimbabwe. C: Harare [Salisbury], between Avondale West and Mabelreign, fl. 23.x.1955, *Drummond* 4916 (BR, K, SRGH). W: Bulawayo, fl. i.1898, *Rand* 189 (BM).

Also known from Swaziland, Lesotho and north-eastern parts of South Africa (NW Province, Mpumalanga, KwaZulu-Natal). Seasonally burned, and frequently also seasonally waterlogged grassland; c.1500 m.

15. **Asclepias nuttii** N.E. Br. in Bull. Misc. Inform. Kew **1898**: 308 (1898). —Goyder in F.T.E.A., Apocynaceae (part 2): 397 (2012). Type: Tanzania, 'between Lake Tanganyika and Lake Rukwa', *Nutt* s.n. (K holotype).

   Stathmostelma nuttii (N.E. Br.) Bullock in Kew Bull. **8**: 55 (1953).

Perennial herb with one or few annual stems arising from a deep-seated vertical tuber; stems 6–30 cm long, simple or branched below, ascending or erect, reddish, minutely pubescent. Leaves subsessile; lamina 3–10(12) × 0.1–0.5 cm, narrowly linear, attenuate both apically and basally, glabrous or with minute indumentum restricted to margin and midrib. Inflorescences solitary and terminal forming umbels of 5–10 spreading or erect flowers; subsessile or pedunculate, peduncle to 2 cm long, reddish, minutely but densely pubescent; bracts filiform, pubescent; pedicels 0.8–1.5 cm long, minutely pubescent. Sepals 3–5 × 0.5–1 mm, lanceolate or narrowly oblong, acute, pubescent. Corolla campanulate, lobes 6–8 × 2–3 mm, ovate-oblong to lanceolate, acute, cream or white tinged with pink and glabrous except towards the margins adaxially, abaxial face green tinged with purple-brown, glabrous or pubescent towards the tip. Corona lobes arising at the top of a gynostegial stipe c.1 mm long, lobes c.2 mm tall, reaching ± to top of column, subcylindrical, purplish pink with white tip. Anther wings 1.5–2 mm long, with a conspicuous notch near the base ± at the point the margins curl under the gynostegium. Stylar head ± flat. Follicles and seeds not seen.

   Malawi. C: Chongoni Forest School, base of Chiwao Hill, fl. 4.ii.1959, *Robson* 1445 (K).

Known from several localities in SW Tanzania, with a single collection from central Malawi. Surprisingly, this species has not been recorded from the Mbala region of Zambia. Seasonally waterlogged 'dambo' grassland; 1500–1800 m.

This species is distinctive with a single terminal inflorescence per branch. The translator arms are differentiated into proximal and distal sections, as in *Stathmostelma*, but the proximal portion is not so conspicuously expanded. The conspicuous notch in the anther wings suggests a possible link to *Asclepias eminens*.

16. **Asclepias breviantherae** Goyder in Kew Bull. **64**: 389 (2009); in F.T.E.A., Apocynaceae (part 2): 399 (2012). Type: Tanzania T7, Rungwe Dist., Poroto Mts E of Kikondo on road to Kitulo, 1.xii.1994, *Goyder et al.* 3872 (K holotype, DSM, EA, PRE, WAG).

Perennial herb with one to several annual stems arising from a napiform tuber; stems 6–12(25) cm long, simple or little-branched, prostrate to erect, reddish, minutely pubescent with indumentum of white hairs. Leaves subsessile; lamina 3–7 × 0.2–1 cm, linear to narrowly lanceolate or oblong, mostly attenuate apically, and narrowing abruptly at the base, margins not revolute, minutely pubescent with stiff white hairs on both upper and lower surfaces, or indumentum restricted to principal veins. Inflorescences extra-axillary forming umbels of 2–7 spreading or erect flowers; peduncles 0.5–4 cm long, reddish with a dense indumentum of white or rusty hairs; bracts filiform, pubescent; pedicels 0.3–1 cm long, pubescent. Sepals (2)2.5–3 × 0.7–1 mm, narrowly oblong to lanceolate, attenuate, generally reddish with white hairs. Corolla campanulate, lobes (3)5–6 × 2–2.5 mm, oblong to narrowly ovate, subacute, adaxial surface pale green or cream with pinkish tinge, glabrous or minutely papillose, abaxial face reddish purple, pubescent. Gynostegium with stipe 0.7–1.5 mm long. Corona lobes attached to and obscuring the gynostegial stipe, c.1.5 mm tall, spreading from the column then curved upwards, somewhat pouched with a thickened fleshy midline and two short rounded auricles apically, cream tinged with red. Anther wings 1 mm long but parallel grooved portion (guide rails) only 0.2 mm long, the remainder flared and curved below the head of the gynostegium. Stylar head flat or domed. Follicles (*Leedal* 5301; *Pawek* 9330) held erect on elongated peduncle, to c.9 cm long, slender, fusiform, minutely pubescent. Seeds c.3 × 2 mm, ovoid in outline, with one plane face and one strongly convex, sparsely verrucose and with a narrow marginal rim; coma c.1.5 cm long.

Subsp. **breviantherae**.

Sepals 2.5–3 mm long. Corolla lobes 5–6 × 2–2.5 mm, oblong to narrowly ovate. Gynostegium with stipe c.1–1.5 mm long, raising the anthers above the level of the corona. Stylar head domed, extending c.1 mm beyond top of anthers but obscured beneath the conspicuous, membranous anther appendages.

Malawi. N: Rumphi Dist., Nyika Plateau, Kasaramba Vitumbi junction, fl. 4.iii.1977, *Pawek* 12446 (K, MO).

Restricted to the Southern Highlands of Tanzania (Kitulo, Poroto and Mbeya massifs) and the nearby Nyika Plateau of northern Malawi. Thin peaty soil over rock in seasonally burned montane grassland; c.2400 m.

Subsp. *minor* Goyder is endemic to the Kitulo Plateau in Tanzania.

17. **Asclepias minor** (S. Moore) Goyder in Kew Bull. **64**: 392 (2009). Type: Angola, between Forte Princesa Amélia and Rio Cubango, *Gossweiler* 2332 (BM holotype).

> *Odontostelma welwitschii* Rendle in J. Bot. **32**: 161, t.344 (1894), not *Asclepias welwitschii* (Britten & Rendle) N.E. Br. in F.T.A. **4**(1): 341 (1902). Type: Angola, Huilla, near Lopollo, *Welwitsch* 4172 (BM holotype, K, LISU).
> *Schizoglossum welwitschii* (Rendle) N.E. Br. in F.T.A. **4**(1): 365 (1902).
> *Odontostelma minus* S. Moore in J. Bot. **50**: 363 (1912).

Perennial herb with slender vertical napiform tuber. Stems single, 6–20 cm long, unbranched, glabrous or shortly pubescent. Leaves spreading to erect, sessile, 3–9 × 0.5–1 mm, narrowly linear, acute, margins strongly revolute, glabrous. Inflorescences extra-axillary with 4–6 flowers at each of the upper nodes. Peduncles 0.5–3 cm long, slender, glabrous or puberulent. Bracts minute, often deciduous, linear, glabrous or rusty puberulent. Calyx lobes 1.5–2 × 0.5–1 mm, lanceolate to oblong, acute, glabrous, tinged with purple. Corolla broadly campanulate, lobed ± to the base, green, often tinged with purple; lobes 4–6 × 2–3 mm, obovate or spathulate, the apex obtuse, reflexed, glabrous. Corona arising at the base of the staminal column, united at the base; lobes with outer face c.1 × 1 mm, with square inflexed shoulders and a fleshy conical projection c.1 mm long on the inner face. Staminal column c.3 mm long, widening gradually to the fertile portion of the anthers. Fruiting pedicel not contorted; peduncle apparently elongating in fruit. Follicles (immature) erect, lanceolate-fusiform, rusty pubescent, only one of the pair developing. Seeds not seen.

Zambia. C: Kabwe [Broken Hill], fl. 3.x.1963, *Fanshawe* 8003 (K, NDO). S: Muckle Neuk, 20 km N of Choma, fl. 28.xi.1954, *Robinson* 994 (K). **Zimbabwe**. C: Harare [Salisbury] Dist., near Makabusi River, fl. 13.x.1972, *Grosvenor* 772A (K, SRGH).

Also known from Angola. Flowering in October and November in seasonally waterlogged grassland; 1200–1500 m.

18. **Asclepias randii** S. Moore in J. Bot. **40**: 255 (1902). —Goyder in F.T.E.A., Apocynaceae (part 2): 401 (2012). Type: Zimbabwe, Harare [Salisbury], xii.1897, *Rand* 194 (BM holotype, K, SRGH).

Perennial herb with one to many annual stems arising from a vertical napiform tuber; stems 6–20 cm long, simple or branched below, ascending or erect, reddish, pubescent with spreading white hairs to c.1 mm long. Leaves subsessile; lamina 2.5–10 × 0.1–0.5 cm, narrowly linear, attenuate both apically and basally, with sparse indumentum of spreading white hairs especially on the margins and midrib. Inflorescences extra-axillary forming umbels of 3–5 erect flowers; peduncles 1–3 cm long, densely pubescent; bracts filiform, pubescent; pedicels 0.7–1 cm long, densely pubescent. Sepals 3–4 × 0.5–1 mm, lanceolate or narrowly oblong, acute, densely spreading-pubescent. Corolla rotate to partially reflexed, lobes 4–6 × 2.5–3.5 mm, oblong to broadly ovate, acute, adaxial surface greenish white tinged with red or brown, glabrous, abaxial face green tinged brown, spreading-pubescent with white hairs. Corona lobes arising at the base of the staminal column, (4)5–8 mm long, ± cucullate basally with upper margins reaching top of column, the mid-line drawn out into a long-attenuate tip, acute or rounded apically, off-white with a reddish fleshy midline, or the entire lobe mottled red, cavity and mid-line of lobe papillose. Anther wings 1.3–1.5 mm long. Stylar head ± flat. Follicles erect, 5–8 × 0.6 cm, narrowly fusiform, minutely but densely pubescent, one or both of the pair developing, the peduncle lengthening in fruit. Seeds c.5 × 4 mm, ± flattened, verrucose and with a slightly inflated marginal rim c.0.5 mm wide.

Zambia. W: Mwinilunga Dist., Dobeka Plain, near Dobeka Bridge, fl. 14.x.1937, *Milne-Redhead* 3665 (K). N: Mbala [Abercorn] Dist., top of Kambole escarpment, fl. & fr. 1.ii.1959, *Richards* 10829 (K). **Zimbabwe**. C: Harare [Salisbury], fl. xii.1897, *Rand* 194 (BM holotype, K, SRGH). E: Mutare Dist., Mutare [Umtali] Golf Course near Range Hill, fl. 22.iii.1960, *Chase* 7407B (K). C/S: Enkeldoorn–Gutu road, 8 km S of Sebakwe River, fl. 4.xii.1960, *Leach & Chase* 10538 (K). **Malawi**. N: Rumphi Dist., Nyika Plateau, Chelinda Bridge, fl. 18.x.1975, *Pawek* 10276 (EA, K, MO, UC, SRGH).

Also recorded from the southern highlands of Tanzania, D.R. Congo, Angola and northern Namibia. Montane grassland and seasonally waterlogged 'dambo' grasslands; (1000)1500–2500 m.

Two Zambian collections (*Richards* 15383 from Kawambwa and *Milne-Redhead* 3175 from Mwinilunga) are anomalous in lacking indumentum on the leaves and corolla.

19. **Asclepias grandirandii** Goyder in Kew Bull. **64**: 394 (2009); in F.T.E.A., Apocynaceae (part 2): 403 (2012). Type: Malawi N, near Chelinda CDC Camp, 26.x.1958, *Robson* 373 (K holotype).

Perennial herb with one to many annual stems arising from a vertical napiform tuber; stems 10–30 cm long, simple or branched below, prostrate to ascending or erect, green, conspicuously pubescent with spreading white hairs. Leaves subsessile; lamina 3–5 × 0.5–2 cm, oblong, triangular-oblong or elliptic, apex acute, base rounded, truncate or weakly cordate, with sparse indumentum of spreading white hairs especially on the margins and midrib. Inflorescences extra-axillary forming umbels of 3–5 erect flowers; peduncles 1–3 cm long, densely pubescent; bracts filiform, pubescent; pedicels 1–2 cm long, densely pubescent. Sepals 3–6 × 0.5–2 mm, linear, lanceolate or narrowly triangular, acute, densely spreading-pubescent. Corolla partially reflexed, lobes 5–8 × 3–4 mm, oblong to broadly ovate, acute, adaxial surface greenish white, glabrous, abaxial face green tinged brown or purple, spreading-pubescent with white hairs. Corona lobes arising at the base of the staminal column, 5–7 mm long, ± cucullate basally with upper margins reaching top of column, the mid-line drawn out into a long-attenuate tip, acute or rounded apically, off-white with a reddish fleshy midline, or the entire lobe mottled red, cavity and mid-line of lobe papillose. Anther wings (1.4)1.5–2 mm long. Stylar head ± flat. Follicles (immature) 6 cm long, erect, fusiform, densely pubescent; seeds not seen.

Zambia. N: Mbala [Abercorn] Dist., top of Chilongowelo escarpment, fl. 27.x.1956, *Richards* 7360 (K). **Malawi**. N: Mzimba Dist., Chimpyai View, Vipya Plateau, 60 km SW of Mzuzu, fl. 22.xi.1975, *Pawek* 10358 (K, MAL, MO, SRGH, UC).

Also known from the Mitumba range running between Kolwezi and Kalemia in SE D.R. Congo, and the Ufipa Plateau in SW Tanzania. Seasonally burned upland grassland or woodland; 1400–2200 m.

This is essentially a broad-leaved version of *Asclepias randii* with longer pedicels and slightly larger flowers. It is at least partially sympatric with *A. randii*, and although the differences seem slight, the overall appearance of the plant is very different – the leaves tend to be shorter and broader in this species than in *A. randii*, and the leaf base is rounded to truncate or weakly cordate, but never attenuate as in the latter species. The geographic range is somewhat narrower than *A. randii*. Pubescence is conspicuous, but the 1 mm long hairs of *A. randii* were not observed.

20. **Asclepias aurea** (Schltr.) Schltr. in J. Bot. **34**: 455 (1896). Type: South Africa, Mpumalanga, near Barberton, 1889, *Galpin* 580 (B† holotype, K, GRA, PRE).

Schizoglossum pedunculatum Schltr. in Verh. Bot. Vereins Prov. Brandenburg **35**: 50 (1893), not *Asclepias pedunculata* (Decne.) Dandy (1952). Type: South Africa, between Kenilworth and Claremont near Cape Town, x.1892, *Schlechter* 351 (B† holotype, BOL).

Gomphocarpus aureus Schltr. in Bot. Jahrb. Syst. **18**(Beibl. 45): 17 (1894).

Gomphocarpus schizoglossoides Schltr. in Bot. Jahrb. Syst. **18**(Beibl. 45): 21 (1894). Type: South Africa, Mpumalanga, Saddleback Mountain near Barberton, *Galpin* 500 (B† holotype, K, PRE, SAM).

Asclepias aurea var. *brevicuspis* S. Moore in J. Bot. **40**: 255 (1902). Type: Zimbabwe, Harare [Salisbury], ix.1898, *Rand* 638 (BM holotype).

Asclepias aurea var. *vittata* N.E. Br. in Fl. Cap. **4**(1): 686 (1908). Type: South Africa, Mpumalanga, Saddleback Mountain near Barberton, *Galpin* 500 (K lectotype, PRE, SAM), designated by Goyder in Kew Bull. **64**: 395 (2009).

Asclepias radiata S. Moore in J. Bot. **50**: 345 (1912). Type: Angola, Kubango River near Forte Princeza Amelia, *Gossweiler* 'with 4210' (BM holotype).

Aidomene parvula Stopp in Bot. Jahrb. Syst. **87**: 21 (1967). Type: Angola, Dist. Nova Lisboa, Canjangu, 2.xi.1959, *Stopp* BO 129 (K holotype, LISC, M).

Perennial herb with one to several annual stems arising from a vertical cylindrical or napiform tuber; stems 8–40 cm long, simple or branched below, ascending or erect, minutely pubescent. Leaves 3–8 × 0.1 cm, filiform with inrolled margins, glabrous. Inflorescences terminal or extra-axillary forming umbels of 3–10 erect flowers; peduncles 3–10(14) cm long, subglabrous; pedicels 0.5–0.8 cm long, subglabrous or minutely pubescent. Sepals 1.5–2 × 0.5–1 mm, ovate to oblong, acute, spreading-pubescent, usually reddish. Corolla rotate to partially reflexed, lobes 2–3.5 × 1.5–2 mm, oblong to broadly ovate, acute, adaxial surface greenish white or yellow, glabrous, abaxial face reddish, spreading-pubescent with white hairs. Corona lobes yellow or white, (2)2.5–3 mm long, ± cucullate basally with upper margins reaching top of column, the fleshy mid-line drawn out into a long-attenuate tip, acute or rounded apically. Anther wings 1 mm long. Stylar head ± flat. Follicles (immature) erect, 5 cm long, narrowly fusiform, minutely pubescent, one of the pair developing. Seeds not seen.

Zambia. N: Mporokoso Dist., Kalungwishi headwaters, fl. & fr. 11.i.2000, *Bingham & Beel* 12113 (K). **Zimbabwe**. W: Matobo Dist., Besna Kobila farm, fl. ix.1956, *Miller* 3667 (K, SRGH). C: Makoni Dist., 1 mile from Eagles Nest towards Rusape, fl. 29.xi.1955, *Drummond* 5070 (K, SRGH). E: Chimanimani Mountains, E side of Bundu plain, fl. 31.xii.1957, *Goodier & Phipps* 242 (K, SRGH).

From Angola and south-eastern D.R. Congo to northern South Africa (NW Province, Limpopo, Mpumalanga and KwaZulu-Natal), Swaziland and Lesotho. Periodically burned montane or seasonally waterlogged grassland; 1000–2000 m.

This species appears most closely allied to *Asclepias randii*, but is much more slender, possessing long slender peduncles, particularly from the lower nodes. The flowers are variable in colour, perhaps deserving infraspecific taxonomic recognition - most South African material has yellow coronas, but some are cream. Zambian, Katangan and possibly Angolan collections have a white corona.

Material from Angola described as *Asclepias radiata* S. Moore appears to belong to this species, but differs in the more hispid indumentum of the stem and leaves. However, Zambian material is more pubescent than most material from further south, so it may be best to regard it just as a local form.

Richards 17269 and *Bingham* 13155, both from the Mwinilunga area of NW Zambia, appear allied to this species, but have longer pedicels and minute flowers with corona lobes lacking the well-developed tongue of *A. aurea*. A third collection from eastern Angola, *Goyder & Gonçalves* 4809, appears to belong to the same taxon.

21. **Asclepias cucullata** (Schltr.) Schltr. in J. Bot. **34**: 455 (1896). Type: South Africa, Mpumalanga, Saddleback Mt., Barberton, 21.ix.1890, *Galpin* 1034 (B† holotype, K).

> *Gomphocarpus cucullatus* Schltr. in Bot. Jahrb. Syst. **18**(Beibl. 45): 17 (1894).
> *Trachycalymma cucullatum* (Schltr.) Bullock in Kew Bull. **10**: 620 (1956).

Perennial herb with 1–3 annual stems arising from a fleshy, cylindrical, vertical tuber; stems 0.1–0.3 m long, mostly unbranched, erect or ascending, minutely pubescent with spreading white hairs. Leaves sessile; lamina 4–10 × 0.1–0.3 cm, narrowly linear, acute, base cuneate, venation obscure, with indumentum of stiff white hairs on both surfaces. Inflorescences terminal or extra-axillary with (2)4–6 erect flowers per umbel, peduncles 1–7 cm long, lower ones markedly longer than the upper ones, erect, pubescent; bracts 0.1–0.5 cm long, filiform, pubescent; pedicels 1–1.5(2) cm long, pubescent. Calyx lobes 2–5 × 1(2) mm, narrowly triangular to lanceolate, acute, pubescent, purple. Corolla subrotate or weakly reflexed, greenish or greyish purple outside, paler within and sometimes with a network of darker veins, pubescent outside, glabrous within; lobes 5–6 × 3–4 mm, ovate, subacute. Corona white turning yellow with age, with purple midline; lobes 2–3 × 3–4 mm, cucullate, adnate to the column to just below the base of the anther wings then free from it to ± half way up the anther wings, upper margins raised proximally, somewhat lower near the mid point where the margins are widely separated, then raised again distally into a subacute tip; interstaminal corona lobes minute. Anther wings 1–1.5 mm long, triangular. Fruiting pedicel contorted to hold follicle erect; follicle 7–8 × 0.6–0.8 cm, narrowly fusiform with an attenuate apex, not inflated, smooth, minutely puberulent. Seeds not seen.

Corona lobes with papillae restricted to narrow band across base of lobe . . . .
. . . . a) subsp. *cucullata*
Corona lobes with papillae distributed over most of the cavity of the lobe; Chimanimani and Vumba . . . . .b) subsp. *scabrifolia*

a) Subsp. **cucullata**.

Zimbabwe. N: Guruve [Sipolilo] Dist., Nyamunyeche Estate, Great Dyke, fl. & fr. 4.x.1978, *Nyariri* 385 (K, SRGH). E: Mare River, Inyanga, fl. 21.x.1946, *Wild* 1522 (K, SRGH).
Known from the Great Dyke and Nyanga highlands of Zimbabwe, the Soutpansberg in South Africa, then following the mountain chain from Pilgrims Rest and the Barberton Mountains through Swaziland and KwaZulu-Natal to the Transkei region of the Eastern Cape. Open grassland; 1400–1800 m.

b) Subsp. **scabrifolia** (S. Moore) Goyder in Kew Bull. **56**: 159 (2001). Type: Zimbabwe, Chimanimani Mountains, 26.ix.1906, *Swynnerton* 1915 (BM holotype).

> *Asclepias scabrifolia* S. Moore in J. Bot. **46**: 297 (1908).

Zimbabwe. E: Vumba Mountains, fl. 30.x.1950, *Chase* 3103 (BM, K, SRGH). **Mozambique**. MS: Between Skeleton Pass and the plateau, fl. 27.ix.1966, *Grosvenor* 198 (K, LISC, PRE, SRGH).
Known only from the Chimanimani and Vumba massifs of eastern Zimbabwe and adjacent regions of Mozambique. Open grassland; 1200–2200 m.
This subspecies can be distinguished from the former by its smaller flowers and shorter corona lobes. The clearest difference, however, concerns the coronal papillae which occur over most of the cavity of the lobe. In subsp. *cucullata*, these papillae are restricted almost entirely to a narrow band across the base of the lobe.

22. **Asclepias pygmaea** N.E. Br. in Bull. Misc. Inform. Kew **1895**: 255 (1895). —Goyder in F.T.E.A., Apocynaceae (part 2): 403 (2012). Type: Tanzania, the lower plateau N of Lake Malawi [Lake Nyassa], *Thomson* s.n. (K holotype).

Perennial herb with one to several annual stems arising from a vertical napiform tuber; stems 6–20 cm long, simple or branched below, ascending or erect, reddish, minutely pubescent with short white hairs. Leaves subsessile; lamina 3–6 × 0.1–0.2 cm, narrowly linear, attenuate both apically and basally, with sparse indumentum of very short white hairs mostly on the margins and midrib. Inflorescences extra-axillary forming umbels of 3–5 erect flowers; peduncles 1–4 cm long, minutely but densely pubescent; bracts filiform, pubescent; pedicels 0.5–1.3 cm long, minutely pubescent. Sepals 2–3 × 0.5 mm, lanceolate or narrowly oblong, acute, densely spreading-pubescent. Corolla somewhat reflexed, lobes 3.5–4 × 1.5–2.5 mm, oblong to broadly ovate, acute, adaxial surface greenish white tinged with pink, glabrous or minutely papillate, abaxial face green or purplish, spreading-pubescent with white hairs. Corona lobes arising from the top of a minute stipe c.0.5 mm tall at the base of the staminal column, 3–3.5 mm long, ± cucullate basally with upper margins reaching top of column, the mid-line drawn out into a long-attenuate tip, acute or rounded apically, off-white with a reddish fleshy midline, or the entire lobe mottled red, cavity and mid-line of lobe papillose. Anther wings 1–1.5 mm long. Stylar head ± flat. Follicles and seeds not seen.

Malawi. N: Nyika Plateau, S slopes of Nganda, fl. 13.ix.1972, *Synge* s.n. in WC 465 (K). Known only from the southern highlands of Tanzania and the Nyika Plateau in Malawi. Montane grassland; 1800–2500 m.

Asclepias pygmaea can be distinguished from *A. cucullata* by the shape of the corona lobes, which are much longer than tall, and the coronal papillae, which are distributed along the entire length of the lobes. It can also be easily confused with *A. randii*. *Asclepias pygmaea* tends to have a more erect habit than *A. randii*, and stem and leaf indumentum is of short curled hairs not the stiff spreading hairs of *A. randii*. The flowers are slightly smaller and the corona has a less drawn-out apex, but are otherwise very similar. However, the pollinarium differs in having clearly differentiated proximal and distal portions to the translator arms, very slender and ± terete distally. The translator arms of *A. randii* are flattened, and broaden gradually towards the pollinia.

23. **Asclepias densiflora** N.E. Br. in F.T.A. **4**(1): 320 (1902). Type: Zimbabwe, between Harare [Salisbury] and Bulowayo, xi.1899, *Cecil* 78 (K holotype).

Perennial herb with 1–3 annual stems arising from a fleshy, cylindrical, vertical tuber; stems 20–40 cm long, generally branched below, spreading below then ascending to erect, pubescent with spreading white hairs. Leaves with pedicels 1–5 mm long; lamina 3–5(8) × 0.5–2 cm, broadly to narrowly triangular, acute, base truncate, venation generally visible, indumentum of somewhat scabrid white hairs on both surfaces. Inflorescences terminal with up to around 35 flowers per umbel, peduncles 6–14 cm long, erect, pubescent; pedicels 8–10 mm long, sparsely to densely pubescent. Calyx lobes 2.5–3 × 1 mm, lanceolate to oblong, acute, pubescent. Corolla reflexed, outside brown or purplish, glabrous, inside white or yellow, minutely pubescent; lobes 5 × 3 mm, ovate, subacute. Corona white cream or yellow; lobes 2 mm long, cucullate, somewhat fleshy and pouched, attached to the column to just below the base of the anther wings, upper margins level with top of column and forming two inward-pointing teeth; interstaminal corona lobes minute. Anther wings 1–1.5 mm long, triangular. Fruiting pedicel contorted to hold follicle erect; follicle 3–4 × 1.5 cm, narrowly ovoid with an attenuate apex, not inflated, ornamented with multiple rows of soft prickles, minutely puberulent. Seeds not seen.

Zimbabwe. W: Plumtree, *Eyles* 8549 (K). C: 12 km SE of Gweru [Gwelo], 14.xi.1966, *Biegel* 1436 (K). E: SW portion of Mutare [Umtali] Golf Course, fl. & fr. 30.xi.1960, *Chase* 7413 (K, NU, SRGH).
Zimbabwe, northern South Africa (Limpopo, NW Province, Gauteng, Mpumalanga, KwaZulu-Natal) and Swaziland. Mixed *Brachystegia* woodland or wooded savanna, often on stony soil, sometimes persisting in cultivated land; 1000–1500 m.

24. **Asclepias adscendens** (Schltr.) Schltr. in J. Bot. **34**: 455 (1896). Type: South Africa, near Barberton, x.1889, *Galpin* 596 (B† holotype, K, PRE).

    Gomphocarpus adscendens Schltr. in Bot. Jahrb. Syst. **18**(Beibl. 45): 16 (1894).

Perennial herb with 1 or more annual stems arising from a fleshy, cylindrical, vertical tuber; stems 15–25 cm long, generally branched below, spreading below then ascending to erect, pubescent with spreading white hairs. Leaves with pedicels 1–3 mm long; lamina 2–4 × 0.4–1 cm, broadly to narrowly triangular, acute, base truncate, venation generally visible, indumentum of somewhat scabrid white hairs on both surfaces. Inflorescences terminal with up to around 20 flowers per umbel, peduncles 2–3 cm long, erect, pubescent; pedicels 8–10 mm long, sparsely to densely pubescent. Calyx lobes 2.5 × 1 mm, lanceolate, acute, pubescent. Corolla reflexed, outside purplish, glabrous, inside white or yellow, minutely pubescent; lobes 4 × 2.5 mm, ovate, subacute. Corona white cream or yellow; lobes 1.5–2 mm long, cucullate, attached to the column to just below the base of the anther wings, upper margins level with top of column; interstaminal corona lobes minute. Anther wings 1.1 mm long, triangular. Fruiting pedicel contorted to hold follicle erect; follicle 5–6 × 1.5 cm, narrowly ovoid with an attenuate apex, not inflated, ornamented with many discrete linear soft prickles, minutely puberulent. Seeds not seen.

Zimbabwe. E: SW portion of Mutare [Umtali] Golf Course, fl. 30.xi.1960, *Chase* 7663 (K, SRGH). **Mozambique**. M: between Boane and Namaacha, fl. 27.viii.1948, *Myre & Carvalho* 124 (LMA).

Also known from Swaziland, Lesotho and the north-eastern provinces of South Africa (Limpopo, NW Province, Gauteng, Mpumalanga, Free State, KwaZulu-Natal). Open grassland, often in stony soil; 1000 m.

Smaller and more branched than the preceding species, habit more decumbent. Peduncles are much shorter in this species than in *Asclepias densiflora*, and the corona appears somewhat less fleshy. The follicle has more discrete processes rather than the irregularly lobed wings of *A. densiflora*.

25. **Asclepias meliodora** (Schltr.) Schltr. in J. Bot. **34**: 455 (1896). Type: South Africa, Limpopo [Transvaal], near Sandloop, 3.ii.1894, *Schlechter* 4373 (B† holotype); Limpopo [Transvaal], N of Potgietersrust, 18.xii.1928, *Hutchinson* 1944 (K neotype), designated by Goyder in Kew Bull. **64**: 397 (2009).

    Gomphocarpus meliodorus Schltr. in Bot. Jahrb. Syst. **20**(Beibl. 51): 33 (1895).

    Asclepias meliodora var. *brevicoronata* N.E. Br. in Fl. Cap. **4**(1): 700 (1908). Type: South Africa, NW Province, near Rustenburg, *Pegler* s.n. in *Bolus* 10553 (K holotype, BOL not seen).

Perennial herb with 1 or more annual stems arising from a long cylindrical rootstock; stems 15–40 cm long, simple or more commonly branched, erect, or spreading below then ascending to erect, pubescent with spreading white hairs. Leaves with pedicels 1–4 mm long; lamina 5–9 × 0.3–1 cm, narrowly oblong, acute, base truncate, venation generally visible, indumentum of somewhat scabrid white hairs on both surfaces. Inflorescences terminal and extra-axillary with up to around 10 flowers per umbel, peduncles 2–3 cm long at upper nodes, longer below, erect, pubescent; pedicels 8–10 mm long, sparsely to densely pubescent. Calyx lobes 4 × 1 mm, lanceolate, acute, pubescent. Corolla reflexed, white, minutely pubescent; lobes 4 × 1.5 mm, ovate, subacute. Corona white; lobes 3–4 mm long and twice as tall as the column, cucullate, erect; interstaminal corona lobes vestigial. Anther wings 1.1 mm long, triangular. Fruiting pedicel contorted to hold follicle erect; follicle 12 × 1.5 cm, fusiform with an attenuate apex, not inflated, smooth, minutely puberulent. Seeds 7–8 × 5 mm, ovate, with a verrucose flattened disc surrounded by an inflated rim; coma 2–3 cm long.

Botswana. SE: Lobatse, *Rogers* 6244 (BOL – cited by Nicholas in unpublished ms). **Mozambique**. M: near Goba, fl. 19.xii.1952, *Myre & Carvalho* 1399 (K, LISC).

Also known from Swaziland and South Africa (Limpopo, NW Province, Gauteng, Mpumalanga, KwaZulu-Natal. Open grassland or scrub woodland, frequently on stony ground; c.1200–1300 m in Botwana, c.200–400 m in Mozambique.

Most collections look like a dwarf version of *Asclepias adscendens*. However, the corona is more elongate and less compressed, and the peduncle is much shorter. Follicles are smooth, lacking wings and spine-like processes. Eastern populations appear taller, and with longer peduncles, particularly those from the lower nodes.

26. **Asclepias albens** (E. Mey.) Schltr. in Bot. Jahrb. Syst. **21**(Beibl. 54): 5 (1896). — Bester & Condy in Fl. Pl. Africa **62**: 110–119 (2011). Types: South Africa, "EM a: prope Geelhoutboom", *Drège* 3414 (K herb. Hook. lectotype, K herb. Benth.), lectotype designated here; *Drège* s.n. (HAL); "EM b: prope Omsamcaba", *Drège* s.n. (K paralectotype); "EM c: Witbergen", *Drège* s.n. (HAL paralectotype).

   Pachycarpus albens E. Mey., Comm. Pl. Afr. Austr.: 214 (1838).
   Xysmalobium albens (E. Mey.) D. Dietr., Syn. Pl. **2**: 902 (1840).
   Gomphocarpus albens (E. Mey.) Decne. in Candolle, Prodr. **8**: 559 (1844).
   Gomphocarpus affinis Schltr. in Bot. Jahrb. Syst. **20**(Beibl. 51): 27 (1895). Type: South Africa, Wilge River, 18.xi.1893, *Schlechter* 3751 (B† holotype, K lectotype, M, PRE, S), lectotype designated here.
   Asclepias affinis (Schltr.) Schltr. in J. Bot. **34**: 455 (1896), not De Wild. in Ann. Mus. Congo Belge, Bot. sér. 5 **1**: 184 (1904).

Perennial herb with 1 or more annual stems arising from a vertical tuber; stems 20–40 cm long, branched near base, spreading below then ascending to erect, pubescent with spreading white hairs. Leaves with pedicels 1–5 mm long; lamina 2–5 × 1.5–3 cm, mostly ovate-triangular but sometimes oblong, acute, base rounded, truncate or weakly cordate, venation generally visible, indumentum of somewhat scabrid white hairs on both surfaces. Inflorescences terminal with up to around 30 flowers per umbel, peduncles 2–10 cm long, erect, pubescent; pedicels 10–25 mm long, pubescent. Calyx lobes 3 × 1 mm, lanceolate, acute, pubescent. Corolla reflexed, purplish outside, white or greenish within, minutely pubescent; lobes 5–6 × 2.5–5 mm, obovate, subacute. Corona white, yellow or greenish brown; lobes 2 mm high and as tall as the column, cucullate but with the fleshy midline ascending or erect and 2–3 mm long; interstaminal corona lobes vestigial. Anther wings 1–1.3 mm long, narrowly triangular. Fruiting pedicel contorted to hold follicle erect; follicle 4–5 × 1.5 cm, narrowly ovoid an attenuate beak, not inflated, smooth or very weakly ridged, minutely puberulent. Seeds not seen.

Mozambique. M: Namaacha, Montes Ponduíni, fl. 20.xi.1966, *Moura* 168 (LMU).
Northern and Eastern provinces of South Africa and Swaziland. Rocky ground; 500 m.

67. **MARGARETTA** Oliv.[39]

Margaretta Oliv. in Trans. Linn. Soc. London **29**: 111, t.76 (1875). —Goyder in Kew Bull. **60**: 87–94 (2005).

Perennial herbs with annual stems arising from a slender to stout vertical tuber; latex milky. Stems usually simple or branched, erect or ascending. Leaves opposite, subsessile; lamina linear to narrowly oblong. Inflorescences terminal and extra-axillary, umbels erect, borne on long or short peduncles. Calyx lobes to the base. Corolla deeply lobed, apex of lobes often revolute, much reduced and smaller and less conspicuous than the petaloid corona. Corona of 5 petaloid lobes arising above the base of the staminal column; erect and somewhat fleshy, the claw cucullate at least at the base, the inner margins generally into a pair of inward- or upward-pointing teeth, with or without a tooth arising from the middle of the cavity, the limb expanded into a flattened oblong to orbicular lobe with entire to dentate margins. Corpusculum ovoid to subcylindrical, black; translator arms short, broad and differentially thickened. Stylar head flat. Fruiting pedicel not contorted; follicles single, erect, ovoid to fusiform. Seeds ovate with a verrucose disc and a narrow inflated rim; coma of silky hairs.

---

[39] by D.J. Goyder

A single variable species in tropical Africa. Very closely allied to *Stathmostelma*, with which it is almost certainly congeneric in any of the likely future generic realignments.

Margaretta rosea Oliv. in Trans. Linn. Soc. **29**: 111, t.76 (1875). —Goyder in F.T.E.A., Apocynaceae (part 2): 404 (2012). Type: Uganda, Unyoro, vii.1862, *Speke & Grant* 531 (K holotype). FIGURE 7.3.**187**.

Perennial herbs with annual shoots arising from a carrot-shaped tuber. Stems erect, to c.60 cm. Leaves to c.19 × c.2.3 cm, lanceolate to narrowly oblong, subcordate at the base, sparsely pubescent on both faces. Inflorescences umbelliform, with 3–10 flowers; peduncles to 7.5 cm long, erect, densely pubescent; pedicels to 1.7 cm long, pubescent. Calyx lobes 2.5–6 mm long, pubescent. Corolla lobed almost to the base, lobes 4–10 × 1–3.5 mm, lanceolate, often revolute apically, glabrous adaxially, abaxial face pubescent. Corona highly variable, petaloid and often brightly coloured, the basal claw with two marginal teeth, a third central tooth present or absent; limb expanded into an entire or toothed dorsiventrally flattened lobe.

Patterns of morphological variation are strongly geographic. None of the characters are absolutely constant, and there are always a few specimens in which one or other of the diagnostic characters is absent. The shape of the corona limb is particularly variable in subsp. *cornetii*. Nevertheless, few individuals are hard to place and the forms can be successfully keyed out. As the forms generally replace each other geographically, subspecific rank is the most appropriate rank at which to recognise these taxa formally.

1. Corona with 3 teeth on the claw, 2 lateral ones and a third centrally . . . .
. . . . .a) subsp. *whytei*
– Corona with 2 teeth on the claw; central tooth absent or vestigial 2
2. Limb of corona ± orbicular, narrowing abruptly into the claw; SW Tanzania. . . .
. . . . b) subsp. *orbicularis*
– Limb of corona oblong or obovate, tapering gradually into the claw; SW Tanzania
. . . . c) subsp. *cornetii*

a) Subsp. **whytei** (K. Schum.) Mwany. in Kew Bull. **51**: 726 (1996). —Goyder in F.T.E.A., Apocynaceae (part 2): 408 (2012). Type: Malawi, Mulanje, *Whyte* 106 (K lectotype, B†), designated by Goyder in Kew Bull. **60**: 91 (2005).
    Margaretta whytei K. Schum. in Engler, Pflanzenw. Ost-Afrikas C: 323 (1895).

Corolla pubescent abaxially in southern populations, glabrous in more northerly populations; adaxial face glabrous to papillose. Corona with 3 teeth on the claw, claw of corona lobes ± same width as limb, corona mauve, generally smaller than in other subspecies.

Zimbabwe. C: Makoni Dist., 3 km E of Nyazura [Inyazura], fl. 19.xi.1960, *Leach & Noel* 10494 (K, SRGH). E: Inyanga Dist., Holdenby Native Purchase Area, fl. 13.xi.1958, *West* 3767 (K, SRGH). **Malawi**. N: Masisi Plateau, fl. ix.1902, *McClounie* 108 (K). S: Blantyre Dist., Ndirande Mountain, fl. 23.xi.1966, *Brummitt et al.* 15169 (K, MAL). **Mozambique**. N: Nampula Dist., Murrupula, fl. 11.i.1961, *Carvalho* 418 (K, LMA). Z: Chiperone Mountain, fl. 3.xii.2006, *Harris et al.* 124 (K, LMA, LMU, MAL). MS: railway between Beira and Manica [Massi Kessi], 1899, *Cecil* 17 (K).
    Also recorded from SE Tanzania. Open *Brachystegia*/*Uapaca* woodland; 300–1500 m.

b) Subsp. **orbicularis** (N.E. Br.) Goyder in Kew Bull. **60**: 91 (2005); in F.T.E.A., Apocynaceae (part 2): 408 (2012). Type: Malawi, Elephant Marsh, North Nyassa, *Scott* s.n. (K lectotype), designated by Goyder in Kew Bull. **60**: 91 (2005).
    Margaretta orbicularis N.E. Br. in Bull. Misc. Inform. Kew **1895**: 256 (1895).
    Margaretta pulchella Schltr. in Fries, Wiss. Ergebn. Schwed. Rhod.-Kongo Exped. 1911-12 1: 266, fig.32, t.18 fig.3 (1916). Type: Zambia, Bwana Mkubwa, viii.1911, *Fries* 491 (UPS holotype).

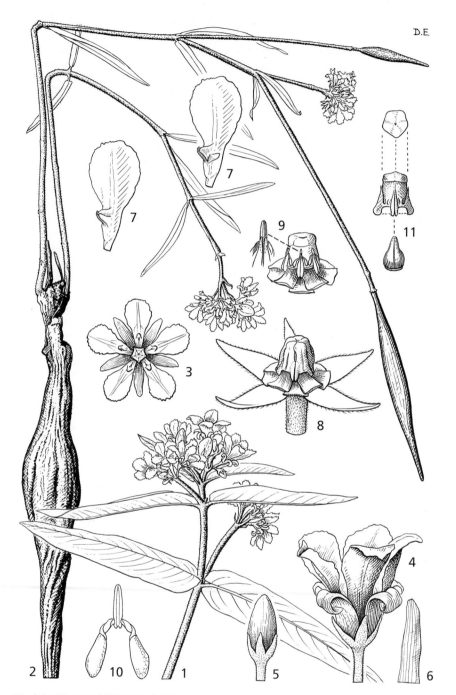

Fig. 7.3.**187**. MARGARETTA ROSEA.   1, flowering shoot (× ²/₃); 2, habit, with tuberous rootstock and follicles (× ²/₃); 3, flower from above (× 1); 4, flower from the side (× 2); 5, flower bud (× 2); 6, corolla lobe (× 2); 7, corona lobes (× 2); 8, flower with both corolla and corona lobes removed to expose gynostegium (× 3); 9, as 8 but with stamens partially removed to show position of pollinaria (× 3); 10, pollinarium (× 14); 11, gynoecium showing paired ovaries, the stylar head and apex of stylar head from above (× 3). 1 from *Richards* 195; 2–11 from *Bullock* 1867. Drawn by D. Erasmus. Reproduced from Flora of Tropical East Africa (2012).

Corolla glabrous to sparsely pubescent. Corona with 2 teeth on the claw, limb of corona ± orbicular, narrowing abruptly into the claw, corona violet, purple, magenta or white.

Zambia. W: Kitwe, fl. 4.xii.1957, *Fanshawe* 4101 (K, NDO). C: Kabwe [Broken Hill], fl. vii.1909, *Rogers* 8290 (K). **Malawi**. N: Mzimba Dist., Rukuru River bridge 30 km N of Edingeni, fl. 27.xii.1975, *Pawek* 10589 (K, MAL). C: Dedza Dist., Sosola Rest House, Mua, fl. 15.xii.1969, *Salubeni* 1437 (K, MAL, SRGH).

Also occurs in SW Tanzania, and the Katanga region of SE D.R. Congo. Burnt, sometimes seasonally waterlogged grassland; 400–1600 m.

c) Subsp. **cornetii** (Dewèvre) Goyder in Kew Bull. **60**: 92 (2005); in F.T.E.A., Apocynaceae (part 2): 409 (2012). Type: D.R. Congo, Katanga, *Cornet* s.n. (BR holotype).

  Margaretta cornetii Dewèvre in Bull. Soc. Roy. Bot. Belgique **34**(2): 90 (1895).
  Margaretta cornetii var. *pallida* De Wild. in Ann. Mus. Congo Belge, Bot., sér. 4 **1**: 108 (1903). Type: D.R. Congo, Lukafu, x.1899, *Verdick* 133 (BR holotype).
  Margaretta verdickii De Wild. in Ann. Mus. Congo Belge, Bot. sér 5 **1**: 183 (1904). Type: D.R. Congo, Lukafu, x.1899, *Verdick* 148 bis (BR holotype).
  Margaretta decipiens Schltr. in Fries, Wiss. Ergebn. Schwed. Rhod.-Kongo Exped. 1911-12 **1**: 265 (1916). Type: Zambia, Kamindas, near Lake Bangweulu, 11.x.1911, *Fries* 987 (UPS holotype).

Corolla densely pubescent abaxially. Corona with 2 teeth on the claw, limb of corona oblong or obovate, tapering gradually into the claw, corona violet, purple, magenta or white.

Zambia. N: Mbala [Abercorn] Dist., Kara Gorge, fl. 14.xi.1956, *Richards* 6980 (K). W: Chingola, fl. 18.x.1955, *Fanshawe* 2540 (K, NDO). C: 12 km E of Lusaka, fl. & fr. 22.ix.1955, *King* 143 (K). S: near Kalomo, fl. 16.xii.1961, *Whellan* 1893 (K; SRGH).

Also known from SW Tanzania and the Katanga region of the D.R. Congo. Burnt grassland; 1000–1800 m.

68. STATHMOSTELMA K. Schum.[40]

Stathmostelma K. Schum. in Bot. Jahrb. Syst. **17**: 129 (1893); in Engler & Prantl, Nat. Pflanzenfam. **4**(2): 239 (1895). —Goyder in Kew Bull. **53**: 577–616 (1998).

Perennial herbs with annual stems arising from a slender to stout vertical tuber; latex milky. Stems 1 to many, simple or branched, erect or ascending, glabrous or minutely pubescent with short, spreading hairs. Leaves opposite, sessile or petiolate; lamina linear to ovate, both upper and lower surfaces pubescent with spreading white hairs or indumentum restricted to margins and midrib below. Inflorescences terminal, extra-axillary or both, umbels erect, borne on long or short peduncles. Calyx lobes to the base. Corolla rotate to strongly reflexed, occasionally broadly campanulate. Corolline corona absent. Gynostegium sessile or with stipe to 3 mm long. Gynostegial corona of 5 lobes arising above the base of the staminal column; erect and somewhat fleshy, frequently weakly pouched towards the base and concave-cucullate for at least some of its length, the inner apical margins generally produced into a pair of inward- or upward-pointing teeth, with or without a tooth arising from the middle of the cavity, glabrous except along the midline of the cavity towards the base of the lobe. Anther wings curved, with or without a contorted basal tail. Corpusculum ovoid to subcylindrical or occasionally subglobose, black; translator arms with a membranous, convex and somewhat contorted proximal portion, and a filiform distal portion broadening abruptly into a short clasping overlap with the pollinium; pollinia flattened, oblong. Stylar head flat. Fruiting pedicel not contorted; follicles single, erect, ovoid to fusiform, beaked, occasionally inflated but generally not, smooth or with weak longitudinal ridges, glabrous to densely pubescent. Seeds ovate with a verrucose disc and a narrow inflated rim; coma of silky hairs.

---

[40] by D.J. Goyder

18 somewhat critical taxa in 14 species. Distributed mostly in eastern parts of tropical Africa, growing mostly at middle altitudes from Ethiopia in the north to Mozambique and Zimbabwe in the south. Very closely allied to *Margaretta*.

1. Leaves at least 4 cm wide . . . . **1.** *spectabile*
– Leaves not more than 3 cm wide 2
2. Anther wings with contorted basal tails; rootstock a narrowly cylindrical tuber; inflorescences mostly terminal . . . . **6.** *pauciflorum*
– Anther wings without contorted basal tails; rootstock a napiform to globose tuber; inflorescences mostly extra-axillary 3
3. Corona lobes white or greenish, frequently speckled purple . . . . . **5.** *gigantiflorum*
– Corona lobes red, orange or yellow 4
4. Corolla red or pink . . . . **2.** *pedunculatum*
– Corolla green or yellow 5
5. Corolla lobes at least 14 mm long; plant generally little branched . . **3.** *fornicatum*
– Corolla lobes less than 12 mm long; plant usually well branched or with many stems from the base . . . . **4.** *welwitschii*

1. **Stathmostelma spectabile** (N.E. Br.) Schltr. in Bot. Jahrb. Syst. **51**: 138 (1913). —Goyder in F.T.E.A., Apocynaceae (part 2): 410 (2012). Type: Malawi, 1891, *Buchanan* 553 (K lectotype), designated by Bullock in Kew Bull. **15**: 196 (1961).

> *Asclepias spectabilis* N.E. Br. in Bull. Misc. Inform. Kew **1895**: 254 (1895); in F.T.A. 4(1): 325 (1902).
> *Stathmostelma odoratum* K. Schum. in Bot. Jahrb. Syst. **28**: 457 (1900). Type: Tanzania, near Lula [Sula], *Goetze* 498 (B† holotype).
> *Stathmostelma pachycladum* K. Schum. in Bot. Jahrb. Syst. **28**: 458 (1900). Type: Tanzania, Iringa, near Kigonsive, i.1899, *Goetze* 531 (B† holotype).
> *Asclepias odorata* (K. Schum.) N.E. Br. in F.T.A. 4(1): 324 (1902).
> *Asclepias pachyclada* (K. Schum.) N.E. Br. in F.T.A. 4(1): 325 (1902).
> *Stathmostelma macropetalum* Schltr. & K. Schum. in Bot. Jahrb. Syst. **33**: 325 (1903). Type: Tanzania, Kilimandjaro, near Ndala, 21.xii.1896, *Trotha* 179 (B† holotype).
> *Asclepias macropetala* (Schltr. & K. Schum.) N.E. Br. in F.T.A. 4(1): 616 (1904).

Perennial herb with several annual stems arising from a slender or stout vertical tuber to 2 m long; stems 0.5–1.5 m long, generally unbranched, erect, densely pubescent with short white hairs. Leaves with petiole 5–20(30) mm long, pubescent; lamina 10–24 × 4–12 cm, oblong or elliptic to ovate, apex acute or obtuse, base rounded to truncate, occasionally somewhat cuneate, both upper and lower surfaces with a dense indumentum of spreading white hairs. Inflorescences extra-axillary with 4–12(15) flowers in an erect umbel, peduncles (1)3–10 cm long, erect, densely spreading-pubescent; pedicels 2–5 cm long, pubescent. Sepals (4)8–16 × 1–2 mm, linear-lanceolate, acute, densely pubescent. Corolla rotate to broadly campanulate, glabrous or pubescent at least towards tip on outer surface, minutely papillose within; lobes 15–20(30) × 6–10 mm, ovate to oblong or oblanceolate, acute or obtuse. Gynostegium shortly stipitate. Corona lobes arising just above the base of the staminal column and adnate to it to the base of the anthers, the free portion longer than the staminal column, 8–12 × 2–4 mm, erect and somewhat fleshy, weakly pouched towards the base and concave-cucullate for most of its length, apex rounded, the inner apical margins produced into a pair of inward-pointing teeth 2–3 mm long, with or without a tooth arising from the middle of the cavity, glabrous except along the midline of the cavity towards the base of the lobe. Anther wings 3–4 mm long, curved; anther appendages c.2–3 mm long, semicircular and inflexed over apex of stylar head. Stylar head flat, white or green. Fruiting pedicel not contorted; follicle erect, ovoid to fusiform, beaked, inflated or not, smooth or with weak longitudinal ridges, densely pubescent. Seeds c.5 × 3 mm, ovate with a verrucose disc and a narrow inflated rim; coma c.4 cm long.

Sepals mostly linear-lanceolate and less than 2 mm wide; corolla lobes strongly replicate . . . . a) subsp. *spectabile*
Sepals ovate, at least 4 mm wide; corolla lobes not replicate . . . . b) subsp. *frommii*

Fig. 7.3.**188**.
A. —STATHMOSTELMA SPECTABILE
subsp. SPECTABILE. A1, flowering shoot;
A2, flower from the side showing spreading
corolla, erect corona, and anther wings.
B. —STATHMOSTELMA PAUCIFLORUM,
inflorescence. Photographed by B. Wursten.

a) Subsp. **spectabile**. FIGURE 7.3.**188A**.

Bracts filiform. Sepals (4)8–16 × 1–2(5) mm, linear-lanceolate to lanceolate, acute. Corolla red or orange; lobes weakly to strongly replicate. Corona lobes red, orange or yellow, sometimes paler at the margins, with a slender or robust tooth to 3 mm long in the cavity. Follicle 10–14 × 1.5–2 cm, fusiform, not inflated, smooth or with weak longitudinal ridges.

Zambia. N: Mbala [Abercorn] Dist., Kawimbe, fl. 29.xii.1958, *Richards* 10374 (K). C: Balmoral Ranch 16 km SW of Lusaka, fl. 11.i.1995, *Bingham & Truluck* 10279 (K). **Zimbabwe**. N: Mtoko, fl. 10.i.1973, *Westwater* s.n. in GH 223127 (K, SRGH). E: Inyanga Dist., Van Niekirk's ruins, Mt. Zima, fl. 21.i.1967, *Chase* 8449 (K, SRGH). **Malawi**. N: Chitipa Dist., 75 km S of Chisenga on Nthalire road, fl. 3.i.1977, *Pawek* 12218 (K, MAL, MO). C: Dedza Dist., Nzoola Village, fl. 13.i.1985, *Patel & Kaunda* 1962 (K, MAL). S: Limbe, fl. 9.i.1970, *Moriarty* 421 (K, MAL).

Also found in Tanzania. Growing among grass in seasonally waterlogged grassland and in *Brachystegia* or mixed deciduous woodland, often in rocky ground or on limestone; 1000–1900 m.

b) Subsp. **frommii** (Schltr.) Goyder in Kew Bull. **53**: 585 (1988); in F.T.E.A., Apocynaceae (part 2): 412 (2012). Types: Tanzania, *Fromm* 89 (B† syntype) & *Fromm* 195 (B† syntype); Ufipa Dist., Lake Sundu, 23.xi.1960, *Richards* 13599 (K neotype, BR), designated by Goyder in Kew Bull. **53**: 585 (1998).

Stathmostelma frommii Schltr. in Bot. Jahrb. Syst. **51**: 139 (1913).

Bracts lanceolate to ovate. Sepals 9–16 × 4–11 mm, ovate to broadly ovate, acute, green or maroon. Corolla yellow, orange or red; lobes not replicate. Corona lobes yellow or orange, occasionally maroon; tooth within cavity sometimes reduced or absent. Follicle c.8 × 3 cm, ovoid, shortly beaked, somewhat inflated or not, smooth.

Zambia. N: Mbala [Abercorn] Dist., Chinakila woodland beyond Loye Flats, fl. 11.i.1965, *Richards* 19472 (K). W: Solwezi-Kafue, Mushama Forest, fl. 10.i.1962, *Holmes* 1434 (K, NDO).

Also occurs in D.R. Congo (Katanga) and southern Tanzania. Open *Brachystegia* woodland or scrub, frequently on limestone or on termitaria; 1200–1900 m.

The only character which can be relied upon to separate the two subspecies is the shape of the sepals, which are broadly ovate in subsp. *frommii* and linear-lanceolate in subsp. *spectabile*. The follicle is commonly inflated in subsp. *frommii* and more slender in the type subspecies, but from the limited material available this difference appears inconsistent. Flower colour does not correlate with other characters and cannot be used to differentiate the taxa.

2. **Stathmostelma pedunculatum** (Decne.) K. Schum. in Bot. Jahrb. Syst. **17**: 132 (1893). —Goyder in F.T.E.A., Apocynaceae (part 2): 413 (2012). Type: Ethiopia, between Adowa and Gondar, 1840, *Quartin-Dillon* s.n. (P lectotype, K photograph), designated by Goyder in Kew Bull. **53**: 586 (1998).

Gomphocarpus pedunculatus Decne. in Candolle, Prodr. **8**: 558 (1844).

Asclepias macrantha Oliv. in Trans. Linn. Soc. London **29**: 111, t.75 (1875). —Brown in F.T.A. **4**(1): 340 (1902). Type: Ethiopia, Sana Dist., near Gadding Gale, 23.vi.1840, *Schimper* s.n. (K holotype, P).

Pachycarpus corniculatus Hochst. in Flora **27**: 101 (1844), nomen nudum.

Gomphocarpus longipes Oliv. in Trans. Linn. Soc. London **29**: 111, t.75 (1875), in error.

Stathmostelma globuliflorum K. Schum. in Engler, Pflanzenw. Ost-Afrikas **C**: 322 (1895). Type: Tanzania/Kenya, Nyika, near Kiyombe, *Volkens* 87 (B† holotype, K).

Asclepias uvirensis S. Moore in J. Bot. **48**: 256 (1910). Type: D.R. Congo, Uvira, shore of Lake Tanganyika, 10.vii.1908, *Kässner* 3162 (BM holotype).

Stathmostelma macranthum (Oliv.) Schltr. in Notizbl. Bot. Gart. Berlin-Dahlem **9**: 27 (1924).

Asclepias pedunculatum (Decne.) Dandy in Andrews, Fl. Pl. Sudan **2**: 401 (1952).

Perennial herb with 1–4 annual stems arising from a slender, vertical, napiform tuber; stems 0.1–0.5(1) m long, simple or occasionally branched below, erect or ascending, minutely pubescent with short white hairs. Leaves sessile or with petiole to 5 mm long; lamina (4)7–15(19) × 0.2–1.3 cm, linear to lanceolate, acute, base truncate or rounded, minutely pubescent with short, stiff hairs on both upper and lower surfaces. Inflorescences terminal or extra-axillary with 2–4(7) flowers in an erect umbel, peduncles (6)8–20(29) cm long, the lower inflorescences with markedly longer peduncles than the upper ones, thus raising all the flowers to approximately the same level, erect, minutely pubescent; bracts caducous; pedicels 2–5 cm long, pubescent. Sepals 2–6 × 1–2 mm, lanceolate to ovate, acute, usually tinged red towards tip, densely pubescent. Corolla rotate or slightly reflexed, pink to orange-red, pubescent at least towards tip on outer surface, minutely papillose towards the base within; lobes 8–17 × 4–6 mm, oblong to obovate, acute, plane. Gynostegium with stipe 1–3 mm long. Corona lobes pink, red or orange with orange or yellow teeth, 4–7 × 1.5–2.5 mm, erect, 1–2 times as tall as column, weakly pouched towards the base on each side of the mid-line, appearing concave-cucullate for most of its length but in fact solid except for the proximal and upper margins, apex rounded distally, the inner apical margins produced into a pair of inward- or upward-pointing teeth 1–2 mm long and with a tooth to 2.5 mm long arising from the middle of the cavity, glabrous. Anther wings 2–3 mm long, curved; anther appendages 1–1.5 mm long, semicircular and inflexed over apex of stylar head. Stylar head flat. Fruiting pedicel not contorted; follicle erect, 8–12.5 × 0.5–1 cm, fusiform, beaked, not inflated, smooth, densely pubescent. Seeds c.5 × 3 mm, ovate with a verrucose disc and a narrow inflated rim.

Mozambique. N: Cabo Delgado Province, 5 km WNW of Quiterajo above Messalo River floodplain, fl. 23.xi.2009, *Goyder & Luke* 6135 (K).

Also known from Cameroon, Sudan, Ethiopia, D.R. Congo (Kivu), Uganda, Kenya and Tanzania. Growing among grass often in seasonally waterlogged grassland; 0–2200 m.

3. **Stathmostelma fornicatum** (N.E. Br.) Bullock in Kew Bull. **8**: 55 (1953). —Goyder in F.T.E.A., Apocynaceae (part 2): 415 (2012). Type: Malawi?, Nyika Plateau, ii-iii.1903, *McClounie* 81 (K holotype).

    Asclepias fornicata N.E. Br. in Bull. Misc. Inform. Kew **1906**: 250 (1906).

Perennial herb to c.0.6 m tall, with a single annual stem arising from a large napiform tuber; stems generally branched below, erect or ascending, glabrous or minutely pubescent with short white hairs. Leaves with petiole 2–5(20) mm long; lamina 8–20 × 0.3–1.3 cm, linear, acute, base truncate or rounded, minutely pubescent with short, stiff hairs particularly on the margins and midrib. Inflorescences terminal or extra-axillary with 2–4(5) flowers in an erect umbel, peduncles 2–14(18) cm long, the lower inflorescences with markedly longer peduncles than the upper ones, erect, minutely pubescent; bracts caducous; pedicels 2–5 cm long, minutely pubescent. Sepals 4–7 × 1–3.5 mm, lanceolate to ovate, acute or subacute, glabrous or pubescent with ciliate margins, green or tinged purple towards apex. Corolla rotate to campanulate, outer face greenish, chrome-yellow within; lobes 14–20 × 4–7 mm, lanceolate to ovate, acute or subacute, plane or somewhat replicate, glabrous on outer surface, minutely papillose within. Gynostegium with stipe 1–2 mm long. Corona lobes yellow or orange, erect and c.1.5 times as tall as column, (5)7–8 × c.3 mm, appearing concave-cucullate for most of their length but in fact solid except for the papillose proximal and upper margins, ± triangular in section above, apex rounded distally, the inner apical margins extending over the head of the column as a pair of broadly oblong, acute or obliquely truncate teeth c.2–3 × 2–3 mm, cavity sinus with or without a well developed central tooth. Anther wings 3–5 × 1 mm, curving somewhat under the anthers; anther appendages c.2 mm long, semicircular and inflexed over apex of stylar head. Stylar head flat. Fruiting pedicel not contorted; follicle erect, c.11 × 0.6 cm, fusiform, beaked, not inflated, smooth, minutely pubescent when young, glabrous at maturity. Seeds c.5 × 3 mm, ovate with a verrucose, minutely pubescent disc and a narrow inflated rim.

Subsp. **fornicatum**.

Corona with 1 or 2 rudimentary fleshy teeth towards the top of the cavity sinus.

Botswana. N: Nata Dist., 25 km towards Jolley's Pan from highway, fl. 24.xii.1996, *Bruyns* 6976 (BOL). **Zambia**. E: Luangwa Valley, Chitungwi Plain, fl. 23.i.1967, *Astle* 5011 (K). S: Machili, fl. 23.xii.1960, *Fanshawe* 6002 (K, NDO). **Zimbabwe**. W: Victoria Falls, fl. 27.xii.1979, *Ncube* 70 (K, SRGH). **Malawi**. N: Nyika Plateau, *McClounie* 81 (K). C: N of Chitala on Kasache road, fl. 12.ii.1959, *Robson* 1572 (BM, K, MAL, SRGH). S: Blantyre, fl. 7.i.1956, *Jackson* 1784 (BR, EA, K, MAL, SRGH).

Also known from Angola. Seasonally waterlogged grassland and mopane- or acacia-woodland; 700–1100(2500) m.

4. **Stathmostelma welwitschii** Britten & Rendle in Trans. Linn. Soc. London, Bot. **4**: 28 (1894). —Goyder in F.T.E.A., Apocynaceae (part 2): 414 (2012). Type: Angola, near Pedras de Guinga, Pungo Andongo, *Welwitsch* 4168 (BM holotype, K, LISU).

   Stathmostelma laurentianum Dewèvre in Bull. Soc. Roy. Bot. Belgique **34**(2): 87 (1895). Type: D.R. Congo, *Laurent* s.n. (BR holotype).

   Asclepias welwitschii (Britten & Rendle) N.E. Br. in F.T.A. **4**(1): 341 (1902).

   Asclepias laurentiana (Dewèvre) N.E. Br. in F.T.A. **4**(1): 342 (1902).

   Stathmostelma chironioides De Wild. & Durand in Bull. Herb. Boissier, sér. 2 **1**: 829 (1901), nomen nudum, not *Gomphocarpus chironioides* Decne. in Candolle, Prodr. **8**: 562 (1844).

Perennial herb with 1 to many annual stems arising from a long, slender, vertical, napiform tuber; stems 0.2–1 m long, simple or branched below, erect or ascending, minutely pubescent with short white hairs. Leaves sessile or with petiole to 10 mm long; lamina 7–16 × 0.2–1.2(1.6) cm, linear to lanceolate, acute, base truncate or rounded, minutely pubescent with short, stiff hairs on both upper and lower surfaces. Inflorescences extra-axillary with 2–7 flowers in an erect umbel, peduncles (4)7–14(20) cm long, the lower inflorescences with markedly longer peduncles than the upper ones, thus raising all the flowers to approximately the same level, erect, minutely pubescent; bracts caducous; pedicels 1.5–4 cm long, minutely pubescent. Sepals 3–6 × 12 mm, lanceolate to ovate, acute, usually tinged red towards tip, pubescent. Corolla rotate or slightly reflexed, yellow, pubescent at least towards tip on outer surface, minutely papillose towards the base within; lobes 9–12 × 6 mm, oblong to obovate, acute, plane. Gynostegium with stipe 1 mm long. Corona lobes yellow or orange, 4–6 × 2–4 mm, erect, slightly longer than column, appearing concave-cucullate for most of their length but in fact solid except for the proximal and upper margins, apex rounded distally, the inner apical margins produced into a pair of teeth 1–2 mm long and with a rounded or acute tooth arising from the middle of the cavity, glabrous. Anther wings 2–3 mm long, curved; anther appendages 1–1.5 mm long, semicircular and inflexed over apex of stylar head. Stylar head flat. Fruiting pedicel not contorted; follicle erect, 5–10 × 0.5–1 cm, fusiform, beaked, not inflated, smooth, densely pubescent. Seeds c.5 × 3 mm, ovate with a verrucose disc and a narrow inflated rim.

Stathmostelma welwitschii can be readily distinguished from *S. pedunculatum* by the more branched growth form and the flowers which are yellow rather than red.

Var. **welwitschii**.

Corona with slender acute tooth in cavity.

Zambia. N: Mporokoso Dist., Kabwe Plain, Mweru Wantipa, fl. 14.xii.1960, *Richards* 13697 (EA, K, SRGH).

Also known from Tanzania, D.R. Congo, Angola and Ethiopia. Seasonally waterlogged grassland; 1000–1650 m.

Bone et al. 410 (K) from the Solwezi Dist. of Zambia appears close to this taxon but differs in the size of floral organs and the more erect habit.

5. **Stathmostelma gigantiflorum** K. Schum. in Bot. Jahrb. Syst. **17**: 129, t.6 A-C (1893).
—Goyder in F.T.E.A., Apocynaceae (part 2): 417 (2012). Type: Bot. Jahrb. Syst. **17**:
t.6, figs.A–C (1893), lectotype designated by Goyder in Kew Bull. **53**: 599 (1998).

   Stathmostelma bicolor K. Schum. in Bot. Jahrb. Syst. **28**: 457 (1900), not *Asclepias bicolor*
Moench (1794). Type: Tanzania, Iringa Dist., Uhehe, Muhinde, *Goetze* 523 (B† holotype, K).

   Asclepias gigantiflora (K. Schum.) N.E. Br. in F.T.A. **4**(1): 326 (1902).

   Asclepias muhindensis N.E. Br. in F.T.A. **4**(1): 344 (1902). Type as for *Stathmostelma bicolor.*

   Stathmostelma nomadacridum Bullock in Kew Bull. **8**: 53 (1953). Type: Tanzania, Ufipa
Dist., Milepa, *Michelmore* 1438 (K holotype).

   Stathmostelma praetermissum Bullock in Kew Bull. **8**: 347 (1953). Type: Kenya, near
Moboloni rock, *Bally* 8380 (K holotype).

   Perennial herb with a single branched annual stem arising from a stout napiform or globose
tuber; stem 0.3–1 m long, branched from near the ground, erect or ascending, minutely pubescent
with short white hairs. Leaves mostly subsessile, occasionally with petiole to 10 mm long; lamina
10–25 × 0.2–1(2) cm, linear or occasionally lanceolate, acute, base cuneate, indumentum
mostly restricted to margins and midvein. Inflorescences terminal and extra-axillary with 3–6
flowers in an erect umbel; peduncles 2–17 cm long, the lower inflorescences generally with
longer peduncles than the upper ones, erect, minutely pubescent; bracts filiform, pubescent;
pedicels 2–5 cm long, minutely pubescent. Sepals 6–12(14) × 2–3(9) mm, triangular to broadly
ovate, acute, subglabrous to densely pubescent. Corolla rotate to slightly reflexed, subglabrous
or pubescent on outer surface, minutely papillose within, white or green outside, white, green
or pink and occasionally speckled purple within; lobes (15)20–30 × 7–15 mm, ovate to elliptic,
acute, somewhat replicate. Gynostegium with stipe c.3 mm long. Corona lobes white or green,
frequently speckled with purple towards upper margin, 8–15 × 4–7 mm, mostly overtopping
column by at least half their length, erect, concave-cucullate for most of their length, weakly
pouched either side of the mid line towards the base, apex rounded to attenuate distally, the
inner apical margins produced into a pair of upward- or inward-pointing acute teeth, inner
margins of lobe papillose, cavity without a median tooth. Anther wings 4–6 mm long, curved, the
basal tails free from the column; anther appendages c.1.5 mm long, semicircular and inflexed
over apex of stylar head. Stylar head flat. Fruiting pedicel not contorted; follicle erect, 10–16
× 0.5–0.8 cm, narrowly fusiform, with a basal stipe, not inflated, smooth, minutely but densely
pubescent. Seeds c.3 × 2 mm, verrucose.

   Zambia. E: Molozi River, Fort Jameson-Lundazi road, fl. & fr. 8.i.1959, *Jackson* 1089
(BM, K, SRGH). **Malawi**. C: North Kasungu, fl. 14.i.1959, *Jackson* 2286 (K, MAL,
SRGH).

   Also reported from Kenya and Tanzania. Seasonally waterlogged grassland on black
clay soils; 500–1800 m.

6. **Stathmostelma pauciflorum** (Klotzsch) K. Schum. in Bot. Jahrb. Syst. **17**: 132 (1893).
—Goyder in F.T.E.A., Apocynaceae (part 2): 421 (2012). Type: Mozambique, Rios
de Sena, *Peters* s.n. (B† holotype, K). FIGURE 7.3.**188B**.

   Gomphocarpus pauciflorus Klotzsch in Peters, Naturw. Reise Mossambique, **6**(Bot., 1): 276
(1861).

   Stathmostelma reflexum Britten & Rendle in Trans. Linn. Soc. London, Bot. **4**: 27, t.6, figs.4–
6 (1894). Type: Malawi, Mulanje, x.1891, *Whyte* s.n. (BM holotype).

   Asclepias reflexa (Britten & Rendle) Britten & Rendle in Trans. Linn. Soc. London, Bot. **4**:
28 (1894). —Brown in F.T.A. **4**(1): 344 (1902).

   Asclepias reflexa var. *longicauda* S. Moore in J. Bot. **47**: 219 (1909). Type: Zambia, Katanino
[Katenina] Hills, *Kassner* 2417 (BM lectotype, K), designated by Goyder in Kew Bull. **53**: 606
(1998).

   Perennial herb with a single annual stem arising from a narrowly cylindrical vertical tuber to
0.3 m or more in length; stems 0.2–1 m long, generally unbranched, erect, pubescent with short
white hairs. Leaves with petiole 0–5(10) mm long; lamina (3)6–14 × 0.1–0.6(1.3) cm, narrowly
linear to linear lanceolate, acute, cuneate or occasionally truncate, sparsely pubescent with stiff

white hairs particularly on margins and midrib below. Inflorescences terminal, sometimes also extra-axillary, with 3–7(12) flowers in an erect umbel, peduncles (5)7–30 cm long, erect, minutely spreading-pubescent; bracts filiform, pubescent, caducous; pedicels 1.5–3 cm long, pubescent and frequently purplish. Sepals 2–4 × 1 mm, ovate-oblong, acute, densely pubescent, purplish. Corolla usually strongly reflexed, red or orange (yellow in parts of southern D.R. Congo), glabrous on outer surface, minutely papillose within; lobes (6)7–10 × 3–4 mm, oblong-ovate, acute, somewhat replicate. Gynostegium sessile. Corona lobes arising at the base of the staminal column, orange or yellow, 7–12(14) × 2 mm, erect and somewhat fleshy, concave-cucullate in the lower half with the inner margins produced into a pair of inward-pointing falcate teeth c.2 mm long arching over the column, the upper half forming an erect or spreading, flattened tongue with a rounded to acute apex, no tooth present within cavity of lobe. Anther wings c.2 mm long with an additional basal tail curled under the anthers; anther appendages c.1 mm long. Stylar head flat. Fruiting pedicel not contorted; follicle erect, c.8 × 0.5 cm, fusiform, not inflated, smooth, glabrous.

Zambia. B: Manjoke, fl. 10.x.1964, *Fanshawe* 8982 (K, NDO). W: 80 km W of Chingola, fl. 18.xii.1963, *Robinson* 6122 (BR, K, SRGH). C: Karubwe Dambo 30 km N of Lusaka, fl. 2.i.1994, *Bingham* 9941 (K). E: Petauke, fl. 17.i.1974, *Fanshawe* 12182 (K, NDO). S: Mazabuka Dist., Siamambo Forest Reserve, fl. 10.i.1960, *White* 6186 (FHO, K, SRGH). **Zimbabwe**. N: Lomagundi Dist., Shinje, SE of Sipolilo, fl. 10.ii.1982, *Brummitt & Drummond* 15846 (K). C: Rusapi, Hislop Z 294, (K). E: Mutare [Umtali] Golf Course, fl. 30.xi.1960, *Chase* 7414 (K, SRGH). **Malawi**. N: Mt Masisi, fl. ix.1902, *McClounie* 95 (K). C: Kasungu National Park, Lingadzi Dambo, fl. 8.ii.1992, *Goyder et al.* 3627 (BR, K, MAL, PRE). S: Zomba, Lake Chilwa road, fl. 17.ii.1971, *Moriarty* 474 (K, MAL). **Mozambique**. N: Nampula Dist., Malema, fl. 20.x.1941, *Hornby* 3467 (K). Z: Mocuba Dist., Namagoa, fl. 6.xi.1948, *Faulkner* Kew 311 (K). MS: 5 km from Moribane towards Matarara do Lucite, fl. 8.viii.1953, *Gomes Pedro* 4239 (K, LMA).

The most widely distributed species of the genus, recorded from coastal Kenya, Tanzania, Mozambique, Malawi, Zambia, southern D.R. Congo and Zimbabwe. Seasonally waterlogged grassland close to sea level near the coast and up to 1500 m inland.

69. **GOMPHOCARPUS** R. Br.[41]

Gomphocarpus R. Br., On the Asclepiadeae: 26 (1810). —Goyder & Nicholas in Kew Bull. **56**: 769–836 (2001).

Shrubby or pyrophytic perennial herbs; stems erect, branched or not, frequently woody below. Root system mostly a fibrous tap root, sometimes somewhat thickened and woody, but never tuberous. Latex white, copious. Leaves generally in opposite pairs, occasionally subopposite or in whorls of 3 or 4. Inflorescences extra-axillary with flowers in a nodding umbel. Sepals free. Corolla lobes united only at the base, spreading or reflexed, abaxial face glabrous or pubescent, occasionally tomentose, adaxial face glabrous or minutely papillose, occasionally pubescent, commonly with a line of long white hairs along the right margin. Corolline corona absent. Gynostegial corona of 5 minute interstaminal lobules near the base of the anther wings and 5 well-developed cucullate staminal lobes attached above the base of the staminal column, with or without teeth or processes on the upper margins and mostly lacking a tooth in the cavity. Margins of anther wings straight, slightly convex or slightly sinuous. Corpusculum ovoid to subcylindrical, brown or black, with or without translucent flanges up the sides; translator arms flattened and geniculate or differentially thickened; pollinia flattened, oblong or oblanceolate to obovate, occasionally subtriangular. Stylar head flat. Fruiting pedicel mostly contorted to hold follicle erect, occasionally straight or weakly sinuous. Follicles generally single by abortion, variable in shape from subcylindrical to ovoid or globose, strongly inflated in some taxa, weakly or not at

---

[41] by D.J. Goyder

all in others, surface frequently ornamented with pubescent filiform processes. Seeds comose, generally ovate, with one convex and one concave face, verrucose, without a marginal rim, or in the *G. glaucophyllus* group with a convoluted margin, or rarely (*G. rivularis*) smooth, and with an inflated rim.

25 taxa in 20 species. Native to drier parts of the African continent and contiguous parts of Arabia, Sinai, Israel and Jordan.

1. Leaves broad, at least at base, ovate to lanceolate, or broadly ovate to suborbicular; base of lamina usually somewhat cordate 2
– Leaves linear to narrowly elliptic 5
2. Leaf margin minutely scabrid . . . . **7.** *swynnertonii*
– Leaf margin smooth 3
3. Corolla lobes at least 14 mm long; upper margin of corona lobes considerably higher than head of column; bracts at least 10 mm long, commonly elliptic or oblanceolate . . . . **9.** *praticola*
– Corolla lobes to 12 mm long; upper margin of corona lobes ± level with head of column; bracts to 10 mm long, linear to filiform 4
4. Corona lobes 2–3 mm long, upper margin straight . . . . **10.** *semiamplectens*
– Corona lobes 4–6 mm long, upper margin toothed . . . . **8.** *glaucophyllus*
5. Upper margin of corona lobes with a tooth at the proximal end (Fig. 7.3.**189A–D**); abaxial face of corolla glabrous or occasionally pubescent 6
– Upper margin of corona lobes lacking a tooth at the proximal end (Fig. 7.3.**189E–K**); abaxial face of corolla tomentose or pubescent 9
6. Upper margin of corona lobe without a distinct notch at base of proximal tooth (Fig. 7.3.**189A, B**) 7
– Upper margin of corona lobe with a distinct notch at base of proximal tooth (Fig. 7.3.**189C, D**) 8
7. Follicle ovoid below, drawn out into a clearly beaked apex; corona lobe oblong, with well developed proximal tooth; characteristic habit generally much branched below . . . . **1.** *fruticosus*
– Follicle subglobose, beak absent or minimal; upper margin of corona lobe sloping downwards away from column, proximal tooth weakly developed; characteristic habit with one principal stem branched above . . . . **2.** *physocarpus*
8. Corona lobes shorter than the staminal column, upper margin denticulate with a vertical cut about half way along separating the proximal tooth; corolla white or pink . . . . **3.** *semilunatus*
– Corona lobes equalling the column, the upper margin entire, and rounded distally, forming an acute angle at junction with proximal tooth; corolla yellow-green . . . . **4.** *kaessneri*
9. Stems slender, flexuous; follicles smooth or somewhat verrucose, but with few or no pubescent filiform processes . . . . **5.** *tenuifolius*
– Stems more robust; follicles densely covered in pubescent filiform processes . . . . **6.** *tomentosus*

1. **Gomphocarpus fruticosus** (L.) W.T. Aiton, Hort. Kew., ed. 2 **2**: 80 (1811). —Goyder in F.T.E.A., Apocynaceae (part 2): 425 (2012). Type: Herb. Linn. 310.33 (LINN lectotype), designated by Wijnands in Bot. Commelins: 48 (1983); South Africa, Northern Cape, Bloeddrif, 12.ix.1968, *Hardy* 2562 (K epitype, PRE, WIND), designated by Goyder & Nicholas in Kew Bull. **56**: 782 (2001). FIGURE 7.3.**189A**.

   Asclepias fruticosa L., Sp. Pl.: 216 (1753). —Brown in F.T.A. 4(1): 330 (1902); in Fl. Cap. 4(1): 691 (1908).

   Asclepias glabra Mill., Gard. Dict., ed. 8: no. 12 (1768). Type: '*Asclepias glabra* Mill. Dict. no. 12' (BM holotype).

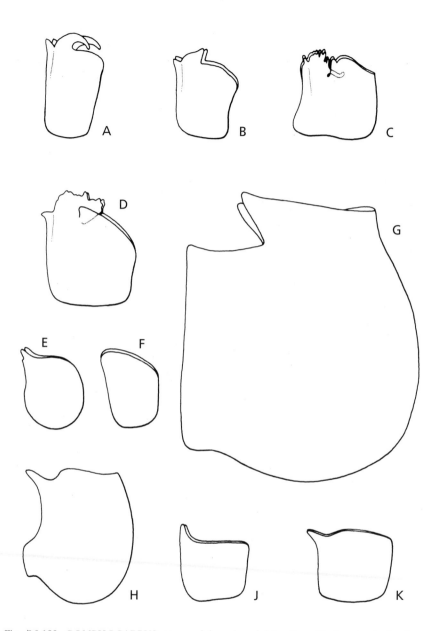

Fig. 7.3.**189**. GOMPHOCARPUS corona lobes viewed laterally, with the staminal column (not illustrated) to the left (× 8). A. —GOMPHOCARPUS FRUTICOSUS, from *Seydel* 3703. B. —GOMPHOCARPUS PHYSOCARPUS, from *Phillipson* 958. C. —GOMPHOCARPUS SEMILUNATUS, from *Goyder et al.* 3710. D. —GOMPHOCARPUS KAESSNERI, from *Richards* 23476. E. —GOMPHOCARPUS TENUIFOLIUS, from *Cribb* s.n. F. —GOMPHOCARPUS TOMENTOSUS, from *Richards* 14638. G. —GOMPHOCARPUS PRATICOLA, from *Sanane* 1415. H.—GOMPHOCARPUS GLAUCOPHYLLUS, from *Pawek* 3959. J.—GOMPHOCARPUS SEMIAMPLECTENS, from *Robinson* 4755. K. —GOMPHOCARPUS SWYNNERTONII, from *Goyder et al.* 3805. Drawn by E. Papadopoulos. Reproduced and adapted from Kew Bulletin (2001).

Asclepias salicifolia Salisb., Prodr.: 150 (1796), illegitimate name. Type as for *G. fruticosus* (L.) W.T. Aiton

?*Asclepias angustifolia* Schweigg., Enum. Pl. Hort. Regiom.: 13 (1812). Type: 'e horto Berolinensi' (KBG†).

?*Gomphocarpus angustifolius* (Schweigg.) Link, Enum. Hort. Berol. Alt. **1**: 251 (1821).

Gomphocarpus cornutus Decne. in Ann. Sci. Nat., Bot., sér. 2 **9**: 324 (1838). Type: Madagascar, *Bojer* s.n. (P holotype).

Gomphocarpus frutescens E. Mey., Comm. Pl. Afr. Austr.: 202 (1838), invalid name – error for *fruticosus*.

Gomphocarpus crinitus G. Bertol., Mem. Reale Accad. Sci. Ist. Bologna **3**: 253, t.20 fig.1 (1851). Type: Mozambique, Inhambane, xii.1848, *Fornasini* s.n. (BOLO holotype, K photograph).

?*Gomphocarpus arachnoideus* E. Fourn., Bull. Soc. Bot. France **14**: 250 (1867). Type: cult. Paris from seeds ex Mexico (P holotype, not seen).

Asclepias cornuta (Decne.) Cordem., Fl. Réunion: 482 (1895).

Asclepias crinita (G. Bertol.) N.E. Br. in F.T.A. **4**(1): 352 (1902).

Shrubby perennial herb 0.5–1.5(3) m tall arising from a tap root; stems erect, much branched from above the base, woody below, densely spreading-pubescent, sometimes shortly tomentose on young shoots. Leaves paired; petiole 1–10 mm long, pubescent; lamina (2.5)4–12 × (0.2)0.3–0.8(1.3) cm, linear or linear-lanceolate, apex acute or attenuate, mucronate, base narrowly to broadly cuneate, margins smooth, plane or somewhat revolute in northern part of the range, coriaceous, sparsely to densely pubescent with soft white hairs on the midrib and margins. Inflorescences extra-axillary with 4–7(12) flowers in a nodding umbel, peduncles 1.5–3(4) cm long, pubescent; bracts filiform, deciduous; pedicels 1–2.5 cm long, pubescent. Sepals 2–5 × 0.6–1.3 mm, lanceolate or triangular, attenuate, pubescent. Corolla reflexed, glabrous outside, minutely papillate and frequently with long white hairs along the right margin within; lobes 5–8 × 3–5 mm, ovate, subacute. Corona lobes attached 1–1.5 mm above base of staminal column, laterally compressed, complicate, 2–4 × 1.5–3 mm, ± as tall as the column, quadrate, the upper margins entire, proximal margins produced into a pair of falcate teeth c.1–1.5 mm long, pointing back along the upper margins of the lobe or curved down into the cavity, cavity without teeth or projections. Anther wings 1.5–2 mm long, margins not curved. Stylar head flat. Fruiting pedicel contorted to hold follicle erect; follicle 4–7 × 1.5–2.5 cm, ovoid, tapering gradually or abruptly into an attenuate beak, strongly or weakly inflated, pubescent, with or without pubescent filiform processes. Seeds 3.5–5 × 2 mm, ovate with one convex and one concave face, verrucose; coma c.3 cm long.

Follicles tapering gradually into an attenuate apex; generally covered with filiform processes. . . . . a) subsp. *fruticosus*
Follicles ± globose at base, narrowing abruptly into a long beak; filiform processes usually absent from fruit . . . . b) subsp. *rostratus*

a) Subsp. **fruticosus**. FIGURE 7.3.**190A**.

Young shoots subglabrous. Corolla white or cream. Corona cream or green; upper margin generally more angular than in subsp. *flavidus*. Follicles ovoid, beaked but not generally long-attenuate, densely covered with filiform processes c.0.6 cm long.

Botswana. N: Boteti River, fl. & fr. 19.xii.1978, *P.A. Smith* 2601 (K, MO, PRE, PSUB, SRGH). SE: 12.9 km from Kanye on Mmathete road, fl. & fr. 8.iii.1991, *Cook et al.* 52 (GAB, K). **Zambia**. N: Mbala [Abercorn] Dist., Lumi Marsh near Kawimbe, fl. & fr. 9.vi.1956, *Richards* 6122 (K). W: Chondwe, Ndola, fr. 2.xii.1972, *Fanshawe* 11674 (K, NDO). C: Chilanga Dist., Quien Sabe, fl. & fr. ix.1929, *Sandwith* 17 (BR, K). S: Choma, fl. 11.x.1955, *Bainbridge* 154/55, (K, FHO). **Zimbabwe**. W: Matopos National Park, Tuli River, fl. ii.1949, *Davies* 246 (K, SRGH). C: Gwebi River near Darwendale, c.50 km W of Harare, fl. 9.v.1989, *Goyder & Adams* 3235 (K, MAL, PRE). E: Nyanga Dist., Nyangombe Falls, fl. 3.xi.1950, *Chase* 3069 (BM, BR, K, SRGH). **Malawi**. C: Ncheu Dist., Chagunda Port, Lake Malawi, fl. 29.ix.1970, *Salubeni* 1496 (K, MAL, SRGH). S:

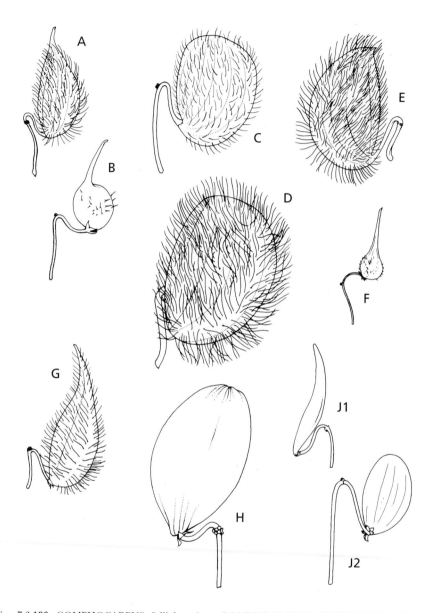

Fig. 7.3.**190**. GOMPHOCARPUS follicles. A. —GOMPHOCARPUS FRUTICOSUS subsp. FRUTICOSUS, from *Herman* 238. B. —GOMPHOCARPUS FRUTICOSUS subsp. ROSTRATUS, from *Long & Rae* 491. C. —GOMPHOCARPUS PHYSOCARPUS, from *Strey* 4859. D. — GOMPHOCARPUS SEMILUNATUS, from *Goyder et al.* 3710. E. —GOMPHOCARPUS KAESSNERI, from *Richards* 13434. F. —GOMPHOCARPUS TENUIFOLIUS, from *Cribb* s.n. G. —GOMPHOCARPUS TOMENTOSUS subsp. TOMENTOSUS, from Richards 14638. H. —GOMPHOCARPUS GLAUCOPHYLLUS, from *Phillips* 4605. J. —GOMPHOCARPUS SEMIAMPLECTENS, J1 narrow form, from *Fanshawe* 1841 and J2 inflated form, from *Gossweiler* 9941. Drawn by E. Papadopoulos. Adapted from Kew Bulletin (2001).

Lower Shire valley, Elephant Marsh, fl. 24.ix.1956, *Robertson* s.n. (K). **Mozambique**. MS: Gorongosa Game Reserve, fl. 27.ix.1953, *Chase* 5091 (BM, NSW, SRGH). GI: Chibuto, Baixo Changana, fl. & fr. 22.viii.1963, *Macedo & Macuácua* 1114 (K, LMA). M: Maputo [Lourenço Marques], between Peter and Costa do Sol, Golf Course, fl. 30.iii.1960, *Balsinhas* 154 (K, LMA).

Native to southern Africa as far north as SW Angola, Botswana, Zambia and Mozambique. Cultivated and widely naturalised elsewhere in warm temperate or dry subtropical areas of the world. Dry sandy soils in open or disturbed places. Frequently on river banks.

Conservation notes: reported by Goyder & Nicholas in Kew Bull. **56**: 784 (2001) as Least Concern.

b) Subsp. **rostratus** (N.E. Br.) Goyder & Nicholas in Kew Bull. **56**: 786 (2001). Type: Botswana, Lake Ngami, *Lugard* 22 (K000234915 lectotype, K), designated by Goyder & Nicholas in Kew Bull. **56**: 786 (2001). FIGURE 7.3.**190B**.

> *Asclepias rostrata* N.E. Br. in F.T.A. 4(1): 331 (1902).
> *Gomphocarpus rostratus* (N.E. Br.) Bullock in Kew Bull. **7**: 410 (1952), in part – excluding Kenyan and Tanzanian collections.

Follicles ± globose at base, narrowing abruptly into a long beak; filiform processes usually absent from fruit.

Caprivi. Eastern Caprivi, Sangwali, fl. & fr. 3.x.1970, *Vahrmeijer* 2160 (K, PRE). **Botswana**. N: floodplain near Xalaba, Boro River, fl. & fr. 10.xii.2007, *A. & R. Heath* 1495 (GAB, K, PSUB). **Zambia**. B: Ilongo, fl. & fr. 16.vii.1962, *Fanshawe* 6937 (K, NDO). C: Chiposhi Island, Lukanga Swamp, fl. & fr. 12.viii.1972, *Verboom* 3439 (K).

Centred on northern Botswana but extending into adjacent regions of Namibia (mostly Caprivi Strip), southern Angola and south west Zambia. Frequent on flood plains and river banks.

Conservation notes: reported by Goyder & Nicholas in Kew Bull. **56**: 786 (2001) as Least Concern.

Subsp. *rostratus* replaces the more widely distributed subsp. *fruticosus* in N Botswana and adjacent territories where it appears to grow in more frequently flooded areas. Occasional populations however, have follicles ± intermediate between the two typical forms (e.g. *Long & Rae* 32 (E, K) from SE Botswana). Other characters, however, correlate over most of the range: the habit is reported as less fruticose than the typical subspecies, upper parts of the plant are more tomentose, and flowers tend to be on the smaller end of the scale for the species.

2. **Gomphocarpus physocarpus** E. Mey., Comm. Pl. Afr. Austr.: 202 (1838). —Goyder in F.T.E.A., Apocynaceae (part 2): 427 (2012). Type: South Africa, by stream near Glenfilling, alt. 500 ft, *Drège* s.n. (K herb. Benth. lectotype, BM, E, K, TCD), designated by Goyder in Kew Bull. **53**: 418 (1998). FIGURES 7.3.**189B**, 7.3.**190C**.

> *Asclepias physocarpa* (E. Mey.) Schltr. in Bot. Jahrb. Syst. **21**(Beibl. 54): 8 (1896); in J. Bot. **34**: 453 (1896).

Shrubby perennial herb to 2.5 m tall arising from a tap root; stems generally single, branching above, woody below, upper parts pubescent with spreading white hairs. Leaves generally paired, but occasionally subopposite or crowded into pseudowhorls; petiole 3–12 mm long; lamina 4–9(12) × 0.5–1.5(2) cm, narrowly oblong to lanceolate, apex acute, base cuneate, margins smooth, subglabrous or sparsely pubescent with soft white hairs, particularly on the midrib and margins. Inflorescences extra-axillary with 5–12 flowers in a nodding umbel, peduncles 1.5–3.5 cm long, ascending, densely spreading-pubescent to tomentose; bracts deciduous; pedicels (1)1.5–2(3) cm long, densely pubescent. Sepals 3–4 × 1 mm, narrowly triangular-attenuate, pubescent. Corolla strongly reflexed, white, glabrous outside, minutely papillate and with long white hairs along the right margin within; lobes 5–8 × 3–4.5 mm, ovate, subacute. Corona lobes attached 1.5–2 mm above base of staminal column, laterally compressed, complicate, 2–3 × 1.5–2(2.5) mm, as tall as the column, quadrate, with a short erect to slightly recurved tooth c.0.5 mm long on the proximal upper margin, cavity of the lobe lacking any form of tooth, white, frequently tinged with pink or purple. Anther wings 1.8–2 mm long, margins very slightly sinuous. Stylar head flat. Fruiting pedicel contorted; follicle to 4–7 cm in diam., globose or subglobose, slightly depressed on one side, not beaked but occasionally somewhat angled apically, strongly inflated, pubescent, densely covered with filiform processes to c.1 cm long. Seeds c.4.5 × 2 mm, ovate with one convex and one concave face, verrucose; coma c.3 cm long.

Zimbabwe. C: National Botanic Garden Harare [Salisbury], fl. & fr. i.1975, *Grosvenor* 845 (K, SRGH). **Mozambique**. GI: Chibuto, Baixo Changane, fl. & fr. 22.viii.1963, *Macedo & Macuácua* 1115 (K, LMA). M: Maputo Elephant Reserve, near Lake Xinguti on track W of Ponta Milibangala, fl. & fr. 2.xii.2001, *Goyder* 5020 (K, LMU).

Native to South Africa, Swaziland and southern Mozambique. Almost certainly introduced elsewhere in Africa. The species is certainly not native elsewhere in the world.

Common in seasonally wet pastures and flood plains, also occurring in disturbed areas.

Conservation notes: reported by Goyder & Nicholas in Kew Bull. **56**: 789 (2001) as Least Concern.

3. **Gomphocarpus semilunatus** A. Rich., Tent. Fl. Abyss. **2**: 39 (1851). —Goyder in F.T.E.A., Apocynaceae (part 2): 428 (2012). Type: Ethiopia, Tigray, *Quartin Dillon* s.n. (P holotype, K). FIGURES 7.3.**189C**, 7.3.**190D**.

Asclepias semilunata (A. Rich.) N.E. Br. in F.T.A. **4**(1): 328 (1902).

Asclepias denticulata Schltr. in J. Bot. **33**: 334 (1895). Type: Uganda, Toro Dist., Ruimi [Wimi], vi.1894, *Scott-Elliot* 7904 (BM lectotype, K), designated by Goyder & Nicholas in Kew Bull. **56**: 791 (2001).

Shrubby perennial herb 1–2.5 m tall arising from a tap root; stems stout, erect, woody below, frequently unbranched, densely pubescent with spreading white hairs above. Leaves paired or in whorls of 4; petiole 2–6 mm long, pubescent; lamina 8–12(15) × 0.5–1.5(2.5) cm, linear or linear-lanceolate, apex acute or attenuate, base cuneate or truncate, margins smooth, revolute, sparsely pubescent with soft white hairs, particularly on the midrib and margins. Inflorescences extra-axillary with 5–9 flowers in a nodding umbel, peduncles 1–4 cm long, ascending, densely spreading-pubescent to tomentose; bracts deciduous; pedicels (1)1.5–2 cm long, densely pubescent, frequently tinged with purple. Sepals 5–9 × 1–2 mm, lanceolate, attenuate, pubescent, purplish. Corolla ± rotate, white or pink, outer surface deeper than the inner, glabrous outside, minutely papillate and frequently with long white hairs along the right margin within; lobes 7–9 × 4–5 mm, ovate, subacute. Corona lobes attached c.1 mm above base of staminal column, laterally compressed, complicate, 3 × 2.5–3 mm, distinctly shorter than the column, quadrate, the upper margins sloping down distally, denticulate and with a vertical cut about half way along, cavity of the lobe with or without a dorsally flattened, entire or bifid tooth, purple or pink with white upper and inner margins. Anther wings c.2 mm long, margins slightly convex; anther appendages c.1 × 0.5 mm, truncate and inflexed over apex of stigma head. Stigma head flat. Fruiting pedicel contorted; follicle to 7 × 5 cm, subglobose/scrotiform, not beaked, generally inflated, pubescent, densely covered with pubescent filiform processes c.1 cm long. Seeds c.3.5 × 1.5 mm, ovate with one convex and one concave face, verrucose; coma c.3 cm long.

Zambia. N: Lake Bangweulu, swamps S of Chafye Island, fl. & fr. 14.ii.1996, *Renvoize* 5603 (K).

Distributed from Nigeria to Ethiopia in the north, then southwards through E Africa to northern Zambia, southern D.R. Congo and Angola.

Seasonally flooded alluvial grasslands and waste places, frequently occurring in large numbers; 300–2600 m.

Conservation notes: reported by Goyder & Nicholas in Kew Bull. **56**: 791 (2001) as Least Concern.

4. **Gomphocarpus kaessneri** (N.E. Br.) Goyder & Nicholas in Kew Bull. **56**: 792 (2001). —Goyder in F.T.E.A., Apocynaceae (part 2): 428 (2012). Type: Kenya, Machakos/ Masai Districts, Kiu, 23.iv.1902, *Kässner* 664 (K holotype, BM). FIGURES 7.3.**189D**, 7.3.**190E**.

Asclepias kaessneri N.E. Br. in J. Bot. **41**: 362 (1903).

Shrubby annual or perennial herb 1–2 m tall arising from a tap root; stems erect, branched above the base, densely spreading-pubescent above. Leaves mostly paired, occasionally in whorls of 4 particularly towards base of stem; sessile or with petiole to 5 mm long; lamina 7–18 × 0.3–1.5 cm, narrowly linear to linear-lanceolate, apex attenuate, margins smooth, weakly to strongly revolute, coriaceous, subglabrous to sparsely pubescent with soft white hairs on both surfaces, particularly on the midrib and margins. Inflorescences extra-axillary with 4–7 flowers in a nodding umbel, peduncles 1–2(3) cm long, pubescent; pedicels 1–4 cm long, variable even within an inflorescence, pubescent. Sepals 4–6 × 1.5 mm, narrowly triangular, acute, pubescent. Corolla reflexed, greenish yellow, marked with brown outside, glabrous or sparsely pubescent outside, minutely papillate and with white hairs along the right margin within; lobes 6–10 × 4–5.5 mm, ovate, acute or subacute. Corona lobes attached 1.5–2 mm above base of staminal column, laterally compressed, complicate, 3.5–4 × 2–3.5 mm, ± as tall as the column, the upper margins acute at margin with the proximal teeth, rounded distally, proximal margins produced into a pair of broad, usually cristate teeth c.1.5–2 mm long, pointing back within the upper margins of the lobe, cavity without teeth or projections, brown or purple except for paler inner margin and tooth. Anther wings 2 mm long, margins ± straight. Stylar head flat. Fruiting pedicel contorted; follicle 4–7 × 2–4 cm, ovoid, rounded apically and without a beak, inflated, densely pubescent and with pubescent filiform processes to c.10 mm long. Seeds c.5 × 2 mm, ovate with one convex and one concave face, verrucose; coma c.3 cm long.

Zambia. N: Musesha, fl. & fr. 15.ix.1958, *Fanshawe* 4843 (K, NDO).

Also known from southern Kenya and central and western Tanzania. Seasonally flooded alluvial grasslands and waste places – perhaps in somewhat drier climatic zones than *G. semilunatus*; 1000 m.

Conservation notes: reported by Goyder & Nicholas in Kew Bull. **56**: 793 (2001) as Least Concern.

Strongly bicoloured flowers with yellow-green corolla and purple corona lobes appear to be diagnostic for this species – both *G. semilunatus* and *G. physocarpus* have a white corolla with pinkish corona lobes. Differences in corona morphology between this species and *G. semilunatus* begin to break down in northern Zambia, with the corona in some collections tending towards the shorter corona lobes of *G. semilunatus*.

5. **Gomphocarpus tenuifolius** (N.E. Br.) Bullock in Kew Bull. **8**: 341 (1953). Type: Zimbabwe, Lee's Farm, Mangwe River, 13.iii.1870, *Baines* s.n. (K holotype). FIGURES 7.3.**189E**, 7.3.**190F**.

Asclepias tenuifolia N.E. Br. in Bull. Misc. Inform. Kew **1895**: 255 (1895).
Asclepias rhodesica Weim. in Bot. Not. **1935**: 381, figs.9–10 (1935). Type: Zimbabwe, Makoni, near 'The Springs', 30.ix.1930, *Fries, Norlindh & Weimarck* 3320 (LD holotype, BM, BR).

Slender shrubby perennial herb 0.1–1.5 m tall, rootstock not seen; stems erect, much branched from base, slender, glabrous below, tomentose above. Leaves paired; sessile; lamina 3–6 × c.0.1 cm, narrowly linear, apex acute, mucronate, margins smooth, strongly revolute, minutely pubescent on both upper and lower surfaces. Inflorescences extra-axillary with 4–5 flowers in a nodding umbel, peduncles 1–1.5 cm long, slender to filiform, pubescent; bracts filiform, deciduous; pedicels 1–1.5 cm long, filiform, pubescent. Sepals 1.5–2 × 0.5–1 mm, triangular, attenuate, pubescent. Corolla reflexed, yellow on upper surface, reddish brown and pubescent outside, glabrous to densely papillate and with long white hairs along the right margin within; lobes 3.5–5 × 2–3 mm, ovate, subacute. Corona lobes attached 1 mm above base of staminal column, laterally compressed, complicate, 1.5–2 × 1.5–2 mm, slightly taller than the column at least proximally, quadrate or D-shaped, the upper margin entire and sloping gently away from the acute, proximal angles of upper margins, lacking any form of tooth within the cavity, yellow or greenish yellow. Anther wings 0.8–1 mm long, triangular but with a curved margin. Stylar head flat. Fruiting pedicel ± straight; follicles 3–3.5 × c.0.6 cm, lanceolate in outline, tapering gradually or abruptly into an attenuate beak, pubescent to tomentose, smooth or with soft prickles along the longitudinal ridges. Seeds not seen.

Zimbabwe. N: Lomagundi Dist., 22 km from Umvukwes to Sipolilo, fl. & fr. 10.ii.1982, *Brummitt & Drummond* (K, SRGH). W: Matopos Hills, fl. 26.ii.1981, *Philcox & Leppard* 8845 (K). C: Harare Dist., Rumani, fl. & fr. 22.ii.1952, *Wild* 3760 (BR, K, SRGH). E: Mutare Dist., Maranki Reserve, fl. 10.ii.1953, *Chase* 4773 (BM, K, SRGH). S: Masvingo Dist., 10 km E of Zimbabwe, fl. iii.1958, *Leach* 5963 (BR, K, SRGH).

Endemic to the central upland belt of Zimbabwe. More or less restricted to granite kopjes; 900–1600 m.

Conservation notes: reported by Goyder & Nicholas in Kew Bull. **56**: 802 (2001) as Least Concern.

6. **Gomphocarpus tomentosus** Burch., Trav. S. Africa **1**: 543 (1822), not *G. tomentosus* (Torr.) A. Gray (1876), illegitimate name. Type: South Africa, Northern Cape Prov., Asbestos Mountains near Kloof village (29°15'S 23°46'E), 16.ii.1812, *Burchell* 2024 (K holotype, TCD). FIGURE 7.3.**189F**.

    Gomphocarpus lanatus E. Mey., Comm. Pl. Afr. Austr.: 202 (1838). Type: South Africa, hills near Hamerkuil, *Drège* s.n. (K lectotype, B†, BM, E), designated by Goyder & Nicholas in Kew Bull. **56**: 807 (2001).

    ?*Gomphocarpus nutans* Klotzsch in Peters, Naturw. Reise Mossambique **6**(Bot., 1): 275 (1861). Type: Mozambique, lower Zambezi, Sena, *Peters* s.n. (B† holotype).

    Gomphocarpus fruticosus var. *tomentosus* (Burch.) K. Schum. in Engler, Pflanzenw. Ost-Afrikas **C**: 322 (1895).

    Asclepias burchellii Schltr. in J. Bot. **33**: 336 (1895). Type as for *G. tomentosus*.

    ?*Asclepias nutans* (Klotzsch) N.E. Br. in F.T.A. 4(1): 352 (1902).

    Asclepias lanata (E. Mey.) Druce in Rep. Bot. Soc. Exch. Club Brit. Isles **4**(Suppl. 2): 605 (1916).

Shrubby perennial herb 0.5–1.5(2) m; stems erect, branched, woody below, tomentose to lanate on young shoots, spreading-pubescent below. Leaves paired; petiole 1–5(10) mm long; lamina 5–12 × 0.2–0.5(0.7) cm, linear, tapering gently to an apex acute, margins smooth, weakly revolute, minutely pubescent on upper surface, tomentose beneath. Inflorescences extra-axillary with 3–6 flowers in a nodding umbel, peduncles 1–2 cm long, pubescent to lanate; bracts filiform, deciduous; pedicels 1–1.5 cm long, pubescent to lanate. Sepals 2–4 × 0.5–1.5 mm, triangular, acute to attenuate, tomentose to sublanate. Corolla reflexed, yellow-brown or green, frequently tinged pink and tomentose outside, glabrous or minutely pubescent within; lobes 5–7 × 3–4.5 mm, broadly ovate, acute. Corona lobes attached 0.5–1 mm above base of staminal column, laterally compressed, complicate, 2.5–4 × 1.5–3.5 mm, ± as tall as the column, narrowly to broadly D-shaped, margins fused only in lower, wider portion of lobe, entire except for minute folds or teeth by proximal end of upper margin, cavity without teeth or projections, lobes white or cream. Anther wings c.2 mm long, ± linear or narrowly triangular, curved slightly towards base. Stylar head flat. Fruiting pedicel contorted to hold follicle erect; follicles ovoid, tapering gradually into a long-attenuate beak, or subglobose, shortly tomentose at least when young and with lines of pubescent filiform processes 6–10 mm long. Seeds c.5 × 2 mm, ovate with one strongly convex and one concave face, verrucose; coma c.3 cm long.

Subsp. **tomentosus**. FIGURE 7.3.**190G**.

Young shoots mostly tomentose. Follicles 5–8 × c.1.5–2 cm, ovoid, tapering gradually into a long-attenuate beak, somewhat inflated.

Botswana. N: Okavango Delta, road between Kaporota and Vumbura camps, fl. & fr. 8.iii.2010, *A. & R. Heath* 1951 (GAB, K, PSUB). SW: track from Tshobokwame to Ramsden, fr. 20.iii.1979, *Skarpe* 337 (K). SE: Seleka Ranch, fl. & fr. 22.ii.1977, *Hansen* 3047 (C, GAB, K, PRE). **Zimbabwe**. W: Lobatsi Pan, fl. & fr. 6.ii.1970, Cannell 104 (K, SRGH). E: Chipinga Dist., road to Sabi Valley experimental station, 400 m from Birchenough Bridge–Chipinga road, fl. & fr. 22.iv.1969, *Biegel* 2955 (E, K, SRGH). **Mozambique**. GI: Between Panda and Mangorro, 6 km from Panda, 7.iv.1959, *Barbosa & de Lemos* 8534 (K, LISC, LMA). M: Maputo Elephant Reserve, near Lake Xinguti on track W of Ponta Milibangalala, fl. & fr. 2.xii.2001, *Goyder* 5019 (K, LMU).

Widespread in southern Africa as far north as southern Angola, Zimbabwe and Mozambique. Growing mostly on sand in open or disturbed areas; 0–2000 m.

Conservation notes: reported by Goyder & Nicholas in Kew Bull. **56**: 809 (2001) as Least Concern.

7. **Gomphocarpus swynnertonii** (S. Moore) Goyder & Nicholas in Kew Bull. **56**: 810 (2001). —Goyder in F.T.E.A., Apocynaceae (part 2): 432 (2012). Type: Zimbabwe, Chimanimani [Melsetter], 6000', 10.x.1908, *Swynnerton* 6092 (BM lectotype), designated Goyder & Nicholas in Kew Bull. **56**: 810 (2001). FIGURE 7.3.**189K**.

Asclepias swynnertonii S. Moore in J. Linn. Soc., Bot. **40**: 142 (1911).
Asclepias nyikana Schltr. in Bot. Jahrb. Syst. **51**: 138 (1913). Type: Tanzania (T7), Kyimbila, *Stolz* 105 (B† holotype, BM, K).
Asclepias stolzii Schltr., invalid name, in sched. *Stolz* 105, type of *A. nyikana*.

Perennial herb arising from stout woody taproot; stems few to many, unbranched, erect, 10–40 cm long, with a dense indumentum of spreading white hairs, or indumentum occasionally restricted to weak vertical bands. Leaves sessile or subsessile, 1.5–5.5(8) × 0.7–3 cm, ovate, acute, slightly cordate at the base, margins slightly revolute at least when dry, veins prominent at c.40° to midrib, glabrous except for the minutely scabrid marginal vein. Inflorescences extra-axillary with 6–13 flowers in a nodding umbel; peduncles 2–4.5 cm long, generally longer than the subtending leaf, pubescent; bracts 5–10 mm long, filiform or occasionally broader, ciliate; pedicels 10–23 mm long, slender, pubescent. Calyx lobes 3–6 × 1.5–2 mm, ovate or oblong, apex acute to rounded, margins ciliate or not. Corolla united at the base for 0.5–1 mm, yellow or greenish yellow, sometimes tinged brownish purple, glabrous or minutely papillate towards base; lobes spreading to reflexed, (4)5–8 × (2)3–5 mm, ovate or elliptic, acute. Corona lobes attached c.1 mm above base of staminal column, laterally compressed, complicate above, solid below line from point of attachment to column to distal end of upper margin, 2–4 × 2–3 mm, quadrate, upper margins ± level with top of column, straight or sometimes weakly toothed at about the middle, produced into a pair of teeth at the proximal end ± reaching head of column, lower margin spreading or ascending. Anther wings c.2 × 0.5 mm; anther appendages c.1 × 1 mm, broadly ovate, obtuse, inflexed over apex of stigma head. Follicles 6–9 × 1.5–2 cm, fusiform or lanceolate in outline, pubescent, usually only one of the pair developing; fruiting pedicel contorted to hold follicle erect. Seeds 5–6 × 4–5 mm, ± suborbicular with a convoluted margin and a darker verrucose disc; coma c.3 cm long.

Zambia. N: Mpika, fl. 6.ii.1955, *Fanshawe* 2006 (K, NDO). W: 50 km E of Mwinilunga on path to Solwezi, fl. & fr. 11.ix.1930, *Milne-Redhead* 1091 (K). E: Nyika Plaeau, near turn-off to Chelinda Camp, fl. & fr. 27.xi.1955, *Lees* 111 (K). **Zimbabwe**. C: Harare [Salisbury], fl. x-xi.1935, *Brain* 6473 (K, SRGH). E: Pungwe source, Nyanga, 19.x.1946, *Wild* 1413 (BR, K, SRGH). **Malawi**. N: Nyika National Park, 1 km W of Chelinda Bridge, 22.i.1992, *Goyder et al.* 3542 (BR, K, MAL, PRE). **Mozambique**. MS: Manica, Chimanimani Mts towards E ridge, 1.xi.2014, *Timberlake et al.* 6035 (BR, K, LMA, SRGH).

In suitable habitats from Angola in the west, through the Katanga plateaus of D.R. Congo, to the highlands of southern Tanzania, Malawi, Zimbabwe and neighbouring territories.

Montane, fire-prone grassland and open *Protea* scrub; (1000)1600–2400 m.

Conservation notes: reported by Goyder & Nicholas in Kew Bull. **56**: 811 (2001) as Least Concern.

This species is variable in size and in the degree of pubescence. A single collection from Mpika and some from the Nyika Plateau approach *G. glaucophyllus* in their larger leaves and in the indumentum reduced to weak bands on the stem and inflorescence.

8. **Gomphocarpus glaucophyllus** Schltr. in Bot. Jahrb. Syst. **18**(Beibl. 45): 19 (1894).
—Goyder in F.T.E.A., Apocynaceae (part 2): 433 (2012). Type: South Africa, Mpumalanga, near Barberton, *Galpin* 663 (B† holotype, K, NH, PRE, SRGH). FIGURES 7.3.**189H**, 7.3.**190H**.

Asclepias glaucophylla (Schltr.) Schltr. in J. Bot. **34**: 455 (1896).

Asclepias lilacina Weim. in Bot. Not. **1935**: 374 (1935). Type: Zimbabwe, between Nyanga and Rusape, 1800 m, 19.ix.1930, *Fries, Norlindh & Weimarck* 3062 (LD holotype, BM).

Perennial herb arising from stout woody rhizomatous rootstock; stems 1–4, robust and somewhat fleshy, unbranched, ascending or erect, 0.3–1 m long, glabrous and commonly glaucous. Leaves sessile or subsessile, (4)7–12 × (1.3)3–6 cm, ovate to lanceolate, acute, cordate at the base, margins smooth, glabrous, generally glaucous with a waxy bloom on both surfaces, veins prominent, sometimes reddish. Inflorescences extra-axillary with (6)9–15 flowers in a nodding umbel; peduncles 1.5–6 cm long, mostly 1/2 to 2/3 length of subtending leaf, glabrous and generally glaucous; bracts 5–10 × 0.5–1 mm, filiform to linear, margins ciliate or not; pedicels 2–3.5 cm long, glabrous, glaucous or not. Calyx lobes 7–10 × 2–4 mm, oblong to broadly ovate, apex obtuse to acute, glabrous, green or glaucous. Corolla united at the base for 1–2 mm, greenish yellow, sometimes purplish on outer face, minutely papillate particularly towards base; lobes strongly or weakly reflexed, 9–12 × 5–8 mm, ovate, acute. Corona lobes attached c.1.5 mm above base of staminal column, similar to *G. swynnertonii* but measuring 4–6 × 3–6 mm, upper margin ± level with head of staminal column. Anther wings 2–2.5 × 0.5–1 mm, triangular, outer margin straight or weakly concave; anther appendages c.1 × 1 mm, broadly ovate, obtuse, inflexed over apex of stigma head. Follicles 5.5–10 × 1.5–3.5 cm, fusiform with occasional longitudinal wings and an acute apex or oblong-elliptic in outline and somewhat inflated with an obtuse apex, glabrous, glaucous, usually only one of the pair developing; fruiting peduncle contorted to hold follicle erect. Seeds c.6 × 5 mm, ± suborbicular with a slightly convoluted, pale and minutely pubescent margin surrounding a verrucose pubescent disc; coma c.4 cm long, ivory.

Zambia. B: Sesheke Dist., *Gairdner* 217 (K). C: Chilanga, fl. 10.ix.1909, *Rogers* 8479 (BOL, K, SRGH). **Zimbabwe**. N: Umvukwe Dist., E of Imshi Mine, fl. 5.xi.1961, *Leach* 11273 (BM, BRLU, K, SRGH). W: Hwange [Wankie] National Park, Main Camp along Dopi Pan road, fl. 8.xi.1968, *Rushworth* 1251 (K, SRGH). C: near Harare [Salisbury], fl. xi-xii.1899, *Cecil* 57 (K). E: Inyanga North, St. Swithins T.T.L. 18 km E of Elim mission, fl. 16.i.1968, *Biegel* 1779 (K, PRE, SRGH). **Malawi**. N: Mzimba Dist., 5 km SSW of Chikangawa, fl. 2.ix.1978, *Phillips* 4169 (K, MAL, MO, WAG). C: Dedza Dist., Chongoni Forest, 15.x.1960, *Chapman* 990 (K, MAL, PRE, SRGH).

Uplands of eastern Africa from Swaziland and adjacent parts of South Africa, through Zimbabwe, Zambia, Malawi and Tanzania, to Rwanda, Burundi and Uganda in the north. Flowering mostly October to December in grassland or open *Brachystegia* or mixed decidous woodland; 700–2400 m.

Conservation notes: reported by Goyder & Nicholas in Kew Bull. **56**: 815 (2001) as Least Concern.

9. **Gomphocarpus praticola** (S. Moore) Goyder & Nicholas in Kew Bull. **56**: 815 (2001), as "*praticolus*"; —Goyder in F.T.E.A., Apocynaceae (part 2): 435 (2012). Type: Angola, Kuelai, near R. Chipumba, 19.ix.1906, *Gossweiler* 3532 (BM holotype, LISC); LISC specimen from 'Sera Pinto, Bie, Munonque, 19.ix.1906'. FIGURES 7.3.**189G**, 7.3.**191**.

Fig. 7.3.**191**. GOMPHOCARPUS PRATICOLA.. 1, rootstock and lower stem (×²/₃); 2, flowering shoot (×²/₃); 3, pollinarium (× 14). 1 from *Bullock* 1877; 2 from *Goyder et al.* 3812; 3 from *Sanane* 1415. Drawn by E. Papadopoulos. Reproduced from Kew Bulletin (2001).

Asclepias praticola S. Moore in J. Bot. **50**: 341 (1912).

Asclepias katangensis S. Moore in J. Bot. **50**: 340 (1912), illegitimate name, not *A. katangensis* De Wild. in Ann. Mus. Congo Belge, Bot. sér. 5 **1**: 187 (1904). —De Wildeman in Contr. Fl. Katanga **2**: 125 (1913). Type: D.R. Congo, Katanga, Lovoi R., ix.1910, *Kässner* 3353 (BM holotype).

Asclepias friesii Schltr. in Fries, Wiss. Ergebn. Schwed. Rhod.-Kongo-Exped. 1911–12 **1**: 267 (1916). Type: Zambia, Mtali, in marsh N of L. Benguela, *Fries* 1177a (UPS holotype, K photograph).

Asclepias moorei De Wild. in Contr. Fl. Katanga: 156 (1921). Type as for *A. katangensis.*

Perennial herb arising from a stout woody rhizomatous roostock; stems 1–3, robust and somewhat fleshy, unbranched, ascending or erect, (0.4)0.6–1 m long, glabrous or rarely with two bands of spreading white hairs along upper portions of stem, glaucous or green. Leaves held erect, sessile, (5)6–12(16) × (1.5)2–5.5 cm, oblong to elliptic or rarely lanceolate, apex acute or obtuse, base cordate, margins smooth, glabrous, green or glaucous on both surfaces, veins conspicous or obscure. Inflorescences extra-axillary with 6–15 flowers in a nodding umbel; peduncles 1.5–4.7(7) cm long, usually much shorter than the subtending leaf, glabrous or softly pubescent; bracts variable in size even within an inflorescence, 10–35 × 0.2–10(15) mm, filiform to elliptic or oblanceolate, acute or obtuse, green or glaucous, glabrous but sometimes ciliate towards tip; pedicels 2.5–5(18.5) cm long, commonly purple, glabrous or with soft spreading hairs. Calyx lobes 5–17(27) × (2)4–10 mm, elliptic to obovate, rarely linear, apex obtuse or rarely acute, glabrous, green, glaucous, purple or brown. Corolla united at base for c.3 mm, green or yellow-green with reddish purple markings outside, papillate towards base; lobes strongly reflexed, (14)16–35 × 7–15 mm, ovate, acute. Corona lobes attached 1–2 mm above base of the staminal column, laterally compressed but with the half of the lobe furthest from the column subcylindrical and solid except for the rim, outer margin of lobe drawn in apically towards column, (5)7–12 × (4)6–9 mm, upper margins generally well above head of staminal column, conspicuously toothed at proximal end and half way along upper margin, brown or purple with greenish yellow upper margin. Anther wings 3–3.5 × 1 mm, outer margin ± straight or weakly curved; anther appendages 1–1.5 × 1–1.5 mm, broadly ovate, obtuse, inflexed above over apex of stigma head. Follicles c.8 × 2 cm, fusiform, inflated, usually only one of the pair developing, glabrous, glaucous; fruiting pedicel contorted to hold follicle erect. Seeds c.7 × 4.5 mm, ± suborbicular with a slightly convoluted, pale and minutely pubescent margin surrounding a verrucose pubescent disc; coma c.4 cm long, ivory.

Zambia. W: E of River Lunga at Mwinilunga, fl. 24.xi.1937, *Milne-Redhead* 3380 (K). N: N'ingi Pans, Mbala, fl. 18.xi.1964, *Richards* 19265 (K, SRGH). **Malawi**. N: Chitipa Dist., Mafinga Hills. Kawumbe Hill, 2 km N of Chisenga, fl. 2.ii.1992, *Goyder et al.* 3601 (BR, K, MAL, MO, PRE).

Contiguous upland areas of northern Malawi, Zambia and southern Tanzania, extending westwards through southern D.R. Congo into Angola. Grassland or open *Brachystegia/Uapaca* woodland on sandy or rocky ground; 1500–2400 m.

Conservation notes: reported by Goyder & Nicholas in Kew Bull. **56**: 816 (2001) as Least Concern.

10. **Gomphocarpus semiamplectens** K. Schum. in Bot. Jahrb. Syst. **17**: 128 (1893). Type: Angola, Pungo Andongo, near Cazella, i.1857, *Welwitsch* 4188 (K lectotype, B†, BM), designated by Goyder & Nicholas in Kew Bull. **56**: 815 (2001). FIGURES 7.3.**189J**, 7.3.**190J**.

Asclepias semiamplectens (K. Schum.) Hiern, Cat. Afr. Pl. **1**: 685 (1898).

Perennial herb arising from stout woody rhizomatous rootstock; stems few to many, moderately slender and not apparently fleshy, unbranched, ascending to erect, 0.3–1 m long, glabrous. Leaves held erect, sessile, 5–9.5 × 2.5–6 cm, ovate or more usually broadly ovate, apex obtuse or acute, base cordate, margins smooth, glabrous, green or glaucous on one or both surfaces, veins conspicuous. Inflorescence extra-axillary with 7–30 flowers in a nodding umbel; peduncles (2.5)4–8(12) cm long, shorter than to much longer than the subtending leaf, glabrous. Bracts

4–8 mm long, filiform, glabrous. Pedicels 15–30 mm long, glabrous. Calyx lobes 3–6 × 1.5–2.5 mm, ovate to triangular, acute or subacute, glabrous but occasionally minutely ciliate on margins. Corolla united at base for c.1 mm, yellow-green, minutely papillate towards base; lobes partially reflexed, 5–8 × (2)3–5 mm, ovate, acute. Corona lobes attached c.1 mm above base of column, similar to *G. swynnertonii* but central cavity open to near base, 2–3 × 2.5–3.5 mm, ± quadrate, upper margin ± level with top of column, flat, level or sloping away from column, yellow. Anther wings c.2 × 0.5 mm, outer margin straight, anther appendages c.1 × 1 mm, ovate, acute, inflexed over apex of stigma head. Follicles 4–8 × 1.5–2.5 cm, ovoid with an acute apex, usually inflated but occasionally slender and not inflated, glabrous, only one of the pair developing; fruiting pedicel contorted to hold follicle erect. Seeds c.5 × 4 mm, pubescent on both the convoluted rim and the verrucose disc; coma c.3 cm long, ivory.

Zambia. N: 15 km E of Kasama, fl. 20.xii.1961, *Robinson* 4735 (K, SRGH). W: Mwinilunga, NE of Dobeka Bridge, fl. 8.xi.1937, *Milne-Redhead* 3156 (K, SRGH). C: Serenje, fr. 23.i.1955, *Fanshawe* 1841 (K, NDO).

Northern and western Zambia, through southern and western D.R. Congo to Angola and Cabinda. Open hillsides or deciduous woodland, frequently on sand; 900–1600 m.

Conservation notes: reported by Goyder & Nicholas in Kew Bull. **56**: 818 (2001) as Least Concern.

70. **PACHYCARPUS** E. Mey.[42]

Pachycarpus E. Mey., Comm. Pl. Afr. Austr.: 209 (1838). —Smith in S. African J. Bot. **54**: 399–439 (1988). —Goyder in Kew Bull. **53**: 335–374 (1998); in Fl. Pl. Africa **58**: 96–103 (2003).

Gomphocarpus sect. *Pachycarpus* (E. Mey.) Decne. in Candolle, Prodr. **8**: 562 (1844). — Schumann in Engler & Prantl, Nat. Pflanzenfam. **4**(2): 236 (1895), in part.

Perennial herbs with annual stems arising from a tuberous rootstock; latex milky. Stems 1 to many, simple or branched below, erect or ascending, pubescent with short spreading hairs. Leaves opposite, petiolate; lamina narrowly lanceolate to broadly ovate (linear in sect. *Campanulati*), generally with prominent secondary veins anastomosing to form an undulating submarginal vein, and an indumentum of scabrid hairs particularly along the margins. Inflorescences terminal and extra-axillary, umbels nodding or erect, sessile or more commonly pedunculate. Calyx lobed to the base. Corolla rotate to campanulate (reflexed in some southern African species). Corolline corona absent. Gynostegial corona of 5 lobes arising near the base of the staminal column (at the top of a gynostegial stipe in *P. medusonema*), spreading or ascending, generally dorsoventrally flattened and usually with a pair of triangular or quadrate appendages arising near the base of the upper surface, but sometimes appearing ± cucullate or pouched, and generally with an apical tongue; upper parts sometimes minutely verrucose or papillate, the remainder of the lobe generally glabrous. Staminal column stout, anther wings vertical or spreading; anther appendages inflexed over stylar head. Corpusculum ovoid, occasionally narrower; translator arms flattened and geniculate, frequently with a short clasping overlap with the pollinia; pollinia flattened, sometimes only weakly, and usually broadest distally. Stylar head truncate, ± level with head of staminal column or slightly raised above it. Follicles single by abortion, held ± erect by contortion of the fruiting pedicel, weakly to strongly inflated, frequently with longitudinal wings or ridges, minutely puberulent or glabrous. Seeds flattened, rarely ovoid, generally differentiated into a verrucose disc and an inflated or convoluted rim; coma of silky hairs.

37 species in tropical and subtropical Africa. Growing mostly in montane grassland and open woodland in the mountains of eastern Africa, and the grasslands of southern Africa from the Transvaal to the eastern Cape.

---

[42] by D.J. Goyder

1. Corolla lobes reflexed; corona lobes terminating in an erect petaloid appendage . . . . **8.** *appendiculatus*
- Corolla lobes not reflexed; corona lobes lacking a petaloid terminal appendage 2
2. Corolla rotate to rotate-campanulate 3
- Corolla campanulate to subglobose 5
3. Peduncles absent or shorter than 3 cm long; rootstock generally a vertical napiform tuber; follicles generally with 4–6 longitudinal wings distally, never strongly inflated . . . . **3.** *concolor*
- Peduncles mostly at least 4 cm long; roots a fascicle of horizontal fusiform, fleshy tubers; follicles often strongly inflated, not ornamented with longitudinal wings 4
4. Corona lobes widest near base, ± as tall as column proximally; anther-wing parallel to axis of column . . . . **1.** *lineolatus*
- Corona lobes narrower at the base than above, ½ to ⅔ height of column proximally; anther-wings angled to give a truncate-conical appearance to the column . . . . **2.** *bisacculatus*
5. Corolla lobes less than 8 mm long . . . . **7.** *firmus*
- Corolla lobes at least 9 mm long 6
6. Corona lobes shorter than or ± as long as column . . . . **6.** *goetzei*
- Corona lobes at least 1.5× as long as column. 7
7. Fertile portion of staminal column raised above corona on a conspicuous stipe; corona lobes without teeth near base. . . . . **4.** *spurius*
- Staminal column not or barely stipitate; corona with erect pair of teeth towards base. . . . . **5.** *chirindense*

1. **Pachycarpus lineolatus** (Decne.) Bullock in Kew Bull. **8**: 333 (1953). —Goyder in F.T.E.A., Apocynaceae (part 2): 437 (2012). Type: Angola, *da Silva* s.n. (P holotype, K photograph). FIGURE 7.3.**192A**.

> *Gomphocarpus lineolatus* Decne. in Ann. Sci. Nat., Bot. sér. 2 **9**: 326 (1838).
> *Asclepias lineolata* (Decne.) Schltr. in J. Bot. **33**: 336 (1895).
> *Asclepias conspicua* N.E. Br. in Bull. Misc. Inform. Kew **1895**: 253 (1895); in F.T.A. **4**(1): 324 (1902). Type: Zambia, Fwambo, *Carson* 12 (K holotype).
> ?*Calotropis busseana* K. Schum. in Bot. Jahrb. Syst. **33**: 323 (1903). —Brown in F.T.A. **4**(1): 615 (1904). Type: Tanzania, Usambara, 9.ix.1900, *Busse* 341 (B† holotype).
> *Asclepias browniana* S. Moore in J. Bot. **47**: 217 (1909). Type: D.R. Congo, Lake Moero, *Kässner* 2806 (BM lectotype), designated by Goyder in Kew Bull. **53**: 345 (1998).
> *Pachycarpus schweinfurthii* (N.E. Br.) Bullock in Kew Bull. **8**: 330 (1953); in F.W.T.A., ed. 2 **2**: 93 (1963).

Perennial herb with annual stems arising from a fascicle of fleshy fusiform tubers; stems 0.4–1.5 m long, 1 to many, usually unbranched, but sometimes branched below, erect or ascending, densely pubescent with spreading white hairs. Leaves with petiole (2)5–13 mm long, pubescent; lamina 4–12(14) × 1–6(9) cm, narrowly lanceolate to broadly ovate, elliptic or oblong, apex obtuse or slightly retuse to acute, base cordate, truncate or rounded, margins scabrid, venation prominent with numerous parallel secondary veins at right angles to the midrib, indumentum of slightly scabrid spreading white hairs on both upper and lower surfaces. Inflorescences terminal and extra-axillary with 4–12 flowers in a nodding umbel, peduncles (2)4–17 cm long, erect or ascending, densely spreading-pubescent; bracts (2)5–14 mm long, filiform, pubescent; pedicels 1–3(4) cm long, pubescent, commonly tinged with purple. Calyx lobes (4)6–11 × (1)2–3 mm, lanceolate, acute, densely pubescent, green or tinged with purple particularly towards the apex and margins. Corolla rotate or saucer-shaped, underside cream or brownish purple, sometimes with one or more longitudinal purple stripes, sparsely to densely pubescent, upper surface cream or pale green, often marked with a network of purple or chocolate-brown lines in the distal half, minutely papillate and with short hairs towards the apex and margins; lobes 8–17

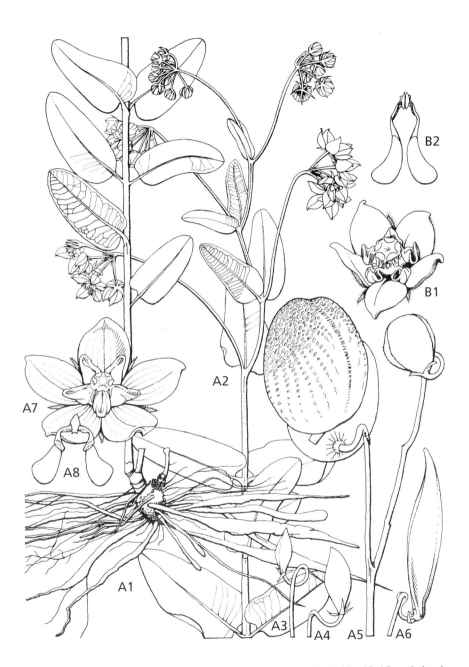

Fig. 7.3.**192**. A. —PACHYCARPUS LINEOLATUS. A1, A2, habit (× ½); A3–A6, variation in follicle age and degree of inflation (× ½); A7, flower (× 1½); A8, pollinarium (× 10). B. — PACHYCARPUS BISACCULATUS. B1, flower (× 1½); B2, pollinarium (× 10). A1, A7, A8 from *Goyder et al.* 3779; A3, A4, A6 from *Dummer* 385; A5 from *Bullock* 2277; B1, B2 from *Keay* in FHI 25897. Drawn by E. Papadopoulos. Reproduced from Kew Bulletin (1998).

× (4)6–9(11) mm, ovate, acute. Corona lobes arising from the base of the staminal column, 7–8 mm long, (1)3–4 mm wide at the base, narrowing gradually towards the apex, solid, fleshy, spreading or ascending from the staminal column, with a pair of triangular or quadrate vertical plates arising from the upper surface, 4–5 mm high at the proximal end reaching to the top of the column, the proximal margins resting against the column, the upper margins meeting to form an enclosed chamber, white or cream with a reddish purple patch on the underside of the fleshy basal plate, often restricted to an area just below the tip, upper parts minutely verrucose or papillate, the remainder of the lobe generally glabrous. Staminal column c.6 mm long, fertile portion forming a drum c.3 mm long and 4–6 mm diam.; anther wings 2–3.5 × 1 mm, extending below the fertile portion for c.1.5 mm, triangular, ± vertical. Stylar head flat, white or green. Follicles and seeds as in *P. bisacculatus.*

Zambia. N: Mbala [Abercorn], fl. 23.xii.1949, *Bullock* 2129 (BR, EA, K, SRGH). W: Ndola, fl. 10.i.1954, *Fanshawe* 659 (BR, EA, K, NDO, SRGH). **Malawi**. N: Nyika Plateau, 12 km N of M1, fl. 2.ii.1978, *Pawek* 13720 (K, MO, MAL). C: Dowa Dist., Kongwe Forest Reserve, fl. 7.iii.1982, *Brummitt et al.* 16377 (K, MAL).

Widespread in tropical Africa from Ivory Coast in the west to Sudan in the east and as far south as Angola and Malawi. Open deciduous woodland or seasonally waterlogged grassland; 700–2500 m.

2. **Pachycarpus bisacculatus** (Oliv.) Goyder in Kew Bull. **51**: 798 (1996); in F.T.E.A., Apocynaceae (part 2): 438 (2012). Type: Tanzania, Kilimanjaro, *Johnston* s.n. (K holotype). FIGURE 7.3.**192B**.

　　Gomphocarpus bisacculatus Oliv. in Trans. Linn. Soc. London, Bot. **2**: 341 (1887).
　　Asclepias lineolata sensu Hutchinson & Dalziel in F.W.T.A. **2**(1): 56 (1931), not Decne.
　　Pachycarpus lineolatus sensu Bullock in Kew Bull. **8**: 333 (1953), not Decne.; in F.W.T.A. ed. 2 **2**: 93 (1963).
　　Pachycarpus bullockii Cavaco in Bull. Mus. Natl. Hist. Nat. sér. 2 **29**: 514 (1958). Type: Angola, NE of Lunda, near Cassai stream, 16.xi.1946, *Gossweiler* 13891 (P holotype not seen, K).

Perennial herb with annual stems arising from a fascicle of fleshy fusiform tubers; stems 0.3–1.5 m long, usually single and unbranched, but sometimes branched below, erect or ascending, densely pubescent with spreading white hairs. Leaves with petiole (3)5–13 mm long, pubescent; lamina 6–12 × 2–7 cm, lanceolate to ovate or oblong, apex obtuse or slightly retuse to acute, base cordate or truncate, margins scabrid, venation prominent with numerous parallel secondary veins at right angles to the midrib, indumentum of slightly scabrid spreading white hairs on both upper and lower surfaces. Inflorescences terminal and extra-axillary with 4–8(14) flowers in a secund umbel, peduncles (2)4–17 cm long, erect or ascending, densely spreading-pubescent; bracts 5–12 mm long, filiform, pubescent; pedicels (1)1.5–3 cm long, pubescent, tinged with purple. Calyx lobes 6–10 × 2–3 mm, lanceolate, acute, densely pubescent, purplish. Corolla rotate or saucer-shaped, underside brownish purple, densely pubescent, upper surface pale green with a network of purple or chocolate brown lines particularly in the distal half, minutely papillate and with short hairs towards the apex and margins; lobes 10–15 × 7–10 mm, ovate, acute. Corona lobes 5–6 mm long, 4–6 mm high at the proximal end and 1/2–2/3 height of the column, 3–4 mm wide, widest point ±1/3 way along lobe, arising from the base of the staminal column and adnate to it to the base of the anthers, cream or more usually pink with a purple tip, spreading and strongly complicate at extreme base with proximal margins parallel and erect reaching to the base or middle of the anthers, distal half or two thirds forming an erect or incurved cone with involute margins, flanges between the base of this apical cone and the proximal margins of the lobes sinuous and dilated to form a pair of shallow pouches, the upper and inward-facing portions of the lobes minutely but densely papillate, the outer faces glabrous. Staminal column 6–8 mm long, fertile portion forming a drum 3–4 mm long and 4–6 mm diam.; anther wings 3 × 1.5 mm, not extending below fertile portion of column, triangular, giving the column a truncate conical appearance. Stylar head flat, white or green. Fruiting pedicel contorted to hold follicle erect; follicle to 12 × 7 cm, ovoid or fusiform, generally inflated, without wings or processes, pubescent. Seeds (immature) 4.5 × 3 mm, ovate with a verrucose disc and a narrow inflated rim; coma c.3 cm long.

Botswana. N: Okavango, Eretsha, fl. 19.xii.1996, *Bruyns* 6943 (K). **Zambia**. B: Kabompo Dist., 110 km S of Mwinilunga on Kabompo road, fl. 25.xii.1969, *Simon & Williamson* 2026 (K, SRGH). N: Luapula Dist., Mbereshi, fl. 11.i.1960, *Richards & Omari* 12316 (K, SRGH). W: Nkana South, fl. 13.xii.1962, *Mutimushi* 245 (K, NDO). C: Chakwenga headwaters, 100 km E of Lusaka, fl. 7.i.1964, *Robinson* 6139 (K, SRGH). E: W of Sasare, fl. 10.xii.1958, *Robson* 891 (BM, K, SRGH). S: Choma SW, fl. 2.i.1955, *Robinson* 1044 (K, SRGH). **Zimbabwe**. N: 8 km from Gokwe, near Sasame village, fl. 23.xii.1962, *Bingham* 345 (K, SRGH). W: Hwangie [Wankie] National Park, Main Camp near Dopi, fl. 16.xi.1968, *Rushworth* 1272 (K, SRGH). C: Umbuma road 16 km from Gwelo, fl. 5.ii.1967, *Biegel* 1882 (K, SRGH). E: Mutare Dist., Burma Farm to Manyera Farm, fl. 30.xii.1959, *Chase* 7235 (BM, BR, K, PRE, SRGH). **Malawi**. N: Karonga Dist., Vinthukhutu Forest Reserve, 3 km N of Chilumba, fl. 6.i.1978, *Pawek* 13558 (K, MAL, MO, UC). C: North Kasungu, fl. & fr. 14.i.1959, *Jackson* 2287 (K, SRGH). S: Micheru Road, Blantyre, fl. 15.i.1970, *Moriarty* 367 (K, MAL). **Mozambique**. N: 25 km E of frontier at Mandimba, fl. 18.xii.1941, *Hornby* 2459 (K). Z: Mocuba Dist., Namagoa, fl. 4.xii.1948, *Faulkner* Kew 406 (BR, K, LISC, SRGH). MS: Espungabera, fl. i.1962, *Goldsmith* 25/62 (BM, BR, K, LISC, PRE, SRGH).

Widespread in tropical Africa from Guinea-Bissau in the west to southern Sudan and Ethiopia in the east and as far south as Angola and Mozambique.

Growing among grass in seasonally waterlogged grassland and in *Brachystegia* or mixed deciduous woodland; 100–2500 m.

Field observations in Zambia and Tanzania by Bullock & Milne-Redhead in Kew Bull. **8**: 333 (1953) indicate that the flowering periods of this species (referred to as *P. lineolatus*) and the closely related *P. lineolatus* (referred to as *P. schweinfurthii*) differ by about six weeks.

3. **Pachycarpus concolor** E. Mey., Comm. Pl. Afr. Austr.: 210 (1838). —Brown in F.T.A. **4**(1): 378 (1902). —Goyder in F.T.E.A., Apocynaceae (part 2): 440 (2012). Type: South Africa, Eastern Cape Prov., King William's Town Division, between Chalumna R. and Kachu (Yellowwood) R., *Drège* s.n. (K lectotype, MEL, P), designated by Smith in S. African J. Bot. **54**: 411 (1988).

Xysmalobium concolor (E. Mey.) D. Dietr., Syn. Pl. **2**: 902 (1840).
Gomphocarpus concolor (E. Mey.) Decne. in Candolle, Prodr.: **8**: 563 (1844).
Asclepias concolor (E. Mey.) Schltr. in Bot. Jahrb. Syst. **21**(Beibl. 54): 6 (1896); in J. Bot. **34**: 452 (1896).

Perennial herb with annual stems arising from a stout, cylindrical, vertical tuber; stems 0.2–1 m long, 1 to several, usually unbranched but sometimes branched below, procumbent to erect even in the same population, densely pubescent with spreading white or rust hairs. Leaves spreading, petiole 1–10 mm long, pubescent; lamina 5–16 × 0.6–4 cm, linear to narrowly lanceolate or triangular, widest at or just above the base then tapering gradually into the acute or subacute apex, or lanceolate to ovate (outside the F.T.E.A. area), base shortly cuneate or truncate, margins scabrid, plane or crisped, indumentum of slightly scabrid spreading white or rusty hairs on both upper and lower surfaces. Inflorescences terminal and extra-axillary with (1)2–7 scented flowers in an erect umbel, sessile or with peduncles to 3 cm long, densely spreading-pubescent; bracts 4–14 mm long, filiform, pubescent; pedicels 1–3 cm long, pubescent. Calyx lobes subequal to strongly unequal, longest 12–16 × 4–8 mm, ovate to lanceolate, acute, pubescent, sometimes reddish towards tip. Corolla rotate or broadly campanulate, underside tinged reddish brown, densely pubescent, upper surface white, cream or pale green, with or without purple markings, glabrous or minutely papillate; lobes 11–15(18) × 8–13 mm, broadly ovate, subacute. Corona lobes dorso-ventrally flattened, arising from the base of the staminal column; basal portion c.3 × 3–6 mm, spreading, rhomboid or continuous with the distal portion, with an erect pair of pyramidal or deltoid obtuse teeth 2–3.5(4.5) × 1.5–3.5 mm arising from the middle of this portion and reaching ± the top of the staminal column; distal portion 4–9(15) × 1.5–2.5 mm, lingulate or slightly spathulate, rounded or truncate, erect or incurved over the head of the staminal column, glabrous except for a minutely but densely papillate central strip on upper side, erect or incurved over the head of the staminal column, entire lobe white or cream, or dull purple at the base with a green tongue. Staminal column truncate conical, 3–4 mm long and 7–8(9) mm diam.; anther wings 3–4 × 1.5–3 mm, triangular with acute or rounded bases. Stylar head flat. Fruiting pedicel contorted to hold follicle erect; follicle to 12 × 3 cm, fusiform, not generally inflated, with 6 narrow longitudinal wings or ridges in the upper half, minutely pubescent. Seeds 5 × 3 mm, ovate and slightly convex on both faces, pubescent, disc not clearly differentiated from rim; rim with transverse ridges; coma c.4 cm long.

Leaves ovate to lanceolate; seasonally burned montane or plateau grasslands at higher elevation; southern Botswana, Zimbabwe, Lebombo mountains. . a) subsp. *concolor*
Leaves linear; seasonally burned coastal dune grasslands at c.40–50 m alt.; Mozambique S of Maputo . . . . b) subsp. *arenicola*

a) Subsp. **concolor**. FIGURE 7.3.**193C**.

Botswana. SE: near Palapye, fl. i.1899, *Lugard* 261 (K). **Zimbabwe**. W: Matopos, fl. 9.ii.1941, *Hopkins* 7950 (K, SRGH). C: 1.5 km from Eagles Nest on Rusape road, fl. 29.xi.1955, *Drummond* 5072 (K, PRE, SRGH). E: Mutare, fl. 17.xi.1951, *Chase* 4995b (BM, SRGH). S: Kyle National Park Game Reserve, Chembezi Gully, fr. 22.v.1971, *Grosvenor* 533 (SRGH). **Mozambique**. M: Namaacha hills, fl. 20.xii.1944, *Torre* 6924 (LISC).

Also known from Swaziland and the northern and eastern provinces of South Africa. Seasonally burned montane or plateau grasslands; 300–1500 m.

b) Subsp. **arenicola** Goyder in Fl. Pl. Africa **58**: 96 (2003). Type: Mozambique, Maputo Elephant Reserve, 2 km W of Ponta Milibangalala, 29.xi.2001, *Goyder* 5015 (K holotype, LMU).

Mozambique. M: Near Ponta Malongane on road to Zitundo, fl. & fr. 27.xi.2001, *Goyder* 5011 (K, LMU).

Known only from the seasonally burned coastal dune grasslands between Maputo and the St. Lucia estuary in northern KwaZulu-Natal; 40–50 m.

4. **Pachycarpus spurius** (N.E. Br.) Bullock in Kew Bull. **8**: 338 (1953). —Goyder in F.T.E.A., Apocynaceae (part 2): 441 (2012). Type: Malawi, Shire Highlands, *Buchanan* 451 (K holotype). FIGURE 7.3.**193A**.

Xysmalobium spurium N.E. Br. in Bull. Misc. Inform. Kew **1895**: 251 (1895).
?*Xysmalobium dolichoglossum* K. Schum. in Bot. Jahrb. Syst. **28**: 456 (1900). Type: Tanzania, Uhehe, Uchungwe Mt, *Goetze* 638 (B† holotype).
Schizoglossum spurium (N.E. Br.) N.E. Br. in F.T.A. **4**(1): 367 (1902).
?*Schizoglossum dolichoglossum* (K. Schum.) N.E. Br. in F.T.A. **4**(1): 367 (1902).
Schizoglossum debeersianum K. Schum. in Bot. Jahrb. Syst. **33**: 323 (1903). Type: D.R. Congo, Haut Marungu, *De Beers* 108 (BR holotype).

Perennial herb with 1–5 annual stems arising from a long cylindrical vertical rootstock, sometimes with a few fusiform laterals; stems 0.8–2 m long, unbranched, erect, densely pubescent with spreading white hairs. Leaves with petiole 3–10(15) mm long, pubescent; lamina 6–13 × 3–8.5 cm, lanceolate to broadly ovate, apex acute or obtuse, base rounded and slightly cordate, margins scabrid, venation prominent with numerous parallel secondary veins, pubescent on both upper and lower surfaces with minute scabrid hairs. Inflorescences terminal and extra-axillary with 4–9 flowers in a nodding or secund umbel, peduncles 2–4.5 cm long, densely pubescent; bracts c.10 mm long, filiform, pubescent; pedicels 1.5–2.3(3) cm long, densely pubescent. Calyx lobes 6–11(14) × 3–4(7) mm, ovate to lanceolate, acute, pubescent. Corolla globose-campanulate, glabrous or minutely papillate within, pubescent outside, reddish purple or occasionally cream outside, white, cream or purple within, often with darker veins; lobes

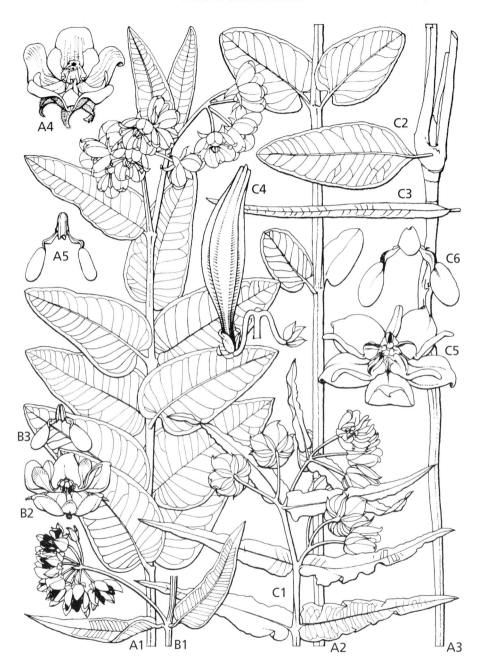

Fig. 7.3.**193**. A. —PACHYCARPUS SPURIUS. A1–A3, habit (× ½); A4, flower (× 1); A5, pollinarium (× 8); all from *Goyder et al.* 3639. B. —PACHYCARPUS CHIRINDENSIS. B1, stem with inflorescence (× ½); B2, flower (× 1); B3, pollinarium (× 8); all from *Pole-Evans* 5291. C. —PACHYCARPUS CONCOLOR subsp. CONCOLOR. C1, upper portion of stem with inflorescence (× ½); C2, C3, leaves (× ½); C4, follicle (× ½); C5, flower (× 1); C6, pollinarium (× 8); C1, C4–C6 from *Grimshaw* 9480; C2 from *Linley* 350; C3 from *Davis* 47. Drawn by E. Papadopoulos. Reproduced from Kew Bulletin (1998).

(11)14–22 × (6)8–14 mm, oblong ovate or obovate, acute or obtuse, recurved at the tip. Staminal column stipitate; corona lobes arising from the base of the stipe, dorsoventrally flattened, spreading below, erect or slightly recurved above, 10–20 × 4–8 mm, spathulate, the dilated apex acute to truncate or emarginate, entire or variously toothed. Anther wings 3–4 × 1–1.5 mm, triangular and convex. Peduncle not elongating in fruit; fruiting pedicel contorted; follicle c.7 × 3–3.5 cm, ovoid and somewhat inflated, with 4 or 6 broad, longitudinal, apically toothed wings, pubescent, not erect. Mature seeds not seen.

Malawi. N: Nyika Plateau, NE of Nganda, fl. 11.iv.1997, *Patel et al.* 5127 (K, MAL). C: Dedza Mountain Forest Reserve, fl. 10.ii.1992, *Goyder et al.* 3639 (BR, K, MAL, PRE) S: Shire Highlands, fl., *Adamson* 292 (E, K).

Occurs in montane regions of Malawi, southern Tanzania and southern D.R. Congo. Montane grassland and *Brachystegia* or *Protea* scrub woodland, sometimes in disturbed sites; (1200)1500–2400 m.

5. **Pachycarpus chirindensis** (S. Moore) Goyder in Kew Bull. **53**: 356 (1998). Type: Zimbabwe, near Chirinda, *Swynnerton* 246 (BM holotype, K). FIGURE 7.3.**193B**.

    Schizoglossum chirindense S. Moore in Journ. Bot. **46**: 295 (1908).

    Schizoglossum gigantoglossum Weim. in Bot. Not. **1935**: 393 (1935). Type: Zimbabwe, Nyanga, 7.xii.1930, *Fries, Norlindh & Weimarck* 36/3 (LD holotype, BM, K photograph).

Perennial herb with 1–3 annual stems arising from a slender vertical tuberous rootstock to c.1 m long; stems 0.3–1 m long, unbranched, ascending or erect, densely pubescent with spreading white hairs. Leaves with petiole 2–5 mm long, pubescent; lamina 7–11 × 1–3.5 cm, lanceolate, tapering gradually to an acute apex, base rounded or truncate, margins scabrid, venation prominent with numerous parallel secondary veins, pubescent on both upper and lower surfaces with minute scabrid hairs. Inflorescences terminal and extra-axillary with 6–18 flowers in a globose umbel, peduncles (0.5)1.5–5.5 cm long, densely pubescent; bracts c.5 mm long, filiform, pubescent; pedicels 1–1.5 cm long, densely pubescent. Calyx lobes 6–10 × 2–3 mm, lanceolate, acute, pubescent. Corolla campanulate, glabrous or minutely papillate within, pubescent outside, reddish purple or occasionally yellow; lobes 8–12 × 5–8 mm, ovate-oblong or obovate, acute, recurved at the tip. Staminal column not or only minutely stipitate; corona lobes reddish brown, dorsoventrally flattened, spreading at the base, ascending or erect above, 7–13 × 2.5–4.5 mm, broadly spathulate with a rounded apex or narrowly obtriangular with a truncate or emarginate apex, inner face with a pair of erect narrowly to broadly triangular plates along the midline of the lobes near the base. Anther wings 1.5–2 mm long, triangular, giving the column a conical appearance. Stigma head c.2 mm diam., not exserted from top of column. Fruiting pedicel contorted to hold follicle erect; follicle 8–12 × 2.5–3 cm, lanceolate-fusiform with 6 longitudinal wings.

Zimbabwe. E: Chimaminani [Melsetter] Dist., near Black Mountain Inn, fl. 21.xi.1956, *Chase* 6250 (K, SRGH). **Mozambique**. MS: Serra da Gorongosa, track to Gogogo peak, fl. 27.ix.1943, *Torre* 5972 (LISC).

Restricted to the east Zimbabwe highlands and adjacent parts of Mozambique. Rocky grassland; 1000–2000 m.

This species is closely allied to *Pachycarpus spurius*, and to *P. macrochilus* (Schltr.) N.E. Br. from the mountains of eastern South Africa and Lesotho. It differs from the latter in its pedunculate inflorescences and the lack of a gynostegial stipe and from the former by the basal corona flanges.

6. **Pachycarpus goetzei** (K. Schum.) Bullock in Kew Bull. **8**: 338 (1953). —Goyder in F.T.E.A., Apocynaceae (part 2): 445 (2012). Types: Tanzania, Uhehe, Kipundi Mts, *Goetze* 674 (B† syntype); Iringa, near Gumbira, *Goetze* 706 (B† syntype); Mbeya Dist., Kitulo-Matamba road, *Bidgood & Congdon* 154 (K neotype, EA, TPRI), designated by Goyder in Kew Bull. **53**: 363 (1998).

Schizoglossum goetzei K. Schum. in Bot. Jahrb. Syst. **28**: 455 (1900). —Brown in F.T.A. 4(1): 371 (1902).

Schizoglossum simulans N.E. Br. in F.T.A. **4**(1): 369 (1902). Type: Malawi, Nyika Plateau, vii.1896, *Whyte* 108 (K lectotype), designated by Goyder in Kew Bull. **53**: 363 (1998).

Perennial herb similar in most respects to *P. eximius* with 1 to several annual stems, rootstock not seen; stems 0.5 to 1.3 m tall, erect, unbranched. Leaves (4.5)7–12 × 1.5–4.5 cm, lanceolate to oblong, acute. Inflorescences with 5–18 sweetly scented flowers in a secund umbel; pedicels 1–1.5 cm long, densely pubescent. Calyx often reddish towards tip. Corolla white or cream, often tinged pink; lobes 9–10 × 5–7 mm. Corona lobes cream or white, ascending or erect with a spreading or slightly recurved tip, 4–5 × 3–5 mm, ± as long as the staminal column, oblong, broadened slightly towards the truncate to 3–toothed apex, inner face with an erect triangular tooth in centre of the lobe c.2 mm long ± reaching the top of the anthers. Anther wings c.2 × 0.5 mm. Fruiting pedicel contorted or not; follicle 6–9 × 2–2.5 cm, narrowly ovoid with 4 broad longitudinal wings, pubescent.

Zambia. N: Mpika, fl. 7.ii.1955, *Fanshawe* 2017 (BR, K, NDO, SRGH). **Malawi**. N: Nyika Plateau, 15 km N of M1, fl. 3.iii.1977, *Pawek* 12420 (BR, K, PRE, SRGH).

Known from southern Tanzania and adjacent parts of Zambia and Malawi. Upland grassland and *Brachystegia* woodland; 1600–2300 m.

7. **Pachycarpus firmus** (N.E. Br.) Goyder in Kew Bull. **53**: 367 (1998); in F.T.E.A., Apocynaceae (part 2): 448 (2012). Type: Angola, Huilla, near Lopollo, *Welwitsch* 4191 (K holotype, BM).

Schizoglossum firmum N.E. Br. in Bull. Misc. Inform. Kew **1895**: 252 (1895); in F.T.A. 4(1): 368 (1902).

Asclepias firma (N.E. Br.) Hiern in Cat. Afr. Pl. **1**: 684 (1898).

Perennial herb with 1 to several annual stems arising from a vertical tuberous rootstock; stems 0.3–0.8 m tall, erect, densely pubescent. Leaves with petiole 2–6 mm long, pubescent; lamina 4–8 × 2–4 cm, lanceolate to ovate or oblong-ovate, apex acute to obtuse or rounded, base rounded or truncate, usually slightly cordate, margins scabrid, venation prominent with numerous parallel secondary veins, indumentum of spreading white hairs on both upper and lower surfaces. Inflorescences terminal and extra-axillary with 5–20 flowers in a globose umbel, peduncles 1–4 cm long, not elongating significantly in fruit, densely spreading-pubescent; bracts 4–10 mm long, filiform, pubescent; pedicels 0.3–0.8 cm long, densely pubescent. Calyx lobes 4–7 × 1–2 mm, lanceolate to ovate, acute, densely pubescent, often purple towards tip. Corolla campanulate, pubescent on outer face at least above, glabrous within, white or pink, marked with deeper pink or maroon; lobes 5–8 × 3–4 mm, ovate, acute, not recurved. Corona lobes arising from the base of the staminal column, white or purple, erect, dorsoventrally flattened and slightly convex, 2–3.5 × 2–3.5 mm, ± quadrate and about as tall as the column, with a subulate apical tooth 2–3 mm long arching over the stylar head, inner face of the lobe with margins inflexed and produced into a pair of fleshy teeth ± level with top of anthers. Anther wings c.1.5 × 0.5 mm. Stylar head flat, 2–3 mm diam., forming a raised pentagonal cushion 1–2 mm above top of anthers. Fruiting pedicel contorted to hold follicle erect; follicle c.6 × 1.5 cm, narrowly ovoid with 4–6 longitudinal, weakly toothed wings, pubescent. Seeds c.4.5 × 2 mm, subovoid but strongly concave and with verrucose or reticulate sculpturing

Malawi. N: Jembya Forest Reserve 16 km SE of Shisenga, fl. 25.i.1989, *Thompson & Rawlins* 6175 (CM, K photograph). C: Chencherere Hill, Dedza, fl. 18.i.1959, *Robson & Jackson* 1240 (K). S: South Kirk Range, Blantryre Dist., fl. 27.i.1956, *Jackson* 1823 (K).

Recorded from Malawi, the southern highlands of Tanzania, D.R. Congo and Angola. Upland grassland or *Brachystegia* woodland; 1500–2000 m.

8. **Pachycarpus appendiculatus** E. Mey., Comm. Pl. Afr. Austr.: 210 (1838). —Brown in Fl. Cap. 4(1): 721 (1908). Type: South Africa, Eastern Cape Province, between Morley and Umtata, *Drège* 4933 (B† holotype, K).

Xysmalobium appendiculatum (E. Mey.) D. Dietr., Syn. Pl. **2**: 902 (1840).
Gomphocarpus appendiculatus (E. Mey.) Decne. in Candolle, Prodr. **8**: 562 (1844).
Gomphocarpus macroglossus Turcz. in Bull. Soc. Imp. Naturalistes Moscou **21**: 259 (1848).
Type: South Africa, near Great Fish River between Kaffir's Drift and Governors Kop, *Ecklon* 34 (KW holotype).

Perennial herb with annual stem arising from a long cylindrical vertical rootstock; stems 0.2–0.6 m long, generally unbranched, erect, minutely pubescent. Leaves with petiole 5–14 mm long; lamina 4–10 × 1.5–5 cm, narrowly to broadly oblong or ovate, apex obtuse, usually shortly apiculate, base acute or rounded, margins scabrid, venation prominent with numerous parallel secondary veins, pubescent on both upper and lower surfaces with minute scabrid hairs. Inflorescences extra-axillary, sessile or subsessile, with 2–4 downward-facing flowers; pedicels 1.5–4 cm long, pubescent. Calyx lobes 6–8 × 3–4 mm, ovate, acute, pubescent. Corolla reflexed or recurved, minutely pubescent within, greenish white, often speckled with red; lobes10–20 × 5–15 mm, oblong or ovate. Staminal column sessile. Corona lobes dull green, dorsoventrally flattened, spreading, 10–23 mm long with an oblong base with a pair of vertical oblong wing-like keels arising from it near the base, the base narrowing abruptly then expanded again into an upturned lanceolate to ovate or suborbicular terminal appendage 6–12 mm long. Anther wings 3–7 mm long, margin strongly curved, the base spreading 5 mm from the column. Peduncle not elongating in fruit; fruiting pedicel contorted; follicle 10–11 × 4–6 cm, subglobose or ellipsoid and somewhat inflated, with 4 or 6 narrow longitudinal wings, glabrous. Mature seeds not seen.

Mozambique. M: Lebombo mountains c.1 km S of Goba Fronteira, fl. 23.xi.2001, *Goyder* 5000 (K, LMU).

Also known from Swaziland and South Africa (Mpumalanga, KwaZulu-Natal, Eastern Cape Provinces). Shallow soil over rock; c.400 m.

71. **PARAPODIUM** E.Mey.[43]

Parapodium E. Mey., Comm. Pl. Afr. Austr.: 221 (1838).
Rhombonema Schltr. in Bot. Jahrb. Syst. **20**(Beibl. 51): 41 (1895).

Stems simple, 1 or more, arising from a tuberous rootstock; latex white. Leaves oblong, linear or lanceolate, margins smooth or undulate and crisped. Inflorescences terminal and extra-axillary, umbelliform, sessile or pedunculate. Corolla with campanulate tube around base of gynostegium, lobed above. Corolline corona of five lobes adnate to the corolla below. Gynostegial corona absent. Anther appendages membranous. Pollinia attached terminally to the flattened translator arms. Stylar head conical. Follicles single by abortion, held ± erect by contortion of the fruiting pedicel, with longitudinal wings or ridges.

Three species in southern Africa.

Parapodium costatum E. Mey., Comm. Pl. Afr. Austr.: 222 (1838). Type: South Africa, Witberg, *Drège* s.n. (B† holotype, K). FIGURE 7.3.**194**.
Rhombonema luridum Schltr. in Bot. Jahrb. Syst. **20**(Beibl. 51): 41 (1895). Type: South Africa, Magaliesberg, 3.xi.1893, *Schlechter* 3610 (B† holotype, PRE0341733-0 lectotype, AMD, K, PRE, S), lectotype designated here.

Perennial herb with annual stem arising from a cylindrical vertical tuberous rootstock; stems 10–30 cm long, stout, simple, erect, minutely pubescent. Leaves with petiole 15–25 mm long; lamina 5–12(14) × 1–4 cm, narrowly to broadly oblong or lanceolate, apex subacute, obtuse or rounded, usually shortly apiculate, base cuneate, rounded or truncate, margins smooth, venation prominent with numerous parallel secondary veins, glabrous or sparsely pubescent except for the midvein beneath. Inflorescences with erect peduncle 0.5–2 cm long; pedicels 8–10 long, pubescent. Calyx lobes 6–8 × 1–2 mm, lanceolate or linear-lanceolate, acute, pubescent.

---

[43] by D.J. Goyder

Fig. 7.3.**194**. PARAPODIUM COSTATUM. 1, habit; 2, tuberous rootstock; 3, gynostegium and corona from above; 4, pollinarium; 5, gynostegium, showing conical stylar head, and corona, from the side. Drawn by Cynthia Letty. Reproduced with permission from Flowering Plants of South Africa (1931).

Corolla brownish green, minutely papillate; tube 2–3 mm deep; lobes 4–6 × 2–3 mm, oblong or ovate, erect with recurved tips. Corona white, adnate to corolla tube basally, with free upper margins, truncate, the lobes contiguous and forming a pentagonal disc around the gynostegium. Staminal column sessile, yellow-green. Anther wings 1.2 mm long, triangular, spreading from the column basally. Follicle (immature), narrowly ovoid, with 4 or 6 narrow longitudinal toothed wings, minutely pubescent. Mature seeds not seen.

Botswana. SE: Lobatsi, fl. xii.1924, *Tapscott* 2467 (K).

Also known from South Africa (NW Province, Gauteng, Limpopo, Mpumalanga, Free State).

72. **XYSMALOBIUM** R. Br.[44]

Xysmalobium R. Br., On the Asclepiadeae: 27 (1810).

Stems simple, 1 or more, arising from a tuberous rootstock; latex white. Leaves ovate to lanceolate or sublinear with lateral veins ± parallel and spreading to almost a right angle with the midrib, margins smooth, undulate or crisped. Inflorescences umbelliform, with erect peduncles. Corolla lobed almost to the base. Corolline corona absent. Gynostegial corona lobes fleshy, generally not strongly compressed either dorsally or laterally. Anthers raised on a variously developed gynostegial stalk formed from the anther filaments; anther appendages membranous. Pollinia attached terminally to the winged translator arms. Follicles inflated or not.

An unsatisfactory assemblage of c.30 species in tropical and southern Africa. Within the Flora region, *X. undulatum* seems only distantly related to the remaining species.

1. Corolla generally bearded at the tip; stems stout; leaves pubescent, margins scabrid to the touch . . . .**1.** *undulatum*
 – Corolla lobes not bearded apically; stems slender; leaves glabrous or pubescent, margins not scabrid 2
2. Corolla strongly reflexed 3
 – Corolla spreading or campanulate, not strongly reflexed 8
3. Leaves linear 4
 – Leaves broader 6
4. Corona lobes minute, entirely adnate to column . . . .**2.** *holubii*
 – Corona larger, free above 5
5. Corona lobes with flattened, suberect outer face, inner face with fleshy projection . . . .**5.** *stocksii*
 – Corona lobes subglobose with fleshy peg or tooth extending up side of staminal column . . . .**4.** *heudelotianum*
6. Corona lobes dorsally flattened or with a flattened outer face . . . . . .**8.** *fraternum*
 – Corona lobes subglobose but sometimes with an upward- or inward-pointing tooth 7
7. Subglobose portion of corona lobes 1–2 mm diameter, reaching to ± half-way up staminal column; translator arms flattened in proximal half, distal half terete . . . .**4.** *heudelotianum*
 – Subglobose portion of corona lobes 3–5 mm diameter, ± reaching top of staminal column; translator arms flattened, becoming broader distally . . . .**10.** *sessile*
8. Inflorescences pedunculate, peduncles 1.5–2.5 cm long; corolla lobes c.3 mm long . . . .**3.** *gramineum*
 – Inflorescences sessile, very rarely with occasional peduncle to 0.5 cm long; corolla lobes at least 4 mm long 9

---

[44] by D.J. Goyder

9. Corolla lobes 10–11 mm long; corpusculum with conspicuous wings . . **11.** *alatum*
– Corolla lobes 4–8 mm long; corpusculum not winged 10
10. Gynostegium with conspicuous stipe c.2.5 mm long; anther wings 0.5–1 mm long, tapering gradually into stipe . . . . **7.** *patulum*
– Gynostegium sessile or subsessile; anther wings (0.6)2–3 mm long 11
11. Anther wings vertical, extending well-below the anther sacs; fleshy apical portion of the corona lobes ± horizontal in the cavities bounded by anther wing tails . . . .**9.** *kaessneri*
– Anther wings wider at base than above, without basal extensions; corona lobes ± erect 12
12. Anther wings distinctly flared at base, outer margin curved . . . . **8.** *fraternum*
– Anther wings narrowly to broadly triangular, outer margin straight . . . .**6.** *rhodesianum*

1. **Xysmalobium undulatum** (L.) W.T. Aiton, Hort. Kew., ed.2 **2**: 79 (1811). —Goyder in F.T.E.A., Apocynaceae (part 2): 450 (2012). Type: "Apocynum Afric. lapathi folio" in Commelin, Hort. Med. Amstelae. Pl. Rar. Exot.: 16, t.16 (1706), iconotype designated by Wijnands in Bot. Commelins: 48 (1983). FIGURE 7.3.**195**.

Asclepias undulata L., Sp. Pl.: 214 (1753).

Xysmalobium angolense Scott Elliot in J. Bot. **28**: 365 (1890). Type: Angola, Huilla, Catumba, fl. i/ii.1860, *Welwitsch* 4171 (BM lectotype, K, LISU), designated here; nr. Huilla and Humpata, *Welwitsch* 4170 (K paralectotype, LISU).

Xysmalobium prismatostigma K. Schum. in Bot. Jahrb. Syst. **17**: 120 (1893). Type: Angola, Malange, fl. xi/xii.1879, *Mechow* 329 (B† holotype, K, M).

Woodia trilobata Schltr. in J. Bot. **33**: 337 (1895). Type: Kenya, Kavirondo, Nandi Range, *Scott-Elliot* 6877 (BM holotype, K).

Xysmalobium trilobatum (Schltr.) N.E. Br. in F.T.A. **4**(1): 306 (1902).

Xysmalobium dispar N.E. Br. in F.T.A. **4**(1): 307 (1902). Type: Malawi, Namasi, fl. 29.v.1899, *Cameron* 4 (K lectotype), designated here; Manganja Hills, near Mt. Soche, 8.iii.1862, *Kirk* s.n. (K paralectotype); Zimbabwe, Leshumo Valley, i.1876, *Holub* 669 & 816 (K paralectotypes).

Xysmalobium barbigerum N.E. Br. in F.T.A. **4**(1): 307 (1902). Type: Angola, Amboella, at the mouth of the R. Kuebe, *Baum* 332 (K holotype, ?B†).

Asclepias leucotricha Schltr. in Warburg, Kunene-Sambesi Exped.: 342 (1903). Type as for *X. barbigerum*.

Xysmalobium leucotrichum (Schltr.) N.E. Br. in F.T.A. **4**(1): 615 (1904).

Robust perennial herb with thick, carrot-like, tuberous rootstock. Stems 1 to several, (0.2)0.5–1.5 m tall, erect, unbranched, renewed annually, stout at least at the base, sparsely to densely pubescent with soft, spreading white hairs, the base of the stem often glabrescent. Leaves shortly petiolate; petiole 1–5(10) mm; lamina 7.5–19(27) × 0.5–8.5(12) cm, very variable in shape, linear or linear-lanceolate to ovate or ovate-lanceolate, apex acute or acuminate, base cordate, somewhat hastate, truncate or broadly rounded, margins crispate, undulate or smooth, often slightly revolute, lateral veins ± parallel, spreading to near 90° to the midrib but ascending slightly near the base, tips of lateral veins anastomosing to form a wavy, submarginal vein, indumentum a sparse to dense covering of spreading white hairs, particularlty on principle veins and near margins. Inflorescences umbelliform, 12–26-flowered, densely pubescent, peduncles arising laterally from the upper nodes, robust, 12–45(90) mm long, up to 3 mm thick. Pedicels slender, (8)14–18 mm. Sepals 4–8(11) mm long, 1–2.5 mm wide, about half as long to as long as the corolla, ovate to lanceolate, apex acute or attenuate, pubescent. Corolla cream or greenish white, often tinged with brown or pink outside, divided almost to the base; lobes 5–10 mm long, 2.5–6 mm wide, ovate, lower 2/3 glabrous, somewhat concave, upper 1/3 variably pubescent, usually lanate but occasionally glabrous or subglabrous, spreading, the apex acute, recurved. Corona very fleshy, ± tetrahedral in shape and attached by one of the apices to the base of the gynostegium, or stalk narrow for most of its length and widening abruptly to the ± triangular outer face, the outer face 2–4 mm wide, 2–3.5 mm high. Gynostegium raised on a distinct stalk

D.E.

Fig. **7.3.195**. XYSMALOBIUM UNDULATUM. 1, habit (× ¾); 2, fruit (× ¾); 3, 4, flowers (× 1½); 5, gynostegium and corona (× 3); 6, gynoecium showing paired ovaries and stylar head (× 3); 7, pollinarium (× 7). 1 from *Milne-Redhead & Taylor* 8231; 2 from *Milne-Redhead & Taylor* 8231A; 3–7 from *Milne-Redhead & Taylor* 1087. Drawn by D. Erasmus. Reproduced from Flora of Ethiopia and Eritrea (2003).

formed from the anther filaments; stalk 1–3(4) mm long, usually only partially obscured by the corona lobes but lobes occasionally overlapping base of anthers when gynostegial stalk short; anther wings jutting out abruptly from column, the anthers 2–3 mm tall. Pollinia 0.75–1.5 mm long, variable in shape from slender pear-shaped to short club-shaped, attached by the tip to the translator arm; translator arms narrowly winged in upper part, 0.5–1 mm long; corpusculum dark brown, ovoid, 0.5 mm long, 0.25 mm wide. Stylar head apex ± level with top of anthers. Fruit a single follicle, somewhat inflated or not, lanceolate or ovate in outline with an obtuse or attenuate apex, the entire surface covered with soft hairy prickles.

Zambia. N: N'Kali Dambo, fl. 8.xi.1954, *Richards* 2159 (K). W: Mwekera forestry training school, 20 km SE of Kitwe, fr. 25.iv.1989, *Goyder et al.* 3111 (K, NDO). C: Serenje Dist., Kundalila Falls, fl. 4.ii.1973, *Strid* 2901 (K). E: Nyika Plateau, fl. 2.i.1964, *Benson in N.R.*486 (K). **Zimbabwe**. C: Greendale, Harare [Salisbury] Dist., fl. 18.ii.1957, *Leach* 5958 (K, SRGH). E: Inyanga Dist., Mare Dam, fl. 4.iii.1969, *Jacobsen* 3729 (K, SRGH). S: 16 km from Masvingo [Fort Victoria], no date, *Cannell* 565 (K, SRGH). **Malawi**. N: Nyika Plateau, Chelinda, fl. 8.xii.1975, *Phillips* 584 (K, MO). C: Chongoni Forestry School, base of Chiwao Hill, fl. 4.ii.1959, *Robson* 1448 (K). S: Zomba Plateau, Mlunguzi Marsh, fl. 1.iii.1979, *Blackmore* 610 (K, MAL). **Mozambique**. Z: Manu Forest, Namuli Mountain, fl. 18.xi.2007, *Harris et al.* 377 (K, LMA).

Widespread across eastern and southern Africa from Ethiopia to South Africa. Fairly common on road or stream banks and in dambos, occasionally in montane grassland; 1000–2400 m.

A very variable species with distinctive local forms. A small-flowered variant with particularly broad leaves is found on the Nyika Plateau and the neighbouring Southern Highlands of Tanzania. Material from Namuli in northern Mozambique is also of this form.

2. **Xysmalobium holubii** Scott Elliot in J. Bot. **28**: 365 (1890). Types: Zimbabwe, Leshumo Valley, *Holub* s.n. (K syntype); Angola, Huilla, near Lopollo, xii.1859, *Welwitsch* 4175 (BM syntype, K).

   Xysmalobium decipiens N.E. Br. in Bull. Misc. Inform. Kew **1895**: 250 (1895). Type: Angola, Huilla, near Lopollo, *Welwitsch* 4175 (K holotype, BM).

   Xysmalobium tenue S. Moore in J. Bot. **50**: 358 (1912). Type: Angola, near Kassuango skirting the summit meadows along the Quiriri [Kuiriri], 14.x.1906, *Gossweiler* 3231 (BM holotype, BR).

Erect perennial herb with one or more slender or robust fusiform tubers; stems usually single and unbranched, 20–60 cm tall, pubescent with short curled hairs at least near the upper nodes. Leaves sessile or subsessile; lamina 6–18 × 0.1–0.4(1) cm, linear, acute, glabrous, subcarnose with prominent midrib and marginal veins. Umbels terminal and extra-axillary at up to 10 nodes, globose, with 20–45 sweetly scented flowers; peduncles (2)5–20 mm long, pubescent, often in one line. Pedicels 4–9 mm long, pubescent, often in one line. Calyx lobes 1.5–2 × 0.5 mm, lanceolate, acute, strongly reflexed. Corolla yellow-green to white, occasionally tinged with pink, strongly reflexed, lobed ± to the base, lobes 3–4.5 × 1–1.5 mm, oblong, the tips concave, acute, glabrous on the upper surface except for the minutely papillate base, lower surface glabrous. Corona lobes minute, adnate to the staminal column and arising in the hollow at the base of the anthers, 0.5–1 × 0.4–1 mm, resembling a narrow downward-pointing tongue when dry, narrowly to broadly ovoid when fresh, free or connected at the base by an acute tooth, yellow-green or brown. Gynostegium 2.5–3 × 1.5–2 mm, cylindrical or weakly conical-truncate. Anther wings extending well-below the anthers, the base somewhat flared, tails to the anther wings contorted or not. Corpusculum 0.3–0.4 mm long, black; translator arms 0.3–0.4 mm long, uniformly flattened for their entire length, contorted distally; pollinia 0.6–0.9 × 0.2–0.3 mm Apex of stylar head raised slightly above anthers, pentagonal, depressed slightly in the centre. Follicles erect, 9–14 × c.0.5 cm, narrowly fusiform, slender attenuate at apex and base, smooth, glabrous.

Zambia. N: 8 km N of Kasama, Misamfu, fl. 7.i.1967, *Anton-Smith* s.n. in SRGH 181841 (K, SRGH). W: Kitwe, fl. 26.x.1955, *Fanshawe* 2557 (K, NDO). S: Namwala, fl. 8.i.1957, *Robinson* 2045 (K). **Zimbabwe**. W: Leshumo Valley, *Holub* s.n. (K). **Malawi**. N: Mzimba Dist., 6 km SW of Mzuzu, fl. & fr. 11.i.1976, *Pawek* 10698 (K, MAL).

Also known from Angola, Gabon and southern D.R. Congo. Found in seasonally waterlogged dambo vegetation; 1000–1500 m.

Most Angolan material has more clearly defined teeth linking the corona lobes than elsewhere in the species. The anther wings are also slightly shorter and more contorted at the base than in other material. These collections represent Brown's *Xysmalobium decipiens*, but differences between them and the rest of *X. holubii* break down in western Zambia and southern D.R. Congo.

Material previously recognised under the name *X. tenue* is smaller in most floral organs than other material with a reduced corona and more slender leaves. I believe it is best treated as a depauperate form of the species.

3. **Xysmalobium gramineum** S. Moore in J. Bot. **40**: 254 (1902). Type: Zimbabwe, Bulowayo, early i.1898, *Rand* 193 (BM holotype, K).

Slender erect herb c.10 cm tall. Leaves shortly petiolate; petiole 0.2–0.4 cm long; lamina 4.5–6.5 × 0.1–0.4 cm, linear, acute, margins revolute. Umbels extra-axillary, shorter than the leaves, minutely pubescent; peduncles 1.5–2.5 cm long with c.10 flowers; pedicels c.5 mm long. Calyx lobes c.2 × 1 mm, triangular, acute, pubescent outside. Corolla united at the base for c.0.5 mm; lobes c.3 × 1.5 mm, ovate, concave below, the margins revolute above, apex sub-acute, reflexed, lower half glabrous on both surfaces, upper half minutely papillate within, pubescent outside. Corona lobes fleshy, c.0.6 mm long, 0.4 mm wide, 0.3 mm thick, spreading outwards from the base of the staminal column and connected to each other by distinct teeth c.0.1 mm long. Staminal column c.1.4 mm long; anther wings 0.8 mm long, ± triangular, widest 0.4 mm from the top of the column. Pollinaria minute, corpusculum 0.2 × 0.06 mm, oblong, black; translator arms c.0.2 mm long, flattened, contorted, c.0.05 mm wide distally; pollinia 0.5 × 0.2–0.3 mm, obovate. Stylar head with conical apex exserted from top of staminal column for c.0.5 mm. Follicles not seen.

Zimbabwe. W: Bulowayo, fl. i.1898, *Rand* 193 (BM, K). Apparently, known only from the type.

4. **Xysmalobium heudelotianum** Decne. in Candolle, Prodr. **8**: 520 (1844). —Goyder in F.T.E.A., Apocynaceae (part 2): 453 (2012). Type: 'Senegambia', *Heudelot* s.n. (P holotype). FIGURE 7.3.**196**.

   Xysmalobium reticulatum N.E. Br. in Bull. Misc. Inform. Kew **1895**: 251 (1895); in F.T.A. **4**(1): 303 (1902). Type: Malawi, Shire Highlands, *Buchanan* s.n. (K holotype).
   Xysmalobium schumannianum S. Moore in J. Bot. **39**: 259 (1901). Type: Kenya, Machakos, 1896, *Hinde* s.n. (BM holotype, K).
   Schizoglossum truncatulum K. Schum. in Engler, Pflanzenw. Ost-Afrikas **C**: 322 (1895). Type: Uganda, Mpororo, Kanjanaberge, 18.iv.1891, *Stuhlmann* 2096 (B† holotype, K).
   Schizoglossum heudelotianum (Decne.) Roberty in Bull. Inst. Franc. Afr. Noire **25**: 1430 (1953).

Erect or occasionally decumbent perennial with an erect carrot-shaped tuber c.3–5 cm long, up to 3 cm wide; stems 1 or few, simple or branched, (10)20–60 cm, pubescent with short curled hairs. Leaves erect or spreading, shortly petiolate, petiole 1–2(5) mm long; lamina (1.5)4–8(17) × 0.2–2(3) cm, ovate to oblong- or linear-lanceolate, rarely linear or obovate, apex obtuse with a short mucro or acute, rarely retuse, the base usually truncate or slightly cordate, occasionally rounded or subcuneate, margins smooth or undulate, secondary and often tertiary venation prominent, often raised, reticulate, with a glabrous or pubescent marginal nerve, remainder of lamina glabrous or with scattered reddish hairs below. Umbels sessile or very rarely pedunculate,

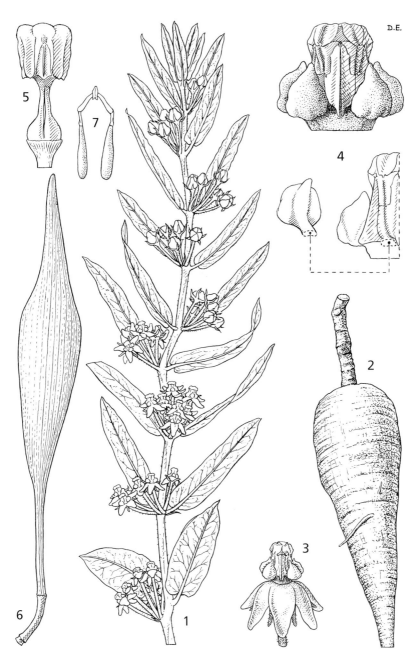

Fig. 7.3.**196.** XYSMALOBIUM HEUDELOTIANUM. 1, flowering stem (× 1); 2, tuber (× 1); 3, flower (× 3); 4, gynostegium and corona, and with one corona lobe detached (× 8); 5, gynoecium showing paired ovaries and stylar head (× 8); 6, follicle (× 1); 7, pollinarium (× 16). 1, 3–7 from *Richards* 3410; 2 from *Dalziel* 693. Drawn by D. Erasmus. Reproduced from Flora of Tropical East Africa (2012).

extra-axillary with 4–10(15) flowers at few or often at many of the nodes, all umbels apparently on one side of the stem. Pedicels 4–15(18) mm long, pubescent. Calyx lobes reflexed slightly by the corolla, c.3 mm long, 1 mm wide, triangular, acute, pubescent. Corolla cream or white to reddish brown, lobed almost to the base, reflexed, lobes 4–6(7) mm long, 2–3.5 mm wide, ovate, acute, glabrous but minutely papillate at base on inner surface, outer surface glabrous or with scattered reddish hairs in the upper half. Corona lobes green or brownish, on short spreading stalks arising near the base of the column, subglobose, 1–2 mm diam. with a smooth or papillate tooth or beak pointing up the side of the anthers and reaching from ½ to the full height of the column. Gynostegium 3–4 mm long, gynostegial stalk up to 1 mm long. Anther wings narrowly triangular giving the column a subcylindric or slightly conical appearance. Stylar head truncate, raised slightly above the tip of the anthers. Corpusculum 0.2–0.3 mm long, ovoid, dark brown; translator arms weakly geniculate at the middle, proximal half winged on both sides or just the lower one, the winged portion (0.2)0.3–0.5 mm long, 0.1–0.2 mm wide, translucent, distal portion (0.2)0.4 mm long, slender, terete or sometimes with a narrow translucent wing; pollinia 0.8–1.2 mm long, 0.2–0.4 mm wide, oblanceolate. Follicles erect, narrowly fusiform, long attenuate at both apex and base, smooth and minutely pubescent.

Zambia. N: Isoka, fl. 21.xii.1962, *Fanshawe* 7203 (K, NDO). C: Mt. Makulu, 20 km. S. of Lusaka, fl. 17.xii.1956, *Angus* 1467 (K; NDO). E: 13 km. E. of Chipata [Fort Jameson], fl. 7.i.1959, *Robson* 1058 (BM, K). S: Machili, fl. 22.xii.1960, *Fanshawe* 6000 (K, NDO). **Zimbabwe**. W: Gwaai River, fl. i.1906, *Allen* 227 (K). C: Rusape, fl. 5.ii.1949, *Munch* 145 (K, SRGH). E: Mutari Dist., Commonage, E. of Palmerston, fl. 4.xii.1951, *Chase* 4229 (BM, K, SRGH). **Malawi**. N: Mzimba Dist., 10 km. E. of Mzambazi, fl. 31.i.1976, *Pawek* 10762 (K, MAL, MO, UC). C: Dedza Dist., Chongoni Forest, fl. 19.xii.1966, *Salubeni* 472 (K, MAL, SRGH). S: Zomba, fl. xii.1900, *Manning* 58 (K). **Mozambique**. N: Marrupa Dist., Murrupula, fl. 16.i.1961, *Fidalgo de Carvalho* 443 (K, LMA).

Scattered records from West Africa to Sudan and Kenya, and south to central and eastern Zimbabwe. Usually found in dambos or other seasonally burned grassland, less commonly in scrub or open woodland; 800–1500 m.

5. **Xysmalobium stocksii** N.E. Br. in Bull. Misc. Inform. Kew **1913**: 302 (1913). — Goyder in F.T.E.A., Apocynaceae (part 2): 455 (2012). Type: Mozambique, Ibo neighbourhood, *Stocks* s.n. (K holotype).

Erect perennial herb with a globose tuber 2–3 cm diam., 2–3 cm below soil surface; stems one or two, 20–50 cm, pubescent with short curled hairs. Leaves erect or spreading, petiolate or sessile, petiole up to c.1 cm long; lamina 5.5–11(19) × 0.2–0.8(1.2) cm, linear with an acute apex, tapering gradually at the base, veins not visible above, inconspicuous or clearly reticulate below, glabrous. Umbels sessile, terminal or extra axillary with 3–8 flowers at each of the upper 1–6 nodes. Pedicels 5–17 mm long, densely pubescent on one side. Calyx lobes 2–4 mm long, 0.5–1.0 mm wide, ovate to lanceolate, acute, reflexed by the corolla. Corolla reflexed, globose in bud, lobed almost to the base; lobes 6–7 mm long, 3–3.5 mm wide, ovate, acute, the underside glabrous or with few scattered reddish hairs towards the tip, upper surface minutely papillate or with white hairs especially near the base of the lobe. Corona lobes suberect, reaching from ½ to about equalling the height of the column, outer face flattened, 2–4 mm long, c.2 mm wide, ovate or obovate, obtuse or rounded at the apex, the inner face with a ± pyramidal swelling c.1.5 mm long in the upper half. Gynostegium c.4 mm long, anther wings narrowly triangular, not spreading markedly at the base, the column appearing ± cylindrical. Corpusculum c.0.3 mm long, ovoid, black; translator arms geniculate, proximal half winged on lower side, the winged portion 0.4–0.5 mm long, 0.2–0.3 mm wide, obovate, distal portion 0.4–0.5 mm long, slender, terete; pollinium 1.1–1.3 mm long, c.0.3 mm wide, oblanceolate. Follicles not seen.

Zambia. N: Kawambwa-Johnson Falls road, fl. 4.xii.1961, *Richards* 15505 (K). **Mozambique**. N: Ibo neighbouhood, *Stocks* s.n. (K). Z: Quelimane Dist., Lugela, Moebede road, fl. 31.xii.1948, fr. iv.1949, *Faulkner* K108 (K).

Also known from eastern Kenya and Tanzania, and Angola. Burned grassland and degraded *Brachystegia* woodland; 300–1100 m.

6. **Xysmalobium rhodesianum** S. Moore in J. Bot. **47**: 215 (1909). Type: Zambia, Chibinga Stream, 21.x.1907, *Kässner* 2079 (BM holotype).

> *Xysmalobium clavatum* S. Moore in J. Bot. **50**: 360 (1912). Types: Angola, in primeval woods between Cuansha and the Mungombe rivulet, fl. 25.xi.1905, *Gossweiler* 2205 (BM lectotype), designated here; Kuiriri, fr. 23.i.1906, *Gossweiler* 3435 (BM paralectotype, K).

Erect perennial herb with short napiform tuber; stems 1–3, 25–40 cm tall, pubescent with short curled hairs. Leaves petiolate or sessile; petiole up to 6 mm long; lamina (2)4–8.5 × 0.2–3 cm, linear to ovate, the lowest leaves shorter and relatively broader than the upper ones, apex obtuse or acute, cuneate at the base, veins prominent, reticulate, glabrous or with scattered reddish hairs on both surfaces. Umbels sessile or rarely with peduncle to 5 mm long, extra-axillary with 5–9 flowers at each of the middle and upper nodes. Pedicels 5–9 mm long, pubescent. Calyx lobes c.3 × 1 mm, ovate, acute, pubescent. Corolla campanulate, yellowish green, lobed almost to the base; lobes c.4.5 × 2.5 mm, broadly ovate, acute, under surface with scattered hairs towards the tip, minutely papillate towards the base on the upper surface. Corona lobes reaching ± to the top of the anthers, fleshy but somewhat flattened dorsally, spreading outwards from the base of the staminal column for c.0.5 mm then becoming erect for c.1–1.5 mm, outer face obovate, rounded at the apex, thickened in the upper half, sometimes toothed at the basal end of the thickening. Gynostegium c.1.5 mm tall, anther wings widest at the base. Pollinarium with corpusculum c.0.3 × 0.1 mm, black, ± cylindrical; translator arms 0.4 × 0.1 mm, translucent, not differentially thickened or contorted; pollinia 0.6–0.7 × 0.3–0.4 mm, triangular to semilunate, widest near the middle. Follicles not seen.

Zambia. N: Mbala [Abercorn] Dist., Sunzu Mt., fl. 9.i.1955, *Richards* 3988 (K). C: 10 km. E of Lusaka, fl. 9.i.1958, *King* 398 (K). E: 80 km. S of Lundazi, fl. xii. 1971, *Williamson* 2086 (K, SRGH). **Malawi**. N: Mzimba Dist., 12 km SW of Chikangawa, fl. 20.i.1979, *Phillips* 4654 (K, MO). C: Kasungu, Kapopo village, fl. 15.i.1959, *Jackson* 2298 (K).

Brachystegia woodland or open grassland; 1200–1800 m.

Very close to *X. patulum* but the gynostegium is not raised on a conspicuous stipe and the translator arms are translucent and not differentially thickened.

7. **Xysmalobium patulum** S. Moore in J. Bot. **47**: 215 (1909). —Goyder in F.T.E.A., Apocynaceae (part 2): 457 (2012). Type: Zambia, under trees at the Katanino [Katinina] Hills, *Kässner* s.n., with 2167 (BM holotype).

Erect perennial herb with short napiform tuber; stems 1 or 2, c.35 cm tall, pubescent with short curled hairs. Leaves subsessile to shortly petiolate; petiole to c.0.3 cm long; lamina 2.5–6 × 0.6–1.3 cm, narrowly oblong to oblanceolate, apex obtuse to subacute, tapering gradually at the base; veins prominent, reticulate, glabrous except for occasional hairs on the marginal nerve. Umbels sessile, extra-axillary, with 5–9 flowers at each of the middle and upper nodes. Pedicels 5–8 mm long, pubescent. Calyx lobes c.2.5 × 1 mm, ovate, acute, pubescent. Corolla campanulate, green tinged with purple, lobed almost to the base; lobes 5 × 2 mm, ovate, acute, pubescent with scattered hairs on lower surface, upper surface minutely papillate especially towards the base. Corona lobes green, reaching ± to the base of the anthers, fleshy but somewhat flattened dorsally, spreading outwards from the base of the staminal column for c.0.5 mm then becoming erect for c.2 mm, the outer face obovate, rounded at the apex, thickened in the upper half, sometimes toothed at the basal end of the thickening. Gynostegium c.2.5 mm tall, the filaments forming an obconic stipe, widest point of anther wings c.0.5 mm from the top of the column, narrowing gradually below into the stipe. Pollinarium with corpusculum 0.4 × 0.3 mm, ovoid, black; translator arms c.0.3 × 0.1 mm, slightly contorted and differentially thickened; pollinia 0.6 × 0.3 mm, triangular, widest around the middle. Apex of stylar head flat, ± level with the anther tips. Follicles not seen.

Zambia. W: 1 km. N of Mwinilunga, fl. 23.xi.1937, *Milne-Redhead* 3351 (K).

Also known from southern Tanzania (T7) and the Katanga region of D.R. Congo. *Brachystegia* woodland; c.1500 m.

This species differs from *Xysmalobium heudelotianum* and other species in the corolla lobes which are not reflexed, the stipitate gynostegium and the form of the anther wings, the spathulate corona lobes and the pollinarium with differentially thickened translator arms. The form of the gynostegium and the pollinarium distinguish it from *X. rhomboideum* N.E. Br.

The type differs slightly from other material in the attitude of the basal portion of the corona lobes, which is more erect, and the upper portion, which is held in a more horizontal position than in other collections of this species. *Pawek* 3112 (MAL), from the Misuku hills of northern Malawi, may also belong to this species.

8. **Xysmalobium fraternum** N.E. Br. in Bull. Misc. Inform. Kew **1895**: 252 (1895); in F.T.A. **4**(1): 305 (1902). —Goyder in F.T.E.A., Apocynaceae (part 2): 455 (2012). Type: Malawi, Shire Highlands, near Blantyre, *Last* s.n. (K holotype).

Erect perennial herb with a subglobose to carrot-shaped tuber; stems usually single, 30–55 cm tall, pubescent with short curled hairs. Leaves petiolate or rarely subsessile; petiole to 8 mm long; lamina 2.5–6.5 × 1.5–3.5 cm, often oblong or ovate in lowest leaves, other leaves obovate, rounded to obtuse or subacute, occasionally slightly retuse, shortly mucronate or not, cuneate at the base, sometimes narrowly so, veins prominent below, reticulate, glabrescent, young leaves with scattered rusty hairs below or on both surfaces. Umbels sessile, extra-axillary, with 4–8 flowers at few or many nodes. Pedicels 5–19 mm long slender, pubescent. Calyx lobes 3–4 mm long, 1–1.5 mm wide, lanceolate, acute, reflexed by the corolla. Corolla campanulate or rotate-campanulate, white or yellowish green, globose in bud, lobed almost to the base; lobes 5 mm long, 2–3 mm wide, ovate to oblong, acute, the underside glabrous or with scattered reddish hairs on the upper half, upper surface minutely papillate or with white hairs near the base of the lobe. Corona lobes about as high as the column, spreading outwards for c.2 mm then arching upwards and inwards to the head of the gynostegium, the outer face broadly ovate, c.4 mm high, 5 mm wide with a subacute apex, the inner face with a pubescent or glabrous ridge running down the inner face from the fleshy apex. Gynostegium 3–4 mm high, anther wings ± triangular, the base spreading outwards from the column. Corpusculum 0.4–0.5 mm long, ovoid, black; translator arms 0.4–0.8 mm long, flattened ± uniformly along the entire length; pollinia 0.6–0.8 × 0.4–0.6 mm, triangular. Apex of stylar head flat, level with or raised slightly above anthers. Follicles erect, c.25 cm long, smooth, pubescent, the lower half forming a narrow 'stipe', the upper half narrowly fusiform.

Zambia. N: Mbala (Abercorn), fl. 22.xii.1949, *Bullock* 2124 (K, NDO). **Zimbabwe**. E: Chipinga Dist., Gungunyana Forest Reserve, fl. i.1962, *Goldsmith* 12/62 (SRGH). **Malawi**. N: Karonga Dist., Vinthukhutu Forest Reserve, 3 km. N of Chilumba, fl. 6.i.1978, *Pawek* 13557 (K, MAL, MO). S: Shire Highlands, near Blantyre, fl. 1887, *Last* s.n. (K). **Mozambique**. Z: Mocuba Dist., Namagoa, 200 km. inland from Quelimane, fl. & fr. xii-i.1943, *Faulkner* 358 (K, PRE).

Also in southern Tanzania. *Brachystegia* woodland; mostly between 1500–1800 m, but collected at c.100 m in Mozambique.

9. **Xysmalobium kaessneri** S. Moore in J. Bot. **46**: 295 (1908). —Goyder in F.T.E.A., Apocynaceae (part 2): 456 (2012). Type: Zambia, Sangolo Spruit, 23.x.1907, *Kässner* 2104 (BM holotype, K).

Erect perennial herb arising from a vertical, woody, napiform tuber. Stems usually single, 40–70 cm tall, densely pubescent with short curled hairs. Leaves sessile or with petiole up to 1 cm long; lamina 3.5–7(9) × (0.7)1–4.5 cm, oblanceolate, obovate, oblong or elliptic, lower leaves occasionally broadly ovate or suborbicular, apex rounded, retuse or, particularly in upper leaves, acute, base rounded to cuneate, margins smooth or slightly crisped, veins prominent especially on lower surface, reticulate, secondary veins at an acute angle to the midrib, marginal nerve present, indumentum of white hairs concentrated mainly on the midrib and major veins. Umbels sessile, extra-axillary, with 5–8 flowers at each of the upper and middle nodes. Corolla campanulate, divided ± to the base, green tinged with purple outside, paler within; lobes c.7 × 3.5 mm, ovate, acute, outer face pubescent, inner surface minutely pubescent. Corona lobes shorter than the staminal column, basal portion of each lobe forming a fleshy claw c.1 × 0.5 mm spreading upwards and outwards from the base of the column, distal portion spathulate, very fleshy, spreading ± horizontally from the cavity between adjacent pairs of anther wings, 2–3 mm long, c.2 mm wide and 1 mm thick, basal margin often produced into a small backward-pointing tongue. Gynostegium c.3.5 mm high, cylindrical, the anther wings ± vertical. Pollinaria robust; corpusculum c.0.5 × 0.4 mm, ovate, black; translator arms c.0.7 × 0.3 mm flattened, dark brown, the distal end ± as wide as the proximal end of the pollinium; pollinia 0.7 × 0.4 mm, oblong-obtriangular. Apex of stylar head c.3 mm diam., flat, ± level with the top of the column. Follicles up to c.30 cm long, 2 cm diam. at the widest point, narrowly fusiform, erect, finely rusty-pubescent, only one of the pair developing.

Zambia. N: 12 km. E of Kasama, fl. 19.i.1961, *Robinson* 4320 (K). C: Chakwenga headwaters, c.100–129 km. E of Lusaka, fl. 7.i.1964, *Robinson* 6140 (K). S: Mapanza, fl. 24.xii.1956, *Robinson* 1998 (K). **Zimbabwe**. E: Watsomba Mt range, fr. 14.iv.1956, *Pole-Evans* 4988 (K) & fl. 17.i.1958, *Pole-Evans* 5349 (K). **Malawi**. N: Nyika Plateau, fr. 11.iii.1978, *Pawek* 14084 (K, MAL). C: Mchinji [Fort Manning], fl. 7.i.1959, *Robson* 1079 (K).

Also known from southern Tanzania and D.R. Congo. *Brachystegia* woodland; 1200–1600 m.

Cameron 133 and *Richards* 15472 from Zambia may belong here, but the corona lobes differ slightly in the shape and orientation of the basal and apical portions. Several collections from D.R. Congo are similarly in this regard, but I have not detected discrete differences, and individuals within a single population may vary in length of anther wing.

10. **Xysmalobium sessile** (Decne.) Decne. in Candolle, Prodr. **8**: 519 (1844). Type: Angola, Benguella Plateau, *da Silva* s.n. (P holotype, K).

  Gomphocarpus sessilis Decne. in Ann. Sci. Nat., Bot. sér. 2 **9**: 325 (1838).

Erect perennial herb with vertical, woody, napiform tuber. Stems usually single, to c.1 m long, pubescent with short curled hairs. Leaves sessile or subsessile; lamina 5–10 × 2–5.5 cm, ovate or sometimes upper leaves oblong, apex rounded to subacute, base rounded to cordate, veins prominent, raised on lower surface, reticulate, secondary veins ± parallel, initially at right-angles to midrib, marginal veins and midrib pubescent, rest of lamina glabrous or with few scattered hairs. Umbels sessile, extra-axillary, with 4–6 flowers at each of the upper and middle nodes. Pedicels 12–25 mm long, pubescent. Calyx lobes 7–10 × 1.5–4 mm, reflexed, linear to lanceolate, acute, pubescent outside. Corolla pink, united for c.1 mm at the base; lobes reflexed, 9–10 × 4.5–6 mm, ovate, acute, outer face pubescent, inner face with fine, short white hairs. Corona lobes about as high as the column, basal portion of each lobe forming a flattened fleshy claw c.2 × 2 mm, spreading from the base of the column, distal portion ± erect, fleshy, 3–5 mm long, 3–4.5 mm wide and up to 3 mm thick, sometimes with a vertical linear depression running down the midline of the ovate to orbicular outer face and a variously developed vertical fleshy ridge or peg arising from the base of the inner face. Gynostegium 4 mm high, conical, the anther wings spreading at c.45° from the column. Pollinaria robust; corpusculum 0.6 × 0.4 mm, ovate, black; translator arms 0.8 × 0.5 mm, oblong-obtriangular, flattened, the distal end wider than the proximal end of the pollinium, dark brown; pollinia 0.8 × 0.5 mm, obovate-oblong. Apex of stylar head flat, c.3 mm diam., ± level with the top of the column. Follicles (immature) narrowly fusiform, erect, with 6 longitudinal wings running the length of the follicle, toothed above, rusty pubescent, only one of the pair developing.

Zambia. W: Mufulira, fl & fr. 18.i.1973, *Fanshawe* 11746 (K, NDO). Ndola, fl. 12.i.1955, *Fanshawe* 1798 (K, NDO). C: Mulungushi River, fl. 20.vii.1907, *Kässner* 2065 (BM, K).

Also known from Angola and the Shaba province of D.R. Congo; 1100–1300 m.

Xysmalobium sessile is very close to *X. rhomboideum* N.E. Br. and *X. andongense* Hiern from Angola and D.R. Congo but is larger in all respects and differs in the possession of reflexed corolla lobes. The corona lobes of all three species share the same basic plan, but are shorter in *X. rhomboideum* and less fleshy in *X. andongense*. The pollinaria of *X. rhomboideum*, however, are quite different from those of *X. sessile*, with very slender translator arms. In *X. andongense* they are slightly broader, but still much narrower than in *X. sessile*. The anther wings of *X. rhomboideum* are more broadly triangular than those of either *X. sessile* or *X. andongense*, giving a more conical look to the column.

11. **Xysmalobium alatum** Goyder, sp. nov. Unique in the genus in possessing a conspicuously winged corpusculum. It can also be distinguished from allied species by the combination of sessile inflorescences and corolla lobes longer that 10 mm. Type: Zambia, Kitwe, Mwambashi, *Mutimushi* 207 (K holotype, NDO).

Erect tuberous perennial herb 20 cm to ? 1 m, pubescent with long white hairs. Leaves petiolate, petiole to 7 mm long; lamina 4–11.5 × 1.7–5 cm, the lowest pair of leaves smaller, ovate, other leaves elliptic or obovate, apex obtuse to rounded or slightly retuse, shortly mucronate or not, cuneate at the base, veins prominent or not below, pubescent with a sparse or dense indumentum of spreading white hairs on both surfaces. Umbels sessile, extra-axillary, with 3–4 flowers at few or many nodes. Pedicels 10–20 mm long, pubescent. Calyx lobes 6–8 × 1.5 mm, lanceolate, acute, pubescent. Corolla campanulate, white or yellowish green, lobed ± to the base; lobes 10–11 × 4–5 mm, ovate, acute, lower surface finely pubescent, upper surface glabrous or minutely papillate towards the base. Corona lobes fleshy, reaching about half way up anthers, spreading outwards for 2–3 mm than arching upwards towards the anthers, the outer face c.6 × 3 mm, broadly ovate and slightly V-shaped in section with a subacute apex. Gynostegium c.5–6 mm high with conspicuous stipe 2–2.5 mm high, c.2 mm diam. at the base, widening to 3 mm immediately below the anthers. Anthers forming a drum c.5 mm diam., c.3 mm high, anther wings ± vertical. Corpusculum c.0.6–0.8 mm long, subglobose, black with a pair of spreading, ± translucent, triangular brown wings c.0.6 × 0.4 mm; translator arms 0.6–0.9 mm long, obtriangular, c.0.1 mm wide at the proximal end, 0.6 mm wide distally, somewhat thickened or contorted near attachment to pollinia; pollinia obovate-oblong, c.1 mm long, 0.4 mm wide proximally, 0.6 mm at the widest point. Apex of stylar head flat, level with top of anthers. Follicles not seen.

Zambia. W: Kitwe, Mwambashi, fl. 4.xii.1962, *Mutimushi* 207 (K, NDO).
Also known from near Kolwezi (*Paterson* 509, K) and Lubumbashi (*Schmitz* 3649 & 5216, BR) in the Katanga region of D.R. Congo. *Brachystegia* woodland; c.1300 m.
The winged corpusculum is unique within the genus.

73. GLOSSOSTELMA Schltr.[45]

Glossostelma Schltr. in J. Bot. **33**: 321 (1895). —Goyder in Kew Bull. **50**: 527–555 (1995).

Slender or robust, erect, perennial herbs with a short, often stout, vertical rhizome and several fusiform, tuberous, lateral roots. Stems usually solitary, occasionally more than one, unbranched, with a line of pubescence running along the stem and alternating at the nodes, or rarely glabrous. Leaves usually petiolate, the petiole weakly channelled; lamina linear to ovate or spathulate. Inflorescences umbelliform, with or without a peduncle; umbels extra-axillary, arising terminally and laterally at the upper 1–4(6) nodes. Corolla medium to large, campanulate, lobed almost to the base. Corolline corona absent. Gynostegial corona of 5 lobes, often fleshy, sometimes dorsally flattened, united very briefly at the base and attached to the gynostegium near the base of the anthers; gynostegium usually with a conspicuous stalk below the attachment of corona lobes. Stylar head truncate, 5-angled, generally ± level with the top of the anthers and

---

[45] by D.J. Goyder

largely obscured by the anther appendages. Corpusculum ovoid, brown or black; translator arms flattened, broadened and geniculate near the attachment of the pollinia, held ± at right angles to the axis of the corpusculum; pollinia pendulous, flattened, falcate-oblong to obtriangular. Fruiting pedicel not contorted; follicles erect, lanceolate or ovate-lanceolate in outline, with an attenuate apex, smooth, glabrous or pubescent, usually only one of the pair developing. Seeds flattened with one face convex, ovate to suborbicular in outline with a narrow convoluted rim surrounding a verrucose disc.

12 species, distributed mostly in a narrow belt of southern tropical Africa from Angola and D.R. Congo to Mozambique and southern Tanzania. Generally found in the 'wetter miombo woodland' vegetation of White (1983) or adjacent regions of afromontane vegetation.

1. Corona lobes forming short fleshy pegs at base of anther wings; pollinia broadest distally . . . . **8.** *brevilobum*
   – Corona lobes erect, of various forms, reaching ± to the top of the staminal column or beyond; pollinia broadest at attachment of translator arms 2
2. Leaves linear 3
   – Leaves lanceolate to ovate or spathulate 5
3. Corolla lobes c.5 mm long; longest leaves < 80 mm long . . . . **5.** *nyikense*
   – Corolla lobes 10–17 mm long; most leaves > 80 mm long 4
4. Corona lobes dorsally compressed for their entire length; neither bulbous nor fleshy at the base but margins sometimes inrolled below; margins ± parallel, not narrowing abruptly above middle of lobe . . . . **1.** *carsonii*
   – Corona lobes not dorsally compressed, or compressed above only; bulbous or fleshy at the base and narrowing abruptly into a tongue half way up . . . .
   . . . . **2.** *lisianthoides*
5. Corona lobes 7–11 mm long, overtopping the head of the column for at least half their length 6
   – Corona lobes 2–5 mm long, not overtopping the head of the column 7
6. Tip of corona lobes dorsally flattened with a downward-pointing apical flap on inner face . . . .**7.** *cabrae*
   – Tip of corona lobes solid, clavate, without an apical flap . . . . **6.** *ceciliae*
7. Corolla lobes 5–7 mm wide; corona lobes c.5 mm long, upper portion produced into a hood over head of column . . . . **4.** *rusapense*
   – Corolla lobes 8–14 mm wide; corona lobes c.3 mm long, not produced into a hood . . . . **3.** *spathulatum*

1. **Glossostelma carsonii** (N.E. Br.) Bullock in Kew Bull. **7**: 415 (1952). —Goyder in F.T.E.A., Apocynaceae (part 2): 458 (2012). Type: Zambia, Tanganyika Plateau, Fife Station, *Carson* s.n. (K holotype). FIGURE 7.3.**197**.

   Xysmalobium carsonii N.E. Br. in Bull. Misc. Inform. Kew **1895**: 250 (1895).
   Asclepias mashonensis Schltr. in J. Bot. **33**: 356 (1895). Type: Zimbabwe, Mashonaland, 1893, *Folliott-Darling* s.n. (B† holotype).
   Schizoglossum lividiflorum K. Schum. in Bot. Jahrb. Syst. **28**: 454 (1900). —Brown in F.T.A. **4**(1): 367 (1902). Type: Tanzania, Iringa [Uhehe] Dist., Matanana Plateau, iii.1899, *Goetze* 741 (B† holotype).
   Gomphocarpus chlorojodina K. Schum. in Bot. Jahrb. Syst. **30**: 383 (1901). Type: Tanzania, Unyiha [Unyika] Plateau, Umalila, 22.x.1899, *Goetze* 1360 (B† holotype, K).
   Schizoglossum carsonii (N.E. Br.) N.E. Br. in F.T.A. **4**(1): 366 (1902). —Weimarck in Bot. Not. **1935**: 393 (1935).
   Schizoglossum chlorojodinum (K. Schum.) N.E. Br. in F.T.A. **4**(1): 366 (1902).
   Schizoglossum kassneri S. Moore in J. Bot. **50**: 362 (1912). Type: D.R. Congo, Kasai Dist., Lubi R., ix.1910, *Kässner* 3303 (BM holotype).

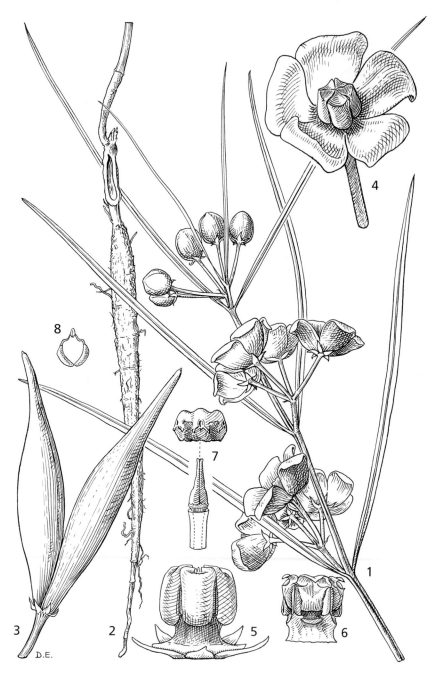

Fig. 7.3.**197**. GLOSSOSTELMA CARSONII. 1, flowering shoot (× 1); 2, rootstock (× 1) (note laterals appear to have been lost from this rootstock); 3, paired follicles (× 1); 4 flower (× 2); 5, gynostegium with corona (× 4); 6, gynostegium with corona removed (× 4); 7, gynoecium showing paired ovaries and stylar head (× 4); 8, pollinarium (× 7). 1 from *Milne-Redhead & Taylor* 7914; 2 from *Milne-Redhead & Taylor* 7914A; 3 from *Cecil* 48A; 4–8 from *Milne-Redhead & Taylor* 1056. Drawn by D. Erasmus. Reproduced from Kew Bulletin (1995).

Asclepias carsonii (N.E. Br.) Schltr. in Fries, Wiss. Erg. Schwed. Rhodesia-Kongo-Exped. 1911-12 **1**: 267 (1916).

Slender erect herb; stems 25–85 cm long. Leaves linear or occasionally narrowly linear-lanceolate, spreading, lowest leaves present at anthesis 1.4–6 × 0.1–0.2(0.4) cm, upper leaves 5.5–20 × 0.1–0.7(0.9) cm, apex acute, tapering gradually into the sessile or subsessile base, glabrous, margins smooth, lateral veins visible only in the widest leaves. Umbels with 3–5(9) spreading or erect flowers. Peduncles (4)16–43(72) mm long, or occasionally absent, pubescent on one side, not lengthening in fruit; bracts 2–6 mm long, filiform, often deciduous. Pedicels 10–17(21) mm long, pubescent on one side. Calyx lobes 3–4 × 1–2 mm, ovate-triangular, acute. Corolla campanulate, lobed to 0.5–1 mm from the base, the lobes generally meeting at the margins when fresh, very variable in colour, usually yellowish green or brown with purple stippling, the stippling denser outside, but sometimes entire corolla cream or maroon; lobes 12–17 × 4.5–11 mm, obovate or obovate-oblong, apex subacute but appearing truncate as tip recurved, glabrous except for the minutely verrucate-papillate apical region and margins. Corona lobes (2)4–6(7) × 1–2.5 mm, creamy white, orange-yellow or occasionally suffused with purple, attached (0)1–2 mm above the base of the gynostegium, dorsally compressed, thin, erect, oblong or obovate-oblong, apex rounded, subacute or, more commonly, variously toothed, the apical portion erect or inflexed over the column for up to half its length, rarely shorter than column, the upper margins reflexed or not, basal portion of the lobes with incurved margins, glabrous or minutely verrucate-papillate. Anther appendages 1–2 mm, broadly ovate, obtuse, sometimes toothed at margin. Corpusculum c.0.3 mm long; translator arms c.0.3 mm long; pollinia c.1 × 0.3 mm, sickle-shaped, tapering towards the base and attached apically to the translator arm. Follicles 11–13 × c.1.1 cm, lanceolate, apex attenuate, glabrous or white-pubescent.

Zambia. N: Mbala Dist., Kawimbe, fl. 26.xii.1967 *Simon et al.* 1582 (K, SRGH). W: Kasempa Dist., 48 km from Chizera on Solwezi road, fl. 26.i.1975, *Brummitt et al.* 14144 (K, NDO). C: Mulungushi R., fl. 19.xii.1907, *Kässner* 2053 (BM, BR, E, K). E: Nyika Plateau, 2 km S of Chowo Forest, fr. 21.i.1992, *Goyder et al.* 3538 (K, NDO). **Zimbabwe**. N: Lomagundi Dist., near Tengwe River, fl. xi.1956, *Davies* 2236 (K, SRGH). C: Gwelo Teachers College, fl. 11.xi.1967, *Biegel* 2311 (K, SRGH). E: Mutare Dist., Honde valley, fl. 28.xi.1948, *Chase* 1552 (BM, BRLU, K, LISC, SRGH). **Malawi**. N: Vipya Plateau, Lwanjati peak E of Champira, fl. 11.i.1975, *Pawek* 8933 (K, MAL, MO, UC). C: Chongoni F.R., fl. 23.xi.1966, *Salubeni* 466 (K, LISC, MAL, SRGH). S: Mt Mulanje, Chambe path up Chapaluka stream, fl. & fr. 22.i.1967, *Hilliard & Burtt* 4589 (EA, K, MAL, SRGH). **Mozambique**. N: Cabo Delgado Province, 35 km W of Palma on Pundanhar road, near turn-off to Nhica do Rovuma, fl. 8.xii.2008, *Goyder et al.* 5092 (K, LMA, LMU, P, PRE).

Also known from D.R. Congo, Rwanda, Burundi, Tanzania and Angola. *Uapaca–Protea* scrub, *Brachystegia* woodland and montane grassland; 1000–2250 m.

2. **Glossostelma lisianthoides** (Decne.) Bullock in Kew Bull. **7**: 416 (1952). Type: Angola, *da Silva* s.n. (P holotype, K).

   Gomphocarpus lisianthoides Decne. in Ann. Sc. Nat., Bot. sér. 2 **9**: 325 (1838).
   Gomphocarpus chironioides Decne. in Candolle, Prodr. **8**: 562 (1844), superfluous name. Type as for *Glossostelma lisianthoides*.
   Xysmalobium dissolutum K. Schum. in Bot. Jahrb. Syst. **17**: 119 (1893). Type: Angola, Mukenge, *Pogge* 1227 (B† holotype, K).
   ?*Schizoglossum violaceum* K. Schum. in Bot. Jahrb. Syst. **17**: 122 (1893). Type: Angola, Sao Salvador, *Büttner* 504 (B†).
   Xysmalobium fritillarioides Rendle in J. Bot. **32**: 162 (1894). Type: Angola, between Mumpulla and Humpata, x.1859, 'Chlorostelma fritillarioides', *Welwitsch* 4179 (BM holotype).
   Asclepias dissoluta (K. Schum.) Schltr., Westafr. Kautschuk-Exped.: 309 (1900). —Brown in F.T.A. **4**(1): 347 (1902).
   Asclepias lisianthoides (Decne.) N.E. Br. in F.T.A. **4**(1): 327 (1902).
   Schizoglossum macroglossum K. Schum. in Bot. Jahrb. Syst. **33**: 324 (1903). Type: Congo, Stanley Pool, x.1895, *deMeeuse* (B† holotype).

Asclepias congolensis De Wild in Ann. Mus. Congo Belge, sér. 5 **1**: 186 (1904). Types: D.R. Congo, Kisantu, 1899, *Gillet* s.n. (BR syntype); Kisantu, 1903, *Gillet* s.n. (BR syntype); 'Bas-Congo', 1902, *Butaye* coll. *Gillet* s.n. (BR syntype).

Asclepias nemorensis S. Moore in J. Bot. **47**: 218 (1909). Type: Zambia, Chibenga Stream, *Kässner* 2078 (BM holotype, K, right hand specimen).

Slender erect herb to 70 cm. Leaves linear, spreading, 2–3 × 0.1–0.2 cm on lower portion of stem, 5.5–12(18) × 0.1–0.4(0.7) cm above, apex acute, tapering gradually into the sessile or subsessile base, glabrous, margins smooth, lateral veins inconspicuous. Umbels with (1)2–4 nodding flowers. Peduncles (2)10–25(35) mm long, apparently lengthening with age, entirely glabrous or pubescent on one side; bracts 2–6.5 mm long, filiform, deciduous. Pedicels (10)16–24(28) mm long, glabrous or pubescent on one side only. Calyx lobes 3–5 × 1–1.5 mm, narrowly triangular, acute. Corolla spreading-campanulate, lobed to c.1 mm from the base, greenish yellow or cream within, suffused with pink or brown outside; lobes 10–13 × 5–7 mm, ovate-oblong, obtuse. Corona lobes 4–5 mm long, attached c.2 mm above the base of the gynostegium, the lower half of the lobes bulbous and ± orbicular when fresh, appearing channelled on the inner surface in dried material, upper half of lobe narrowed abruptly to an oblong tongue, c.2 × 1 mm, truncate or shortly bifid at the apex and inflexed over the head of the gynostegium. Anther appendages membranous, c.1.5 × 1.5 mm, ovate. Corpusculum 0.3 mm long; translator arms c.0.3 mm long, extremely slender; pollinia sickle-shaped, c.0.7 × 0.3 mm, attached apically to the translator arms. Follicles 8–11 × 0.8–1.2 cm, slender, lanceolate with an attenuate apex, glabrous or puberulent. Seeds c.5 × 4 mm, ovate; coma 3.5–4 cm long.

Zambia. N: Mbala Dist., Mpukutu Forest, Chinakila, fl. 14.i.1965, *Richards* 19526 (K). W: Mwinilunga Dist., E of Dobeka Bridge, fl. 5.xi.1837, *Milne-Redhead* 3110 (BR, K). C: Chakwenga Headwaters, 100–129 km E of Lusaka, fl. 1.xii.1963, *Robinson* 5870 (K, SRGH). E: Chipata, fl. iii.1952, *Benson* NR20 (BM, SRGH). **Malawi**. N: Mzimba Dist., Mzuzu, Marymount, fl. 26.ii.1975, *Pawek* 9110 (K).

Also known from Angola, Gabon and D.R. Congo. Deciduous woodland (miombo) or grassland; (500)1200–1500 m.

The differences in corona morphology between this species and the previous one are readily apparent in fresh or spirit material, but the corona of *G. lisianthoides* distorts badly in dried material, the bulbous base in particular becoming flattened, making identification of some herbarium material difficult. Additional characters of value in distinguishing between the species include the generally less showy corolla of *G. lisianthoides*, which tends to have narrower lobes and a less deeply campanulate form than in *G. carsonii*.

3. **Glossostelma spathulatum** (K. Schum.) Bullock in Kew Bull. **7**: 414 (1952). — Goyder in F.T.E.A., Apocynaceae (part 2): 460 (2012). Types: Angola, Cuango, *Mechow* 539a (B†, K lectotype), designated by Goyder in Kew Bull. **50**: 537 (1995); Congolo, *Buchner* 611 (B† paralectotype), Malandsche, *Mechow* 356 (B† paralectotype, M, W). FIGURE 7.3.**198A**.

Schizoglossum spathulatum K. Schum. in Bot. Jahrb. Syst. **17**: 120 (1893).
Gomphocarpus spathulatus (K. Schum.) Schltr. in J. Bot. **33**: 269 (1895).
Xysmalobium bellum N.E. Br. in Bull. Misc. Inform. Kew **1895**: 69 (1895); in F.T.A. 4(1): 311 (1902). Types: Malawi, *Buchanan* 603 (K) & Blantyre, *Buchanan* 43 (K syntype, E); Manganja Hills, *Kirk* s.n. (K syntype); Zambia, Fwambo, *Carson* 62 (K syntype).
Xysmalobium spathulatum (K. Schum.) N.E. Br., F.T.A. 4(1): 312 (1902).
?*Xysmalobium mildbraedii* Schltr. in Bot. Jahrb. Syst. **51**: 134 (1913). Type: D.R. Congo, Kimuenza, x.1910, *Mildbraed* 3759 (B† holotype).

Erect herb to c.50 cm. Leaves petiolate or occasionally sessile, petiole (2)5–15(20) mm long; lamina spathulate, obovate, oblanceolate, lanceolate or oblong, spreading or ascending, lowest leaves present at anthesis 2–3.5 × 0.3–1.2 cm, upper leaves 2.5–8(10) × (1.1)2–4(4.8) cm, apex obtuse, rounded or retuse, apiculate, the apices of lower leaves generally more acute than those of upper leaves, tapering gradually or abruptly at the base into the petiole, glabrous or sparsely hairy beneath, midrib channelled on upper surface, lateral veins clearly visible but neither impressed nor raised, numerous, parallel to each other and at c.90° to the midrib. Umbels with 2–4(8) flowers. Peduncles pubescent at least on one side, 2–20(35) mm long; bracts deciduous, often of two types, one 2–4 mm long, filiform, the other 5–7(12) × 1–1.5 mm, narrowly triangular with ciliate margins. Pedicels 11–25(31) mm long, pubescent on one side. Calyx lobes 6–10 × (2.5)4–5(6.5) mm, ovate, acute or subacute. Corolla campanulate, lobed to 1–2 mm from the base, green or maroon with reddish brown, yellow or white markings within; lobes 13–20(25) × 8–14 mm, obovate, apex rounded or obtuse, recurved, glabrous. Corona lobes white or yellowish, c.3 mm long, 2 mm wide, reaching the top of the column or exceeding it slightly, subglobose but with a flattened inner face and three short, inward-pointing teeth near the top, attached to the staminal column 3–5 mm above the base of the gynostegium. Anther appendages c.1.5 × 2 mm, broadly ovate. Corpusculum c.0.5 mm long; translator arms c.0.5 mm long; pollinia 0.7–1 × c.0.5 mm, falcate-oblong. Follicles c.7 × 2 cm, ovate-lanceolate, puberulent. Seeds c.7 × 6 mm, suborbicular.

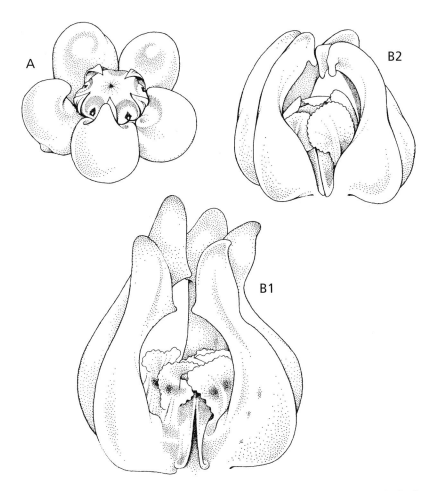

Fig. 7.3.**198**. A. —GLOSSOSTELMA SPATHULATUM, gynostegium with corona (× 4), from *Cribb et al.* 11382. B. —GLOSSOSTELMA CECILIAE. B1, gynostegium and "Mbala type" corona (× 4); B2, gynostegium and "general type" corona (× 4), B1 from *Lawton* 567, B2 from *Pole-Evans* 5044. Drawn by D. Erasmus. Reproduced and adapted from Kew Bulletin (1995).

Zambia. B: Kaoma Dist., Luampa Mission, fl. 22.ii.1952, *White* 2114 (BR, K, FHO). N: Mpika, fl. 15.ii.1955, *Fanshawe* 2066 (K, NDO). W: 21 km NW of Kaoma on road to Kasempa, fl. 28.ii.1996, *Harder et al.* 3596 (K, MO, NDO). C: Protea Hill Farm, 13 km SE of Lusaka, fl. 2.ii.1995, *Bingham* 10355 (K). E: 10 km E of Chipata, fl. 12.i.1959, *King* 454 (SRGH). S: Kalomo, fl. 10.ii.1965, *Fanshawe* 9173 (NDO, SRGH). **Zimbabwe**. N: Mwami, K34 farm, fl. 23.i.1948, *Bates* s.n. in GH 18617 (K, SRGH). C: Makabusi area, fl. 20.xii.1969, *Linley* 356 (K, SRGH). E: Nyanga village, fl. 6.ii.1965, *Chase* 8255 (BR, K, SRGH). **Malawi**. C: Lilongwe, Bunda forest, fl. 1.iii.1962, *Adlard* 418 (MAL, SRGH). S: Blantyre Dist., Ndirande F.R., 3 km N of Limbe, fl. 15.ii.1970, *Brummitt* 8586 (K, MAL, SRGH). **Mozambique**. N: 18 km from Mutuáli on Malema road, fl. 25.ii.1954, *Gomes e Sousa* 4224 (K). Z: Guruè, 6 km from Nintulo, fl. 10.ii.1964, *Torre & Paiva*, 10524 (LISC). T: Muatize, 10 km from Zóbuè, fl. 11.i.1966, *Correia* 374 (LISC).

Also occurs in Angola, D.R. Congo and Tanzania. Deciduous woodland (miombo) or open grassland; (650) 1500–1900 m.

4. **Glossostelma rusapense** Goyder in Kew Bull. **50**: 540 (1995). Type: Zimbabwe, Makone Dist., 1.5 km from Eagles Nest on Rusape road, 29.xi.1955, *Drummond* 5069 (K holotype, SRGH).

Erect herb to c.20 cm. Leaves petiolate or subsessile; petiole 0–2 mm long; lamina 2.5–5 × 0.5–1 cm, lanceolate or oblong, apex acute, tapering gradually or abruptly at the base, glabrous except for the midrib and margins beneath. Umbels with 2–4 flowers. Peduncles absent or up to 2 mm long; bracts deciduous, 1–3 mm long, filiform. Pedicels 8–17 mm long. Calyx lobes c.4 × 1.5 mm, ovate or triangular, acute. Corolla campanulate, lobed to 1–2 mm from the base, green speckled with reddish brown markings within; lobes c.13 × 5–7 mm, glabrous except for the puberulous apex and margins. Corona lobes attached to the gynostegium 2 mm above the base of the corolla, c.5 mm long, 2 mm wide, swollen and appearing slightly pouched at the base, with two short lateral teeth on the inner surface c.2 mm above the base, the upper part narrowed gradually into a dorsally flattened tongue arched over the head of the column. Anther appendages c.1.5 × 2 mm, broadly ovate. Corpusculum c.0.5 mm long, ovoid, black; translator arms c.0.5 mm long, broadly flattened; pollinia c.1 × 0.5 mm, falcate-oblong. Follicles not seen.

Zimbabwe. C: Makone Dist., 1.5 km. from Eagles Nest on Rusape road, 29.xi.1955, *Drummond* 5069 (K holotype, SRGH).

Only known from this region. Recently burned grassland; 1400–1600 m.

The corona has a broadly similar form to that of *G. spathulatum* but is somewhat less fleshy and is more inflated above.

5. **Glossostelma nyikense** Goyder in Kew Bull. **50**: 545 (1995). Type: Malawi, Nyika Plateau, near Chelinda C.D.C. Camp, 28.x.1958, *Robson* 452 (K holotype, LISC, PRE, SRGH). FIGURE 7.3.**199**.

Erect herb with 1 or 2 stems 10–20 cm long. Leaves sessile; lamina 2.5–7(8) × 0.2–0.3 cm, linear, apex acute, slightly fleshy, lateral veins obscure. Umbels with 4–10 flowers. Peduncles 1–5 mm long, pubescent on one side; bracts 1–3 mm, filiform. Pedicels 5–12 mm long, pubescent on one side. Calyx lobes 2–3 × 1 mm, triangular-ovate, apex attenuate. Corolla greenish yellow or brown; lobes c.5 × 2 mm, ovate-oblong, apex acute or subacute. Corona lobes attached to the gynostegium c.1 mm above the base of the corolla, c.2 mm long, ovoid, with no teeth or projections on the inner face, slightly exceeding the top of the column. Corpusculum c.0.4 mm long, subcylindrical, brown; translator arms c.0.3 mm long, flattened and contorted distally; pollinia 0.5 × 0.3 mm, broadly oblong and rounded at the base. Follicles (very immature) densely rufous pubescent.

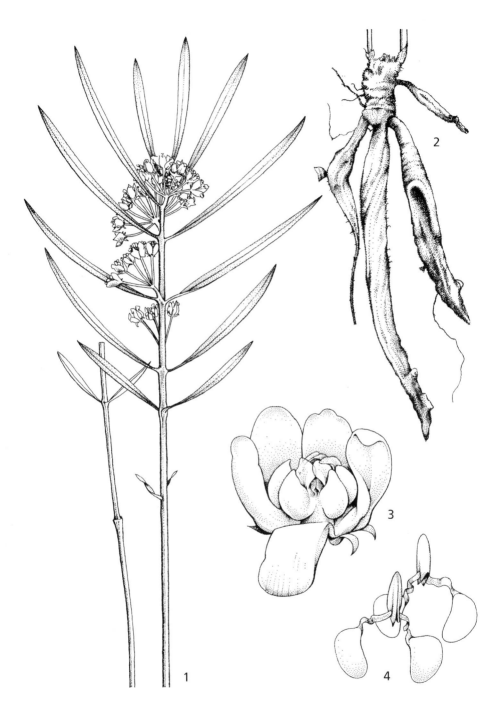

Fig. 7.3.**199**. GLOSSOSTELMA NYIKENSE. 1, flowering stem (× ²/₃); 2, rootstock (× ²/₃); 3, flower (× 4); 4, pollinaria (× 22). All from *Robson* 452. Drawn by E. Papadopoulos. Reproduced from Kew Bulletin (1995).

Zambia. E: Lundazi Dist., Nyika Plateau, fl. ix.1968, *Williamson* 1009 (SRGH). **Malawi**. N: Nyika Plateau, fl. 25.x.1971, *Moriarty* 723 (MAL).

Known from just six collections from the Nyika Plateau. Montane grassland; 2000–2250 m.

This species is close to both *Glossotelma mbisiense* Goyder and *G. spathulatum*, differing principally in its smaller flowers, narrow leaves, and the lack of any teeth or projections on the corona lobe.

6. **Glossostelma ceciliae** (N.E. Br.) Goyder in Kew Bull. **50**: 547 (1995). —Goyder in F.T.E.A., Apocynaceae (part 2): 462 (2012). Type: Zimbabwe, Mashonaland, at Harare [Salisbury], *Cecil* 60 (K holotype). FIGURE 7.3.**198B**.

　　Xysmalobium ceciliae N.E. Br. in F.T.A. **4**(1): 310 (1902).

Robust erect herb; stems 20–90 cm long. Leaves petiolate or sessile, petiole (0)3–6 mm long; lamina narrowly oblong or oblanceolate, spreading or ascending, lowest leaves present at anthesis 2.5–3.5 × 0.7–0.9 cm, upper leaves (3.5)5–8.5 × 0.9–2.1 cm, apex acute, obtuse or rounded, apiculate, rounded at the base or more commonly tapering gradually into the petiole, glabrous except on the margins, midrib channelled on upper surface, lateral veins clearly visible but neither impressed nor raised, numerous, parallel to each other and at c.90° to the midrib. Umbels with (1)3–5 flowers. Peduncles pubescent, 1–12 mm long or absent; bracts deciduous, often of two types, one 5–6 mm long, filiform, the other 5–9 × 1 mm, narrowly triangular with ciliate margins. Pedicels 16–25(30) mm, pubescent on one side. Calyx lobes 6–10 × 2–4 mm, ovate or triangular, acute. Corolla campanulate, lobed to 2–3 mm from the base, green or cream within, tinged brown or reddish brown outside; lobes 23–28 × 9–19 mm, obovate, apex acute or obtuse, recurved, glabrous. Corona lobes green or white, attached to the staminal column 2–4 mm above the base of the gynostegium, 7–10 mm long, 3–4 mm wide at the slightly swollen base, laterally compressed but not flattened above, falcate, the upper half arched over the head of the column, dilated slightly at the truncate, clavate apex. Anther appendages 2–3 × 2–3 mm, broadly ovate. Corpusculum c.0.3 mm long, ovoid; translator arms c.0.5 mm long; pollinia 1.25 × 0.5 mm, falcate-oblong. Follicles 11–24 × 1–1.3 cm, lanceolate to long-fusiform, attenuate at both ends, puberulent. Seeds c.5 × 4 mm, ovate, convex on one face, the disc verruculose, the rim convoluted.

Zambia. N: Lunzua P.F.A., fl. 8.iv.1959, *Lawton* 567 (K, NDO). W: Mwinilunga aerodrome, fl. 30.xi.1937, *Milne-Redhead* 3457 (K). **Zimbabwe**. C: Marondera, fl. i.1967, *Davy* s.n. in GH175488 (SRGH). E: Chimanimani Dist., Cashel, fl. 26.xi.1947, *Chase* 674 (BM, SRGH). S: Victoria Dist., Glenlivet, fl. xii.1955, *Leach* 5955 (SRGH).

Also known from Angola, D.R. Congo and Tanzania. Deciduous woodland (miombo); c.1300–1600 m.

Material from within the Flora area falls into two imperfectly delimitable types. Corona lobes of material from around Mbala in northern Zambia have a characteristic apex with the flattened face held ± vertically (Fig. 7.3.**197B1**). In most other collections the flattened face is held ± horizontally or the tip of the corona lobe is rounded (Fig. 7.3.**197B2**). The Mbala collections flower from March to May in contrast to the rest of the species which flowers between October and December.

7. **Glossostelma cabrae** (De Wild.) Goyder in Kew Bull. **50**: 548 (1995). —Goyder in F.T.E.A., Apocynaceae (part 2): 463 (2012). Type: D.R. Congo, Valley of the Tawa, 3.x.1902, *Cabra-Michel* 52 (BR holotype).

　　Asclepias cabrae De Wild. in Ann. Mus. Congo Belge, Bot. sér. 5 **1**: 185 (1904).

　　Xysmalobium speciosum S. Moore in J. Bot. **47**: 216 (1909). Type: Zambia, Lisanga Spruit, 26.xii.1907, *Kässner* 2144 (BM holotype, K).

Erect herb to c.50 cm. Leaves petiolate or sessile, petiole (0)1–5 mm long; lamina oblong, obovate-oblong or rarely linear, spreading or ascending, lowest leaves present at anthesis 1.5 × 0.2 cm, upper leaves 3.5–8(20) × (0.4)1.5–3.1 cm, apex obtuse, rounded or retuse, apiculate, tapering gradually at the base, glabrous or with short, sparse hairs beneath, midrib channelled on upper surface, lateral veins clearly visible but neither impressed nor raised, numerous, parallel to each other and at c.90° to the midrib. Umbels with 2–4 flowers. Peduncles pubescent, 1–12 mm long, or absent; bracts deciduous, 5–6 mm long, filiform. Pedicels 18–25 mm long, pubescent on one side. Calyx lobes 5–6 × 1–1.5 mm, triangular, acute. Corolla campanulate, lobed to 2–3 mm from the base, green or cream with reddish brown markings within; lobes 23–26(28) × 9–14(20) mm, obovate, apex acute or obtuse, recurved, glabrous except on the margins. Corona lobes green, attached to the staminal column 3–6 mm above the base of the gynostegium, 9–11 mm long, 2 mm wide at the base, the lower half of the lobe somewhat compressed laterally, falcate, arched over the head of the column and narrowed into the base of the erect, dorsally flattened, slightly fleshy, spathulate upper half, the apex rounded or truncate, with one or occasionally two narrowly triangular downward pointing flaps up to 2 mm long on the ventral face. Anther appendages c.1.5 × 2 mm, broadly ovate. Corpusculum c.0.3 mm long; translator arms c.0.5 mm long; pollinia c.1 × 0.25 mm, falcate-oblong. Follicles not seen.

Zambia. W: Mwinilunga Dist., 12 km NW of Kalene Mission, fl. 11.xi.1962, *Richards* 17181 (K). **Mozambique**. Z: Zambezia, Altomolócuè, fl. 29.xi.1967, *Torre & Correia* 16287 (LISC).

Also known from Angola, D.R. Congo and Tanzania. Deciduous woodland (miombo) or open grassland; 1200–1500 m.

As in *Glossostelma ceciliae*, the degree of inflation of the corona lobes of *G. cabrae* is variable although the lobes are always more slender than in *G. ceciliae*. One collection from Mozambique (*Torre & Correia* 16287 (LISC)) is anomalous in having linear leaves up to 20 cm long and only 0.4–0.8 cm wide. In all other respects, however, it is typical of the species.

8. **Glossostelma brevilobum** Goyder in Kew Bull. **50**: 551 (1995). —Goyder in F.T.E.A., Apocynaceae (part 2): 463 (2012). Type: Malawi, Nkhata Bay Dist., Vipya Plateau, 40 km SW of Mzuzu, 12.xi.1972, *Pawek* 5954 (K holotype, CAH, MAL, MO, UC).

Erect herb with stems 12–40 cm long. Leaves sessile or petiolate, petiole to 5 mm long; lamina spathulate, obovate or narrowly oblong to ovate, lowest leaves present at anthesis c.1.3–2 × 0.4–1 cm, upper leaves (2)3–7 × (0.5)1–3.6 cm, apex rounded, obtuse or subacute, apiculate, the apices of lower leaves generally more acute than those of upper leaves, rounded at the base or tapering gradually into the petiole, glabrous or sparsely hairy beneath, midrib channelled on upper surface, lateral veins clearly visible but neither impressed nor raised, numerous, parallel to each other and at c.90° to the midrib. Umbels with up to 9 flowers. Peduncles pubescent, c.1 mm long. Pedicels 8–13 mm long, pubescent on one side. Calyx lobes 2–3 × 1–1.5 mm, ovate or triangular, acute. Corolla campanulate, lobed to 1 mm from the base, green or brown; lobes 5–8 × (2)4–5 mm, ovate, apex acute or subacute, glabrous. Corona lobes attached near the base of the gynostegium, c.1 mm long, 0.5 mm wide, forming fleshy outward-pointing pegs briefly united into an annulus at the base. Anther appendages c.1.5 × 2 mm, broadly ovate. Corpusculum 0.5–0.8 mm long, ovoid, brown; translator arms 0.4–0.7 mm long; pollinia obtriangular, c.0.75 × 0.5–0.6 mm. Follicles not seen.

Malawi. N: Nkhata Bay Dist., Vipya Plateau, 40 km SW of Mzuzu, fl. 12.xi.1972, *Pawek* 5954 (K holotype, MAL).

Also known from Tanzania, D.R. Congo and Burundi. Montane grassland; 1500–1800 m.

This species is close to *Glossostelma spathulatum* and its allies, but the gynostegial stalk and corona lobes are much reduced.

74. **SCHIZOSTEPHANUS** Benth. & Hook.f.[46]

Schizostephanus Benth. & Hook.f., Gen. Pl. **2**: 762 (1876). —Liede in Bot. J. Linn. Soc. **114**: 81 (1994).

Scandent shrubs or twiners with succulent stems; latex clear or yellowish, not milky. Inflorescences extra-axillary, many-flowered, with the flowers scattered along an indeterminate axis. Corolla rotate, the lobes fused basally. Corolline corona absent. Gynostegial corona fused for most of its length into a tube surrounding and obscuring the gynostegium. Gynostegium stipitate. Anther connectives with apical appendages. Pollinaria with a pair of pendent pollinia. Apex of stylar head flat. Follicles paired, or single by abortion, winged longitudinally. Seeds smooth, winged.

Two species in tropical Africa. According to evidence presented by Liede in Ann. Missouri Bot. Gard. **88**: 657–668 (2001) its affinities lie most closely with *Cynanchum* and *Pentarrhinum*.

Schizostephanus alatus K. Schum. in Bot. Jahrb. Syst. **17**: 139 (1893). —Goyder in F.T.E.A., Apocynaceae (part 2): 466 (2012). Type: Ethiopia, Mai-Mezano, Djeladjeranne, *Schimper* 1687 (K lectotype, B†), designated by Goyder in F.T.E.A., Apocynaceae (part 2): 466 (2012). FIGURE 7.3.**200**.

Cynanchum validum N.E. Br. in F.T.A. **4**(1): 398 (1903), new name, not *Cynanchum alatum* Wight & Arn. (1834). Type as for *Schizostephanus alatus*.

Scandent shrub or twiner to 8 m; stems thick and fleshy, glabrous except on very young shoots. Leaves with petiole 2–6 cm long, glabrous; lamina 5–10(12) × 4–6(10) cm, broadly ovate to suborbicular, apex acute to shortly attenuate, base weakly to strongly cordate, glabrous except on principal nerves beneath. Inflorescences axes to 10(13) cm long, somewhat fleshy, subglabrous or minutely pubescent; pedicels 4–6 mm long, minutely pubescent. Calyx lobes 0.5–1 mm long, ovate, rounded to subacute, minutely pubescent. Corolla united at the base, rotate, greenish yellow or the lower half reddish purple; lobes 2.5–4 × 1–1.5 mm, oblong or oblong-ovate, truncate and somewhat twisted apically, pubescent with long slender hairs in basal half of lobes adaxially, abaxial face glabrous. Gynostegium raised on a stipe c.1–1.5 mm long. Corona forming a fluted cylinder surrounding and obscuring the gynostegium, c.2 mm long with longitudinal ridges on the inner face, divided apically into ten erect, acute lobes c.0.5–1 mm long arising in both staminal and interstaminal positions. Anther wings 0.3–4 mm long. Apex of stylar head not exserted from staminal column. Pollinaria with corpusculum c.0.3 mm × 0.2 mm with thin lateral wings; translator arms c.0.15 mm long, flattened and geniculate; pollinia c.0.4 mm long, ovoid, attached apically to translator arms. Follicles mostly in widely divergent pairs, 5–6 cm long, fusiform, with three longitudinal wings c.0.5 cm wide running the entire length of the follicle, smooth, glabrous. Seeds flattened, c.6 × 3 mm, smooth and with a marginal rim c.0.5 mm wide; coma 1.5–2 cm long.

Zimbabwe. C: Marondera [Marandellas], fr. 5.iv.1950, *Wild* 3262 (K, SRGH). E: Mutare [Umtali] Dist., SW side of Murahwa's Hill Commonage, fl. 17.ix.1964, *Chase* 8174 (K, SRGH). **Malawi**. S: Mt Mulanje, Likhabula valley, Nasato stream, fl. 8.iv.1986, *Chapman & Chapman* 7391 (K, MO).

Eastern Africa from Somalia, Ethiopia, Sudan and Kivu in eastern D.R. Congo to NE parts of South Africa. Mostly on granite outcrops in dry woodland or forest; 1000–1600 m.

---

[46] by D.J. Goyder

Fig. 7.3.**200**. SCHIZOSTEPHANUS ALATUS. 1, flowering shoot (×²/₃); 2, inflorescence (× 2); 3, paired follicles from above, single follicle from beneath (× ²/₃); 4, flower (× 3); 5, gynostegium largely obscured by the cupular corona (× 10); 6, stipitate gynostegium (× 14); 7, pollinarium (× 20). All from *Greenway* 10727. Drawn by Margaret Tebbs. Reproduced from Flora of Tropical East Africa (2012).

75. **PENTARRHINUM** E. Mey.[47]

Pentarrhinum E. Mey., Comm. Pl. Afr. Austr.: 199 (1838). —Liede & Nicholas in Kew
Bull. **47**: 475 (1992). —Liede in Ann. Missouri Bot. Gard. **83**: 283 (1996); in Syst.
Bot. **22**: 347 (1997).

Twining herbs with perennial rootstock and slender annual stems; latex white. Leaves
herbaceous, petiolate, cordiform, colleters conspicuous at the base of the lamina; small leaves
on short axillary shoots frequently giving the appearance of stipules. Inflorescences extra-
axillary, cymose, pedunculate. Corolla lobes fused basally, recurved. Corolline corona absent.
Gynostegial corona generally consisting of a short tube and prominent free staminal parts, flat
or laterally conduplicate, apically either conduplicate or provided with a horn-like ornament
projecting towards the gynostegium; if corona fused for much of its length, then with vertical
'suture marks'. Gynostegium sessile. Anther connectives with apical appendages. Pollinia
pendulous. Stylar head flat. Follicles mostly single, ellipsoid, rather thick-walled, with or without
protuberances of variable length and density. Seeds ovate, papillate or pubescent; with or without
a marginal wing; comose.

Five species in tropical Africa. *Pentarrhinum* has recently been absorbed within an
expanded concept of *Cynanchum* by Bruyns *et al.* in S. African J. Bot. **112**: 399–436
(2017), but to maintain consistency with treatments in neighbouring Flora regions I
have maintained it as separate in the current work.

Corona yellow, papillose, fleshy, expanded abruptly into a flat-topped apex with a
    conspicuous inward-pointing horn . . . .**1.** *insipidum*
Corona white, smooth, thin and with inrolled margins and apex forming a tooth on
    the inner face. . . . . **2.** *abyssinicum*

1. **Pentarrhinum insipidum** E. Mey., Comm. Pl. Afr. Austr.: 200 (1838). —Goyder in
F.T.E.A., Apocynaceae (part 2): 469 (2012). Type: South Africa, Uitenhage Div.,
Enon, *Drège* 2220 (B† holotype, K lectotype), designated by Liede & Nicholas in
Kew Bull. **47**: 484 (1992). FIGURE 7.3.**201**.
    Cynanchum insipidum (E. Mey.) Liede & Khanum in Taxon **65**: 479 (2016).

Slender perennial twining to c.8 m, stems minutely pubescent, arising at intervals from a
long, horizontal rhizome; latex white. Abbreviated leafy shoots often present in leaf axils. Leaves
with petiole 1–5 cm long, minutely pubescent; lamina 3–8 × 2–6 cm, ovate to broadly triangular,
apex acute or attenuate, base strongly cordate, minutely pubescent. Inflorescences extra-axillary,
simple, very rarely branched distally, initially ± umbelliform, eventually with flowers scattered
along an axis; peduncle 2–8 cm long, minutely pubescent; rhachis to c.2 cm; bracts 0.5–1.5 mm
long, linear, pubescent; pedicels 10–20 mm long, minutely pubescent. Sepals 1–2 mm long,
oblong or lanceolate, rounded or acute, subglabrous to pubescent. Flower buds subglobose.
Corolla yellow-green or white, frequently tinged brown or purple, divided almost to the base;
lobes strongly reflexed, 2.5–4.5 × 1–3 mm, ovate to oblong, glabrous but with ciliate margins.
Corona yellow, sometimes brown or purple apically and perhaps becoming darker with age,
divided almost to the base; lobes ascending, c.2 mm long, slightly taller than the staminal
column, fleshy and subcylindrical below, broadening abruptly to form a flattened apex with
a well-developed tooth or horn pointing towards the gynostegium, papillose. Gynostegium
subsessile. Anther wings c.0.7 mm long. Corpusculum c.0.2 mm long, rhomboid; translator arms
c.0.2 mm long; pollinia c.0.5 mm long, attached apically to translator arms. Follicles occurring
singly or in pairs reflexed alongside the pedicel, 5–8 × 1.5 cm, lanceolate in outline, tapering to
a weakly clavate tip, with few to many irregular winged processes. Seeds flattened, c.6 × 4 mm,
ovoid, verrucose and with a crenate margin; coma c.3 cm long.

---

[47] by D.J. Goyder

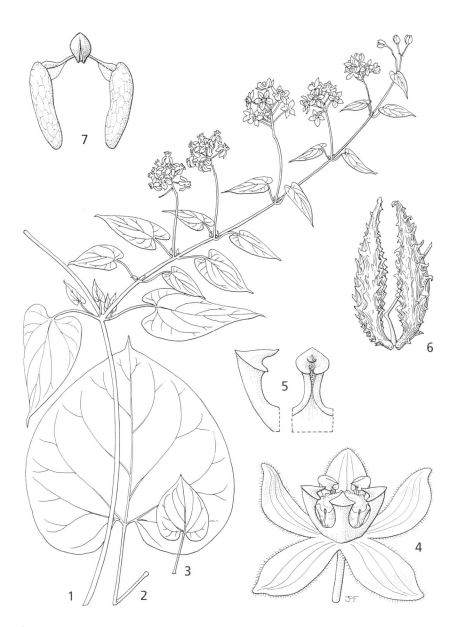

Fig. 7.3.**201**. PENTARRHINUM INSIPIDUM. 1, habit (× ½); 2, 3, leaves (× ½); 4, flower (× 4); 5, corona lobe, lateral and ventral views (× 1); 6, paired follicles (× ½); 7, pollinarium (× 50). 1, 3, 4 from *Drummond & Hemsley* 3062; 2 from *Polhill & Paulo* 726; 3 from *Milne-Redhead & Taylor* 11310; 6 from *Bally* 9110; 7 from *Pope* 86. Drawn by M. Fothergill. Reproduced from Kew Bulletin (1992).

Caprivi. Linyanti, fl. 24.xii.1958, *Killick & Leistner* 3146 (K, PRE). **Botswana**. N: Horseshoe Bend or the Kwando River, fl. 1.ii.1978, *P.A. Smith* 2329 (K, SRGH, PSUB). SE: Gabarone, Broadhurst, fl. 20.ii.1977, *Hansen* 3041 (C, GAB, K, PRE, SRGH, UPS). **Zambia**. B: Machili, fl. 14.iv.1961, *Fanshawe* 6492 (K, NDO). C: Blue Lagoon National Park, Shamukobo road, fl. 26.iv.2000, *Bingham* 12212 (K). S: 41 km from Choma on Namwala road, fl. 8.iii.1997, *Zimba et al.* 1042 (K, MO, NDO). **Zimbabwe**. W: Bullima Mangwe [Bululima Mangwe] Dist., Simukwe River 8 km downstream from Mount Jim, fl. 11.iv.1974, *Ngoni* 380 (K, SRGH). C: Que Que Dist., 25 km along Gokwe road to NW Skipper Mine, fl. 16.iii.1966, *Wild* 7549 (K, SRGH). E: Mutare [Umtali] Dist., Odzi Rover, fl. 13.ii.1963, *Chase* 7957 (K, SRGH). **Malawi**. C: Chongoni Forest Reserve, fl. 17.ii.1968, *Salubeni* 966 (K, MAL, SRGH). S: Lengwe Game Reserve, fl. 22.iv.1970, *Hall-Martin* 622 (K). **Mozambique**. MS: Moribane Forest between Dombe and Sussendenga, fl. 6.iv.2010, *Burrows & Burrows* 11663 (BNRH, K, LMA, PRE). GI: Gaza, between Magul and Macia, fr. 1.vi.1959, *Barbosa & Lemos* 8554 (K, LMA). M: Marracuene, Costa do Sol, fl. & fr. 18.iv.1961, *Macuácua* 91 (K, LMA).

Widely distributed over Eastern and southern Africa from Sudan, Eritrea and Somalia in the north to South Africa and Namibia. Open grassland, coastal dunes and *Brachystegia* (miombo) woodland, occasionally also found in forest; sea level to 1800 m.

2. **Pentarrhinum abyssinicum** Decne. in Candolle, Prodr. **8**: 553 (1844). —Goyder in F.T.E.A., Apocynaceae (part 2): 469 (2012). Type: Delessert, Icon. Sel. Pl. **5**: t.80 (1846) lectotype, designated by Liede & Nicholas in Kew Bull. **47**: 481 (1992).

Cynanchum ethiopicum Liede & Khanum in Taxon **65**: 478 (2016), new name, not *Cynanchum abyssinicum* K. Schum. (1895). Type as for *Pentarrhinum abyssinicum.*

Slender perennial twining to 5 m, stems simple or branched, minutely pubescent; latex white; rootstock not known. Abbreviated leafy shoots often present in leaf axils. Leaves with petiole 1.5–5 cm long, minutely pubescent; lamina 3–10 × 1.5–6 cm, ovate to broadly triangular, apex acute or attenuate, base weakly to strongly cordate, shortly pubescent or glabrous. Inflorescences extra-axillary, simple or irregularly branched, umbelliform (most forest forms) or the flowers scattered singly or in pairs along the inflorescence axes (more open habitats); peduncle 1–6 cm long, minutely pubescent or subglabrous; rhachis to c.3 cm; bracts c.1 mm long, linear, glabrous or pubescent; pedicels 3–20 mm long, glabrous or minutely pubescent. Sepals 1–3 mm long, triangular, acute, minutely pubescent. Flower b uds ovoid or subglobose. Corolla greenish white or pale yellow, sometimes with purple veins, divided almost to the base; lobes reflexed or not, 2–6 × 1–2 mm, ovate to oblong, frequently with somewhat revolute margins, entirely glabrous or pubescent on margins and abaxial face. Corona white, divided almost to the base; lobes spreading or erect, 1.5–2.5 mm long, slightly longer than the staminal column, thin or somewhat fleshy, the lateral and apical margins inrolled forming a weakly to strongly developed apical tooth on the inner face. Gynostegium subsessile. Anther wings 0.5–1 mm long. Corpusculum c.0.2 mm long, ovoid; translator arms c.0.2 mm long; pollinia c.0.5 mm long, attached apically to translator arms. Follicles occurring singly or in pairs, 5–6 × 1 cm, lanceolate in outline, tapering to a weakly clavate tip, smooth or with irregular soft processes. Mature seeds not seen.

Corolla less than 10 mm diameter, lobes fully reflexed at anthesis, margins generally glabrous; mostly in grassland or open woodland . . . . a) subsp. *abyssinicum*
Corolla more than 10 mm diameter, lobes fully not reflexed at anthesis, margins ciliate; occurring in wet forest . . . . b) subsp. *angolense*

a) Subsp. **abyssinicum**.

Corolla less than 10 mm diam., lobes fully reflexed at anthesis, margins generally glabrous.

Zambia. N: Mbala Dist., Kalala village, fl. 7.iii.1968, *Sanane* 74 (K). C: 50 km E of Mumbwa, fl. 11.iv.1966, *Robinson* 6935 (K). S: Zimba, fl. 13.iii.1963, *Lawton* 1047

(K, NDO). **Zimbabwe**. W: Shangani, fl. 2.iii.1955, *Goldsmith* 109/55 (K, SRGH). S: Chiredzi Dist., Triangle Suagar Estate, fl. 4.iii.1970, *Mavi* 1075 (K, SRGH). **Malawi**. C: Kasungu–Lilongwe road, fl. 25.ii.1961, *Richards* 14476 (K).

Scattered localities across tropical Africa from Cameroon and Eritrea to Namibia and Zimbabwe. Grassland, open scrub or deciduous woodland, less often in forest; 1100–2200 m.

b) Subsp. **angolense** (N.E. Br.) Liede & Nicholas in Kew Bull. **47**: 482 (1992). —Goyder in F.T.E.A., Apocynaceae (part 2): 471 (2012). Type: Angola, Icolo e Bengo Dist., near Lagoa de Foto, *Welwitsch* 4240 (K lectotype, BM), designated by Liede & Nicholas in Kew Bull. **47**: 482 (1992).

Pentarrhinum abyssinicum var. *angolense* N.E. Br. in F.T.A. **4**(1): 379 (1902).
Cynanchum ethiopicum subsp. *angolense* (N.E. Br.) Liede & Khanum in Taxon **65**: 478 (2016).
Pentarrhinum insipidum sensu Hiern, Cat. Afr. Pl. **1**: 687 (1898); sensu Bullock in F.W.T.A. ed. 2, **2**: 90 (1963), not E. Mey.

Corolla more than 10 mm diam., lobes fully not reflexed at anthesis, margins ciliate.

Zambia. C: Mumbwa, *Macaulay* 1137 (K). **Malawi**. N: 3 km SE of Chikangawa, fl. 31.vii.1978, *Phillips* 3654 (K, MO). C: Dedza Mountain, road to main peak, fl. 10.iv.1980, *Blackmore et al.* 1214 (K, MAL).

Also known from Cameroon, D.R. Congo, Uganda, Kenya, Tanzania, Angola and Namibia. Wet forest; 1400–1800 m.

76. CYNANCHUM L.[48]

Cynanchum L., Sp. Pl.: 212 (1753). —Liede in Bot. Jahrb. Syst. **114**: 503 (1993); in Ann. Missouri Bot. Gard. **83**: 283 (1996).

Sarcostemma R. Br., Prodr. Fl. Nov. Holland.: 463 (1810).
Bunburia Harv., Gen. S. Afr. Pl.: 416 (1838).
Colostephanus Harv., Gen. S. Afr. Pl.: 417 (1838).
Cyathella Decne. in Ann. Sci. Nat., Bot. sér. 2, **9**: 332 (1838).
Cynoctonum E. Mey., Comm. Pl. Afr. Austr.: 215 (1838), illegitimate name, not J.F. Gmel. (1791).
Endotropis Endl., Gen. Pl.: 591 (1838), illegitimate name, not Raf. (1825), not Raf. (1838).
Sarcocyphula Harv., Thes. Cap. **2**: 58 (1863).
Perianthostelma Baill., Hist. Pl. **10**: 247 (1890).
Flanagania Schltr. in Bot. Jahrb. Syst. **18**(Beibl. 45): 10 (1894).

Synonymy relates to mainland Africa only.

Leafy or leafless perennial plants; stems herbaceous or succulent, twining or not; latex white or very occasionally clear or yellowish. Leaves, when present (i.e. not reduced to scales), with colleters at the base of the lamina. Abbreviated short leafy shoots present the leaf axils of several species giving the appearance of stipules. Inflorescences extra-axillary, sessile or pedunculate, umbelliform or clustered along an axis that develops with age. Corolla fused basally, glabrous or occasionally pubescent. Corolline corona absent. Gynostegial corona in one or two whorls; inner whorl where present of free, fleshy, staminal lobes; outer whorl a variously developed tube of partially or completely fused staminal and interstaminal parts surrounding the gynostegium. Gynostegium sessile or stipitate. Pollinia pendant in the anther cells. Stylar head ± flat, conical, or with rostrate apical appendage. Follicles generally developing singly by abortion. Seeds flat, winged or not, smooth or sculptured, glabrous or pubescent; with a coma of white hairs.

About 300 mostly Old-World species distributed across tropical and subtropical regions, with centres of diversity in Madagascar and southern China.

---

[48] by D.J. Goyder

1. Stems and leaves absent at time of flowering; inflorescence emerging direct from ground . . . . **11.** *praecox*
–   Stems clearly present at time of flowering; inflorescence borne on leafy or leafless shoots 2
2. Leaves reduced to minute scales; stems semi-succulent and photosynthetic . . . 3
–   Leaves well-developed; stems not succulent and not obviously photosynthetic . 7
3. Corona in a single series forming a tube around the gynostegium; fleshy staminal lobes absent . . . . **1.** *gerrardii*
–   Corona in two series – a short tube or annulus around the base of the gynostegium, and discrete fleshy lobes on the backs of the anthers (former *Sarcostemma* spp.) 4
4. Dwarf clump-forming perennial with stems no more than 15 cm long and 3 mm thick . . . . **5.** *oresbium*
–   Plants larger; stems more robust 5
5. Plants twining or scrambling to several metres; main stem thicker than other stems (but not generally apparent in herbarium material) . . . . **2.** *viminale*
–   Plants erect or scrambing, but never twining, and generally less than 1 m in height; lacking an obvious main stem 6
6. Stems arching and rooting at the nodes . . . . **3.** *doleriticum*
–   Stems not as above, erect or scrambling . . . . **4.** *mulanjense*
7. Herb or subshrub less than 30 cm tall, stems not twining, leaves narrowly linear . . . . **12.** *orangeanum*
–   Stems twining to several metres; leaves broader 8
8. Mature leaves more than twice as long as wide, generally auriculately lobed at the base. 9
–   Mature leaves generally less than twice as long as wide, if longer, then rounded at the base. 10
9. Flowers minute, corolla lobes 1.5–2 mm long; veins green; plants of riverine vegetation. . . . .**8.** *schistoglossum*
–   Flowers larger, corolla lobes 6–7 mm long; leaves with red or purple veins beneath; montane forest. . . . . **10.** *rungweense*
10. Abaxial surface of leaves pubescent. 11
–   Abaxial surface of leaves glabrous 12
11. Stylar head with a filiform, bifid appendage exserted beyond the stamens for 2–3 mm; corolla reddish or white with red stripes; montane or submontane forest. . . . . **7.** *umtalense*
–   Stylar head without an appendage exserted beyond the stamens; corolla green; coastal dunes . . . . **13.** *obtusifolium*
12. Leaf base deeply cordate; corona tubular in lower half, fluted above, extending into 10 acuminate lobes . . . . **6.** *mossambicense*
–   Leaf base rounded; corona cupular, upper margin undulate or crenulate, not strongly lobed . . . . **9.** *ellipticum*

1. **Cynanchum gerrardii** (Harv.) Liede in Taxon **40**: 117 (1991). —Goyder in F.T.E.A., Apocynaceae (part 2): 476 (2012). Type: South Africa, KwaZulu-Natal, Tugela, *Gerrard* 1321 (TCD holotype, BM).

    Sarcocyphula gerrardii Harv., Thes. Cap. **2**: 58, t.191 (1863).
    Cynanchum sarcostemmatoides K. Schum. in Engler, Pflanzenw. Ost-Afrikas **C**: 323 (1895); in Engler & Prantl, Naturl. Pflanzenfam. 4(2): 252 (1895). Type: Tanzania, Tanga, Amboni, vi.1893, *Holst* 2706 (K lectotype), designated by Liede in Bot. Jahrb. Syst. **114**: 515 (1993).
    Cynanchum edule Jum. & H. Perrier, Compt. Rend. **152**: 1016 (1911); in Rev. Gen. Bot. **22**: 258 (1911). Type: not known; synonymy after Liede in Bot. Jahrb. Syst. **114**: 515 (1993).
    Cynanchum aphyllum sensu R. Br. in Mem. Wern. Nat. Hist. Soc. **1**: 51 (1810).

Cynanchum tetrapterum sensu Bullock in Kew Bull. **10**: 624 (1956), not *Sarcostemma tetrapterum* Turcz. (1848), not *Cynanchum tetrapterum* (Turcz.) Bullock (1956).

Slightly succulent twining perennial to c.4 m with white latex; stems much branched, glaucous, glabrescent, corky below. Leaves reduced to minute scales to c.1 mm long, frequently subopposite. Inflorescences sessile, extra-axillary, forming a subumbelliform cluster of flowers along a congested rachis, 2–4 flowers open at one time; rhachis to 5 mm long; bracts to 0.5 mm long, broadly deltoid, pubescent; pedicels 2–5 mm long, glabrous. Sepals 0.3–0.7 mm long, ovate, glabrous or pubescent. Flower buds 1–2.5 mm long, globose, ovoid or subcylindrical, the lobes not contorted in bud. Corolla rotate, united only at the base, lobes 1.5–2.5 × 0.7–1.5 mm, oblong or ovate, green, glabrous. Corona white, united into a tube 0.5–1 mm long ± as tall as the staminal column; staminal portions of the tube very variable in length, extending for a further 0.1–1 mm with a linear apical portion; adaxial appendage absent, but staminal portion of tube thicker than interstaminal ones; interstaminal portions of the tube truncate to acute, occasionally as long as interstaminal portions. Anther wings 0.2–0.4 mm long. Corpusculum c.0.15 mm long, ovoid; translator arms c.0.1 mm long, flattened; pollinia attached subapically to translator arms, c.0.4 × 0.15 mm, ovoid, round in section. Stylar head obscured by anther appendages. Follicles occurring singly, 8–12 × 0.6–0.8 cm, fusiform, tapering to a subacute tip. Seeds 5–6 × 2–3 mm, ovate-convex, without a marginal rim; coma 2–2.5 cm long.

Subsp. **gerrardii**.

Sepals 0.3–0.5 mm long. Flower buds 1–1.5(2) mm long. Corona ± as long as gynostegium, fused for most of its length, staminal portions sometimes extended as linear or narrowly oblong teeth.

Zambia. C: Lusaka Dist., Kafue Gorge, fl. 8.iii.1971, *Fanshawe* 11191 (K, NDO). S: Mapanza West, fl. 22.ii.1954, *Robinson* 549 (K). **Zimbabwe**. N: Whindale ranch, Mhangula [Mangula], fl. 24.iii.1969, *Leach et al.* 14328 (K, SRGH). W: Hwange [Wankie] Dist., near Inyanatue sale pens, fl. 29.iii.1963, *Leach* 11630 (K, SRGH). S: Mwenezi [Nuanetsi] Dist., near Kapatensis 65 km NE of Malvernia, fl. 25.iv.1962, *Drummond* 7718 (K, SRGH). **Mozambique**. N: Cabo Delgado, 5 km NNW of Quiterajo on track along N margins of Namacubi Forest, fr. 23.xi.2009, *Goyder et al.* 6131 (K, LMA). Z: 30 km N of Quelimane, fl. 20.viii.1962, *Wild* 5877 (K, SRGH). MS: Between Divinhe and Cherinda, fl. 2.ix.1961, *Leach* 11258 (K, SRGH). GI: Inhambane, Pomene, fr. 24.ix.1980, *Jansen et al.* 7498 (K, LMU). M: Matatuine Dist., Licuati Forest Reserve, fl. 19.vi.2009, *Timberlake* 5707 (K, LMA).

Widespread in eastern and southern Africa, tropical Arabia, Madagascar and the Mascarene islands. Dry scrub, often on rock; from sea level to c.1300 m.

2. **Cynanchum viminale** (L.) Bassi in Bononiensi Sci. Inst. Acad. Comment. **6**(Opusc.): 17 (1768). —Goyder in F.T.E.A., Apocynaceae (part 2): 479 (2012), as *C. viminale* (L.) L. Types: Alpino, De Plantis Aegypti: 190, t.53 (1640), lectotype, designated by Liede & Meve in Bot. J. Linn. Soc. **112**: 2 (1993); *Bassi* s.n. in Herb. Linn. 308.1 (LINN epitype), designated by Liede & Meve in Bot. J. Linn. Soc. **118**: 47 (1995).
    Euphorbia viminalis L., Sp. Pl.: 452 (1753).
    Cynanchum aphyllum L., Syst. Nat., ed. 12 **3**: 235 (1768), superfluous name, based on *Euphorbia viminalis* L.
    Sarcostemma viminale (L.) R. Br. Prodr. Fl. Nov. Holland.: 464 (1810).
    Sarcostemma 'Taxon 3' of B.R. Adams & R.W.K. Holland in Cact. Succ. J. (Los Angeles) **50**: 108, 168 (1978).

An extremely variable succulent perennial with white latex; lacking rhizomes or stolons, but occasionally some stems rooting at the tips; stems erect, scrambling or twining to several metres, glabrous or minutely pubescent. Leaves reduced to minute scales. Inflorescences umbelliform. Sepals c.1 mm long. Corolla cream yellow or green, united only at the base, lobes spreading, 4–6 × 2–2.5 mm, ovate-oblong, glabrous. Corona gynostegial, in two series, white; outer corona forming a fluted tube; inner corona lobes staminal, attached to the back of the anthers, and ±

reaching top of the gynostegium, ovoid, fleshy. Anther wings 0.5–0.7 mm long. Corpusculum c.0.25 mm long, ovoid-subcylindrical; translator arms slender, c.0.15 mm long; pollinia c.0.3 × 0.1 mm, oblanceolate, somewhat flattened. Stylar head obscured by anther appendages. Follicles occurring singly, fusiform, tapering to an attenuate apex. Seeds flattened, discoid.

Subsp. **suberosum** (Meve & Liede) Goyder in Kew Bull. **63**: 472 (2008); in F.T.E.A., Apocynaceae (part 2): 480 (2012). Type: South Africa, Mpumalanga, Komatipoort [Transvaal, Komati Poort], *Kirk* 97 (K holotype).

Sarcostemma viminale subsp. *suberosum* Meve & Liede in Bot. J. Linn. Soc. **120**: 35 (1996).
Sarcostemma viminale sensu B.R. Adams & R.W.K. Holland in Cact. Succ. J. (Los Angeles) **50**: 108, 167 (1978).

Plants scrambling for several metres, with one or occasionally more main stems, much branched, the branching frequently opposite and regular, glabrous. Inflorescences mostly sessile at the nodes, but in addition some frequently as at the tips of lateral shoots; pedicels 5–9 mm long, minutely pubescent. Corolla lobes c.5 × 2 mm. Gynostegium (including corona) ± as tall as broad. Outer corona c.1 mm long; inner corona lobes c.1 mm long. Anther wings c.0.7 mm long, the margins extended basally to form short tail-like appendages.

Botswana. N: 56 km E of Makalamabedi, fl. 21.iii.1965, *Wild & Drummond* 7213 (K, SRGH). **Zambia**. N: Mbala [Abercorn], Lake Chila, fl. 6.xi.1958, *Robson & Fanshawe* 504 (K). S: Mabviya, Nanzhila, Kafue National Park, fl. 11.iv.1964, *Mitchell* 25/13 (K). **Zimbabwe**. N: Sebungwe, fl. vi.1956, *Davies* 1999 (K, SRGH). W: Matopos, fl. iii.1918, *Eyles* 966 (K). C: 61 km W of Harare on road to Bulowayo, 9.v.1989, *Goyder & Adams* 3230 (K, MAL, MO, PRE). E: Sabi Valley, Chimanimani [Melsetter] Junction, fl. 3.i.1964, *Pole-Evans* 6688 (K). S: Chitsa's Kraal, on way to Sabi River Gorge, Gonarezhou Game Reserve, fl. 1.vi.1970, *Ngoni* 149 (K, SRGH). **Malawi**. N: Rumphi Gorge, fl. 9.iii.1975, *Pawek* 9141 (K, MAL, MO, SRGH, UC). S: western slopes of Mt Mchese, fr. 10.vi.1987, *Chapman & Chapman* 8583 (K). **Mozambique**. N: Cabo Delgado, hunting concession between Pundanhar and Nangade, 18.xi.2009, *Goyder et al.* 6086 (K, LMA). MS: Mt Zembe, S of Chimoio [Vila Pery], fl. 15.vi.1959, *Leach* 9117 (K, SRGH). GI: Inhambane, Massinga, Pemene, fl. 11.vii.1981, *de Koning & Hiemstra* 8973 (K, LMU). M: Maputo, road towards Goba, Macanda, fl. 13.iv.1979, *Schäfer* 6729 (K, LMU).

Widespread across tropical and southern Africa. Dry sandy soils or rocky ground; sea level to c.1600 m.

This is the most common and widespread form of the species in Africa.

3. **Cynanchum doleriticum** Goyder, sp. nov. Similar in habit to *Cynanchum resiliens* with stems arching and rooting at the nodes, but flowers with slightly more spreading corolla lobes and with basal projections to the anther wings; it differs also in habitat preference, shallow soil over doleritic domes. Type: Zimbabwe, 22 km S of Mutare [Umtali] on Birchenough Bridge road, fl. 20.xi.1960, *Leach & Chase* 10496 (K holotype, SRGH).

Succulent perennial with white latex; stems arching and rooting in contact with the soil, very regularly and stiffly branched, glabrescent. Leaves reduced to minute scales. Inflorescences at the tips of short lateral shoots, umbelliform; pedicels c.5 mm long, pubescent. Sepals c.1 mm long, ovate-triangular, pubescent. Flower buds c.5 mm long, ovoid, the lobes not contorted in bud. Corolla united only at the base, lobes spreading to suberect, 5–6 × 2 mm, ovate to oblong and plane to somewhat replicate, glabrous. Gynostegial corona, in two series; outer corona forming a fluted tube c.1 mm long, the staminal portions slightly longer than the interstaminal sections; inner corona lobes staminal, attached to the back of the anthers, and ± reaching top of the gynostegium, ovoid, fleshy, 1–1.5 mm long. Anther wings 0.7–0.9 mm long, drawn out basally to form short tails. Stylar head obscured by anther appendages. Follicles occurring singly, 5–9 × 0.5 cm, fusiform, tapering to an attenuate apex. Seeds flattened, discoid, c.6 × 2 mm, brown; coma c.1 cm long, white.

Zimbabwe. E: 22 km S of Mutare [Umtali], fl. 20.xi.1960, *Leach & Chase* 10496 (K holotype, SRGH).

Known from several collections on outcropping domes of doleritic rock in the Rusape–Mutare–Nyanga area. A single collection (*Fanshawe* 9217) from Machili in SW Zambia may also belong to this species; 1000–1900 m.

Cynanchum resiliens, to which these collections have been referred in the past, is restricted to coastal Kenya where it inhabits a sandy, non-rocky substrate. The morphological differences seem slight, but the anther wing tails suggest affinities with the widespread and highly variable *C. viminale* rather than *C. resiliens*, and the habitat preference is also anomalous for *C. resiliens*.

4. **Cynanchum mulanjense** (Liede & Meve) Liede & Meve in Kew Bull. **67**: 753 (2012). Type: Malawi, Mt Mulanje, Chitakale stream, 27.xi.1985, *Chapman & Chapman* 6982 (MO holotype).

> *Sarcostemma mulanjense* Liede & Meve in Novon **2**: 223 (1992).
> *Cynanchum viminale* subsp. *mulanjense* (Liede & Meve) Goyder in Kew Bull. **63**: 471 (2009).

As for *C. viminale*, but stems not twining, forming dense patches of erect or scrambling, glabrous stems; gynostegium taller than broad.

Malawi. N: S Vipya forest, c.25 km SW of Mzuzu, fl. 6.ii.1992, *Goyder et al.* 3623 (BR, K, MAL). C: Dedza, Chiwao Hill, fl. 21.xi.1966, *Salubeni* 462 (K, MAL, SRGH). S: Mt Mulanje, slopes of Namisili above Sombani Basin, fl. 16.v.1981, *Chapman & Patel* 5705 (K, MAL). **Mozambique**. N: Nampula, Mecubúri, Serrão Chinga, fl. 1.vi.1968, *Macedo et al.* 3327 (K, LMA). Z: Mt Namuli, top of Licungo valley, fl. 26.xi.2007, *Harris et al.* 473 (K, MAL, LMA).

Known only from Malawi and northern Mozambique. On rocks outcrops; 1400–2400 m.

Collections from central and northern Malawi lack the elongated gynostegium typical of the species, but are otherwise indistinguishable from it.

5. **Cynanchum oresbium** (Bruyns) Goyder in Kew Bull. **63**: 471 (2009). Type: Mozambique, Nampula, mountains SW of Ribáuè, 19.xi.2000, *Bruyns* 8607 (BOL holotype).

> *Sarcostemma oresbium* Bruyns in Kew Bull. **58**: 433 (2003).

Clump-forming succulent perennial; latex colour not recorded. Stems erect to spreading, 3–15 cm tall, 1.5–3 mm thick, sometimes rhizomatous, constricted at the nodes, glabrous or subglabrous, green or suffused with pink. Leaves reduced to minute scales, caducous. Inflorescences sessile, umbelliform, at or near tip of stem; pedicels 10–12 mm long, erect but decurved above so that flowers are nodding; sepals c.1 mm long. Corolla campanulate, divided almost to the base, glabrous, outside cream suffused with pink, inside cream suffused with pink towards base; lobes 5–7 mm long, 2.5–3 mm broad, forming a tube around the gynostegium then spreading slightly towards the tips. Gynostegial stipe conspicuous at least in dried material, 2 mm long; gynostegial corona at the base of the stipe, c.1 mm tall, tubular, with a denticulate-emarginate rim; staminal corona lobes c.0.8 mm long ovoid, fleshy, adpressed to backs of anthers. Apex of stylar head shortly exserted beyond the anther-appendages Follices and seed not seen.

Mozambique. N: Mt Wesse, 35 km E of Malema, fl. 18.xii.2000, *Bruyns* 8597 (BOL, E, K).

Known only from two localities in Nampula Province – note that the locality above was cited incorrectly as "W of Malema" by Bruyns in Kew Bull. (2003). Crevices of steep granite cliffs or in shallow soil on granite domes; 700–900 m.

6. **Cynanchum mossambicense** K. Schum. in Engler, Pflanzenw. Ost-Afrikas **C**: 323 (1895). Type: Mozambique, Quelimane [Quillimane], *Stuhlmann* 843 (K lectotype, B†, HBG), designated by Liede in Ann. Missouri Bot. Gard. **83**: 325 (1996).

   Cynanchum complexum N.E. Br. in Bull. Misc. Inform. Kew **1895**: 256 (1895). Type: Mozambique, Shupanga forest, x.1887, *Scott* s.n. (K lectotype), designated by Liede in Ann. Missouri Bot. Gard. **83**: 325 (1996) citing the locality as "Mapanga".

   Slender twining perennial to several metres; latex white. Stems glabrous or occasionally pubescent. Abbreviated leafy shoots not present in leaf axils. Leaves with petiole 1–3 cm long, glabrous or sparsely pubescent; lamina 2–5 × 1–4 cm, broadly ovate-triangular, apiculate, base deeply cordate with rounded auricles, glabrous. Inflorescences extra-axillary, umbelliform, with 5–12 flowers; peduncle 10–25 mm long, glabrous or sparsely pubescent; pedicels 5–7 mm long, glabrous or sparsely pubescent. Sepals 1–1.5 mm long, ovate, glabrous. Flower buds 5–7 mm long, ovoid. Corolla rotate, united only at the base, lobes 6–7 × 1.5–2 mm, oblong, white, cream or yellow, glabrous. Corona white, 5–6 mm tall, exceeding and totally obscuring the gynostegium, tubular in lower half, fluted above and extending apically into 10 acuminate lobes. Anther wings c.1 mm long. Follicles 5–7 cm long, with an attenuate beak. Seeds 5–6 × 2–3 mm, flattened, ovate, brown; coma c.2 cm long.

   Zimbabwe. N: Darwin Dist., upper reaches of Nyarandi River, fl. 27.i.1960, *Phipps* 2426 (K, SRGH). S: Gonarezhou [Gona-re-zhou], Chipinda Pools, N bank of Lundi River, fl. 28.v.1971, *Grosvenor* 554 (K, SRGH). **Malawi**. S: Shire Valley above the Cataracts, fl. & fr. viii.1861, *Kirk* s.n. (K syntype). **Mozambique**. Z: 30 km N of Quelimane, fl. 20.viii.1962, *Wild* 5882 (K, SRGH). MS: near Gondola, fl. & fr. 16.vi.1957, *Pole-Evans* 5226 (K). GI: Vilanculos, fl. 27.viii.1968, *Balsinhas* 1314 (LMA).

   Also known from South Africa and Swaziland. Coastal dunes and riverine vegetation; 0–1000 m.

7. **Cynanchum umtalense** Liede in Ann. Missouri Bot. Gard. **83**: 341 (1996). Type: Zimbabwe, Chirinda Forest, i.1962, *Goldsmith* 1/62 (K holotype, BR, FT, P, SRGH).

   Slender twining perennial to several metres; latex white. Stems pubescent with spreading indumentum. Abbreviated leafy shoots not present in leaf axils. Leaves with petiole 1–3.5 cm long, pubescent; lamina 4.5–8 × 1.5–6 cm, oblong to ovate or elliptic, apex rounded to subacute, apiculate, base truncate to weakly cordate, pubescent on both surfaces. Inflorescences extra-axillary, umbelliform; peduncle 5–25 mm long, pubescent; pedicels 5–12 mm long, pubescent. Sepals 1–2 mm long, triangular, densely pubescent. Flower buds 5–6 mm long, ovoid-conical, the corolla lobes contorted in bud. Corolla rotate, united only at the base, lobes 5–6 mm long, narrowly triangular, reddish, or white with three longitudinal red stripes, outer face glabrous, upper surface pubescent. Corona white, cupular, c.2 mm long, enclosing the staminal column, staminal portions extended into erect lobes 0.5–1 mm long. Anther wings 0.7–0.8 mm long. Stylar head with a filiform appendage exserted beyond the stamens for 2–3 mm, apically bifid. Follicles lanceolate in outline with a long tapering beak, densely pubescent. Seeds 5–7 × 3–4 mm, flattened, ovate; coma c.2 cm long.

   Zimbabwe. E: Vumba Mts, fl. 28.iv.1957, *Chase* 6465 (B, FT, K, SRGH). **Malawi**. N: 5 km S of Chikangawa, fl. 10.vii.1978, *Phillips* 3515 (K, MAL, MO, SRGH, WAG). C: Dedza Mtn, fl. 5.iv.1978, *Pawek* 14228 (BR, K, MAL, MO, WAG, UC).

   Restricted to the highlands of eastern Zimbabwe and Malawi. Margins of montane and sub-montane forest; 1100–1800 m.

8. **Cynanchum schistoglossum** Schltr. in J. Bot. **33**: 271 (1895). —Goyder in F.T.E.A., Apocynaceae (part 2): 490 (2012). Type: South Africa, KwaZulu-Natal, Stanger, Phoenix, iv.1895, *Schlechter* 7090 (B neotype, AMD, BM, GZU, K, S, US), designated by Liede in Ann. Missouri Bot. Gard. **83**: 337 (1996).

   Cynanchum brevidens N.E. Br. in Bull. Misc. Inform. Kew **1895**: 257 (1895). Type: Congo, ix.1863, *Burton* s.n. (BM holotype).

Cynanchum brevidens var. *zambesiaca* N.E. Br. in Bull. Misc. Inform. Kew **1895**: 257 (1895). Type: Mozambique, Expedition Island, vii.1838, *Kirk* s.n. (K holotype).

Cynanchum vagum N.E. Br. in Bull. Misc. Inform. Kew **1895**: 257 (1895). Type: D.R. Congo, Stanley Pool, 26.viii.1888, *Hens* 77 (K holotype, BR, L).

Cynanchum minutiflorum K. Schum. in Bull. Soc. Roy. Bot. Belgique **37**: 123 (1898), illegitimate name. Type as for *C. vagum*.

Cynanchum dewevrei De Wild. & T. Durand in Ann. Mus. Congo Belge, Bot. sér. 2 **1**(2): 42 (1900). Type: D.R. Congo, Mwanana Toumbwé, 27.vii.1890, *Dewèvre* 904 (BR lectotype), designated by Liede in Ann. Missouri Bot. Gard. **83**: 337 (1996).

Slender twining perennial to c.4 m, latex white; stems branched, pubescent with spreading white hairs. Abbreviated leafy shoots frequently present in leaf axils giving the appearance of broadly ovate stipules. Leaves with petiole 1–2.5 cm long, densely pubescent; lamina 3.5–5(7.5) × 1.5–3(4) cm, ovate, lanceolate or oblong, apex acute to attenuate, base weakly to strongly cordate, pubescent on both surfaces. Inflorescences extra-axillary, forming a subumbelliform cluster of flowers along a rachis which develops with age, 3–10 flowers open at one time; peduncle 3–6 mm long, pubescent; rhachis 1–2(15) mm long; bracts c.1 mm long, triangular, pubescent; pedicels 3–8 mm long, slender, pubescent. Sepals 0.5–1(1.5) × c.0.5 mm, ovate or lanceolate, pubescent. Flower buds 1–1.5 mm long, globose, the corolla lobes not contorted in bud. Corolla rotate or campanulate, lobes united only at the base, 1.5(2) × 0.7–1 mm, ovate, yellowish green or cream, glabrous. Corona white, united into a fluted tube c.0.5–1 mm long, ± as tall as the staminal column; staminal portions of the tube each extending for a further 0.7–1 mm into an oblong or attenuate lobe, generally with angular shoulders at the base of the lobe, adaxial appendage absent; interstaminal portions not or only barely extending beyond the mouth of the corona tube, thinner than the staminal portions. Anther wings c.0.4 mm long. Corpusculum c.0.15 mm long, ovoid; translator arms c.0.1 mm long, flattened; pollinia attached laterally to translator arms, c.0.2 mm long, ovoid. Stylar head obscured by anther appendages. Follicles occurring singly, 4.5–7 × 0.5 cm, narrowly lanceolate in outline, tapering gradually to an obtuse or slightly clavate tip, smooth or longitudinally ridged, brown, pubescent. Seeds c.4.5–6 × 3.5–4 mm, ovate, smooth, brown, with a paler marginal rim; coma 1.5–2 cm long.

Botswana. N: Northern Dist., Mojeye area, fl. 26.iii.1975, *P.A. Smith* 1297 (K, SRGH). **Zambia**. C: Mt Makulu research Station near Chilanga, fl. 24.iii.1962, *Angus* 3078 (K, FHO, MRSC). E: Luangwa Valley, Musandile, Nsefu, fl. 12.iv.1968, *Phiri* 158 (K). S: Kasungulu, fl. v.1910, *Gairdner* 546 (K). **Zimbabwe**. W: Victoria Falls, S bank of Zambesi, fl. v.1915, *Rogers* 13125 (BOL). **Malawi**. N: Kondowe to Karonga, fl. & fr. vii.1896, *Whyte* s.n. (K). C: Namitete River on Lilongwe–Chipata [Fort Jameson] road, fl. 5.ii.1959, *Robson* 1464 (K). **Mozambique**. Z: Mocuba Dist., Namagoa, fl. 13.vii.1949, *Faulkner* 459 (K). T: Zambesi River, Baroda Province, Msusa, fr. 21.vii.1950, *Chase* 2230 (K, SRGH). MS: Gorongosa National Park, track from Chitengo camp towards Pungué River, fl. & fr. 24.vii.2006, *Goyder & Timberlake* 4093 (K, LMU).

Also recorded from D.R. Congo, Rwanda, Burundi, Uganda, Kenya, Tanzania, Angola, Namibia and South Africa (Northern Province and KwaZulu-Natal). Twining through grasses and scrub vegetation, riverine margins; 100–1600 m.

9. **Cynanchum ellipticum** (Harv.) R.A. Dyer in Mem. Bot. Surv. South Africa **17**: 138 (1937). Type: South Africa, near Grahamstown, *Bunbury* s.n. (K holotype).

Cynanchum capense Thunb., Prodr. Pl. Cap.: 47: (1794), illegitimate name, not L.f., Suppl.: 168 (1782). —Liede in Bot. Jahrb. Syst. **114**: 511 (1993). Type: *Thunberg* s.n. (UPS 6289 holotype).

Bunburia elliptica Harv., Gen. S. Afr. Pl.: 417 (1838).

Slender twining perennial to 3 m; latex white. Stems glabrous. Abbreviated leafy shoots (pseudo-stipules) frequently present in leaf axils. Leaves with petiole 1–2 cm long, glabrous; lamina 2–5 × 1–3 cm, ovate to oblong or elliptic, apex rounded to subacute and shortly apiculate, base rounded or truncate, rarely very weakly cordate, glabrous. Inflorescences extra-axillary, subumbelliform on a rachis which develops with age; peduncle 5–10 mm long, glabrous; pedicels

4–7 mm long, glabrous. Sepals c.1 mm long, glabrous. Flower buds 1.5–2 mm long, ovoid, the corolla lobes not contorted in bud. Corolla rotate, united only at the base, lobes 2–2.5 mm long, oblong, white, glabrous. Corona white, united into a broad tube 1.5–2 mm long, exceeding the staminal column and completely obscuring it, upper margin undulate or crenulate, not strongly lobed. Anther wings c.0.5 mm long. Stylar head not exserted beyond anthers. Follicles 4.5–6 cm long, narrowly fusiform, the apex drawn out into a long beak; seeds 6–7 × 3 mm, flattened, ovate; coma c.2.5 cm long.

Mozambique. GI: Xai-Xai [João Belo], Praia Sepulveda, fl. & fr. 13.viii.1957, *Barbosa & Lemos* 7824 (K, LMA). M: Ilha de Inhaca, near Biological station, fl. 19.vi.1973, *Correia & Marques* 2877 (K, LMU).

Also recorded from eastern South Africa. In the Flora area restricted to coastal dune scrub or forest, but extending inland to higher elevations in South Africa; 0–50 m.

10. **Cynanchum rungweense** Bullock in Kew Bull. **10**: 622 (1956). —Goyder in F.T.E.A., Apocynaceae (part 2): 483 (2012). Type: Tanzania, Mbeya, below Mporoto, Inkuyu, 17.iii.1932, *St. Clair-Thompson* 846 (K holotype). FIGURE 7.3.**202**.

Slender twining perennial to 6 m with white latex; stems branched, glabrous or pubescent. Abbreviated leafy shoots usually present in leaf axils. Leaves with petiole 0.5–3.5 cm long, sparsely pubescent; lamina 3.5–7 × 1–3 cm, oblong, apex apiculate, base deeply cordate with rounded auricles, dark green above, glaucous beneath with red-purple veins, sparsely pubescent adaxially, glabrous beneath. Inflorescences extra-axillary, forming a subumbelliform cluster of flowers along a simple or branched rachis which develops with age, 6–9 flowers open at one time; peduncle 6–35 mm long, glabrous or pubescent; rhachis to 15 mm long; bracts minute; pedicels 10–20 mm long, glabrous or pubescent. Sepals c.2 × 1–1.5 mm, ovate, glabrous or pubescent. Flower buds 5–7 mm long, ovoid, the corolla lobes not contorted in bud. Corolla rotate, united only at the base, lobes 6–7 × 2–2.5 mm, oblong, yellow-green, frequently tinged with purple, glabrous. Corona white or cream at the base, yellow-green above, united into an urceolate tube 4–5 mm long, exceeding the staminal column and completely obscuring it; mouth of tube irregularly dentate; adaxial appendages absent. Anther wings c.1 mm long. Corpusculum c.0.3 mm long, obovoid; translator arms c.0.5 mm long, flattened and geniculate; pollinia attached apically to translator arms, c.0.6 × 0.25 mm, narrowly obovoid. Stylar head obscured by anther appendages. Follicles not seen.

Zambia. E: Nyika Plateau, S of Zambian Rest House, 21.v.1989, *Goyder et al.* 3269 (K, MAL, NDO). **Malawi**. N: Viphya Plateau, SW of Dam 111 on the Chosi road, fl. 24.ii.1976, *Phillips* 1273 (K, MAL, MO).

Also known from the Southern Highlands of Tanzania. Montane forest; 2000–2400 m.

11. **Cynanchum praecox** S. Moore in J. Bot. **40**: 256 (1902). —Goyder in F.T.E.A., Apocynaceae (part 2): 481 (2012). Type: Zimbabwe, Harare [Salisbury] Dist., valley of Mazoe River, ix.1898, *Rand* 512 (BM holotype, K).

   Cynanchum pygmaeum Schltr. in Bot. Jahrb. Syst. **51**: 140 (1913). Types: Cameroon, *Ledermann* 2226 & 2230 (B† syntypes); Bamenda, c.3 miles from Kumbo on Oku road, 15.ii.1958, *Hepper* 2011 (K neotype), designated by Liede in Ann. Missouri Bot. Gard. **83**: 331 (1996).

Low pyrophytic perennial from slender vertical rootstock which sometimes clearly arises from a horizontal rhizome, latex white; leafy shoots developing after flowering, stems pubescent with white hairs. Abbreviated leafy shoots not present in leaf axils. Leaves sessile; lamina 2.5–6 × 0.5–1.2 cm, linear to ovate, apex acute, base cuneate, glabrous or sparsely pubescent. Inflorescences forming a subumbelliform head of 5–25(50) flowers at ground level, irregularly branched; bracts 1–2 mm long, ovate, pubescent; pedicels 5–13(20) mm long, pubescent. Sepals 1–1.5 × 0.3–0.5 mm, ovate or oblong, pubescent. Flower buds c.3 mm long, conical, the corolla lobes contorted in bud. Corolla rotate, united only at the base, lobes 3–5 × 1 mm, oblong, somewhat twisted

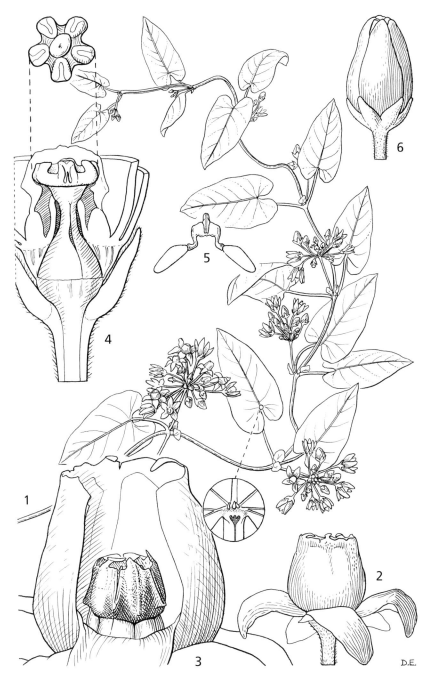

Fig. 7.3.**202**. CYNANCHUM RUNGWEENSE. 1, habit (× ²⁄₃); 2, flower with tubular corona completely obscuring the gynostegium (× 5); 3, flower with corona partially deflected to expose gynostegium (× 11); 4, flower with stamens half cut away to expose gynoecium of two ovaries and a stylar head, the latter also viewed from above (× 11); 5, pollinarium (× 18); 6, flower bud (× 8). 1 from *Richards* 6737; 2–5 from *Richards* 6807; 6 from *Stolz* 1983. Drawn by D. Erasmus. Reproduced from Flora of Tropical East Africa (2012).

distally, olive-green, yellow-brown or brownish purple, glabrous. Corona white or cream, united into a fluted tube 1–1.5 mm long ± as tall as the staminal column, glabrous or papillose; staminal portions of the tube each extending for a further 1–1.5 mm into an oblong lobe with a rounded apex, channelled on the inner face, and continuing along the inner face of the tube as paired longitudinal ridges; adaxial appendage absent; interstaminal portions of the tube ending in a broad, recurved lobe c.0.5 mm long. Anther wings c.0.7 mm long. Corpusculum c.0.2 mm long, ovoid; translator arms c.0.1 mm long, flattened; pollinia attached laterally to translator arms, c.0.2 mm long, ovoid. Stylar head conical, shortly exserted from anther appendages. Follicles (immature) occurring singly, pubescent.

Zambia. C: Muchinga Province, Mutinondo Wilderness Area, 11.x.2015, *Dreschler* s.n. (K). N: Kara road c.5 km from Kawimbe, fl. 29.viii.1956, *Richards* 6027 (K). **Zimbabwe**. C: Harare [Salisbury], Avondale, Lomagundi road, fl. 6.x.1955, *Drummond* 4893 (K, SRGH). **Malawi**. N: Vipya Plateau, 5 km along Vipya Link road, fl. 25.ix.1972, *Pawek* 5814 (K, MAL, MO).

Scattered localities across tropical Africa from Sierra Leone to Zimbabwe. Flowering in bare ground in burnt upland or montane grassland, anticipating the early rains; 1500–2300 m.

12. **Cynanchum orangeanum** (Schltr.) N.E. Br. in Fl. Cap. **4**(1): 745 (1908). Type: South Africa, Orange River near Bethulie, fl. xii.1892, *Flanagan* 1502 (SAM lectotype, B†, BOL), designated by Liede in Bot. Jahrb. Syst. **114**: 531 (1993).

   Flanagania orangeana Schltr. in Bot. Jahrb. Syst. **18**(Beibl. 45): 10 (1894).

Perennial herb with annual stems from a slender vertical rootstock arising from a horizontal rhizome; latex white. Stems erect or ascending, 10–15 cm tall, pubescent with crisped white hairs. Abbreviated leafy shoots not present in leaf axils. Leaves sessile; lamina 2–7 × 0.5 cm, linear, densely pubescent. Inflorescences extra-axillary, umbelliform, subsessile, of 2–4 flowers; pedicels 3–4 mm long, pubescent. Sepals 1–1.5 mm long, triangular, pubescent. Flower buds c.3 mm long, broadly ovoid, the corolla lobes not contorted in bud. Corolla rotate, united only at the base, lobes 3–4 × 1.5 mm, ovate with an attenuate apex and recurved margins, off-white, perhaps tinged with brown or purple outside, glabrous or with a few hairs only. Corona white or cream, united into a fluted tube 1–1.5 mm long ± as tall as the staminal column, glabrous; staminal and interstaminal portions of the tube each extending for a further 0.5 mm into a broadly triangular lobe to form 5 outward-folded lobes and 5 inward-folded lobes. Anther wings c.0.4–0.5 mm long. Stylar head conical, only shortly exserted from anther appendages. Follicles 4–7 cm long, fusiform, beaked. Seeds 5–6 mm long, flattened, ovate; coma 1.5–2 cm long.

Botswana. N: Groot Laagte (East) River valley, fr. 17-18.iii.1980, *P.A. Smith* 3234 (K, PSUB, SRGH). SW: Central Kalahari Game Reserve, Kalahari Plains Camp, fl. 30.xi.2011, *A. & R. Heath* 2184 (GAB, K).

Also in Namibia and South Africa. Recorded from sandy soil on flood plain margin, and in clay overlying calcrete in fossil lake bed; c.1000 m.

13. **Cynanchum obtusifolium** L.f., Suppl. Pl.: 169 (1782). —Brown in Fl. Cap. **4**(1): 749 (1908). Type: *Thunberg* s.n. (UPS 6311 holotype).

   Cynoctonum brownii Meisn. in London J. Bot. **2**: 546 (1843). Type as for *Cynanchum obtusifolium* L.f.
   Cynanchum obtusifolium var. *pilosum* Schltr. in Bot. Jahrb. Syst. **18**(Beibl. 45): 10 (1894). —Brown in Fl. Cap. **4**(1): 750 (1908). Type: South Africa, near Kei mouth, 1892, *Flanagan* 382 (SAM lectotype, BOL, GRA), designated by Liede in Bot. Jahrb. Syst. **114**: 527 (1993).
   Vincetoxicum obtusifolium (L.f.) Kuntze, Revis. Gen. Pl. **3**(2): 200 (1898).

Twining perennial to 3 m; latex white. Stems glabrous or pubescent. Abbreviated leafy shoots (pseudo-stipules) frequently present in leaf axils. Leaves semisucculent; petiole 1–3.5 cm long, glabrous or pubescent; lamina 2–5 × 1.5–4.5 cm, oblong or suborbicular, apex rounded or emarginate, shortly apiculate, base rounded or truncate to weakly cordate, glabrous or densely pubescent beneath. Inflorescences extra-axillary, sessile, umbelliform; pedicels 3–5 mm long, pubescent. Sepals c.2 mm long, pubescent with white hairs. Flower buds 3–4 mm long, ovoid, the corolla lobes not contorted in bud. Corolla rotate, united only at the base, lobes 4–5 mm long, oblong, green, sparsely pubescent outside. Corona white, united into a tube 1–1.5 mm long, about as tall as the staminal column, with 5 erect lobes 1–1.5 mm long opposite the stamens. Anther wings c.0.5 mm long. Stylar head not exserted beyond anthers. Follicles 4–6 cm long, fusiform, the apex attenuate; seeds 6–8 × 3–5 mm, flattened, ovate; coma c.3 cm long.

Mozambique. M: Ponta Mamoli, on coast E of Zitundo, fl. 26.xi.2001, *Goyder* 5004 (K, LMU).

Distributed along the eastern and southern coasts of South Africa as far as the Cape Peninsula. In Mozambique occurring only in dune forest margins; close to sea level.

77. VINCETOXICUM Wolf[49]

Vincetoxicum Wolf, Gen. Pl.: 130 (1776). —Liede-Schumann & Meve in Phytotaxa **369**: 131 (2018).
   Tylophora R. Br., Prodr. Fl. Nov. Holland.: 460 (1810).
   Pentatropis Wight & Arn. in Wight, Contr. Bot. India: 52 (1834).
   Sphaerocodon Benth. in Bentham & Hooker f., Gen. Pl. **2**: 772 (1876).
   Pleurostelma Baill., Hist. Pl. **10**: 266 (1890). —Bullock in Kew Bull. **10**: 612 (1956).
   Podostelma K. Schum. in Bot. Jahrb. Syst. **17**: 133 (1893).
   Tylophoropsis N.E. Br. in Gard. Chron., ser. 3 **16**: 244 (1894).
   Strobopetalum N.E. Br. in Bull. Misc. Inform. Kew **1894**: 335 (1894).
   Microstephanus N.E. Br. in Bull. Misc. Inform. Kew **1895**: 249 (1895); in F.T.A. **4**(1): 288 (1902).
   Diplostigma K. Schum. in Engler, Pflanzenw. Ost-Afrikas **C**: 324 (1895).
   Goydera Liede in Novon **3**: 265 (1993).

Vines or pyrophytic herbs; latex clear but sometimes becoming cloudy on exposure to air. Inflorescences extra-axillary, frequently lax and with many flower clusters along an extended axis, more rarely few-flowered. Corolla deeply five-lobed, normally adaxially with indumentum. Corolline corona absent. Gynostegial corona consisting of five discrete lobes adnate to the staminal column below the anther wings, or vestigial (*V. cernuum*). Anther connectives with apical appendages. Pollinia more or less horizontal, rarely pendant or erect, globose to ovoid or cylindrical, minute. Stylar head flat or capitate, but beaked in *V. cernuum*. Follicles usually occurring singly, smooth, wingless; seeds smooth, winged, comose.

About 150 species with its centre of diversity in Old World tropics and subtropics.

Molecular evidence demonstrates that north-temperate *Vincetoxicum* has evolved independently more than once out of the more tropical *Tylophora* (Liede-Schumann *et al.* in Taxon **61**: 803–825 (2012); Liede-Schumann & Meve in Phytotaxa **369**: 129–194 (2018)). *Vincetoxicum* takes nomenclatural priority for the enlarged genus, which in its expanded form also includes a number of small genera centred on the western Indian Ocean (*Blyttia*, *Goydera*, *Pentatropis* and *Pleurostelma*), and others with a more Asian distribution.

1. Stylar head produced beyond the anthers into a beak; gynostegial corona lobes vestigial . . . . **1**. *cernuum*
 – Stylar head flat or capitate, not beaked; gynostegial corona of five discrete lobes 2
2. Pyrophytic herbs with erect stems, or stems or scrambling to no more than 60 cm 3
 – Slender or robust twiners, usually to several metres 4

[49] by D.J. Goyder

3. Leaves when present held stiffly erect; flowers minute, corolla lobed ± to the base . . . . **3.** *congolanum*
– Leaves not as above; flowers larger, corolla generally united for at least half of its length . . . . **2.** *caffrum*
4. Corolla green with minute velvety indumentum adaxially . . . . **4.** *anomalum*
– Corolla pink or purple, if green, then glabrous or with slender hairs adaxially – never densely velvety-pubescent 5
5. Inflorescences 2 to many times as long as subtending leaf, with 5–20+ sessile clusters of flowers distributed at regular intervals along the inflorescence axes; leaves deeply cordate at base . . . . **9.** *sylvaticum*
– Not as above 6
6. Leaves weakly to strongly cordate at the base, at least 5 cm long; corolla at least 1 cm across, lobes broad, c.3–4 mm wide and with a clearly displaced apex . . . . **5.** *conspicuum*
– Leaves rounded to cuneate at the base, if weakly cordate, then leaves less than 3 cm long; corolla not as above 7
7. Inflorescences sessile . . . . **8.** *stenolobum*
– Inflorescences pedunculate 8
8. Leaves 3.5–6 cm wide; whole plant glabrous . . . . **6.** *apiculatum*
– Leaves less than 3 cm wide; at least the young stems generally minutely pubescent 9
9. Leaves ovate to lanceolate with an apiculate apex; corona narrowing abruptly into a free apical appendage . . . . **7.** *iringensis*
– Leaves not as above; corona lobes adnate to staminal column for their entire length, lacking a free apical appendage 10
10. Leaves linear or linear-lanceolate, less than 3 mm wide . . . . **12.** *gracillimum*
– Leaves ovate, oblong or triangular, broader than 5 mm 11
11. Anther wings c.0.2 mm long; northern Malawi . . . . **10.** *tenuipedunculatum*
– Anther wings c.0.6 mm long; Eastern Highlands of Zimbabwe, Gorongosa . . . . **11.** *monticola*

1. **Vincetoxicum cernuum** (Decne.) Meve & Liede in Phytotaxa **369**: 137 (2018). Type: Tanzania, Pemba, *Bojer* s.n. (P holotype). FIGURE 7.3.**203**.

> *Astephanus cernuus* Decne. in Ann. Sci. Nat., Bot. sér. 2 **9**: 342 (1838).
> *Astephanus ovatus* Decne. in Ann. Sci. Nat., Bot. sér. 2 **9**: 342 (1838). Type: Madagascar, *Commerson* s.n. (P holotype).
> *Astephanus arenarius* Decne. in Candolle, Prodr. **8**: 507 (1844). Type: Madagascar, Vohémar, [collector not listed] 100 (P holotype).
> *Astephanus recurvatus* Klotzsch in Peters, Naturw. Reise Mossambique **6**(Bot. 1): 274 (1861). Type: Mozambique (B† holotype).
> *Pleurostelma grevei* Baill., Hist. Pl. **10**: 266 (1890). Type: Madagascar, Morondava, Be-Kapaké, *Grevé* 273 (P holotype).
> *Microstephanus cernuus* (Decne.) N.E. Br. in Bull. Misc. Inform., Kew 1895: 249 (1895); in F.T.A. **4**(1): 288 (1902).
> *Pleurostelma cernuum* (Decne.) Bullock in Kew Bull. **10**: 612 (1956). —Goyder in F.T.E.A., Apocynaceae (part 2): 464 (2012).

Woody herb or shrub, prostrate or twining to 2 m; stems glabrous, pubescent along one line with curved white hairs or occasionally pubescent all round. Leaves somewhat fleshy, the petiole (2)4–8(13) mm long, pubescent along the upper surface; leaf-blade oblong or elliptic to ovate or occasionally rhomboid, 10–40(55) mm long, 4–15(25) mm wide, glabrous or pubescent, apex acute, obtuse or rounded, commonly with a short apiculus, base generally acute, but occasionally obtuse or rounded, margins flat or revolute; lateral veins generally in two pairs beneath, forming an acute angle with the midrib and ending almost parallel with the margin.

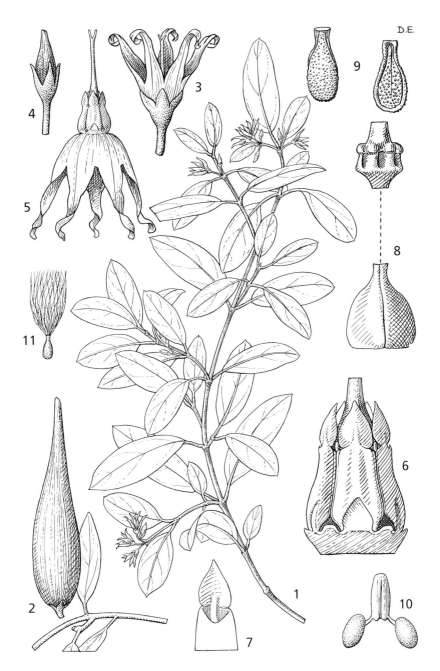

Fig. 7.3.**203**. VINCETOXICUM CERNUUM. 1, habit (× 1); 2, fruit (× 1); 3, flower (× 8); 4, flower bud (× 8); 5, flower with corolla folded back to expose gynostegium (× 10); 6, gynostegium including base of stylar head appendage (× 36); 7, stamen from within (× 36); 8, gynoecium, showing paired ovaries and base of stylar head (× 36); 9, seed, dorsal and ventral views (× 4); 10, pollinarium (× 40). 1, 3–10 from *Hildebrandt* 178; 2 from *Faulkner* 707. Drawn by D. Erasmus. Reproduced from Flora of Tropical East Africa (2012).

Inflorescences 1–4(6) flowered, axillary, umbelliform, glabrous to densely pubescent; peduncles to 7 mm, or absent; bracts subulate, 0.5–1 mm; pedicels 2–6 mm. Sepals ovate, acute, glabrous or pubescent, hyaline, at least at the margin. Corolla 4–8 mm long, white or cream, glabrous outside, the tube campanulate, ± 2 mm, enclosing the staminal column, sparsely pubescent within; lobes contorted, 2–6 mm, glabrous, tapering gradually to a blunt, slightly offset apex, with small coronal pockets formed at the sinus of 2 lobes within the tube. Staminal column ± 1 mm, forming the base of a narrow cone continued upward by the stylar head appendage. Gynostegial corona inconspicuous, formed of 5 triangular ridges barely projecting from the lower half of each anther and linked by minute pockets at the base of the column, alternating with the anthers. Anther appendages white, ovate, acute and erect-connivent around the base of the apical portion of the style. Stylar head projecting beyond the staminal column as a subulate beak, ± 2 mm long, sub-entire or deeply bifid. Follicles 40–60 mm long, ± 10 mm wide, reddish brown, usually only one of the pair developing. Seeds ± 4 mm long, 1.5 mm wide, reddish brown, crowned with an ivory plume of hairs.

Mozambique. N: Ilha de Moçambique, fl. ix.1887, *Scott* s.n. (K).

Also known from coastal Somalia, Kenya, Tanzania Madagascar and the Mascarenes. Bare ground or thickets on sand dunes or coral rock near the coast; sea level to 100 m.

2. **Vincetoxicum caffrum** (Meisn.) Kuntze in Rev. Gen. Pl. **2**: 424 (1981). Type: South Africa, 'Port Natal', viii.1839, *Krauss* 85 (TUB lectotype, BM), designated by Liede-Schumann & Meve in Phytotaxa **369**: 136 (2018). FIGURE 7.3.**204**.

   Tylophora caffra Meisn. in London J. Bot. **2**: 542 (1846). —Goyder in F.T.E.A., Apocynaceae (part 2): 498 (2012).

   Sphaerocodon obtusifolium Benth. in Hooker's Icon. Pl. **12**: 78, t.1190 (1876). —Brown in F.T.A. **4**(1): 412 (1903). Type: Malawi, Shire River near Moramballa, *Kirk* s.n. (K holotype).

   Sphaerocodon natalense Benth. in Hooker's Icon. Pl. **12**: 79 (1876). Type: South Africa, KwaZulu-Natal, *Gerrard* 1797 (K holotype).

   Vincetoxicum caffrum (Meisn.) Kuntze, Revis. Gen. Pl. **2**: 424 (1891).

   Gongronema welwitschii K. Schum. in Bot. Jahrb. Syst. **17**: 145 (1893). Type: Angola, *Welwitsch* 4196 (BM holotype, K, LISC).

   Sphaerocodon caffrum (Meisn.) Schltr. in J. Bot. **33**: 339 (1895).

   ?*Sphaerocodon platypodum* K. Schum. in Ann. Mus. Congo Belge, Bot. ser. 4 **1**: 225 (1903). Type: D.R. Congo, Lukafu, ii.1900, *Verdick* 411 (BR holotype).

Perennial with one or more annual erect or trailing stems to c.60 cm arising from a tuberous rootstock; stems minutely but densely pubescent. Leaves herbaceous; petiole 1–8 mm long, densely pubescent; lamina 2–9 × (0.7)1–4.5 cm, generally oblong-elliptic, sometimes ovate-oblong or obovate-oblong, apex rounded to subacute, frequently shortly apiculate, base rounded to truncate, frequently slightly asymmetric, indumentum sometimes restricted to main veins, subglabrous to sparsely pubescent adaxially, denser and more prominent abaxially. Inflorescences extra-axillary, solitary at each node, sparsely to densely pubescent; peduncles (to 1st cluster of flowers) 0.1–4 cm long; bracts 1–5 mm long, filiform to narrowly oblong, densely pubescent; pedicels 3–13 mm long, densely pubescent. Calyx lobes 2–7 × 0.5–2 mm, oblong, pubescent. Corolla globose in bud, campanulate when open, generally united for half to 2/3 of its length, rarely more deeply divided, green tinged with red or purple outside, deep reddish purple within, the centre occasionally pale, glabrous or sparsely pubescent at least towards the tip abaxially, subglabrous or sparsely pubescent within; extremely variable in size, 2–14 mm long; lobes 2–6 × 1.5–6 mm, oblong to broadly triangular, apex rounded, but margins sometimes somewhat revolute making the apex appear acute. Gynostegium subsessile or stipitate, 1.5 mm long. Corona forming a cream, papillose cylinder around the staminal column; free lobes 0.3–0.4 mm long, fleshy, suborbicular, maroon or purple, radiating from the top of the column just below the base of the anthers. Anther wings falcate, 0.2–0.4 mm long. Apex of stylar head flat. Follicles mostly occurring singly, c.4–6 × 1.5–2 cm, ovoid, smooth, with thick fleshy walls. Seeds 6–10 × 5–7 mm, oblong to ovate, winged; coma 0.7–2 cm long.

Fig. 7.3.**204**. VINCETOXICUM CAFFRUM. Main image, flowering shoot; 1, flower bud; 2, gynostegium with corona; 3, pollinarium. Drawn by M. Smith. Reproduced from Curtis's Botanical Magazine (1903).

Anther wings 0.3–0.4 mm long, generally with a basal tail; corolla 2–14 mm long; sprawling to erect subshrub with few stems; widely distributed across tropical and subtropical Africa . . . . .a) subsp. *caffrum*
Anther wings 0.2 mm long, lacking a conspicuous basal tail; corolla 2–4 mm long; a virgate subshrub mostly restricted to Kalahari sands of western Zambia and southern Angola . . . . .b) subsp. *melananthum*

a) Subsp. **caffrum**.

Anther wings 0.3–0.4 mm long, generally with a basal tail; corolla 2–14 mm long; sprawling to erect subshrub with few stems.

Zambia. B: Kataba, fl. 12.xii.1960, *Fanshawe* 5965 (K, NDO). N: Mbala [Abercorn] Distr., Sumbawanga road 12 km from Kawimbe, fl. 30.i.1957, *Richards* 8009 (K). W: SW of Matonchi Farm, Mwinilunga Distr., fl. 20.xi.1937, *Milne-Redhead* 3330 (K). C: between Undaunda and Rufunza 135 km E of Lusaka, fl. 2.i.1972, *Kornas* 770 (K). S: Mapanza, Choma Distr., fl. 27.xi.1958, *Robinson* 2938 (K, PRE). **Zimbabwe**. C: Harare [Salisbury], fl. 1.xii.1936, *Eyles* 8841 (K, SRGH). E: Vumba, Burma Valley, foothills of Himalayas, fl. 7.xii.1961, *Wild & Chase* 5563 (K, SRGH). **Malawi**. N: Vinthukhutu Forest Reserve 3 km N of Chilumba, fl. 6.i.1978, *Pawek* 13556 (K, MAL, MO, SRGH, UC). C: Mua-Livulezi Forest reserve, fl. 21.i.1959, *Robson & Jackson* 1284 (K). S: Bongwe Forest Reserve, Mulanje Distr., fl. 12.i.1983, *Tawakali & Nachamba* 128 (K, MAL). **Mozambique**. Z: 50 km NE of Mopeia [Mopeia Velha] on road to Quelimane, fl. 7.xii.1971, *Pope & Muller* 537 (K, SRGH).

Widespread across tropical and subtropical Africa from Sierra Leone to southern Sudan in the north, and from northern Namibia to KwaZulu-Natal in the south. Mixed deciduous woodland frequently on rocky slopes, or seasonally burned grassland; 1000–2400 m.

A polymorphic subspecies occurring across tropical Africa. Plants from open habitats tend to have smaller leaves and reduced corolla dimensions in comparison with plants from more wooded environments. They are sometimes also erect, whereas woodland plants often have a more sprawling habit. In the Flora region, two collections from Inyanga in eastern Zimbabwe (*Grosvenor* 778 and *Chase* 3563) have very small flowers, but as the plants have a sprawling habit and anther wing tails are present, these collections are assigned to this subspecies rather than subsp. *melananthum*.

b) Subsp. **melananthum** (N.E. Br.) Goyder, comb. et stat. nov.

Sphaerocodon melananthus N.E. Br. in F.T.A. **4**(1): 412 (March 1903). Type: Angola, Cuando Cubango, Rio Cuito, 'by the River Quito, below the River Longa', 10.xii.1899, *Baum* 526 (K holotype, BR, HBG, M, W).
Gymnema melananthum K. Schum. in Warburg, Kunene-Sambesi Exped.: 344 (May 1903). Type: Angola, Cuando Cubango, Rio Cuito, 'Unwelt des Kuito, auf Sandboden, 11.xii.1899', *Baum* 526 (B† holotype, K).
Sphaerocodon angolensis S. Moore in J. Bot. **47**: 219 (1909). Type: Angola, Kutchi, xii.1905, *Gossweiler* 4124 (BM holotype, K).

Anther wings 0.2 mm long, lacking a conspicuous basal tail; corolla 2–4 mm long; virgate subshrubs.

Zambia. S: Choma, Mapanza, 27.xi.1958, *Robinson* 2938 (K).
Mostly restricted to Kalahari sands of southern Angola, southern and western Zambia and possibly also adjacent parts of northern Botswana and Namibia; 1000–1200 m.

A distinctive small-flowered variant which is almost always erect and shrubby. It is treated here as a regional subspecies of *V. caffrum*.

3. **Vincetoxicum congolanum** (Baill.) Meve & Liede in Phytotaxa **369**: 139 (2018).
Type: Gabon, Ogôoué, 1884, *Thollon* 141 (P00151988 lectotype, P), designated by
Liede-Schumann & Meve in Phytotaxa **369**: 139 (2018).

> *Nanostelma congolanum* Baill., Hist. Pl. **10**: 248 (1890). —Brown in F.T.A. **4**(1): 411 (1903).
> *Tylophora orthocaulis* K. Schum. in Bot. Jahrb. Syst. **23**: 235 (1896). —Brown in F.T.A. **4**(1):
> 410 (1903). Type: *Afzelius* 138 (not traced).
> *Schizoglossum glanvillei* Hutch. & Dalziel in F.W.T.A. ed. 1 **2**: 58 (1931); in Bull. Misc.
> Inform. Kew **1937**: 340 (1937). Type: Sierra Leone, Kulufaga, in Sambaia Chiefdom, fl.
> 27.iv.1929, *Glanville* 192 (K lectotype, BM), designated here.
> *Tylophora congolana* (Baill.) Bullock in Kew Bull. **9**: 585 (1955); in F.W.T.A. ed 2 **2**: 96 (1963).

Perennial with one or more annual erect stems to c.100 cm arising from a cluster of thickened
roots; stems green, somewhat flattened and striate at least when dry, glabrous or minutely but
pubescent in one or two lines. Leaves herbaceous or reduced to scales, sessile; lamina of well-
developed leaves held erect, 2.5–6 × 0.5–1.5 cm, lanceolate or narrowly elliptic, apex acute,
base cuneate, glabrous. Inflorescences extra-axillary, solitary or paired at the nodes, minutely
pubescent, with clusters of pedicellate flowers at intervals along the inflorescence axis; pedicels
2–5 mm long, pubescent. Calyx lobes c.1 mm long, glabrous. Corolla ovoid in bud lobed ± to
the base; lobes c.2 mm long, reddish purple, glabrous. Gynostegium on a short stipe 0.5 mm
long. Corona lobes minute, fleshy, maroon or purple. Anther wings 0.1–0.15 mm long. Apex of
stylar head not exserted. Follicles occurring singly or in pairs, c.4–6 cm long, narrowly fusiform,
smooth, glabrous. Seeds not seen.

Zambia. C: Chakwenga Headwaters, 100–129 km E of Lusaka, fl. 10.i.1964, *Robinson*
6204 (K).

Also recorded from Guinea Conakry, Sierra Leone, Cameroon, Gabon, Central
African Republic, the D.R. Congo and Angola. Habitat recorded as "seasonally damp
bog"; c.1050 m.

This collection is anomalous in possessing only scale leaves – all other material of
Vincetoxicum congolanum I have examined has conventional leaves along most of the
stem, with scale leaves restricted just to the basal nodes. However, florally, and in
all other respects the Zambian collection appears identical to the others, so I have
treated it as an aberrant material of this species. Detailed habitat notes are lacking
for most collections, but the only mention of seasonally damp conditions is for the
Zambian plant. More material is needed to make any further judgement on the
identity of this taxon.

4. **Vincetoxicum anomalum** (N.E. Br.) Meve & Liede in Phytotaxa **369**: 132 (2018).
Types: South Africa, KwaZulu-Natal, 'Buck-bush, near Durban,' *Gerrard* 1320 ex
herb. *McKen* 4 (K lectotype, TCD), lectotype designated by Goyder in F.T.E.A.,
Apocynaceae (part 2): 501 (2012); 'Buck-bush, Umgeni,' *Gerrard* 1320 (BM
paralectotype, K, TCD).

> ?*Pentarrhinum coriaceum* Schltr. in J. Bot. **32**: 357 (1894). Type: South Africa, KwaZulu-
> Natal, *Gerrard & McKen* s.n. (B† holotype).
> *Tylophora anomala* N.E. Br. in Fl. Cap. **4**(1): 766 (1908). —Burrows in Pl. Nyika Plateau: 75
> (2005). —Goyder in F.T.E.A., Apocynaceae (part 2): 501 (2012).
> *Cynanchum chirindense* S. Moore in J. Bot. **46**: 305 (1908). Type: Zimbabwe, Chirinda
> Forest, ii.1906, *Swynnerton* 137 (BM holotype, K).
> *Cynanchum papillosum* Weim. in Bot. Not. **1935**: 398 (1935). Type: Zimbabwe, Inyanga,
> above Pungwe River, 17.xii.1930, *Fries, Norlindh & Weimarck* 3879 (LD holotype).
> *Tylophora* sp. B in Agnew, Upland Kenya Wild Fl., ed. 2: 181 (1994).

Twining to several metres, stems to 3 mm diam., glabrous or pubescent. Leaves semi-succulent,
leathery and frequently with prominent venation when dry; petiole 0.2–2 cm long, glabrous or
pubescent; lamina (3)4–13(17) × (1.2)2–7.5 cm, ovate to oblong or suborbicular, apex rounded
to acute, frequently apiculate or mucronate, base weakly to strongly cordate, occasionally

rounded or truncate, generally with a pair of colleters, completely glabrous or with indumentum of minute hairs on principal veins. Inflorescences extra-axillary, sessile or pedunculate, or with both at the same node, axis of pedunculate inflorescences frequently continuing beyond the first cluster of flowers, forming further clusters at intervals, axes generally simple or occasionally branched; peduncles (to 1st cluster of flowers) to 7 cm long, glabrous or minutely pubescent; bracts minute, triangular, ciliate; pedicels 3–11(20) mm long, glabrous or pubescent. Calyx lobes (0.5)1–1.5 mm long, narrowly to broadly triangular, pubescent or more rarely glabrous. Corolla green or cream, sometimes tinged with purple, densely velvety-pubescent with white hairs on upper surface, glabrous beneath, generally rotate, the lobes united at the base, but in occasional plants the lobes are united to about half-way, resulting in campanulate or even urceolate corollas; corolla lobes c.2–3 × 1–2 mm, triangular to ovate, apex rounded. Staminal corona lobes fleshy, erect, c.0.8–1.2 mm high, almost as tall as the gynostegium and free from it for much of their length, oblong or ovate in outline with a truncate apex, slightly flattened dorsoventrally. Anther wings 0.2 mm long. Stylar head flat. Follicles mostly occurring singly, to c.12 × 1 cm, narrowly lanceolate in outline, smooth or slightly winged. Seeds 6–10 × 3.5–5 mm, oblong to ovate, winged, glabrous or pubescent; coma 2.5–3.5 cm long.

Zambia. N: Mutinondo Wilderness Area, Choso Pool, 21.ii.2015, *Osborne et al.* 1022 (K, UZL). **Zimbabwe**. E: Cecil Hotel garden, Mutare [Umtali], fl. 9.ii.1965, *Chase* 8258 (K, SRGH). S: Gwasha Forest, Mt Buhwa, fl. 23.vii.1965, *West* 6685 (K, SRGH). **Malawi**. N: Wilindi Forest, Misuku Hills, fl. xi.1954, *Chapman* 255 (K, MAL). C: Nchisi Mountain, fl. 3.v.1980, *Blackmore et al.* 1388 (K, MAL). S: Zomba Plateau, Chingwe's Hole Forest, fl. 17.xii.1983, *Dowsett-Lemaire* 1108 (K). **Mozambique**. N: Mueda Plateau, 13.xii.2003, *Luke et al.* 10070 (EA, K). MS: Buzi, Inhafenga, Mabala R., fl. 28.vi.1964, *Carvalho* 747 (K, LMA). GI: Namagoa, Mocuba, fl. viii.1943, *Faulkner* Pretoria 140 (K, PRE). M: Licuati Forest Reserve c.20 km W of Bela Vista, fl. 7.xii.2001 *Goyder* 5035 (K, LMU).

Widely distributed across eastern Africa from Uganda and Kenya to southern Mozambique and KwaZulu-Natal. Morphologically similar material from Cameroon and Bioko appears separate in DNA analyses. Moist riverine forest or thicket; (0)1400–2200 m.

Meve in Kew Bull. (1999) suggested that *Pentarrhinum coriaceum* Schltr. might be conspecific with *Vincetoxicum anomalum*, although Schlechter's diagnosis makes no mention of the conspicuous indumentum on the adaxial face of the corolla, which throws some doubt on this view. Were *P. coriaceum* proved to be conspecific, its publication predates not only *V. anomalum*, but also *Tylophora coriacea* Marais (a new name for *T. laevigata* Decne.).

The inflorescence of *Vincetoxicum anomalum* is hugely variable in its architecture, varying from sessile subumbelliform clusters of flowers, through typical *Vincetoxicum* pedunculate zig-zag inflorescences with clusters of flowers on the angles of a single shoot, to many-branched inflorescences in a single leaf axil. Local forms are distinctive but are not given formal taxonomic recognition here as they do not correlate with other characters, and intermediates occur between the main inflorescence types.

5. **Vincetoxicum conspicuum** (N.E. Br.) Meve & Liede in Phytotaxa **369**: 139 (2018). Types: Angola, Golungo Alto, near Sobato Bumba, iii.1855, *Welwitsch* 4214 (K lectotype, BM, LISC), lectotype designated by Bullock in Kew Bull. **9**: 583 (1955); Angola, Golungo Alto, River Delamboa, *Welwitsch* 4215 (K paralectotype, BM, LISC).

    Tylophora conspicua N.E. Br. in Bull. Misc. Inform. Kew **1895**: 257 (1895); in F.T.A. **4**(1): 405 (1903). —Goyder in F.T.E.A., Apocynaceae (part 2): 502 (2012).

Twining to several metres, stems glabrous or minutely pubescent. Leaves not succulent; petiole 1.5–5 cm long, minutely pubescent; lamina to c.13(16) × 6(10) cm, elliptic to slightly obovate or oblong, apex mostly attenuate, base ± truncate to deeply cordate, glabrous. Inflorescences extra-axillary, with several spiral clusters of flowers arising at the angles of a weakly zig-zag axis, inflorescence axes subglabrous or minutely pubescent; peduncles (to 1ˢᵗ branching point) 3–5.5 cm long; bracts broadly triangular, pubescent; pedicels 5–9 mm long, minutely pubescent. Calyx lobes 1–1.5 mm long, ovate or triangular, apex acute or rounded, minutely pubescent, frequently with ciliate margins. Corolla dark reddish purple, glabrous, rotate, the lobes united at the base for c.2 mm; corolla lobes 3–4 × 3–4 mm, oblong, apex displaced, rounded. Staminal corona lobes maroon or purple, fleshy, subovoid, 0.5–0.8 mm long, adnate to the staminal column and reaching to the base of the anthers. Anther wings 0.2 mm long. Stylar head flat. Follicles occurring singly, 7–9 × 1 cm, narrowly lanceolate in outline, smooth. Seeds 8 × 5 mm, ovate, irregularly winged, verrucose; coma 1.5 cm long.

Zambia. N: Shiwangandu, fl. 5.ii.1955, *Fanshawe* 1991 (K, NDO). **Zimbabwe**. E: Gungunyana Forest Reserve, fl. ii.1964, *Goldsmith* 6/64 (K, SRGH). **Malawi**. N: Nkhata Bay Secondary School, 8 km S of Nkhata Bay, fl. 21.xii.1975, *Pawek* 10417 (K, MAL, MO). **Mozambique**. MS: Buzi River forests SE of Espungabera, fl. xi.1962, *Pole-Evans* 6143 (K).

Widely distributed across tropical Africa from Liberia to Kenya and Angola to Zimbabwe. Moist forest; 500–1000 m.

6. **Vincetoxicum apiculatum** (K. Schum.) Meve & Liede in Phytotaxa **369**: 132 (2018). Type: Tanzania, Pangani, *Stuhlmann* I: 848 (B† holotype, HBG lectotype, K (fragment, as I: 884)), lectotype designated by Meve & Liede in Phytotaxa **369**: 132 (2018).

   Tylophora apiculata K. Schum. in Engler, Pflanzenw. Ost-Afrikas **C**: 325 (1895), not Schltr. (1907). —Brown in F.T.A. **4**(1): 407 (1903). —Goyder in F.T.E.A., Apocynaceae (part 2): 503 (2012).

Twining to several metres, whole plant glabrous. Leaves not succulent; petiole 1–3 cm long; lamina 5–12 × 3.5–6 cm, elliptic to oblong or ovate, apex rounded and apiculate, or acute to attenuate, base rounded or very weakly cordate. Inflorescences extra-axillary, appearing much branched with several clusters of flowers arising at the angles of a strongly zig-zag axis; peduncles (to 1ˢᵗ branching point) 1–5 cm long; bracts minute; pedicels 5–15 mm long, very slender. Calyx lobes 0.5–1 mm long, triangular, apex acute. Corolla pale pink or mauve, glabrous, rotate; lobes 1.5–3 mm long, ovate, apex rounded. Gynostegium 0.5 mm long. Staminal corona lobes green, fleshy, c.0.4 mm long, adnate to the column for their entire length but thicker basally and radiating from the staminal column. Anther wings 0.1 mm long. Stylar head flat. Follicles occurring singly, 6–7 × 1 cm, narrowly lanceolate in outline, smooth. Seeds not seen.

Zimbabwe. E: Gungunyana Forest Reserve, fl. iii.1966, *Goldsmith* 1/66 (K, SRGH). **Mozambique**. MS: Buzi River, fr. 16.viii.1961, *Pole-Evans* 6141 (K) & fl. *Pole-Evans* 6341 (K). M: Maputo Elephant Reserve between Lake Xinguti and Ponta Milibangalala, fl. 2.xii.2001, *Goyder* 5022 (K, LMU).

Also known from coastal Kenya and Tanzania. Moist coastal or riverine forest; 0–1100 m.

Very similar to *Vincetoxicum conspicuum*, but generally more delicate, with rounded not deeply cordate base to the leaf, a more open, often branched, glabrous inflorescence, with smaller paler flowers and a shorter gynostegium.

7. **Vincetoxicum iringensis** (Markgr.) Goyder, comb. nov. Type: Tanzania, Iringa, Lupembe, near Ditimi, 2.iv.1931, *Schlieben* 557 (B† holotype, BM lectotype, BR), designated by Liede-Schumann & Meve in Phytotaxa **369**: 140 (2018).

   Pentarrhinum iringense Markgr. in Notizbl. Bot. Gart. Berlin-Dahlem **11**: 404 (1932).
   Tylophoropsis erubescens Liede & Meve in Kew Bull. **49**: 752 (1994). Type: Tanzania, Mufindi, forest opposite Lugoda Tea Estate factory entrance, 11.v.1968, *Renvoize & Abdallah* 2112 (K000305324 lectotype, K), designated by Goyder in Kew Bull. **60**: 613 (2006).
   Tylophora erubescens (Liede & Meve) Liede in Taxon **45**: 206 (1996). —Burrows in Pl. Nyika Plateau: 75 (2005).

Tylophora iringensis (Markgr.) Goyder in Kew Bull. **60**: 613 (2006). —Goyder in F.T.E.A., Apocynaceae (part 2): 503 (2012).

Twining to several metres, stems minutely pubescent. Leaves not succulent; petiole 3–6 mm long, pubescent; lamina 1–2.3 × 0.4–1.5 cm, ovate to lanceolate, apex acute, usually apiculate, base rounded, with a pair of minute colleters, sub-glabrous with indumentum restricted to margins and principal veins. Inflorescences extra-axillary, pedunculate, generally with a single cluster of 2–4 flowers; peduncles 4–5 mm long, glabrous, slender; bracts minute, triangular; pedicels 4–5 mm long, glabrous. Calyx lobes c.1 mm long, ovate or triangular, pubescent. Corolla maroon or purple, upper surface of the lobes densely pubescent with white hairs, glabrous beneath, rotate, the lobes united at the base; corolla lobes c.1–2 × 1–1.5 mm, ovate, apex acute or subacute. Staminal corona lobes red, fleshy, oblong, spreading horizontally from the gynostegium, with a free, erect, ligulate appendage 0.5–0.8 mm long, about as tall as the column, apex recurved. Anther wings c.0.15 mm long. Stylar head domed, cream. Follicles occurring singly, c.8 × 1 cm, narrowly lanceolate in outline with a long attenuate beak, smooth.

Malawi. N: Nyika National Park, Zovochipolo Forest, fl. 2.iv.2000, *Winter & Burrows* 4134 (K, MAL, PRE).

Also known from the Southern Highlands of Tanzania – apparently restricted to montane forests around the northern end of Lake Malawi. Moist evergreen montane forest; c.2000 m.

Meve & Liede in Phytotaxa **369**: 140 (2018) united this species with *Pentarrhinum coriaceum* Schltr., a species known only from KwaZulu-Natal. *P. coriaceum* has floral and leaf dimensions twice those of *Vincetoxicum iringensis* and I maintain the taxa as separate species.

8. **Vincetoxicum stenolobum** (K. Schum.) Meve & Liede in Phytotaxa **369**: 163 (2018). Type: Tanzania, 'Usambara', Doda, vi.1893, *Holst* 2977a (B† holotype, K lectotype), designated by Liede-Schumann & Meve in Phytotaxa **369**: 163 (2018).

   Astephanus stenolobus K. Schum. in Engler, Pflanzenw. Ost-Afrikas **C**: 321 (1895).
   Tylophora stenoloba (K. Schum.) N.E. Br. in Bull. Misc. Inform. Kew **1895**: 257 (1895); in F.T.A. **4**(1): 409 (1903). —Goyder in F.T.E.A., Apocynaceae (part 2): 507 (2012).

Slender twiner to c.3 m, stems minutely pubescent or subglabrous. Leaves semi-succulent; petiole 2–4(6) mm long, pubescent adaxially; lamina 2–3.5 × 0.7–1.7 cm, elliptic, apex rounded and with a short apiculus, base rounded, glabrous except for the midrib and basal margins, a minute colleter present at base of lamina. Inflorescences extra-axillary, mostly sessile, rarely (*Vollesen* MRC 3102) with peduncle to 1.3 cm; bracts minute, triangular; pedicels filiform, 1.5–2.5 cm long, glabrous. Calyx lobes 0.5–1 mm long, narrowly to broadly triangular, glabrous. Corolla pink or purple, rotate, united at the base for c.1 mm; lobes spirally twisted in bud, spreading at anthesis, 6–12 mm long, narrowing from a triangular base into a weakly to strongly twisted filiform limb with a somewhat spathulate apex, pubescent with minute scattered white hairs at least towards the base adaxially, otherwise glabrous. Gynostegium 0.5 mm long. Staminal corona lobes maroon or purple, fleshy, ovoid, c.0.3 mm long, adnate to the staminal column for most of its length. Anther wings c.0.1 mm long. Stylar head flat. Follicles occurring singly, 3.5–5 × 0.7–0.8 cm, narrowly lanceolate in outline with a long slender beak, smooth. Seeds c.7 × 4 mm, ovate, with a paler marginal wing c.1 mm wide, minutely pubescent on both faces; coma 2 cm long.

Mozambique. N: Quiterajo, head of Río Muenhe below Namacubi Forest, fl. 26.xi.2008, *Goyder & Crawford* 5060 (K, LMA, LMU, P).

Also known from coastal Kenya and Tanzania. Dry evergreen coastal or riverine forest; 0–100 m.

9. **Vincetoxicum sylvaticum** (Decne.) Kuntze in Revis. Gen. Pl. **2**: 425 (1891). Type: Gambia. "In sylvis Gambiae haud procul Cicae", 1827, *Leprieur* s.n. (G-DC holotype).

   Tylophora sylvatica Decne. in Ann. Sc. Nat., Bot. sér. 2 **9**: 273 (1838). —Brown in F.T.A. **4**(1): 407 (1903). —Goyder in F.T.E.A., Apocynaceae (part 2): 508 (2012). Type: Senegambia,

Leprieur (P holotype).
?*Tylophora adalinae* K. Schum. in Engler & Prantl, Nat. Pflanzenfam. **4**(2): 286 (1895). Type not designated.

Twiner to several metres, stems minutely pubescent when young, becoming glabrescent. Leaves not succulent; petiole 1–3 cm long, pubescent adaxially; lamina 3–9 × 1.5–5 cm, ovate to oblong, apex acute and somewhat attenuate, base cordate, rarely truncate, with minute colleters, glabrous or sub-glabrous with minute indumentum restricted to margins and principal veins. Inflorescences extra-axillary, up to c.20 cm long with clusters of minute flowers scattered at intervals of 1–3 cm along the simple or branched axes; peduncles (to 1ˢᵗ cluster of flowers) 2–5 cm long, minutely pubescent; bracts minute, narrowly triangular; pedicels 1–5 mm long, minutely pubescent. Calyx lobes 0.5–0.75 mm long, ovate or triangular, pubescent. Corolla reddish brown or dull purple, glabrous, rotate, the lobes united at the base; lobes 1.5–2.5 × 1–1.5 mm, triangular, ovate or oblong, acute. Gynostegium 1 mm long. Staminal corona lobes maroon, c.0.5 mm high, adnate to the staminal filaments to the base of the anthers and lacking a free apical appendage, broadest near the base of the column forming a fleshy subtriangular tubercle. Anther wings 0.25 mm long. Stylar head flat. Follicles mostly occurring singly, if paired then held at 180°, 6–9.5 × c.0.5 cm, narrowly fusiform, smooth, glabrous. Seeds 7–9 × 3 mm, flattened, ± elliptic with a narrow marginal rim which is irregularly toothed at the end opposite the 1.5–2 cm long coma.

Zambia. W: Kitwe, fl. 25.x.1969, *Mutimushi* 3818 (K, NDO).
Widely distributed across Central and West Africa, it is also known from southern Sudan, SW Ethiopia, western provinces of Uganda, Kenya and Tanzania, and Angola. Riverine and ground-water forest, mushitu; 1200–1400 m.

10. **Vincetoxicum tenuipedunculatum** (K. Schum.) Meve & Liede in Phytotaxa **369**: 165 (2018). Type: Congo (Brazzaville), Loango, near Povo Zala, Chinchocho Dist., 17.xi.1874, *Soyaux* 163 (B† holotype, K lectotype, M), designated by Liede-Schumann & Meve in Phytotaxa **369**: 165 (2018).

   Tylophora tenuipedunculata K. Schum. in Bot. Jahrb. Syst. **17**: 144 (1893). —Brown in F.T.A. **4**(1): 409 (1903). —Goyder in F.T.E.A., Apocynaceae (part 2): 509 (2012).
   Tylophora gracilis De Wild. in Ann. Mus. Congo Belge, Bot. sér. 5 **1**(2): 194 (1904). Type: D.R. Congo, Kisantu, 1899, *Gillet* 268 (BR lectotype), designated by Liede-Schumann & Meve in Phytotaxa **369**: 165 (2018); *Gillet* 572 (not traced, paralectotype); *Gillet* 1710 (BR paralectotype).
   Tylophora congoensis Schltr. in Bot. Jahrb. Syst. **38**: 51 (1905). Type: D.R. Congo, Stanley Pool near Kinshasa, vi.1899, *Schlechter* 12551 (B† holotype, Schlechter (1905: fig.9) iconolectotype), designated by Liede-Schumann & Meve in Phytotaxa **369**: 165 (2018).

Slender twiner to c.3 metres, stems subglabrous to uniformly but minutely pubescent. Leaves not succulent; petiole 0.6–2 cm long, pubescent with hairs restricted mostly to groove along adaxial surface, occasionally subglabrous; lamina 2–6 × 1–2 cm, ovate to oblong or narrowly triangular, apex acute or attenuate, base rounded to truncate and with minute colleters, sub-glabrous. Inflorescences extra-axillary, pedunculate, with one or more clusters of flowers along the axis; peduncles (to 1ˢᵗ cluster of flowers) 1.5–3.5 cm long, glabrous, slender; bracts minute, pubescent; pedicels 0.6–2 cm long, glabrous. Calyx lobes 0.5–1 mm long, triangular, glabrous. Corolla pale to deep pink, sparsely pubescent or subglabrous adaxially, glabrous beneath, rotate, united at the base only, lobes 1.5–3.5 × 0.5–1.5 mm, ovate to triangular, apex acute to subacute. Gynostegium 0.6–0.8 mm long. Staminal corona lobes fleshy, ovoid, 0.4–0.6 mm high, reaching the base of the anthers and adnate to the staminal column for their entire length. Anther wings 0.2 mm long. Stylar head flat. Follicles 4–6 × c.0.5 cm, narrowly lanceolate in outline with a long slender beak, smooth; seeds not seen.

Malawi. N: Nkhata Bay Dist., 8 km E of Mzuzu, Roseveare's, fl. 6.x.1973, *Pawek* 7336 (K, MAL, MO, UC).
Also known from Congo (Brazzaville), Cameroon, Uganda, Kenya and Tanzania. Riverine forest; 1200 m.

11. **Vincetoxicum monticola** Goyder, sp. nov. Differs from *Vincetoxicum tenuipedunculatum* in its anther wings measuring c.0.6 mm long, 2–3 times as long as for *V. tenuipedunculatum*. Type: Zimbabwe, Bunga Forest Botanical Reserve, 3 km from Vumba [Bvumba] Botanical Garden, fl. 9.ii.1997, *Harder et al.* 3817 (K holotype, MO, SRGH).

Slender twiner, stems glabrous or unifariously pubescent. Leaves not succulent; petiole 0.6–1.2 cm long, pubescent with hairs restricted mostly to groove along adaxial surface; lamina 2.5–4 × 0.6–1.6 cm, ovate to lanceolate, tapering gradually into a long-attenuate and somewhat apiculate apex, base rounded to very shallowly cordate, with minute colleters, glabrous except for margins and midvein above. Inflorescences extra-axillary, pedunculate, generally with a single clusters of flowers; peduncles 2–3.5 cm long, glabrous, slender; bracts minute; pedicels 0.6–0.7 cm long, glabrous. Calyx lobes 0.7–1 mm long, triangular, subglabrous. Corolla pink, red, purplish or yellow, sparsely pubescent or subglabrous adaxially, glabrous beneath, rotate, united at the base only, lobes 3.5–4.5 × 1–1.5 mm, triangular to oblong, apex subacute or rounded. Gynostegium 1 mm long. Staminal corona lobes fleshy, ovoid, 0.6 mm high, reaching midway up the anthers-wings and adnate to the staminal column for their entire length. Anther wings 0.6 mm long. Stylar head flat. Follicles (immature) 2.5 × c.0.3 cm, narrowly fusiform with a long slender beak, smooth; seeds not seen.

Zimbabwe. E: Elephant forest, Vumba, fl. 20.i.1955, *Chase* 5473 (K, LMA, SRGH). **Mozambique**. MS: Serra da Gorongosa, monte Nhandore, fl. 6.v.1964, *Torre & Paiva* 12298 (LISC).

Not known elsewhere. The two localities are c.150 km apart. Montane forest; 1400–1800 m.

This species is a distinct local variant, clearly allied to the more widely distributed *Vincetoxicum tenuipedunculatum*. *Tylophora* sp. B of F.T.E.A. is another local variant, this time restricted to the Taita Hills in Kenya, and can be readily distinguished by the basal tails to the anther wings.

12. **Vincetoxicum gracillimum** (Markgr.) Meve & Liede in Phytotaxa **369**: 147 (2018). Type: Tanzania, Morogoro, Uluguru-Gebirge, 10.xii.1932, *Schlieben* 3067 (B† holotype, K lectotype, BR, EA, F, LISC, MO, NY, PRE), designated by Liede-Schumann & Meve in Phytotaxa **369**: 147 (2018).

Tylophora gracillima Markgr. in Notizbl. Bot. Gart. Berlin-Dahlem **13**: 284 (1936).

Slender twiner to c.3 m, stems uniformly spreading-pubescent. Leaves not succulent; petiole 1–3.5 mm long, pubescent with spreading hairs; lamina 0.7–1.6 × 0.2–0.8 cm, linear-lanceolate to broadly ovate, apex acute or occasionally rounded, minutely apiculate, base rounded to subcuneate and with minute colleters, margins minutely undulate or crisped, minutely pubescent with any indumentum ± restricted to margins and principal veins. Inflorescences slender, extra-axillary, pedunculate, with very open branching; peduncles (to 1st branching point) 0.5–2 cm long, glabrous or minutely spreading pubescent; bracts minute; pedicels 0.3 cm long, glabrous. Calyx lobes 0.5–1 mm long, ovate to narrowly triangular, glabrous or minutely pubescent. Corolla greenish white to deep pink or purple, glabrous, rotate, united at the base only, lobes 1–2 × 0.7–1 mm, ovate, apex rounded to subacute. Gynostegium purple or green, 0.3 mm long. Staminal corona lobes fleshy, ovoid, 0.2–0.3 mm long, radiating from the staminal column and adnate to it to the base of the anthers. Anther wings 0.1 mm long. Stylar head flat. Follicles usually single, occasionally paired and held at ± 180°, 4–5 × c.0.3 cm, narrowly fusiform with a long slender beak, smooth; seeds not seen.

Malawi. N: Mughesse Forest, Misuku Hills, 12.ix.1977, *Pawek* 12992 (K, MAL, MO). Also known from the Uluguru mountains, Uzungwa Escarpment and Lulanda forest in Tanzania. Moist montane forest; 1500–1800 m.

INDEX TO BOTANICAL NAMES

Accepted names in roman, synonyms in *italic*. **Bold** page numbers indicate main entries of accepted names (names with description and in the keys) and illustrations.

sect. Chamaesiphon H. Huber **96**, 159
sect. Convolvuloides (H. Huber) Bruyns
93, 113
sect. Coreosma H. Huber **95**, 140
sect. *Duvalia* (Haw.) Bruyns 196
sect. *Hoodia* (Decne.) Bruyns 190
sect. *Huernia* (R. Br.) Bruyns 202
sect. Laguncula H. Huber **94**, 126, 133, 138
sect. *Astropegia* H. Huber 94
sect. *Orbea* (Haw.) Bruyns 212
sect. Phalaena H. Huber **93**, 106
sect. *Piaranthus* (R. Br.) Bruyns 200
sect. Pseudoceropegiella Bruyns **94**, 115
sect. Psilopegia H. Huber **96**, 158
sect. Radicantiores Bruyns **95**, 141
sect. Speciosae (H. Huber) Bruyns **94**, 116
sect. *Stapelia* (L.) Bruyns 192
sect. Stenatae Bruyns **93**, 111
sect. *Tavaresia* (Welw.) Bruyns 227
sect. *Tridentea* (Haw.) Bruyns 198
sect. Umbraticolae (H. Huber) Bruyns **94**,
121
sect. *Riocreuxia* sensu Huber 86
ser. *Abyssinicae* H. Huber 94
ser. *Africanae* H. Huber 95
ser. *Arabicae* H. Huber 93
ser. *Campanulatae* H. Huber 96
ser. *Convolvuloides* H. Huber 93
ser. *Mirabiles* H. Huber 94
ser. *Multiflorae* H. Huber 95
ser. *Nigrae* H. Huber 94
ser. *Ringentes* H. Huber 94
ser. *Speciosae* H. Huber 94
ser. *Stenanthae* H. Huber 95
ser. *Umbraticolae* H. Huber 94
subsect. *Aristolochioides* (H. Huber) Bruyns
93
subsect. *Junceae* (H. Huber) Bruyns 93
abyssinica Decne. 102, 105, 128, **126, 127**,
129, 132
var. *songeensis* H. Huber 129
acacietorum Dinter 149
achtenii De Wild. **131**
subsp. achtenii 102, **127**, 132
subsp. adolfi (Werderm.) H. Huber 106,
127, 132
var. *achtenii* 138
achtenii s. lat. 132
adamsiana M.G. Gilbert vi, 98, **161, 165**
adolfi Werderm 132
var. *gracillima* Werderm. 132
affinis Vatke 118
albipilosa (Peckover) Bruyns 99, 164, **178**
aloicola M.G. Gilbert vi, 101, 103, **108, 110**
ampliata E. Mey. 103, **151, 158**
var. *madagascariensis* Lavranos 158
var. *oxyloba* H. Huber 158
angiensis De Wild. 113

apiculata Schltr. 107
arachnoidea (P.R.O. Bally) Bruyns 111
archeri P.R.O. Bally 107
arenaria R.A. Dyer **96**, 147, 148, 149
arenarioides M.G. Gilbert vi, 101, **142, 147**,
148
arnotii (Baker) Bruyns 99, **172**
banforae (J.-P. Lebrun & Stork) Bruyns 170
barberae (Hook.f.) Bruyns 98, **173, 181**
barbertonensis N.E. Br. 155, 156
barbertonensis 156
barklyana Bruyns 228
bequaertii De Wild. 128
bikitaensis (Peckover) Bruyns 100, **160**
boerhaaviifolia Schinz 149
bonafouxii K. Schum. 102, 105, **127, 130**
var. *linearifolia* Stopp 128
boussingaultiifolia Dinter 146
brachyceras Schltr. 143
breviflora (Schltr.) Bruyns 98, **161, 166**
subsp. flavida (Schltr.) Bruyns 166
brevipedicellata (Turrill) Bruyns 99, **161**,
171, 172
buchananii (N.E. Br.) Bruyns 99, **185, 189**
bulbosa Roxb 152
burchelliana Bruyns **96**, 100
burchellii (K. Schum.) H. Huber 88
subsp. *profusa* (N.E. Br.) H. Huber 88
caffrorum Schltr. 154
var. *dubia* N.E. Br. 154
calcarata N.E. Br. 113
carnosa E. Mey. 104, **114, 120**
cataphyllaris Bullock 102, 138, **151, 157**
caudata (N.E. Br.) Bruyns 215
subsp. *rhodesiaca* (L.C. Leach) Bruyns
216
chimanimaniensis M.G. Gilbert vi, 103,
114, 121
chipiaensis Stopp 122
chlorantha (Schltr.) Bruyns 100, **159, 161**
chlorozona (E.A. Bruce) Bruyns 100, **182**
chrysochroma H. Huber 88
cimiciodora Oberm. 101, **141, 142**
circinata (E. Mey.) Bruyns 98, **162, 163**, 164
claviloba Werderm. 105, **135, 136, 137**
collaricorona subsp. mutongaensis 154
conrathii Schltr. 104, **151, 156**
constricta N.E. Br. 146
cordifolia M.G. Gilbert vi, 103, **114, 119**
crassifolia Schltr. 104, **142, 143**
criniticaulis Werderm. 113
var. *copleyae* (E.A. Bruce & P.R.O. Bally)
H. Huber 144
crispata N.E. Br. 143
cupulata (R.A. Dyer) Bruyns 99, 101, 174,
173, 175
currorii (Decne.) Bruyns 190
subsp. *lugardii* (N.E. Br.) Bruyns 192

longidens N.E. Br. 221
longipedicellata (A. Berger) N.E. Br. 195
melanantha Schltr. 217
paradoxa (I. Verd.) P.V. Heath 222
rogersii L. Bolus 227
schinzii 192
tapscottii I. Verd. 224
umbracula (M.D. Hend.) P.V. Heath 224
unicornis C.A. Lückh. 192, **194**
youngii N.E. Br. 194
Stapeliopsis E. Phillips 212
STATHMOSTELMA K. Schum. 59, 63, 260, 273, **283**
bicolor K. Schum. 289
chironioides De Wild. & Durand 288
fornicatum (N.E. Br.) Bullock 284, **287**
  subsp. fornicatum **288**
frommii Schltr. 286
gigantiflorum K. Schum. 284, **289**
globuliflorum K. Schum. 286
laurentianum Dewèvre 288
macranthum (Oliv.) Schltr. 287
macropetalum Schltr. & K. Schum. 284
nomadacridum Bullock 289
nuttii (N.E. Br.) Bullock 273
odoratum K. Schum 284
pachycladum K. Schum. 284
pauciflorum (Klotzsch) K. Schum. 284, **285, 289**
pedunculatum (Decne.) K. Schum. 284, **286**
praetermissum Bullock 289
reflexum Britten & Rendle 289
spectabile (N.E. Br.) Schltr. **284**
  subsp. frommii (Schltr.) Goyder 284, **286**
  subsp. spectabile 284, **285, 286**
verdickii De Wild. 263
welwitschii Britten & Rendle 284, **288**
  var. welwitschii **288**
STENOSTELMA Schltr. 59, 63, **243**
capense Schltr. **243, 244**
corniculatum (E. Mey.) Bullock 243, **244**
eminens (Harv.) Bullock 272
STOMATOSTEMMA N.E. Br. 3, **18**
monteiroae (Oliv.) N.E. Br. **19, 20**
pendulina Venter & D.V. Field **19**
Strobopetalum N.E. Br. 349
Stultitia E. Phillips 212
paradoxa I. Verd. 222
tapscottii (I. Verd.) E. Phillips 224
umbracula M.D. Hend. 224
Swynnertonia S. Moore 84
Syzigium 158
TACAZZEA Decne. 3, **21**
africana (Schltr.) N.E. Br. 48
amplifolia S. Moore 26
apiculata Oliv. 21, **22, 23**
bagshawei S. Moore 22

  var. *occidentalis* Norman 22
conferta N.E. Br. 21, **24**
floribunda K. Schum. 24
galactagoga Bullock 24
kirkii N.E. Br. 22
oleander S. Moore 24
rosmarinifolia (Decne.) N.E. Br. 21, **24**
salicina Schltr. 24
thollonii Baill. 22
venosa Decne. subsp. *rosmarinifolia* (Decne.) Bullock 24
welwitschii Baill. 22
TAVARESIA Welw. 59, 61, **227**
angolensis Welw. 227
barklyi (Dyer) N.E. Br. 227, **228**
grandiflora (K. Schum.) A. Berger 228
  var. *recta* Van Son 228
meintjesii R.A. Dyer 227
thompsoniorum van Jaarsv. & R. Nagel 227
TELOSMA Coville 64, **80**
africana (N.E. Br.) N.E. Br. **80, 81**
africanum (N.E. Br.) Coville 82
unyorensis S. Moore 82
Tenaris E. Mey. 92
rubella E. Mey. 97
rostrata N.E. Br. 97
simulans N.E. Br. 97
volkensii K. Schum. 97
bikitaensis (Peckover) J.E. Victor & Nicholas 160
chlorantha Schltr. 159
filifolia (Schltr.) N.E. Br. 160
schultzei (Schltr.) E. Phillips 183
Terminalia 19
sericea 221
Toxocarpus
brevipes (Benth.) N.E. Br. 58
lujaei (De Wild. & T. Durand) De Wild. 58
parviflorus (Benth.) N.E. Br. 58
Trachycalymma (K. Schum.) Bullock 260
buchwaldii (Schltr. & K. Schum.) Goyder 270
cristatum (Decne.) Bullock 268
cucullatum (Schltr.) Bullock 277
fimbriatum (Weim.) Bullock 269
foliosum (K. Schum.) Goyder 272
graminifolium (Wild) Goyder 269
pulchellum (Decne.) Bullock 271
pulchellum sensu Bullock 272
Traunia K. Schum. 69
TRIDENTEA Haw. 59, 61, **198**
marientalis **199**
  subsp. marientalensis **199, 200**
Triodoglossum Bullock 48
Tylophora R. Br. 349
adalinae K. Schum. 359
anomala N.E. Br. 355
apiculata K. Schum. 357

FAMILIES OF VASCULAR PLANTS REPRESENTED IN THE FLORA ZAMBESIACA AREA

PTERIDOPHYTA
(Flora Zambesiaca families and family number. Published 1970)

Actiniopteridaceae		Gleicheniaceae	9	see Adiantaceae	18
see Adiantaceae	18	Grammitidaceae	20	Polypodiaceae	21
Adiantaceae	18	Hymenophyllaceae	15	Psilotaceae	1
Aspidiaceae	27	Isoetaceae	4	Pteridaceae	
Aspleniaceae	23	Lindsaeaceae	19	see Adiantaceae	18
Athyriaceae	25	Lomariopsidaceae	26	Salviniaceae	12
Azollaceae	13	Lycopodiaceae	2	Schizaeaceae	10
Blechnaceae	28	Marattiaceae	7	Selaginellaceae	3
Cyatheaceae	14	Marsileaceae	11	Thelypteridaceae	24
Davalliaceae	22	Oleandraceae		Vittariaceae	17
Dennstaedtiaceae	16	see Davalliaceae	22	Woodsiaceae	
Dryopteridaceae		Ophioglossaceae	6	see Athyriaceae	25
see Aspidiaceae	27	Osmundaceae	8		
Equisetaceae	5	Parkeriaceae			

GYMNOSPERMAE
(Flora Zambesiaca families and family number. Volume 1(1) 1960)

Cupressaceae	3	Cycadaceae	1	Podocarpaceae	2

ANGIOSPERMAE
(Flora Zambesiaca families, volume and part number and year of publication)

Acanthaceae			Aristolochiaceae	9(2)	1997
tribes 1–5	8(5)	2013	Asclepiadaceae		
tribes 6–7	8(6)	2015	see Apocynaceae part 2	7(3)	2020
Agapanthaceae	13(1)	2008	Asparagaceae	13(1)	2008
Agavaceae	13(1)	2008	Asphodelaceae	12(3)	2001
Aizoaceae	4	1978	Avicenniaceae	8(7)	2005
Alangiaceae	4	1978	Balanitaceae	2(1)	1963
Alismataceae	12(2)	2009	Balanophoraceae	9(3)	2006
Alliaceae	13(1)	2008	Balsaminaceae	2(1)	1963
Aloaceae	12(3)	2001	Barringtoniaceae	4	1978
Amaranthaceae	9(1)	1988	Basellaceae	9(1)	1988
Amaryllidaceae	13(1)	2008	Begoniaceae	4	1978
Anacardiaceae	2(2)	1966	Behniaceae	13(1)	2008
Anisophylleaceae			Berberidaceae	1(1)	1960
see Rhizophoraceae	4	1978	Bignoniaceae	8(3)	1988
Annonaceae	1(1)	1960	Bixaceae	1(1)	1960
Anthericaceae	13(1)	2008	Bombacaceae	1(2)	1961
Apocynaceae	7(2)	1985	Boraginaceae	7(4)	1990
subfam. Apocynoideae	7(2)	1985	Brexiaceae	4	1978
subfam. Asclepiadoideae	7(3)	2020	Bromeliaceae	13(2)	2010
subfam. Periplocoideae	7(3)	2020	Buddlejaceae		
subfam. Rauvolfioideae	7(2)	1985	see Loganiaceae	7(1)	1983
subfam. Secamonoideae	7(3)	2020	Burmanniaceae	12(2)	2009
Aponogetonaceae	12(2)	2009	Burseraceae	2(1)	1963
Aquifoliaceae	2(2)	1966	Buxaceae	9(3)	2006
Araceae	12(1)	2011	Cabombaceae	1(1)	1960
Araliaceae	4	1978	Cactaceae	4	1978

Caesalpinioideae				tribes 24–26	10(3)	
see Leguminosae	3(2)	2006		tribe 27	10(4)	2ᴜ
Campanulaceae	7(1)	1983		Guttiferae	1(2)	19ᴜ
Canellaceae	7(4)	1990		Haloragaceae	4	1978
Cannabaceae	9(6)	1991		Hamamelidaceae	4	1978
Cannaceae	13(4)	2010		Hemerocallidaceae	12(3)	2001
Capparaceae	1(1)	1960		Hernandiaceae	9(2)	1997
Caricaceae	4	1978		Heteropyxidaceae	4	1978
Caryophyllaceae	1(2)	1961		Hyacinthaceae	-	-
Casuarinaceae	9(6)	1991		Hydnoraceae	9(2)	1997
Cecropiaceae	9(6)	1991		Hydrocharitaceae	12(2)	2009
Celastraceae	2(2)	1966		Hydrophyllaceae	7(4)	1990
Ceratophyllaceae	9(6)	1991		Hydrostachyaceae	9(2)	1997
Chenopodiaceae	9(1)	1988		Hypericaceae		
Chrysobalanaceae	4	1978		see Guttiferae	1(2)	1961
Colchicaceae	12(2)	2009		Hypoxidaceae	12(3)	2001
Combretaceae	4	1978		Icacinaceae	2(1)	1963
Commelinaceae	-	-		Illecebraceae	1(2)	1961
Compositae				Iridaceae	12(4)	1993
tribes 1–5	6(1)	1992		Irvingiaceae	2(1)	1963
Connaraceae	2(2)	1966		Ixonanthaceae	2(1)	1963
Convolvulaceae	8(1)	1987		Juncaceae	13(4)	2010
Cornaceae	4	1978		Juncaginaceae	12(2)	2009
Costaceae	13(4)	2010		Labiatae		
Crassulaceae	7(1)	1983		see Lamiaceae, Verbenacaeae		
Cruciferae	1(1)	1960		Lamiaceae		
Cucurbitaceae	4	1978		Viticoideae, Pingoideae	8(7)	2005
Cuscutaceae	8(1)	1987		Lamiaceae		
Cymodoceaceae	12(2)	2009		Scutellaroideae-		
Cyperaceae	14	2020		Nepetoideae	8(8)	2013
Dichapetalaceae	2(1)	1963		Lauraceae	9(2)	1997
Dilleniaceae	1(1)	1960		Lecythidaceae		
Dioscoreaceae	12(2)	2009		see Barringtoniaceae	4	1978
Dipsacaceae	7(1)	1983		Leeaceae	2(2)	1966
Dipterocarpaceae	1(2)	1961		Leguminosae,		
Dracaenaceae	13(2)	2010		Caesalpinioideae	3(2)	2007
Droseraceae	4	1978		Mimosoideae	3(1)	1970
Ebenaceae	7(1)	1983		Papilionoideae	3(3)	2007
Elatinaceae	1(2)	1961		Papilionoideae	3(4)	2012
Ericaceae	7(1)	1983		Papilionoideae	3(5)	2001
Eriocaulaceae	13(4)	2010		Papilionoideae	3(6)	2000
Eriospermaceae	13(2)	2010		Papilionoideae	3(7)	2002
Erythroxylaceae	2(1)	1963		Lemnaceae		
Escalloniaceae	7(1)	1983		see Araceae	12(1)	2011
Euphorbiaceae	9(4)	1996		Lentibulariaceae	8(3)	1988
Euphorbiaceae	9(5)	2001		Liliaceae sensu stricto	12(2)	2009
Flacourtiaceae	1(1)	1960		Limnocharitaceae	12(2)	2009
Flagellariaceae	13(4)	2010		Linaceae	2(1)	1963
Fumariaceae	1(1)	1960		Lobeliaceae	7(1)	1983
Gentianaceae	7(4)	1990		Loganiaceae	7(1)	1983
Geraniaceae	2(1)	1963		Loranthaceae	9(3)	2006
Gesneriaceae	8(3)	1988		Lythraceae	4	1978
Gisekiaceae				Malpighiaceae	2(1)	1963
see Molluginaceae	4	1978		Malvaceae	1(2)	1961
Goodeniaceae	7(1)	1983		Marantaceae	13(4)	2010
Gramineae				Mayacaceae	13(2)	2010
tribes 1–18	10(1)	1971		Melastomataceae	4	1978
tribes 19–22	10(2)	1999		Meliaceae	2(1)	1963